# Material Behavior and Physical Chemistry in

# LIQUID METAL SYSTEMS

# Material Behavior and Physical Chemistry in

# LIQUID METAL SYSTEMS

Edited by
## H. U. BORGSTEDT

Center for Nuclear Research
Karlsruhe, Federal Republic of Germany

PLENUM PRESS • NEW YORK AND LONDON

Library of Congress Cataloging in Publication Data

Main entry under title:
  Material behavior and physical chemistry in liquid metal systems.

  "Proceedings of a conference on Material behavior and physical chemistry in liquid metal
systems, held March 24–26, 1981, in Karlsruhe, Federal Republic of Germany"—Verso t.p.
  Includes bibliographical references and index.
  1. Liquid metals—Congresses. 2. Chemistry, Metallurgic—Congresses. I. Borgstedt, H. U.,
1930–
TN689.2.M36                                669'.94                              82-3680
ISBN 978-1-4684-8368-0          ISBN 978-1-4684-8366-6  (eBook)        AACR2
DOI 10.1007/978-1-4684-8366-6

Proceedings of a conference on Material Behavior and Physical Chemistry in Liquid Metal
Systems, held March 24–26, 1981, in Karlsruhe, Federal Republic of Germany

©1982 Plenum Press, New York
Softcover reprint of the hardcover 1st edition 1982
A Division of Plenum Publishing Corporation
233 Spring Street, New York, N.Y. 10013

# PREFACE

The international seminar "Material Behavior and Physical Chemistry in Liquid Metal Systems" was organized by the Institute of Materials and Solid State Research of the Karlsruhe Nuclear Research Center (Karlsruhe, Federal Republic of Germany). The seminar was held at the Nuclear Engineering School of the center on March 24-26, 1981.

The aim of the seminar was to give metallurgists, chemists, and physicists working in different areas of the science and technology of liquid metals an opportunity to discuss the basic work and the need for further work in this field. Since the seminar was held near one of the laboratories which for the last few years has been engaged in liquid alkali metal studies, participants also had an opportunity to observe modern equipment for liquid alkali metal research.

Interest in the application of liquid metals as working fluids in energy production, conversion, and storage is increasing. The technology has already demonstrated its high standards, which make possible the operation of large sodium-cooled fast reactors. Past conferences have shown, however, that there is still a lack of basic knowledge and understanding. Therefore, the aim of the present seminar was to discuss basic work in detail, and most of the papers contributed to this objective.

The book contains the 56 original contributions and one discussion report to the seminar, divided into ten topics:
- sodium corrosion and mass transfer
- impurities in sodium
- lithium corrosion
- materials behavior
- lead corrosion
- chemical reactions
- analytical chemistry
- physical chemistry
- solubility in alkali metals
- interstitial elements transfer

v

Sodium corrosion and mass transfer were treated in thirteen papers and one discussion report. The contributions deal with detailed corrosion phenomena of steels in sodium and with the deposition of corrosion products. Problems of corrosion and mass transfer modeling are discussed, and the need for more basic data on solubilities and reaction enthalpies is demonstratèd.

The purification of sodium by cold traps and methods of estimating the impurity contents were reported to have reached a high standard for both laboratory and plant application.

The contributions to the topic of lithium corrosion indicated increased efforts in basic and applied studies. Corrosion by lithium can cause severe compatibility problems and has to be evaluated extensively.

The impact of liquid alkali metals on material behavior was discussed from the point of view of testing methods, and some results of creep rupture and fatigue tests were reported.

The increasing interest in the use of lead and lead alloys in research and technology was not fully reflected by the three papers on the topic of lead corrosion--a surprisingly small number. A wide field of research seems to be still open regarding compatibility problems of lead and its alloys.

Under the heading of chemical reactions, both thermodynamic and kinetic aspects of hydrogen reactions with oxygen and carbon in sodium were discussed.

Four analytical chemistry contributions gave information on in-line monitoring of oxygen, hydrogen, and carbon and on laboratory methods of estimating metallic and nonmetallic impurities in sodium.

The physical chemistry contributions comprise reports on methods and results of investigations on liquid and solid lithium alloys and their physical properties, together with kinetic and thermodynamic studies of other alkali metal systems.

Actual results of solubility studies of metals and nonmetals in lithium and sodium were contributed, indicating the increasing importance of this type of basic study of solubility in alkali metals.

Exchange of carbon and nitrogen was pointed out to be one of the major compatibility problems of materials in alkali metals causing interstitial elements transfer.

All in all, the book gives an overview of a broad cross section of current work in liquid metal science and technology, and may help to increase the reader's understanding of the phenomena connected with the application of liquid metals in energy technology.


Karlsruhe, August 27, 1981                    H.U. Borgstedt

# CONTENTS

WELCOME AND INTRODUCTORY REMARKS

Kurt Anderko

Kernforschungszentrum Karlsruhe
7500 Karlsruhe
Fed. Rep. of Germany

I have the great pleasure to open the Seminar on <u>Material Behaviour and Physical Chemistry in Liquid Metal Systems</u> and to welcome you cordially at Karlsruhe. The attendance list shows guests from the following countries:

Austria, Belgium, CSSR, Germany, France, India, Italy, Netherlands, Spain, United Kingdom, USA.

The idea to invite you to this seminar was nucleated by two local events: the replacement of the first German-built thermal reactor FR2 by the fast sodium-cooled reactor KNK II as a research tool, and secondly the completion of two sodium loops for material studies on the KfK-campus. The idea grew to critical size for a positive decision by the encouragement given to us by many of you. This external impetus certainly is due to the facts that liquid metal technology is in fast development and such a meeting, which concentrates on the materials behaviour, will be valuable in offering to the research man or manager a timely chance to gain a representative overview on the state of the art. Having this in mind we intended to avoid parallel sessions. This could not be fully reached; however, the programme committee did its best to minimize the problem of overlapping themes.

There is no question that the liquid metal community can look upon its achievements within the last decade with pride. On the much disputed energy field liquid metal use is a technical reality and, in addition, offers bright outlooks into new and challenging applications. Mention should be made here of the fusion reactor, of the so-called topping cycles in thermal power stations and of intense neutron sources.

Allow me now, as a critical admirer of liquid metal technology, to make some rather sketchy comments on a selection of liquid metal fields. To begin with sodium it may stated that liquid sodium technology is more advanced then our basic knowledge of thermodynamic and kinetic aspects of alloy and compound formation. Thus, systematization of phase relations (phase diagrams, existence diagrams) of multinary systems especially with non-metals appears to be a rewarding task in the near future. This certainly also holds for the improvement of corrosion models allowing even more reliable calculations on LMFBR fuel elements. To further promote the reliability and safety of sodium cooling, the influence of impurity elements on the mechanical properties of reference (and of alternative) materials has to be studied. This, of course, should include fracture mechanics methods like crack propagation under fatigue and creep loodings of base and welded material.

Turning to lithium it may be remarked that it forms binary compounds with approximately the same number of other elements as does sodium ($\sim$ 30). This tie is reflected in the equal number of papers on lithium and sodium physical chemistry aspects at this seminar. Liquid lithium is a technological newcomer and therefore quantitative corrosion data (as reaction rates or solubilities) are still sparse. However, it appears that at this seminar remarkable progress will be seen in this field. The technological interest in liquid lithium is centered in the fusion field. However, it seems that liquid lithium as a coolant and breeder substance is not only challenged by the combination of He-cooling plus solid breeder, but also by highly alloyed liquid lithium. The designers of fusion reactor systems – be it Tokamak-, Mirror- or Inertia-Type – appear to prefer Pb-Li eutectics: $Pb_{38}Li_{62}$ (95 Wt-% Pb) or $Pb_{83}Li_{17}$ (99.3 Wt-% Pb). The advantages are to be found in the neutron multiplication effect of Pb and in two safety-related aspects: the relatively weak reaction with water and the low active tritium inventory in such a blanket. The compatibility between those liquid Pb-Li-alloys and prospective structural materials (austenitics or martensitic 9-12 % chromium steels) remains to be established, however.

A modern application of pure lead (or of a lead-bismuth alloy) will possibly be the liquid metal target of the so-called Spallation Source. In this device high energy protons (600 MeV) split up target atoms, producing a high density neutron beam. The liquid metal target will be heated up to about 450 $^\circ$C, which may lead to compatibility problems with the structural materials.

Returning to alkali metals, the intention of improving the efficiency of thermal power stations to over 50 % with the help of potassium deserves mentioning. Potassium possesses a boiling point (775 $^\circ$C) which is lower than that of sodium and is therefore being developed as a working medium between 500 and 900 $^{\circ}$C propelling a potassium vapor turbine.

The compatibility of liquid rubidium with steel had to be ascertained within the project of reprocessing LWR fuel. In this process radioactive Krypton-85 is released which will be long-term stored in high pressure steel cylinders. By decay rubidium is formed and heated up to around 150 $^{\circ}$C. This rather exotic example may draw your attention to the fact that in considering areas of technological interest, we met nearly the complete series of liquid alkali metals. Rightly, they will stand in the center of our interest within the next three days.

Ladies and gentlemen, I hope now that we will have a successful meeting in a relaxed atmosphere and that you will enjoy the personal contacts and discussions as well as your stay in the city of Karlsruhe and in its lovely neighbourhood.

# MASS TRANSFER OF STAINLESS STEEL IN PUMPED SODIUM LOOPS AND ITS EFFECT ON MICROSTRUCTURE

Alan W. Thorley, Anthony Blundell, J. Alan Bardsley

Risley Nuclear Power Development Laboratories
Risley
United Kingdom

## INTRODUCTION

In the operation of sodium cooled reactors the mass transfer behaviour of corrosion products requires consideration from the standpoint of maintenance procedures if one is dealing with activated constituents from stainless steel fuel cladding (1,2,3), and also from the thermal hydraulic standpoint if in addition we are dealing with spalled oxides, entrained metal swarf or fragments from pipework surfaces. The paper highlights, using suitable photomicrographs, the type of corrosion which can occur in various parts of a loop and the form of the corrosion products which can collect in the circuit when the loops are operated with sodium containing 10 and 25 ppm oxygen. It also indicates how difficult it is to achieve a mass balance in sodium loops when the stainless steel pipework adds significantly to the corrosion product burden in the system.

## MATERIALS

Types 321 and 316 stainless steels have been used in the experiments and details of composition are given in Table 1. Annular specimens 12 mms long and internal diameter 5.8 mms and 18.2 mms (narrow and wide bore respectively) were machined from bar.

Table 1.  Analysis (wt %) of stainless steels used in tests

| Material | Fe | Cr | Ni | Ti | Mo | Si | Mn | C |
|---|---|---|---|---|---|---|---|---|
| 321 | 75.3 | 16.25 | 8.1 | 0.4 | – | 0.7 | 1.6 | 0.07 |
| 316 | 70.0 | 15.9 | 12.1 | – | 2.0 | 0.6 | 1.3 | 0.04 |

EXPERIMENTAL

The experimental programme has utilised two pumped sodium loops similar to those described previously (4). Specimens made from Type 316 and 321 stainless steels were inserted in various positions around the loop (Fig. 1) with wide bore specimens adjacent to and upstream of the narrow bore specimens. With the exception of the main holder specimens which can be removed at any time during a test run all other specimens were welded into the rig pipework in special holders. Each holder contained a batch of 6 specimens of one size. After the specimens were inserted in the loop the sodium temperature was maintained at 400 °C while the loop sodium was purified using the cold trap. Once the sodium was clean its oxygen level was adjusted to the required value by raising or lowering the cold trap temperature; when this had been obtained the main loop sodium was then raised to the required test temperature. Sampling to determine the oxygen level (4) of the sodium was carried out periodically and over the two year period the loops operated at nominal oxygen levels of 10 and 25 ppm. For details of experimental conditions see Table 2.

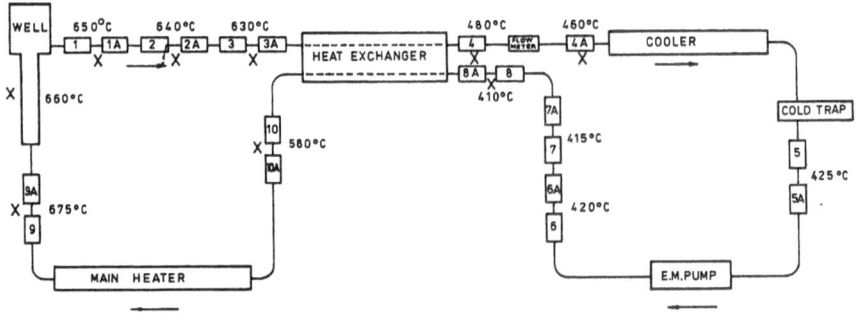

X POSITIONS EXAMINED

Fig. 1. Specimen Layout and Temperature Distribution in Loop.

After the appropriate exposure period the sodium was frozen in the loop and the specimen holders were removed by cutting adjacent pipework. The specimens plus solid sodium were then transferred to a distillation flask where the sodium was removed by low tempera- ture distillation at 300 °C. Some specimens were then removed for cleaning and weighing while others were transferred in the distillation flask to an evacuable glove box where under an atmosphere of pure argon they were immersed in diffusion pump oil previously treated with hot calcium chips to remove any residual moisture. These specimens and certain non-oil treated specimens were then sectioned in the glove box and their exposed bore surfaces scraped with a tool to remove the corrosion products. These products were then sealed into quartz phials for subsequent X-ray examination.

Table 2.   Corrosion-deposition Test Conditions

| Period | Time Weeks | Temperature max. ($^o$C) | min. | Cold Trap Temperat. ($^o$C) | Velocity Wide bore (m/sec) | Narrow bore | Re No Wide bore | Narrow bore (x $10^{-4}$) | Oxygen Level (ppm) |
|---|---|---|---|---|---|---|---|---|---|
| 1 | 23 | 660 | 420 | 200 | 0.4 | 4.0 | 0.3 | 0.9 | 25 |
| 2 | 16 | 660 | 410 | 220 | 0.4 | 4.0 | 0.3 | 0.9 | 25 |
| 3 | 35 | 675 | 410 | 220 | 0.4 | 4.0 | 0.3 | 0.9 | 25 |
| 4 | 25 | 665 | 445 | 230 | 0.6 | 6.0 | 0.5 | 1.4 | 25 |
| 5 | 26 | 655 | 425 | 235 | 0.4 | 4.0 | 0.3 | 0.9 | 25 |
| 1 | 19 | 652 | 428 | 145 | 0.4 | 4.0 | 0.3 | 0.9 | 10 |
| 2 | 13 | 651 | 492 | 151 | 0.4 | 4.0 | 0.3 | 0.9 | 10 |
| 3 | 58 | 650 | 495 | 165 | 0.4 | 4.0 | 0.3 | 0.9 | 10 |
| 4 | 26 | 650 | 467 | 160 | 0.4 | 4.0 | 0.3 | 0.9 | 10 |
| 5 | 9 | 650 | 440 | 160 | 0.6 | 6.0 | 0.5 | 1.4 | 10 |

Metallographic examination of specimens was carried out using standard optical microscopy techniques. Nickel plating or vacuum impregnation methods using Araldite resin were used on all specimens as a means of retaining corrosion products. Conventional methods of grinding followed by polishing were employed, special attention being paid to edge preparation both at the corrosion and deposition surfaces. Certain specimens were examined in both the polished and etched conditions. Etching being used to highlight various microstructural changes such as ferrite layers, carbides, and grain boundaries. In support of the metallographic work, electron probe microanalysis (EPMA) was undertaken on most of the specimens. The specimen used for this analysis was either taken from the same annular section as the metallographic specimen or the metallographic specimen itself was used. The area to be analysed by EPMA was defined by appropriate microhardness indentations. Additional analyses involving scanning electron microscopy with energy dispersive facilities (EDAX) and X-ray diffraction were used to analyse both specimen surfaces and deposits from the various loop positions.

RESULTS

The results from the various types of investigation are illustrated in Figs. 2-3. Fig. 2 illustrates the structural behaviour

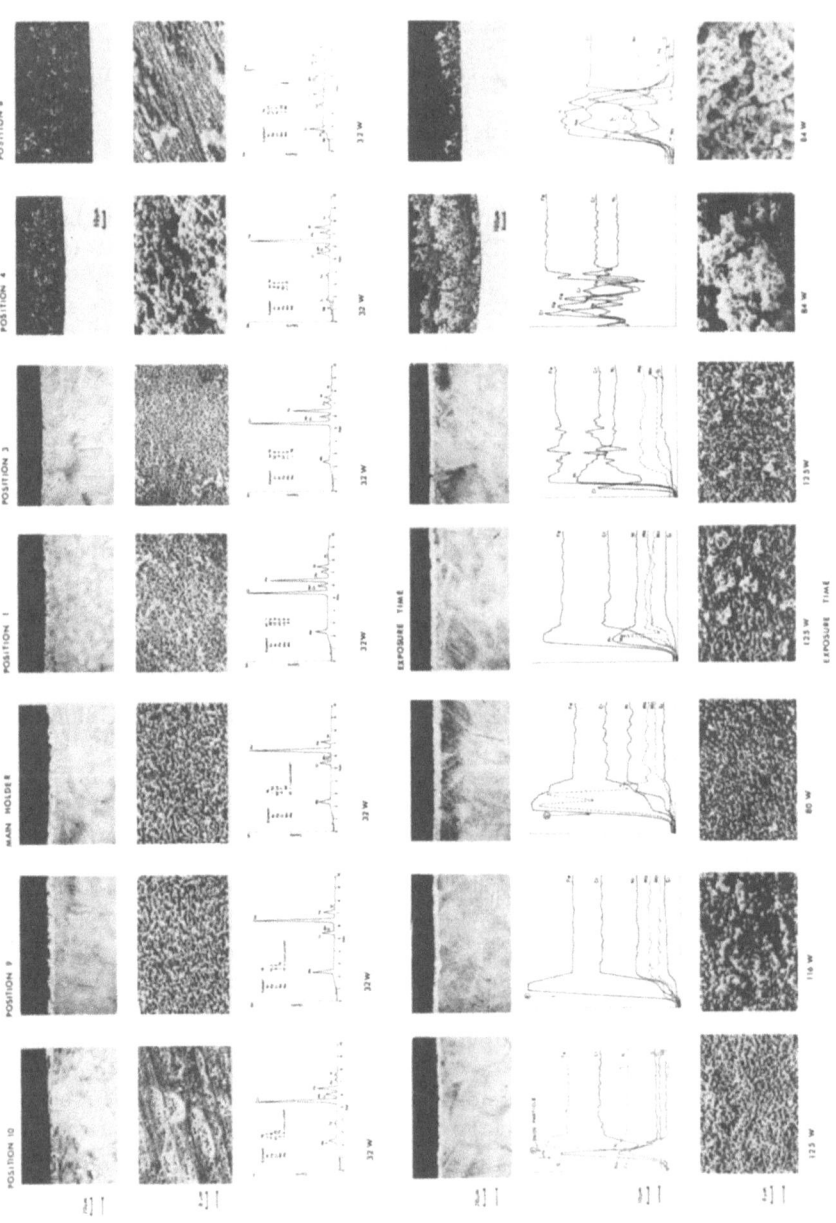

Fig. 2. Structural and Compositional Changes Observed on Type 316 Steel in Various Loop Positions; NA Velocity 3M/Sec; $O_2$ Level 10 PPM

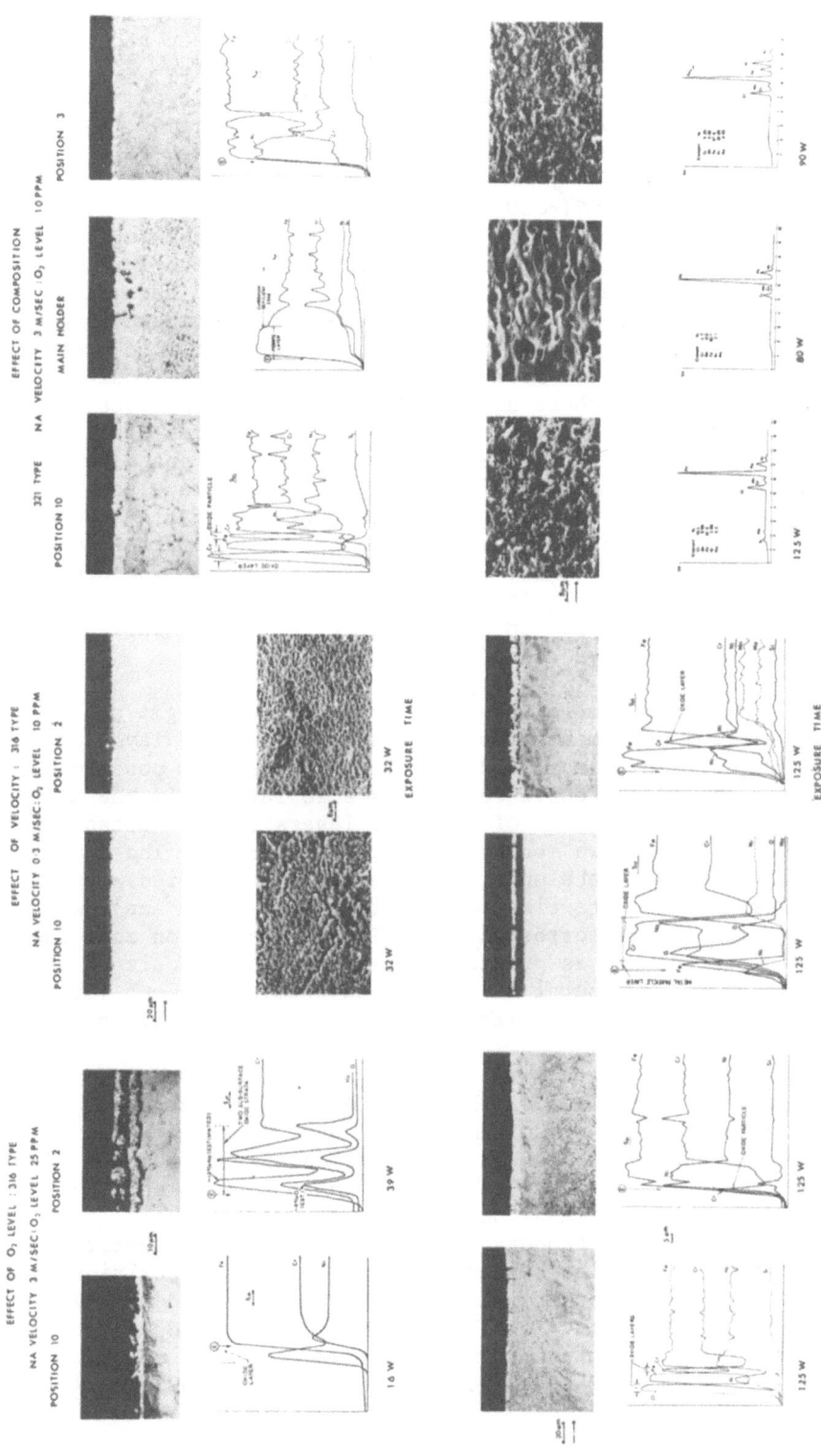

Fig. 3. Effect of Experimental Variables on Surface Structure and Composition of Stainless Steels

of Type 316 stainless steel after exposure to sodium over timescales
of ~ 30 and ~100 weeks at a sodium velocity of 3 m/sec and an
oxygen level of ~10 ppm in 7 positions in the loop (see Fig. 1).
The effect of experimental variables such as oxygen level, sodium
velocity and metal compositions are illustrated in Fig. 3. Again
selected positions have been chosen to illustrate these effects.

Over the short term (~30 weeks) ferrite layers have formed
on the surfaces of specimens in the highest temperature region of
the loop. Surface enrichment of Mo has also occured and formation
of $M_6C$ carbides has been identified in position 9 and the main
holder. Upstream of the maximum temperature position there is
evidence of internal oxidation and surface modification in position
10 due to oxidation of Cr. Downstream of the maximum corrosion
position (Nos 1 and 3) there are isolated regions of surface oxide,
a modified layer of austenite and a more general region of sigma
phase on the surface of the steel. The presence of σ phase has been
identified by differential etching techniques, point analysis using
EPMA and EDAX and by X-ray diffraction methods (see Table 3 and
Fig. 2). In the sections of the loop subject to deposition, positions
4 and 8 are covered with oxide and deposited material. The deposited
material is in the form of steel particles of average size 2 μm,
which are slightly enriched in Ni and deficient in Cr, while the
oxide has been identified as sodium chromite.

With increasing time of exposure to sodium the sigma phase has
been removed from specimens in positions 1 - 3 and the layer of
modified austenite has increased in depth. In upstream position 10
a distinct modified austenite region has also formed. In the main
holder itself, the existence of ferritic layers is well established
and in some instances two modified layers are present. The upper
layer is ferrite while the underlying layer is a modified austenite.
$M_6C$ carbides are also starting to form in the upstream and downstream position in the corrosion zone. In the deposition zone the
only significant change is greater accumulation of deposit in
position 4, and more general oxidation and element pick-up (Si.Ti.Mn)
in the minimum temperature position in the loop (ie position 8).

Increasing the oxygen level of the sodium from 10 to 25 ppm
causes significant break-up of the surface in positions upstream
and downstream of the maximum temperature part of the loop during
short term exposure to sodium at sodium velocities of 3 m/sec. The
surface break-up is caused by the formation of surface and sub-
surface sodium chromite which undercuts the surface layers (see
Fig. 3). In the maximum temperature part of the system ferrite
layers are present and the structural changes in this position and
in the deposition regions are similar to those observed in the
10 ppm oxygen level tests. Notable differences in the 25 ppm tests
are more extensive oxidation in most parts of the loop and the
formation of sodium ferrate as well as sodium chromite in the 425 $^{\circ}$C
position.

Table 3. Carbide and Oxide Phases Observed in Various Loop Positions. Type 316 Steel Unless Stated Otherwise. (X-ray and EPMA data).

| Position | Carbides | | | | Oxides | | | | Other Phases |
| --- | --- | --- | --- | --- | --- | --- | --- | --- | --- |
| | 10ppm $O_2$ | | 25ppm $O_2$ | | 10ppm $O_2$ | | 25ppm $O_2$ | | |
| | Low Vel. | High Vel. | Low Vel. | High Vel. | Low Vel. | High Vel. | Low Vel. | High Vel. | |
| 10 | $M_6C$ (125) | $M_{23}C_6$ (32) | | $M_{23}C_6$ + $M_6C$ (16) | | | | | α and/or γ phases were observed on all specimens. |
| 9 | $M_{23}C_6$ (32) $M_{23}C_6$ + $M_6C$ (116) | $M_{23}C$ (32) $M_6C$ (116) + $M_{23}C_6$ | | $M_{23}C_6$ (30) $M_{23}C_6$ + $M_6C$ (125) | $NaCrO_2$* | | | $NaCrO_2$ | |
| Main Holder | | | | $M_6C$ (32) | | | | $NaCrO_2$(p) | |
| 1 | $M_{23}C_6$ (32) | $M_{23}C_6$ (32) $M_{23}C_6$ + $M_6C$ (125) | $M_{23}C_6$ (12) | $M_{23}C_6$ + $M_6C$ (39) | $NaCrO_2$* (125 Type 321) | | $NaCrO_2$ (p) (39) | $NaCrO_2$ (39) | σ phase high and low velocity 10ppm $O_2$. |
| 2 | $M_{23}C_6$ (32) $M_{23}C_6$(125) | $M_{23}C_6$ + $M_6C$ (125) | $M_{23}C_6$ (12) | $M_6C$ (39) $M_{23}C_6$ + $M_6C$ (125) | $NaCrO_2$* (125) | $NaCrO_2$ (125) | | $NaCrO_2$(125) | |
| 3 | $M_{23}C_6$ (32) $M_{23}C_6$(123) | $M_{23}C_6$ (32) | | $M_6C$ (39) $M_{23}C_6$ + $M_6C$ (125) | | | | $NaCrO_2$ (Type 321) (p) (74) | σ phase low velocity specimens (10ppm $O_2$) |
| 4 | $M_{23}C_6$ (32) | $M_{23}C_6$ (32) | | $M_7C_3$ (39) $M_6C$ + $M_{23}C_6$(125) | $NaCrO_2$ (32) | $NaCrO_2$* (Type 321 + 316) ( ) | | | |
| 8 | | $M_{23}C_6$ (32) $M_{23}C_6$ (84) | | $M_7C_3$ | $NaCrO_2$ (32) $NaCrO_2$* | | | $NaCrO_2$ (p) Type 321 (102) | Oxide containing Si + Ti 10ppm $O_2$ high velocity. |

Key:
* - EPMA
(p) - probable
{ } - Time in weeks

With increasing time of exposure to sodium at the 25 ppm $O_2$ level oxidation products are removed from the corrosion zone and these are transported into the deposition region. However some sodium chromite still exists in the upstream and downstream positions. In the maximum temperature part of the corrosion zone ferrite layer formation, although extensive on Type 321 steels (see later), has not been observed on the Type 316 steels. The indications are that the layer has been corroded away leaving behind a layer rich in $M_6C$ carbides and some oxide. In the deposition zone of the loop there are no significant changes.

There are distinct differences between the two velocity regimes. In the corrosion zone most of the low velocity specimens after 6 months exposure show evidence of retained oxide and surface breakup and in the 10 ppm tests sigma phase is also apparent on the downstream specimens. With increasing time of exposure there is still retention of oxide on certain specimens while others are tending to form Cr depleted modified austenite. The general impression is that the low velocity specimens are reflecting over the long term, conditions which have occurred in the high velocity specimens over the shorter term (compare Figs. 2 - 3). In the deposition zone the low velocity specimens have not attracted any particles and the only feature in this region is surface oxidation where some of the oxides are multi-layer in appearance.

In the main holder position ferrite layer formation, cavity formation and surface pitting are much more extensive in Type 321 steels at both oxygen levels. Surface modifications to form Fe-Ni rich austenite also seems more advanced in Type 321 materials and internal oxidation appears more widespread. In the deposition area of the loop material composition does not appear to be a factor and the amount of material deposited on both types of steel is roughly about the same. See Fig. 3 for details of cavity formation.

The main alloying elements which have been removed from the hot zone of the loop are Ni, Cr, Mn, Si and Mo. Cr loss is mainly caused through surface oxidation in the early stages of loop operation, and possibly some dissolution at the lower oxygen level. Mn and Si may also be involved in the oxidation process, in the early stages of loop operation, while Mo seems to be removed at a much slower rate through the formation of $M_6C$ carbide. The depths to which the various elements are removed in Type 316 steels is consistent with depths of ferrite layer formation, although in certain high temperature positions selected elements can be removed to greater depths to form underlying regions of modified austenite. By contrast, in Type 321 steels, Si and Cr losses have been observed to greater depths and the reason for this is currently being investigated. Detailed surface analysis also shows that there is an interchange between Ti and Mo in the high temperature position of the loop at the 10 ppm oxygen level. Ti moves from Type 321 to

Type 316 steels whereas Mo moves in the opposite direction. In the deposition part of the loop the main region where the various elements seem to collect is at the minimum temperature position, position 8. In this position accumulations of $Si$, Mn and Ti have been identified along with Na, Cr and oxygen. Nickel has also been identified in this region and possibly in downstream positions 1-3.

In most regions of the loop high resolution metallography has identified the existence of oxides on the various specimen surfaces even at oxygen levels of $\sim$ 10 ppm. The oxide is continuous in certain areas whereas in other areas it is patchy and adherent. The most notable oxide is sodium chromite ($NaCrO_2$) which appears in most regions away from the maximum temperature corrosion zone. However there is the possibility of more complex oxide formation in position 8 where elements such as Ti, Mn and Si have collected. It is to be noted sodium ferrate ($Na_2O)_2$. FeO was observed in the 25 ppm tests in the cold region of the loops at a temperature of 425 $^{o}$C, but not in the 10 ppm tests. For details see Table 3.

DISCUSSION

From an analysis of the experimental findings the behaviour of the stainless steel materials in the two sodium loops can be related to processes involving corrosion, surface and sub-surface oxidation and the deposition of corrosion products.

In keeping with observations from other loop systems (5,6,7,8) maximum corrosion losses occur in the high temperature parts of both loops and ferrite layers are formed more or less instantaneously in the high velocity highest temperature parts of the circuits. Release of soluble elements such as Ni and Cr provides thermodynamic driving forces for their deposition in lower temperature regions and it seems in the studies at 10 ppm oxygen level that some of the Cr deposits immediately downstream of the corrosion zone to form an Fe-Cr alloy which has the characteristics of sigma phase. Nickel on the other hand seems to move to lower temperature positions before it deposits, although a small amount of deposition in positions 1-3 cannot be completely discounted.

With increasing time of exposure to sodium two effects are noted which have a bearing on the behaviour of the materials in downstream positions. First the formation of ferritic layers in the hot zone restricts nickel and chromium supply to downstream regions as the release of these elements becomes solid state controlled through the ferrite, and secondly, oxide films of the sodium chromite form both on the surface and sometimes in the sub-surface of the steel. This oxidation process causes two effects to occur; first it causes surface break-up and undercutting of the steel and secondly it extracts Cr from the surface leaving behind a ductile region of transformed austenite which is enriched in Fe

and Ni and deficient in Cr. The picture therefore emerging in the
corrosion zone of the loop is that a number of material transport
processes are occurring more or less simultaneously in this part
of the circuit and although steady state corrosion conditions may
have been archieved in the high temperature regions, this is taking
much longer in downstream positions.

The formation of sodium chromite in the various loop positions
is of technological interest because of its importance in tribological
applications (9). The thermodynamic stability of the oxide is dependent on the chemical activity of the Cr available for reaction and
the chemical potential of the oxygen in the sodium. In practice
both the thermodynamic and mechanical stability of the oxide phase
and the diffusion behaviour of Cr in stainless steel are of importance. For example if the layer is removed by chemical reduction
or erosive forces, further formation of the oxide may not occur
because chromium supply to the surface is insufficient to support
bulk oxide formation. These effects are typified in Fig. 4 which
shows the appearance of those regions in the loop where the various
oxides exist at various times and oxygen levels for the two velocities. This graph which is similar to the one provided by Campbell
(9) indicates that although theoretically sodium chromite should
form at most temperatures, the final state of the surface depends
on many factors such as loop position, oxygen level, and sodium
velocity. The formation of sodium ferrate $(Na_2O)_2FeO$ at temperatures
of 425 °C is considered to be of more fundamental significance
however and in this context there might be good reason for reconsi-
dering the free energy of formation of this oxide as a function of
temperature (10).

Turning to the steel material deposited in position 4, this
is thought to be related to material removed from positions 1, 2, 3
and possibly the higher temperature positions. The particle size
and composition of the particles plus the slight nickel enrichment
are for example consistent with the size and composition of the
eroded material from these positions. The pattern of build-up of
corrosion products in position 4 is also consistent with release
from higher temperature positions at the two oxygen levels (Fig. 5).
For example, increased build-up and breakaway occurs very early in
the test at the 25 ppm oxygen level as more material is released
from upstream and downstream positions in the corrosion zone, whereas, at the 10 ppm level, gradual build-up of deposits occurs in the
initial stages until breakaway of corrosion products occurs in the
hot zone. Once this happens, there is large influx of particles into
position 4, followed by breakaway of deposits to downstream positions.
It is to be noted that the deposit formed in the initial stages at
the 10 ppm oxygen level is less dense and therefore much thicker
than the deposit formed over the same period in the 25 ppm tests,
although the weight deposited at the 10 ppm oxygen level is about
half that deposited at 25 ppm.

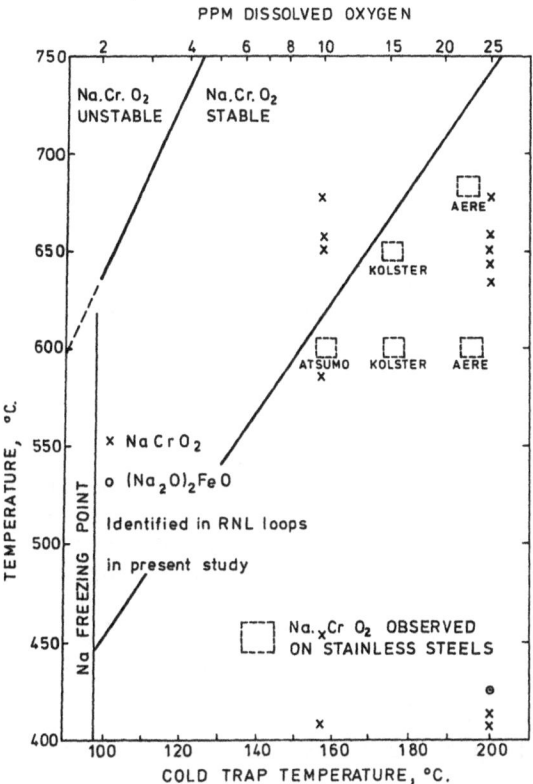

Fig. 4. Observations of No. Cr. $O_2$

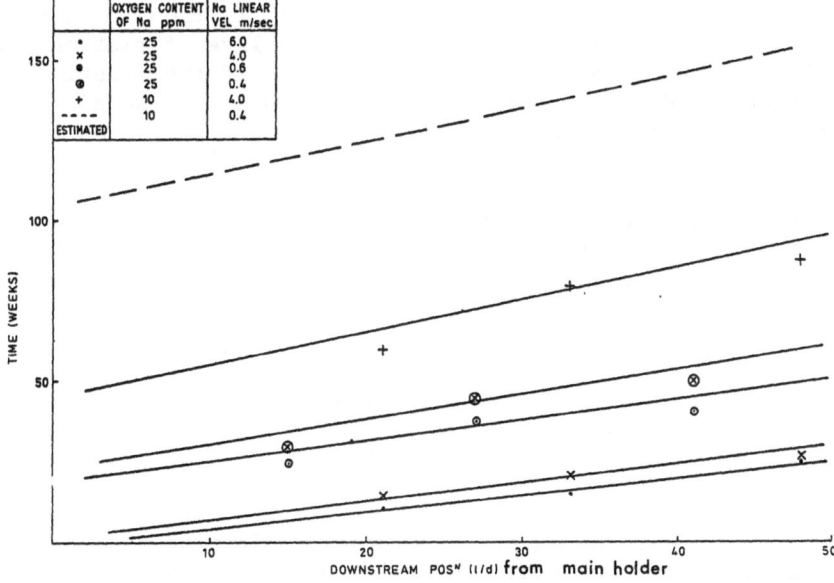

Fig. 5. Time for Downstream Specimens to Enter Weight Loss Regime:
Small No Corrosion Loops.

SUMMARY AND CONCLUSIONS

Microscopic evidence and supporting analytical techniques have
shown that in pumped sodium loops operated at oxygen levels of 10
and 25 ppm, the loop may be roughly divided into zones where corrosion, oxidation or deposition occurs.

In the high temperature zone of the loop where corrosion occurs,
ferrite layer formation, due to the removal of alloying elements
such as Cr Ni and Mn, is a general feature. The depths to which
these layers form depends on temperature, oxygen level, sodium velocity and material composition. Surface pitting which has been observed in the ferrite of Type 321 steels is a new finding at RNL
and requires further investigation.

Away from the high temperature corrosion zone surface oxides
of the sodium chromite type are a feature in positions upstream and
downstream of the maximum temperature positions. Both surface and
sub-surface oxidation can occur in these regions and the process
causes the removal of Cr from the surface layers and also break-up
of surface material. The effect is more apparent at the 25 ppm
oxygen level during short term exposure and significant amounts
of material are transported to the cold zone of the loop during this
period. At the 10 ppm oxygen level it seems longer times are required
to remove the same amount of material and in the low velocity
sections of the loop the oxidised material is still intact after
2 years operation.

Once the oxidised material has been removed the transformation
of the exposed modified austenite to ferrite depends on the rate of
release of nickel from upstream positions. It is thought that once
this becomes solid state controlled, ferrite layer formation will
gradually spread to lower temperature regions if the kinetics for
pre-oxidation and nickel removal are favourable. If the nickel is
not removed and the oxygen levels are maintained it is possible
sub-surface oxidation of the steel may continue if the oxygen has
access to Cr rich regions. It is interesting to note that these
processes are typical of those already reported many years ago for
short term corrosion of steels in the hot zone of a corrosion loop
(10). The only difference is that the corrosion of downstream
material is a function of time dependent processes which are occurring upstream in the loop.

The corrosion products which have deposited in the cooler parts
of the loop are primarily steel particles which have been removed
from the oxidised surfaces in the corrosion zone. The time at which
bulk material deposits depends on the oxygen level of the sodium.
At the 25 ppm oxygen level the bulk of the material deposits in the
short term while at the 10 ppm level the development of mechanically
unstable layers and subsequent deposition of bulk material takes

longer for the same loop operating conditions of temperature and velocity.

It has been found that steel particles do not deposit in the wide bore specimens, suggesting that their mass and the tangential velocities are insufficient to drive them to the walls of the conduit. Also it has been found that in the absence of the narrow bore specimens, which act as efficient particle separators, particles do not deposit on the surfaces of loop pipework. They do however deposit in the slow flow shell side of the regenerative heat exchanger (Fig. 1). Limited experience also indicates that particles do not collect in narrow bore specimens if the loop is operated at oxygen levels of 5 ppm or less, suggesting again that sodium chromite is an important precursor for the removal of particulate material. These factors concerning particle behaviour are of technological imprtance to the design and operation of reactor components.

The difficulties of achieving a mass balance in experiments where oxidised material is breaking away from specimen surfaces has already been identified in a larger loop at RNL operating at oxygen levels of $\sim$ 7 ppm (13). In this loop which models the PFR primary circuit, breakaway of material was observed at the entrance of the heat exchanger section. Such effects not only make the determination of a mass balance difficult but they also confuse the analytical data relating to element transfer, especially as the oxide films can concentrate and transport elements such as Cr, Mn and Si. Removal of elements by the oxidation process also causes the relative enrichment of elements in the substrate which again is not connected with true mass transfer processes. On the practical side the inclusion of new pipework into the loop during any part of the experiments aggravates the problem.

REFERENCES

1. W. F. Brehm, J. C. McGuire, R. P. Colburn, H. P. Mafei, W. H. Olsen, 2nd Int.Conf.on Liquid Metal Technology in Energy Production, Richland, Washington, April 1980.
2. G. Menken, U. Quandt, J. Ebbernik, 2nd Int.Conf.on Liquid Metal Technology in Energy Production, Richland, Washington, April 1980.
3. I. H. Newson, K. T. Claxton, R. W. Dawson, P. Hawtin, 2nd Int.Conf.on Liquid Metal Technology in Energy Production, Richland, Washington, April 1980.
4. A. W. Thorley, A. C. Raine, The Alkali Metals. Special Publication No. 22, The Chemical Society, 1967.
5. P. Roy, D. Dutina, Section 3 GEAP 10394.
6. S. A. Shiels, C. Bagnall, A. R. Keeton, R. F. Witkowski, R. P. Anantatmula, 2nd Int.Conf.on Liquid Metal Technology in Energy Production, Richland, Washington, April 1980.

7.  H. U. Borgstedt, 2nd Int.Conf.on Liquid Metal Technology in Energy Production, Richland, Washington, April 1980.
8.  B. H. Kolster, 2nd Int.Conf.on Liquid Metal Technology in Energy Production, Richland, Washington, April 1980.
9.  C. S. Campbell, M. W. J. Lewis, 2nd Int.Conf.on Liquid Metal Technology in Energy Production, Richland, Washington, April 1980.
10. P. Gross, G. L. Wilson, Jnl.Chem.Soc.(A) 1970, p. 1913.
11. A. W. Thorley, J. A. Bardsley, Proceedings of Royal Microscopical Society, Vol. 1, Pt. 3, 448 (1966).
12. A. W. Thorley, C. Tyzack, A. Blundell, A. C. Raine, 2nd Int.Conf.on Liquid Metal Technology in Energy Production, Richland, Washington, April 1980.
13. A. W. Thorley, C. Tyzack, B. Longson, A. C. Raine, Int. Conf.on Liquid Metal Technology in Energy Production, Champion, Pennsylvania, April 1976.

# THE CORROSION OF IRON IN SODIUM AND THE INFLUENCE OF

# ALLOYING ELEMENTS ON ITS MASS TRANSFER BEHAVIOUR

Alan W. Thorley

United Kingdom Atomic Energy Authority
Risley
United Kingdom

## INTRODUCTION

In order to establish what factors affect the corrosion behaviour of iron and ironbase alloys in liquid sodium simple experiments have been conducted in pumped loop and static sodium environments at the Risley Nuclear Laboratories (RNL) using pure iron and iron-base alloys containing different amounts of oxidising elements. The main purpose of the experiments has been to try and establish the role of oxygen in the corrosion process and how certain alloying elements can change the corrosion mechanism from one of surface corrosion to one of internal oxidation. This paper reports the findings from these experiments and highlights those factors which have to be taken into consideration in the formulation of corrosion models for constructional steels in sodium.

## EXPERIMENTAL

### Loop Studies

The corrosion behaviour of pure iron specimens in sodium has been studied using a simple loop made from mild steel (Fig. 1). The loop was operated with a minimum temperature differential ($\Delta T$) across the circuit and the loop design was such that the circuit had no cold-trap, valves or dump tank. The reasons for this approach were respectively: to eliminate the possible transfer of oxidising elements such as Cr from stainless steel to the pure iron surfaces, secondly to prevent side reactions involving the formation of complex oxides such as $(Na_2O)_2$, FeO which are known to form on the cold side of the standard RNL test loop (1), and thirdly to eliminate cold spots which can become competitive sinks

19

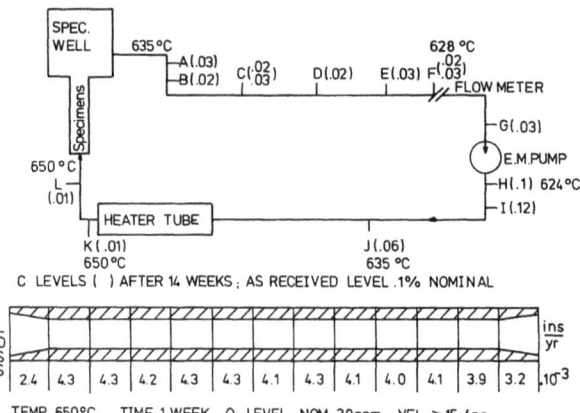

Fig. 1a: Schematic arrangement of mild steel loop

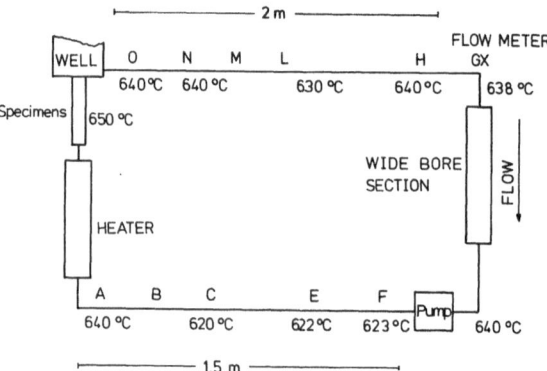

Fig. 1b:   Temperature distribution around the loop
           & analytical data at various positions

for the oxygen in the sodium. All these factors can affect the oxygen level and therefore frustrate the objective of the experiment which was to study the effect of corrosion of pure iron specimens in the loop on the measured oxygen level of the sodium.

In parallel with the mild steel loop a second loop of identical design was constructed from Type 321 (18.8.Ti) stainless steel. (Fig. 1). The corrosion specimens used in this loop were made from Type 316 stainless steel. Both loops contained sodium of distilled quality and oxygen was added in the form of sodium monoxide ($Na_2O$) to the sodium before the start of each test. During the test the oxygen level of the sodium was monitored by taking samples of sodium from the loop which were then analysed for oxygen using the distillation method (2). Specimens were exposed to sodium in the hottest part of the loop at 650 $^{\circ}$C and periodically removed from the loop for weighing and examination of the corroded surfaces.

In order to establish whether the corrosion products released from the pure iron specimens were transported as soluble or insoluble material in the sodium, additional tests involving filters and particle separators were undertaken. The filters were made from sintered iron and molybdenum powder (size 5 to 50 μm) while the particle separators were made from small conduits inside which Mo wires were spirally wound to act as turbulence promoters. These units were located in the removable holder, and the iron specimens were welded into the loop pipework upstream of the filters. For this stage of the programme a dump tank was attached to the rig.

To establish whether pure iron corrodes in a simple loop at the same rate as it does in a loop of more complex geometry (3) a third series of tests was carried out at different sodium velocities and oxygen levels so that the effect of these experimental parameters on corrosion rate could be directly compared with rates obtained on other circuits under the same conditions. For these investigations the mild steel pump duct was replaced with a stainless steel unit and a stainless steel flow meter was inserted in the loop to provide a more accurate measurement of sodium velocity.

Turning to the corrosion of the specimens themselves, two factors were considered important. The first was the influence of grain boundaries on surface corrosion and the second was the effect of alloying elements such as Al, Cr, Mo and Si on the nature and structure of the corroding surface. To investigate these factors a single crystal of pure iron was tested in flowing sodium alongside polycrystalline materials and for the second investigation comparisons were made between 3 groups of material, namely: Fe containing small additions of Al, binary Fe-Cr alloys, and commercial Fe Cr Mo Si alloys.

Finally, in order to establish whether the hot side stainless steel components of a test loop contribute soluble iron to the sodium before the sodium enters the corrosion zone, the hot side components of standard figure-of-eight test loop (2), including the heater, were tungsten plated up to the inlet of the specimen holder. The loop was then operated under conditions which allowed direct comparison with corrosion data from other loops which were not tungsten plated.

## Capsule tests

In support of the loop studies capsule tests were used to study the solubility of iron in sodium at different temperatures and oxygen levels, secondly to equilibrate corrosion products removed from the mild steel pipework with sodium of known oxygen content, and thirdly to establish whether oxide phases exist in iron wires after exposure to sodium ferrate powder at 650 $^{\circ}$C and 800 $^{\circ}$C.

In the solubility tests capsules made from pure iron were cleaned and hydrogen fired before filling with sodium of known oxygen content. Certain capsules were rotated and tilted in a furnace during the thermal treatments while others remained static. After heat treatment the capsules were quenched in a mixture of ice and water to prevent iron segregating to the walls of the capsule while other capsules which contained inverted alumina crucibles in the gas space above the sodium were inverted to drain the sodium into the crucibles before quenching. Samples of sodium for analysis were obtained by coring the solid sodium across its diameter with a copper tool, or by dissolving the contents of the alumina crucible. For the reaction studies the corrosion products were equilibrated at 650 $^{\circ}$C in contact with sodium containing 25 ppm oxygen. After equilibration the capsules were quenched in liquid $N_2$. Sodium was then vacuum distilled from the residue at 300 $^{\circ}$C after which the residues were sealed into quartz capillary tubes in an evacuable glove box for subsequent examination by X-rays.

## RESULTS

## Loop studies

From an analysis of the measured oxygen levels and corrosion rates of the pure iron specimens obtained in the mild steel loop (Fig. 2) it appears that the corrosion of the specimens does not significantly influence the oxygen level of the sodium over time-scale of 14 days or longer. For example tests carried out at oxygen levels of   80 ppm (at 80 ppm the formation of a corrosion product based on the assumed formation of FeO should have caused measurable changes in the oxygen level of the sodium) did not indicate any

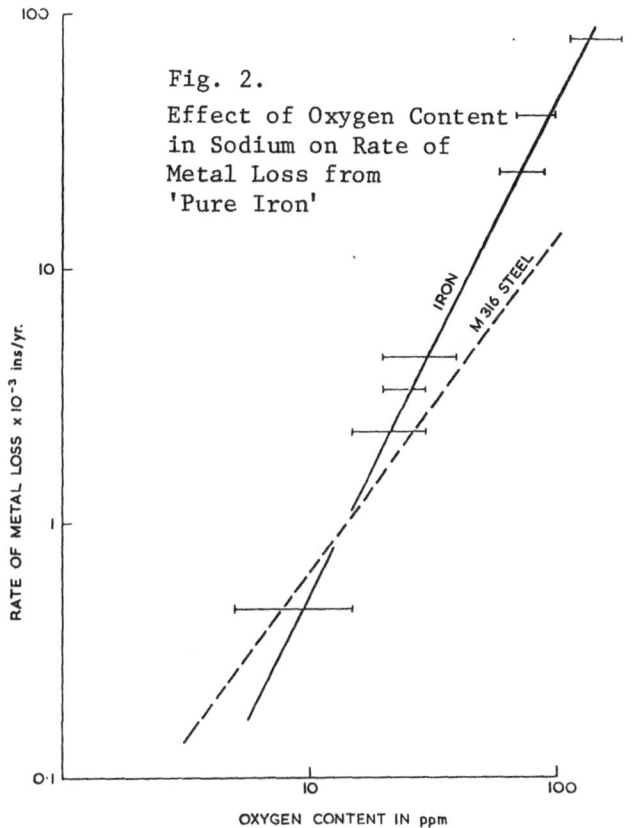

Fig. 2.
Effect of Oxygen Content
in Sodium on Rate of
Metal Loss from
'Pure Iron'

change in the oxygen level of the sodium, neither did tests at
lower $O_2$ levels (Fig. 2). In contrast the oxygen level in the stainless steel loop decreased significantly from similar starting levels
in times of less than 14 days. Examination of the specimens from
both loops indicated significant corrosion had occured and in the
stainless steel loop an exact mass balance was obtained when corrosion losses were compared with the amounts of steel material
deposited in the magnetic flow meter.

Examination of the mild steel loop pipework surfaces showed
that in the hot side of the loop the surfaces were clean and typical
of corroded surface while in the cooler parts debris had collected
on the surfaces; some of the debris contained what appeared to be
single crystals of iron (Fig. 3). Analysis of the loop surfaces
indicated some decarburisation of the mild steel pipework surface,
whereas the pipework in the stainless steel loop was internally
damaged through the formation of sodium chromite (Fig. 4).

From an inspection of the filter assemblies neither filter
assembly collected significant amounts of corrosion products. Some
α iron was detected in the vicinity of the Mo wire but this was not
considered to be significant.

As polished        x 500        As Polished        x 500

Surface of hot side pipework    Surface of lower temperature
between heater and well         pipework

                    x 90                              x 360

Iron deposits in flow meter    Deposits on surface of pipework

Fig. 3:   Isothermal mild steel loop studies

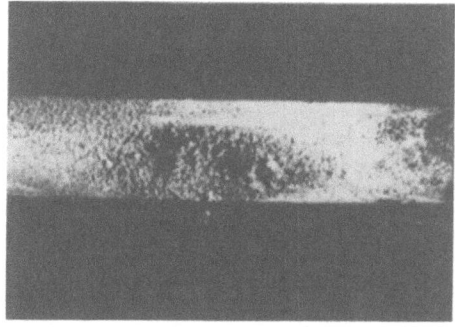

X-ray analysis = 90 % metallics
(70 % α + 30 % γFe)
10 % NaCrO$_2$

Chemical analysis in w/o

Flow meter deposit adhering   x1
to pipework.

| Fe | Cr | Ni | Na |
|------|------|-----|-----|
| 46.4 | 10.2 | 8.7 | 5.4 |

Microprobe analysis (area count)

| Fe | Cr | Ni |
|----|----|----|
| 58 | 33 | 8 |

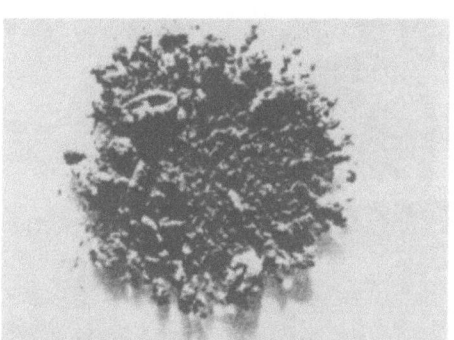

Extracted deposit from flow   x2
meter (Stage 1) position GX.

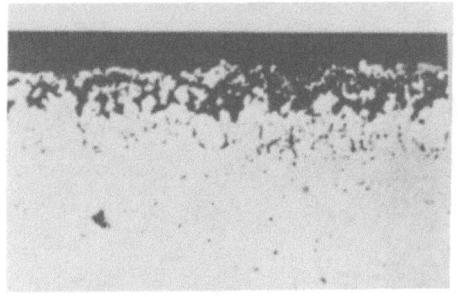

X-ray analysis of material from
pipework surface adjacent to flow
meter section = 50 % metallics
(90 % α + 10 % γFe) + 50 % NaCrO$_2$

AS POLISHED                 x500

Surface of pipework in flow
meter section showing break-
up and oxide phase.

Fig. 4:   Isothermal loop studies. Analysis of products from stain-
          less steel loop.

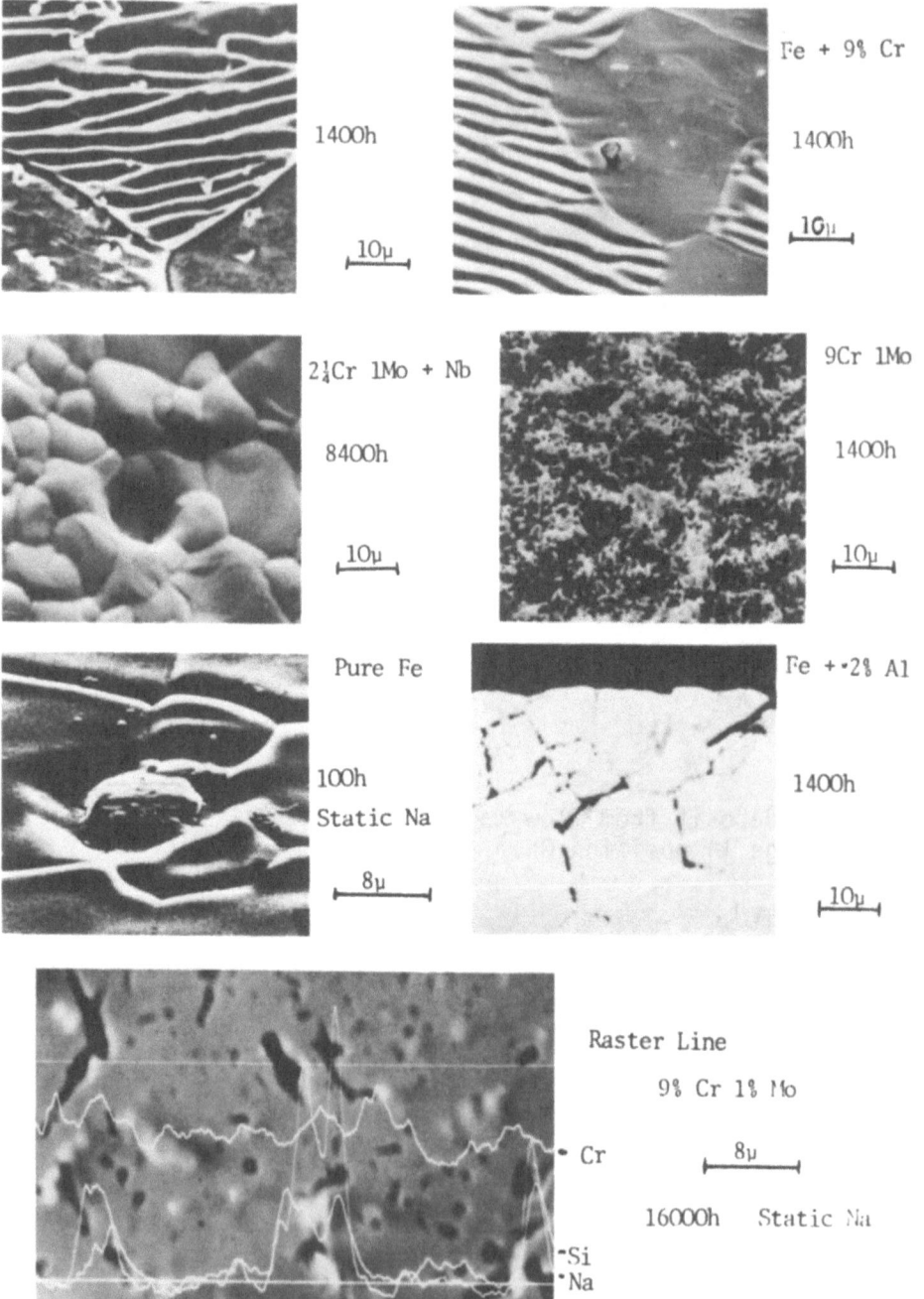

Fig. 5:  Effect of exposure to sodium on surface structure and cross
         section of certain Fe base alloys. $O_2$ level $\sim$ 25 ppm.
         Temp. 600 - 650 $^O$C.

These tests showed that within the limits of measurement the corrosion values obtained at different oxygen levels and sodium velocities were in good agreement with values obtained from the standard RNL test loops (3). On removal of the flow meter after the tests it was found that iron particles had collected in this part of the circuit (Fig. 3).

Neither of these experiments highlighted significant changes in corrosion behaviour. The corrosion rate of the single crystal of iron was if anything slightly higher than polycrystalline material at temperatures of 650 °C and oxygen levels of 25 ppm, whereas the rate of corrosion of Type 316 specimens in the tungsten plated loop was the same as in other circuits for the same experimental conditions. Over the period of test the tungsten had not collected any Fe Ni or Cr from the sodium.

These studies revealed that small amounts of Al (0.2 %) added to iron promotes the formation of a grain boundary phase after exposure to sodium containing 25 ppm $O_2$. This phase is presumably oxide. Corroded surfaces of the various Fe-Cr binary alloys showed no difference in appearance after exposure to flowing sodium at 600 °C. Compare Fe and Fe-9Cr, Fig. 5. Additions of small amounts of Mo (1 %) to commercial alloys promotes the formation of a refined grain structure and a network of surface $M_6C$ carbides in the higher Cr alloys. Si, like Al, is associated with structural changes in the subsurface of the steel. See Fig. 5 for details.

## Capsule tests

The measured iron levels which were obtained at different temperatures and oxygen levels are illustrated in Fig. 6. The figure shows that at the temperatures of interest to this paper (650 °C) oxygen does not appear to affect the solubility level of iron in sodium up to oxygen levels of 100 ppm. Although there is a large scatter in the data the values are in reasonable agreement with certain values determined by Thompson (4) at high oxygen levels.

The equilibration studies did not reveal any change in structure due to interaction with oxygen bearing sodium at high temperature. The corrosion products after test were identified as α-iron.

Interaction between iron wires and sodium ferrate in the absence of free sodium caused a lattice expansion of $\sim 60$ % in the surface of the wire. This expansion could be accounted for by the formation of a non-stoichiometric $FeO_{1-x}$ structure due to interaction with the sodium. The structure was a simple NaCl type with lattice dimensions similar to MgO.

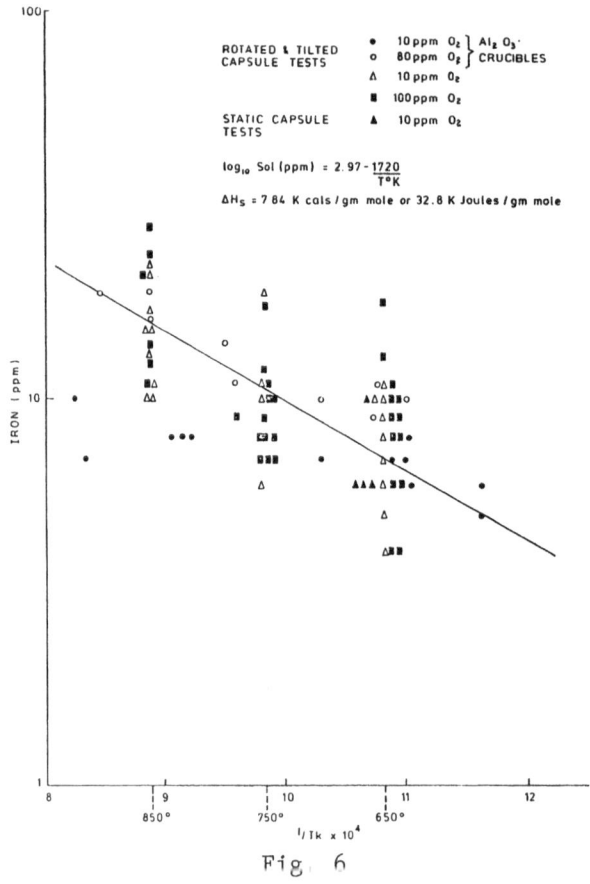

Fig. 6

## DISCUSSION

From a comparison of the behaviour of oxygen in the mild steel
and stainless steel loops it is evident that the behaviour of oxygen
in the mild steel system is more difficult to understand than the
stainless steel system. In the stainless steel loop the formation
of thermodynamically stable oxides, eg sodium-chromite, in the
pipework materials accounts for most of the observed reductions
in oxygen level and the fact that corroded stainless steel particles
deposit in the flowmeter is not surprising in the light of experimental data presented in ref 1.

The surprising finding concerning the mild steel loop experiments is that although the corrosion of pure iron has a greater
oxygen dependency than stainless steel there is no obvious reduction
in the oxygen level of the sodium, suggesting the corrosion products
do not form chemically stable oxides. This conclusion would seem to
be further supported by the ancillary experiments which indicate
that the corrosion debris removed from the loop remains as α-iron

after reaction with sodium containing 25 ppm $O_2$ and oxygen levels up to 100 ppm do not appear to influence the level of soluble iron in sodium.

In trying to establish whether dissolution or oxidation type processes are involved in the corrosion process, the evidence for dissolution seems to centre on 3 observations. One is the appearance of the iron surface after exposure to static and dynamic sodium environments (Fig. 5), the second is the absence of insoluble material in the filter assemblies (the possibility that particles <2 µm diameter may have passed through the filters cannot be ignored (1)), and the third is the absence of oxidation products in the accumulated depris removed from the pipework surfaces of the mild steel loop. Of these 3 observations the appearance of the specimen surfaces probably provides the strongest evidence for dissolution. The surfaces exhibit distinct grain boundary grooving and isolated cracking at triple points after exposure to static sodium at 650 $^o$C (Fig. 5); and after exposure to dynamic sodium they have all the characteristics of an electrolytically etched specimen with active corrosion sites at grain boundaries and the edges of certain crystallographic planes. However, despite these observations it has to be remembered that corrosion rates determined on single crystal specimens are the same as those found on poly-crystalline materials.

Looking at those factors which do not support simple dissolution of Fe in sodium as a major parameter in the corrosion process, consideration has to be given to the following:

a) Iron solubility measurements do not indicate a solubility dependence on oxygen level at oxygen levels of interest (4) despite the fact that oxygen is involved in the corrosion process. However the possibility that solubility measurements are not strictly relevant to flowing sodium systems where the corrosion products are removed from the corrosion zone has to be borne in mind.

b) The determined solubility levels for iron in sodium are in most instances too high to sensibly relate corrosion rates to iron levels in the sodium unless presaturation or fluid resistance effects are incorporated in the mass transfer equation. On the basis of the experimental work using the tungsten plated loop pre-saturation does not appear to be a significant factor.

c) Conversion of the apparent or observed activation energy (3) for the corrosion process of iron based alloys in sodium to a more fundamental value based on constant oxygen activity and not con-centration gives a value of $\sim$ 30 KCals/gram mole. Summation of processes which are influenced by temperature such as solubility, viscosity and diffusion in the liquid phase is not sufficient to account for this relatively high value.

The possibility that an oxidation step may be involved in the
mass transfer process which would accommodate some of the discrepancies identified in items b and c above, does not seem to be
supported by the experimental evidence unless the amount of oxidation product formed in the experiments is too small to be identified.
Tests involving iron wires and sodium ferrate were set up specifi-
cally to look for oxide phases in the wire and although a defect
$FeO_{1-x}$ phase was produced in its surfaces, this finding is probably
not too surprising in that one might expect to find an intermediate
phase based on FeO in the binary $Fe-FeNa_4O_3$ system. Further tests
are in hand to establish the range of non-stoichiometry across the
defect FeO phase as it develops in the wire and to ascertain to
what depth sodium is involved.

In the absence of other experimental evidence to support an
oxidation process an assessment was made of the Na-O-Fe system to
establish whether there were any other factors which might influence
the corrosion process. This assessment revealed (Fig. 7) little or
no evidence of ternary oxides in the Fe-Na part of the system,
although  there is substantial evidence of ternary oxides in the
region contained by the pseudobinary $FeO-Na_2O$ (5,6,7). The Fe-O
binary is also well established and although the principal oxides
are characterised by non-stoichiometry, especially FeO, little is
known about the effect of sodium on defect or sub-oxide structures
based on $FeO_{1-x}$. In the Na rich corner the solubilities of $O_2$ and
Fe have been reported for the respective binary Na-O and Na-Fe
liquid solutions (2,4). Interactions between the soluble species
in regions above the liquidus shelf have not been established
however. In the Fe-rich corner the region of interest centres around
the solid solubility of $O_2$ in the α γ δ allotropes. Solubility
levels have been reported (8) for all three phases and although
the terminal solubility value at 650 $^{\circ}$C ( 0.2 ppm) is lower than
those normally quoted for oxygen in steels, consideration has to
be given to the effect of impurities and the possibility that the
high values do not represent true saturation of the solid solution.
Thus if the steel can still take up oxygen from the sodium then
the oxygen will partition between the steel and the sodium to form
sub-oxides in the steel.

The two well-established ternary oxides shown in the Figure are
$(Na_2O)_2$, FeO or $FeNa_4O_3$ and $NaFeO_2$. According to experimental and
thermodynamic data however neither of these oxides are thermodynamically stable in sodium of low oxygen content, see Fig. 8 and refs
(9) (10) (11). Spinel oxides of the $FeCr_2O_4$ type which are more
thermodynamically stable than FeO (Fig. 8) may favour the formation
of $FeNa_4O_3$ in stainless steel loop systems and this point is being
examined at RNL.

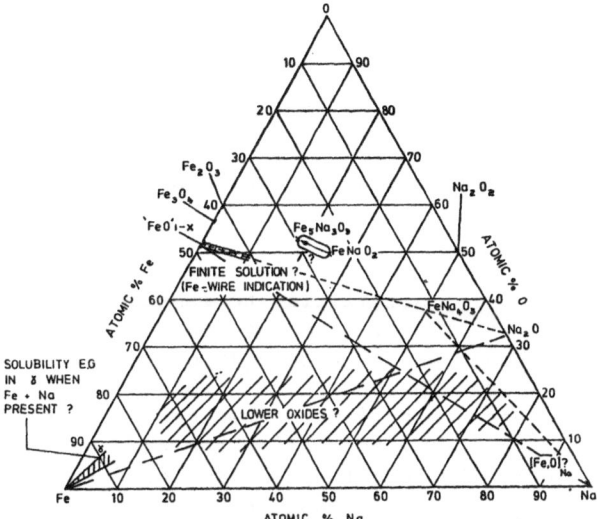

Fig. 7.   The Fe – Na – O System
(Tentative, Schematic)

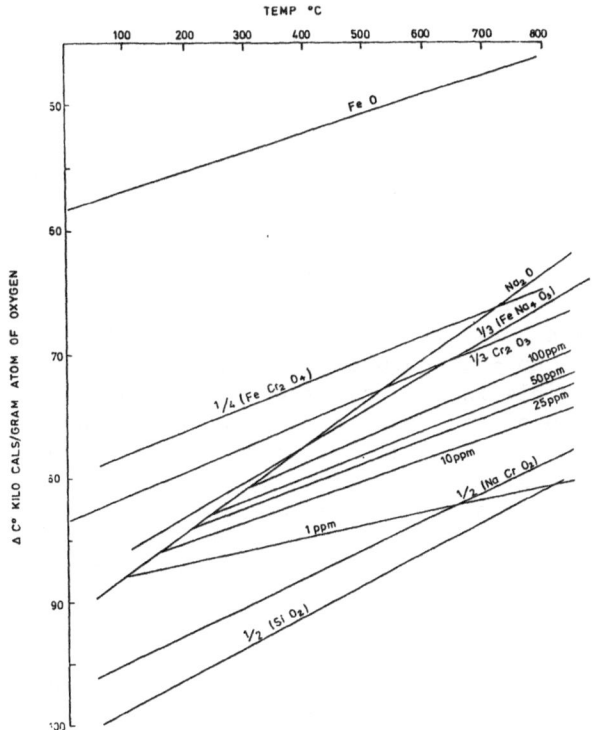

Fig. 8.   Free Energy of Certain Standard State Oxides and
Na-O Solutions as a Function of Temperature

Therefore, summarising this part of the paper, the experimental evidence so far obtained indicates first that the release of iron to the sodium is affected by the oxygen content of the sodium (Fig. 2) and secondly, the process involves very little uptake of oxygen unless the corrosion product is thermodynamically unstable in sodium over a small decrement of $\Delta T$. If this is so then this would imply that the oxygen acts purely as a carrier for the iron in the sodium (12).

In trying to establish the role oxygen plays in the corrosion process there are two possibilities to consider. One is that the oxygen interacts with the iron in the sodium to form a solvated complex probably based on FeO or $FeNa_4O_3$ or secondly surface oxidation also plays a part in the removal process. The formation of a solvated complex has attractions on two counts. First, according to the Law of Mass Action, increasing the chemical activity of the oxygen in the sodium to form the complex would lower the activity of iron in the sodium and increase the driving force for iron transport from the metal and, secondly it introduces a chemical step in the process bearing in mind that mass transfer equations based on iron solubilities cannot be reconciled with existing corrosion data. The possibility that other phases may exist in the Na-O-Fe system requires a more detailed study in the region of the liquides shelf to identify the existence of any new phases and to measure interactions between soluble oxygen and iron when oxygen is added to the sodium.

Effect of alloying elements on oxygen behaviour in sodium

Although earlier work has shown (3) that the corrosion rates of the major alloying elements of stainless steel (Fe Ni Cr) have different oxygen dependency terms eg n = 2 for Fe, 1 for Cr and 0 for Ni rich alloys, it has been found from experience that because of the non-stoichiometric nature of the corrosion process it is not possible in most cases to predict an 'n' value based on the composition of the original alloy. Exceptions to this may be found in the nickel rich alloys where the dilution of Fe by Ni may, with increasing nickel content, change the value of n from 2 to 0, as long as other factors are not involved. However in the more complex nickel-rich alloys cognisance has to be taken of the effect of other alloying additions, especially in alloys such as PE16 and Incoloy 800 which are known to be oxygen sensitive materials with n values much greater than zero (3).

Other examples where predictions of corrosion behaviour based on original composition can be misleading are seen in Fe-Cr binary experiments, and are also in work reported in ref. 13. In the Fe-9Cr binary for example it could be anticipated that Cr, because of its greater thermodynamic affinity for oxygen, will be protective to iron, and n would change from 2 to say 1. In practice this is not so because all the alloys tested in high velocity sodium tended towards a common Cr content through selective removal of Cr from the specimen surface either through the formation of sodium chromite or dissolution processes. Thus 'n' for all three binaries still approximates to 2 after some initial period of exposure to sodium.

From a comparison of the structural changes which occur in the various alloys the effect of Mo in the commercial alloys appears to be of particular significance (14). The general impression from the current series of tests is that Mo besides being a good solid solution strengthener also produces a more refined and stable grain size (Fig. 5) in both unstabilised and stabilised 2 1/4Cr1Mo alloys and 9Cr1Mo steels. It also produces a network of surface $M_6C$ carbides in the 9Cr1Mo alloys. In addition it seems to impart some of its good corrosion resistance in sodium (15) to the iron layer in that it removes active corrosion sites. However in practice 1 % Mo additions are not sufficient to affect the oxygen dependency term for 2 1/4Cr1Mo steel, and n for both 2 1/4Cr1Mo and 9Cr1Mo (preliminary values) approximates to 2 (16). Mo additions such as the 3 % level used in Type 316 stainless steel could, in principle, have a more beneficial effect.

One of the major concerns relating to the effects of alloying additions to iron base alloys is that oxidising elements such as Al, Nb, Ti and Si may change the surface corrosion process into a more damaging one involving internal oxidation. Experiments have shown that when sinks for oxygen such as Al exist in the matrix of the Fe the more mobile oxygen enters the steel to cause internal damage and possible becomes the precursor for other reactions if the oxygen level of the sodium is high enough (16). Whether Si can be involved in a similar fashion in the high Cr ferritic steels is currently under consideration at RNL. $SiO_2$ is stable at all practical oxygen levels in sodium (Fig. 8) and the question centres around whether the influence of Ni and Cr in the respective austenitic and ferritic materials coupled with the greater mobility for interstitial movement in the ferritic alloys is responsible for producing interactions in the ferritic alloys which so far have not been observed in austenitic steels. For details of other damaging mechanisms see refs. 17, 18.

## The effect of alloying on corrosion behaviour

A knowledge of the effect of alloying elements on the surface structure of steels exposed in sodium is of importance if experimental variables, such as temperature, sodium velocity and downstream position are to be sensibly related to the more fundamental aspects of corrosion behaviour in sodium. Although pure iron and the low alloy steels do not appear to present any problems, because after initial exposure their structure and surface composition remain sensibly constant, the high chromium content stainless and ferritic

steels have been known to retain oxide films (NaCrO$_2$) for
exceptionally long times under certain conditions of test and
although weight losses are recorded, this may be more associated
with the removal of oxide than true corrosion of the surface.

Generally speaking, these problems are more apparent at low
temperature, < 550 °C, low velocities and high oxygen levels
($\sim$ 25 ppm). Experience has shown for example, that in measurements
of 1/d effects on corrosion (downstream position) at sodium
velocities of 2-3 m/sec, oxide films are retained at low velocities
down the 1 metre length of specimens although the inlet specimen
surfaces have been transformed to ferrite. Similar findings have
also been observed in the Risley Large Mass Transfer Loop (19)
when corrosion losses have been measured at different temperatures
along a specimen length to obtain an estimate of the apparent
activation energy of the corrosion process; once again oxide films
were retained in the lower temperature positions. As oxide formation
has also been observed in low velocity situations (1) this brings
into question whether the recorded changes in weight as a function
of velocity at the lower levels represents diffusional transport
of material through a boundary layer or whether they are simply
recording loss of damaged or oxidised material. Investigations are
therefore continuing at RNL to establish a relationship between
surface composition and structure and the experimental findings
so that the corrosion behaviour of certain constructional steels
can be sensibly evaluated.

CONCLUSIONS

Experimental work undertaken in small simple loops made from
mild steel where there were no competing sinks for oxygen has shown
that the corrosion of iron in sodium does not lower the oxygen of
the liquid metal. This contrasts with the behaviour of oxygen in
sodium contained in stainless steel circuitry where the oxygen
level was effectively reduced through the formation of sodium
chromite in the hot stainless steel pipework.

The corrosion rate of iron in sodium as a function of oxygen
level at 650 °C is the same in all types of loop geometry
irrespective of material of construction (mild steel or stainless
steel) or the magnitude of the $\Delta$T. The determined oxygen dependency
term 'n' approximates to two in all these systems and this value
suggests that iron is more sensitive to the oxygen level of the
sodium compared with stainless steels and nickel-rich alloys which
have oxygen dependency terms of 1.5 and 0 respectively. However,
although the corrosion of iron is sensitive to oxygen level, no
oxidation products have been identified in the simple mild steel
loop tests. One positive conclusion from the tests is that the
corrosion of iron in sodium is not influenced by oxidation and
precipitation of compounds such as sodium ferrate on the cold
side of a test loop.

Other factors which might influence the corrosion process such as the effect of oxygen on the solubility of iron, pre-saturation effects before the sodium enters the corrosion zone and grain boundaries appear to have little influence on the corrosion behaviour of iron-base materials in sodium.

The tentative conclusion from the experimental work is that iron corrodes by a dissolution process through the formation of a solvated chemical complex based on FeO or $FeNa_4O_3$. Experimental evidence suggests that the complex is formed transiently perhaps becoming thermodynamically unstable over a small decrement of temperature.

Metallographic and SEM analyses of specimens of iron-chromium alloys containing various levels of chromium indicate that the appearance of their corroded surfaces is similar in all cases after exposure to sodium irrespective of their initial chromium content. This is attributed to the selective removal of chromium from the steel during the initial stages of exposure to flowing sodium. The finding is typical of effects observed in stainless steels after exposure to sodium at high temperatures, or for longer durations at lower temperatures and also in more complex nickel alloys. From an academic standpoint these findings highlight the importance of understanding how iron corrodes in sodium.

The addition of molybdenum to iron-chromium alloys causes a change in the grain structure and the removal of active corrosion sites, while additions of aluminium and silicon to iron or iron-chromium-molybdenum alloys respectively can be associated with internal damage at 600 to 650 $^{\circ}$C if the oxygen level of the sodium is high enough.

The possibility that higher chromium content steels may retain oxide films for some indefinite period of time during exposure to sodium has to be taken into consideration when evaluating the effect of certain experimental parameters on the corrosion behaviour of high chromium content steels in sodium.

REFERENCES

1.  A. W. Thorley, A. Blundell, A. Bardsley, Mass transfer of stainless steel in pumped loops and its effect on micro-structure. Paper this Symposium.
2.  A. W. Thorley and A. C. Raine, The Alkali Metals Int. Symp., Nottingham, July 1966, Special Publication 22, Chem. Society, pp. 374 - 392.
3.  A. W. Thorley and C. Tyzack, Liquid Alkali Metals Proc. Int. Conf. BNES Nottingham, 1973, pp. 257 - 273.
4.  W. P. Stanaway and R. Thompson, Solubility of materials in sodium. 2nd Int.Conf. on Liquid Metal Tech. in Energy

Production, Richland, USA, 1980.

5.  A. M. El Balkji et al. J. Solid State Chem., 1976, 18(3)
    pp. 293-7.

6.  A. M. El Balkhi et al. Comptes Rendus (C), 1977, 285,
    120-131.

7.  A. Tschudy, H. Kessler and A. Hatterer, Liquid Alkali
    Metals Proc. Int. Conf. BNES, Nottingham, 1973, pp.209
    -212.

8.  J. H. Swisher, E. T. Turkodgen, Trans.Met.Soc. of AIME,
    Vol. 239, April 1967, pp. 426 - 431.

9.  C. C. Addison, M. G. Barker, A. J. Hooper, Jnl. of Chem.
    Soc. Dalton Trans. 1972, pp. 1017 - 1019.

10. G. W. Horsley, Jnl. of Iron & Steel Inst., Jan. 1956,
    pp. 43 - 48.

11. P. Gross, G. W. Wilson, Jnl. Chem. Soc. (A), 1970, 1913.

12. B. H. Kolster, Jnl. of Nuclear Mtls. 55 (1975) pp.155-168.

13. C. Bagnall, R. E. Witkowski, Metal Progress 1979 pp.50-51.

14. S. Shiels et al. 2nd Int.Conf.on Liquid Metal Tech. in
    Energy Production, Richland, USA, April 1980.

15. A. W. Thorley, C. Tyzack, Corrosion of molybdenum in sodium.
    RNL Internal document.

16. C. Tyzack, A. W. Thorley, Proc. Int. Conf. on Ferritic
    Steels for Fast Reactors, BNES London, Paper 39,
    pp. 39-1 to 39-14.

17. A. Crouch, The growth and stability of sodium chromite
    and its influence on corrosion. 2nd Int.Conf. on
    Liquid Metal Tech. in Energy Production, Richland,
    USA, April 1980.

18. A. J. Hooper, Proc.Int.Conf. on Ferritic Steels for Fast
    Reactors, BNES London, 280, 1977.

19. A. W. Thorley et al. 2nd Int.Conf. on Liquid Metal Techn.
    in Energy Production, Richland, USA, April 1980.

# THE DEPOSITION BEHAVIOUR OF Fe, Cr, Ni, Co and Mn IN STAINLESS STEEL SODIUM LOOPS

Benjamin H. Kolster, Johan v.d. Veer and Leo Bos

Metaalinstituut TNO
Apeldoorn
The Netherlands

## INTRODUCTION

In the literature the corrosion of stainless steel in liquid sodium is extensively reported. Less is known about the deposition behaviour of the elements from stainless steel, that is Fe, Cr, Ni and Mn. It is conceivable that the deposition behaviour of these elements as a function of the relevant parameters provide information with respect to the corrosion and mass transport mechanism of the steel. This was the main reason to study the deposition behaviour of the above mentioned elements. In addition the deposition behaviour of Co was studied, which together with Mn are the main radionuclides released by the canning materials. The parameters taken into account were the temperature, the oxygen level and the downstream position. As only isothermal loop systems were at our disposal, a driving force different from a thermal force had to be created. To that end various metals as Co, Ni and Fe (in two runs Mn as well) were taken as specimens to getter the steel elements. Thus, the elements to be gettered are released by the loop system, except for Co, and by the specimens, except for Cr and in most runs for Mn.

After each run the weight change of the specimens and the chemical composition of the deposit was determined. In addition the deposited layer was examined by means of optical microscopy, SEM and electron microprobe analysis. This paper only deals with the weight change results.

## EXPERIMENTAL

Table 1 shows the condition with respect to temperature and oxygen level of the 10 runs carried out, the symbols 11 and 12 refer

to the loop systems. The temperature dependence is determined at
four temperatures, at 15 ppm oxygen and in loop system 12. The
oxygen dependence is determined at 7 oxygen levels, at 700 °C and in
loop systems 11 and 12. In each run Co, Ni and Fe specimens were
exposed, except for the runs at 14 and 26 ppm, for which Co, Ni and
Mn specimens were used. The specimens were spread over three channels,
each channel containing 9 platelets of one metal. The platelets of
33 x 10 x 2 mm were situated downstream to each other in order to
determine a possible downstream effect (1).

Table 1.   The temperatures and oxygen levels for the runs in loop
           systems 11 and 12.


                              Oxygen level (ppm)

| Temp. (°C) | 3 | 6 | 10 | 14 | 15 | 20 | 26 |
|---|---|---|---|---|---|---|---|
| 700 | 11 | 12 | 11 | 11 (Mn) | 12 | 11 | 12 (Mn) |
| 600 | | | | | 12 | | |
| 500 | | | | | 12 | | |
| 400 | | | | | 12 | | |

       The flow velocity in the specimen holders was about 6 m/s for
all runs under consideration. The exposure time for each run was
about 2.000 h.

       Following on the exposure the weight change and the metallo-
graphic appearance of the deposit was determined. Furthermore the
chemical composition of the deposit was analysed. To that end each
specimen was entirely dissolved and by means of atomic absorption
the deposited elements were analysed. In the same way the starting
getter materials were analysed.

RESULTS

       For the following description of the results one should realize
some constraints with respect to the values analysed. For instance,
the amount of Co deposited is very small. If the getter material
contains Co as an impurity, as for Ni, the resulting amount of Co
deposited is calculated from the difference of two small numbers,
generally giving a poor reliability. In these cases the results are
not considered in the following. The same holds for specimens that
have been used for metallographic investigations before chemical
analysis. For each run the average of the maximum number of specimens was considered.

It may be stated that none of the deposited metals shows a downstream dependent deposition rate. This was observed at each temperature and for each oxygen level applied. A representative example is given in Fig. 1, showing the deposition rate of Fe, Mn and Cr as a function of downstream position $L/D_h$ on Ni as the getter material (L = distance from entrance, $D_h$ = hydraulic diameter).

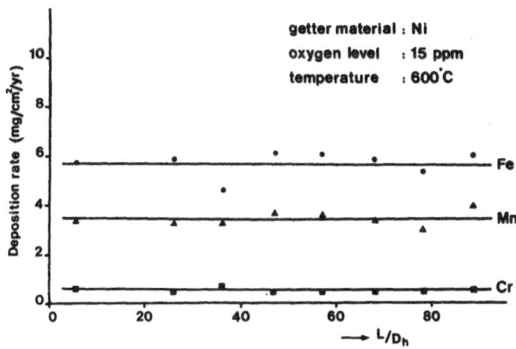

Fig. 1   Deposition rate for various elements on Ni as a function of downstream position

Considering the deposition rate as a function of oxygen level one should realize that the results refer to two loop systems (cf. Table 1). Hence, the question is whether the loop system itself affects the results. To that end we plotted the average weight change for each getter material as a function of oxygen level (Fig. 2). Obviously, the deposition rate on Fe and Co exceeds the corrosion rate for these materials, whereas the reverse holds for Ni. Moreover, the weight change for the Fe and Co specimens show a continuous course, indicating no effect of the loop system on the deposition behaviour for these metals.

Nickel, however, shows two different plots referring to loop system 11 and 12. This must be due to the influence of the loop system, either on the corrosion rate of Ni or on the deposition rate of the elements gettered. From the following results it may be concluded that the deposition rate for the individual elements does not show any difference between the loop systems 11 and 12, meaning that the effect in Fig. 2 reflects a distinct corrosion rate for Ni in the two systems. The reason for this is not clear.

The deposition curves for the various elements are given in Fig. 3 - 6. It should be noted that, unlike Fig. 1, the curves now refer to one deposited metal on the various getter materials.

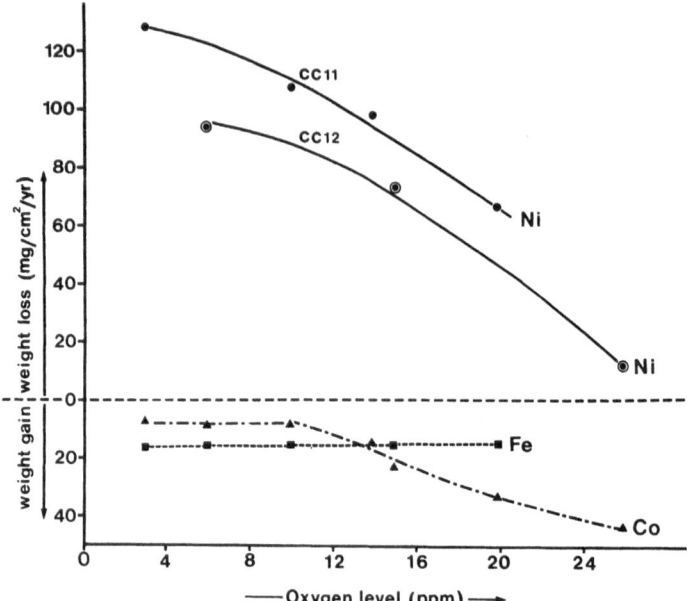

Fig. 2   The accumulated weight change for the getter materials as
         a function of oxygen level

   The deposition of Mn obviously is not dependent on oxygen level
(Fig. 3). Nickel serves as a much more effective getter than Fe and
Co. The high values for 14 and 26 ppm oxygen are due to the simul-
taneous exposure of Mn specimens (Table 1).

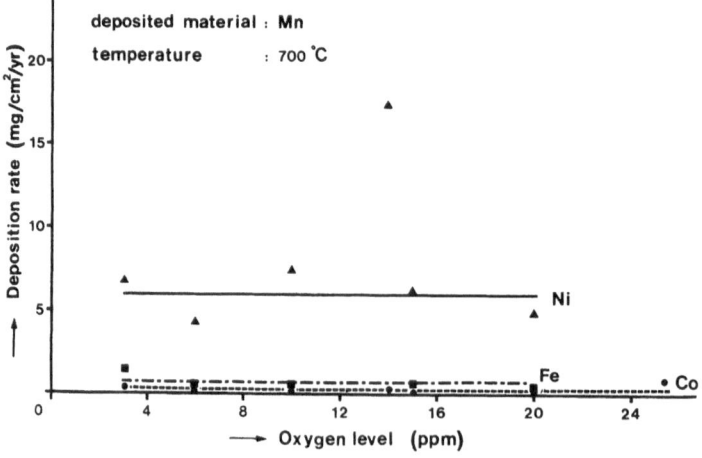

Fig. 3   The deposition of Mn on various getter materials as a
         function of oxygen level

The deposition of Ni on Fe shows an oxygen independent behaviour (Fig. 4). For Co as a getter material two regions can be distinguished, below 14 and above 15 ppm oxygen, both being oxygen independent but at a different level.

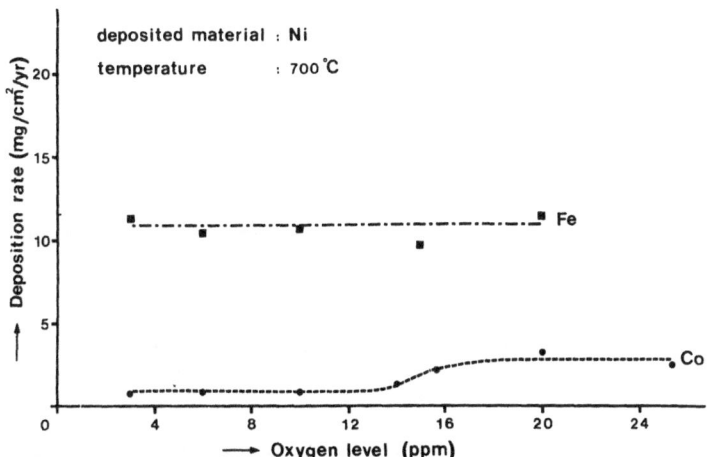

Fig. 4   The deposition of Ni on various getter materials as a function of oxygen level

The deposition of Cr shows a similar trend as Ni on Co (Fig. 5). Up to 14 ppm oxygen a weak or no oxygen effect can be observed. Subsequently an increase in deposition rate between 14 and 15 ppm oxygen can be noticed up to a level that remains the same for higher oxygen levels. The deviating position for the Co getter at 26 ppm is reflected by a quite different composition and morphology of the deposit. So, in this respect this point is suspect and not involved in the curve.

The most remarkable deposition behaviour shows Fe (Fig. 6). At low oxygen levels a slight or no oxygen effect can be observed. A considerable increase in deposition rate takes place between 14 and 15 ppm oxygen. Unlike the behaviour of Cr, with increasing oxygen level the deposition rate still increases, albeit at a lower slope than between 14 and 15 ppm oxygen.

The influence of temperature on the deposition rate is given in Figs. 7 - 11. The curves reflect the distinct affinity between getter and deposited material. In this respect it is worthwhile to consider the difference between Ni and Fe/Co in their ability to getter Mn, indicating that Ni may be used as a getter to remove selectively radio active Mn.

Apart from these qualitative consideration we calculated the activation energy for deposition. The results are given in Table 2.

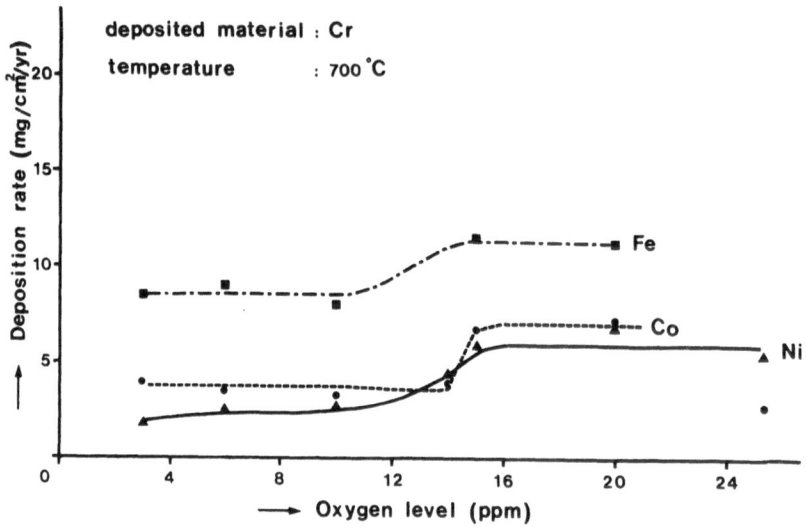

Fig. 5   The deposition of Cr on various getter materials as a
         function of oxygen level

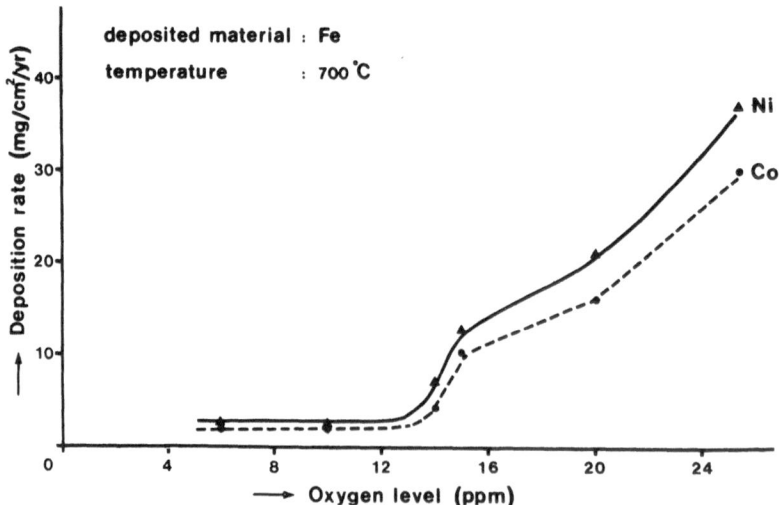

Fig. 6   The deposition of Fe on various getter materials as a
         function of oxygen levels

Table 2.   Activation energies for deposition

| Deposited material | Substrate material | Measuring points | Activation energy (kcal/mol) | Correlation coefficient | Average act. energy (kcal/mol) |
|---|---|---|---|---|---|
| Fe | Ni | 4 | 16.8 | 0.996 | 16.8 |
|    | Co | 4 | 16.8 | 0.998 | |
| Mn | Ni | 4 | 14.6 | 0.988 | 15.9 |
|    | Fe | 4 | 17.2 | 0.963 | |
|    | Co | 4 | 8.9 | 0.743 | |
| Cr | Ni | 3 | 36.4 | 0.998 | 32.5 |
|    | Fe | 4 | 28.6 | 0.889 | |
| Ni | Fe | 4 | 27.6 | 0.959 | 27.9 |
|    | Co | 3 | 28.2 | 0.913 | |
| Co | Fe | 4 | 5.0 | 0.941 | 5.0 |

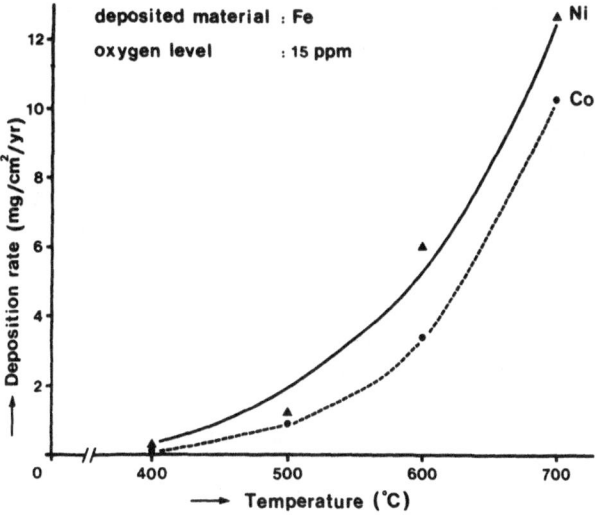

Fig. 7   Deposition of Fe on various getter materials as a function
of temperature

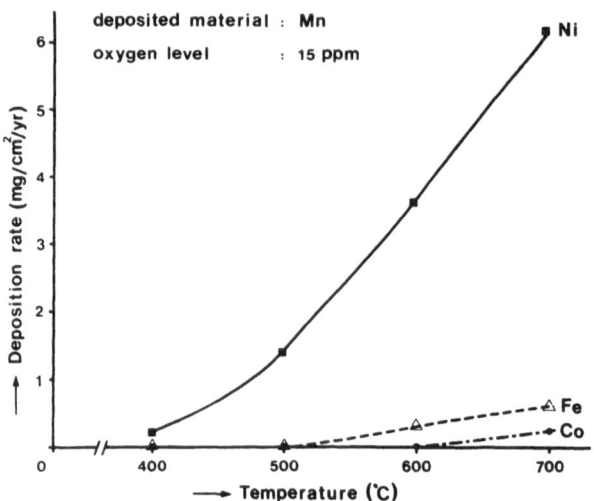

Fig. 8   Deposition of Mn on various getter materials as a function
         of temperature

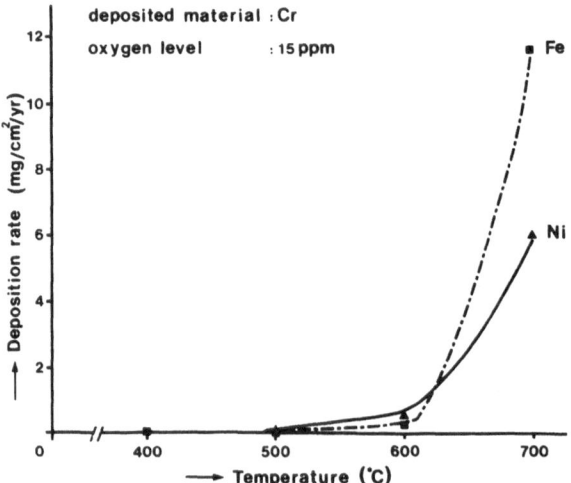

Fig. 9   Deposition of Cr on various getter materials as a function
         of temperature

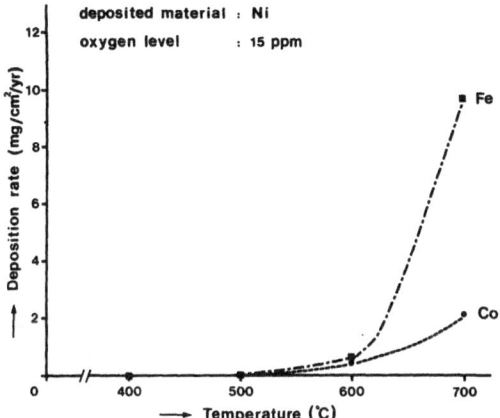

Fig. 10   Deposition of Ni on various getter materials as a function
          of temperature

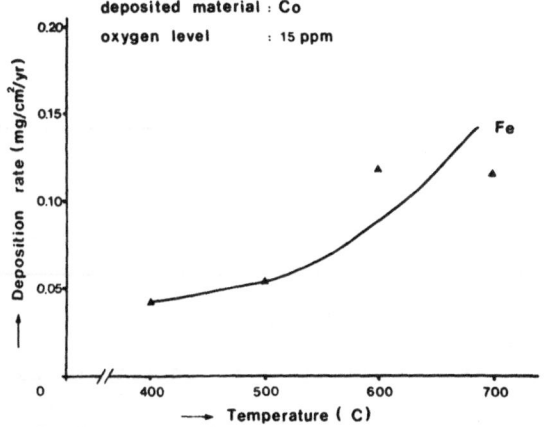

Fig. 11   Deposition of Co on Fe as a function of temperature

     The figures show a remarkable agreement for each deposited
element, irrespective of the getter material. That is to say, if
the correlation coefficient indicates a high reliability. The only
exception is shown by Mn on Co. For the calculation of the average
activation energy the letter value is not taken into account. The
physical meaning of the difference in activation energy for the
deposited materials needs further study and will be given in due
time.

DISCUSSION

     The most interesting result to point out concerns the oxygen
dependence of the deposition of Fe and Cr. As the deposition be-

haviour reflects also the corrosion characteristics, the question
arises whether the deposition results are in line with the corrosion
phenomena observed till now. To that end the characteristic deposition behaviour of Cr and Fe is schematically represented in Fig.
12. Three regions may be distinguished:

    - up to an oxygen level of $x_1$ the deposition of Fe and Cr is
similar, that is weakly dependent or independent on oxygen level.

    - at $x_1$ a sharp increase in deposition rate occurs for both
elements, up to $x_2$,

    - at $x_2$ a second inflexion point can be observed, giving rise
to a diminished increase for Fe deposition, whereas for Cr the deposition rate is no longer affected by the oxygen level.

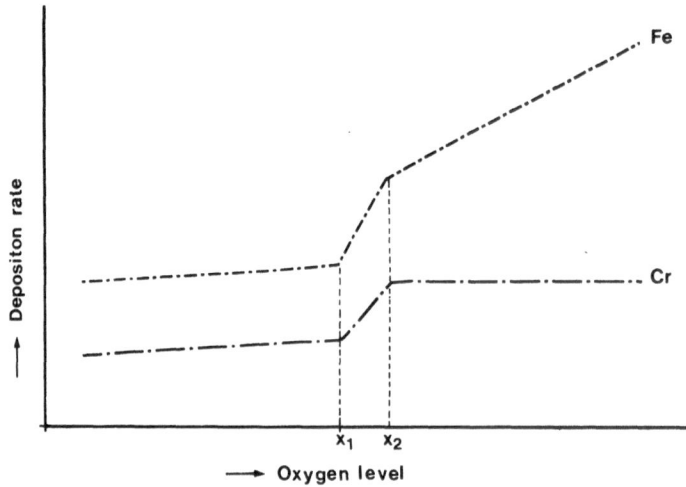

Fig. 12   Deposition of Fe and Cr as a function of oxygen level

    As far as the region up to $x_1$ is concerned the results reflect
the low oxygen corrosion behaviour for steel. This corresponds with
the situation of ferritization of the stainless steel. It should be
emphasized that under steady state conditions Fe, and only Fe,
determines the corrosion rate of the steel, in agreement with the
model proposed previously (2). It is not the diffusion of Cr and
Ni through the ferritic layer that determines the corrosion rate of
steel, as suggested by a number of authors. Having stated this, it
can be understood that the deposition rate of Cr is similar to the
one for Fe up to an oxygen level $x_1$. The very weak oxygen dependence
is in agreement with previous results (2).

    The inflexion point $x_1$ represents a sudden increase of the
corrosion of Fe. We do not know an explanation for this phenomenon,
but we have experimental indications from corrosion experiments
with Fe in a Mo-loop system that such a sudden increase occurs.

This behaviour is still investigated and will be published. The point to make here is that if such an increase takes place for Fe, Cr again will follow with a higher corrosion rate. We think that this is indeed reflected in the deposition behaviour, that is up to an oxygen level $x_2$.

At the inflexion point $x_2$ another phenomenon occurs. At a certain oxygen level it may be anticipated that sodium chromite will be saturated, as we have noticed in another isothermal loopsystem (3). Whether the formation of $NaCrO_2$ is according to a two step mechanism, as proposed in (3) or according to the following reaction is not relevant in this respect:

$$Cr + Na + 2O \rightleftarrows NaCrO_2$$

The important point is that the solubility product of $NaCrO_2$ will be exceeded at a specific oxygen level, after which $NaCrO_2$ will precipitate. However, according to the model for $NaCrO_2$-transfer in Na (4) this compound will only be precipitated on steel substrates, that is on the walls of the loop system. The Cr deposited on the Fe, Ni and Co specimens is metallic Cr, in equilibrium with $NaCrO_2$, according to the above mentioned reaction. So, if the solubility product for $NaCrO_2$ is reached, the concentration of Cr in Na remains constant. Consequently, the deposition rate for Cr will be constant. This happens at the oxygen level $x_2$, reflecting a deposition rate which is no longer oxygen dependent. Having realized this, it may be understood why the deposition rate for Fe proceeds with a slower rate compared to the one at oxygen levels between $x_1$ and $x_2$. At oxygen levels $> x_2$ oxygen is consumed by Cr, leaving less oxygen for the reaction with Fe. In other words the available amount of oxygen for Fe corrosion is less than the overall oxygen level.

CONCLUSIONS

The difference in deposition rate on various getter materials is only very pronounced for Mn, in the sense that Ni, unlike Fe and Co, is very effective in gettering Mn.

The activation energy for deposition is independent on the getter material and shows a very high reliability. The difference in activation energy (Co: 5, Mn: 16, Fe: 17, Ni: 28 and Cr: 33 kcal/mol) needs further study in terms of its physical significance.

None of the deposited elements show a downstream dependent deposition rate.

The oxygen dependence for Cr and Fe deposition reveal three regions of which the low oxygen regimes reflect the corrosion behaviour of Fe as the rate determining component, whereas the third region reflects the formation of sodium chromite in concentrations exceeding the solubility product.

ACKNOWLEDGEMENT

   This work was performed as part of the SNR development program.
It was sponsored by the Project Group for Nuclear Energy TNO. The
authors are much indebted to D. Schakelaar of the Central Technical
Institute TNO for the chemical analyses.

REFERENCES

   1.  B. H. Kolster and L. Bos, Proc. Intern. Conf. on Liquid
       Metal Technology in Energy Production, Champion, PA,
       (1976) p. 368.
   2.  B. H. Kolster, Proc. 7th Int. Conf. on Metallic Corrosion,
       Rio de Janeiro (1978), p. 1524.
   3.  P. L. F. Rademakers and B. H. Kolster, J. Nucl. Mat. 97,
       (1981), 309.
   4.  B. H. Kolster, J. Nucl. Mat. 55, (1975), 155.

# CORROSION RATE AND DOWN STREAM EFFECT IN LIQUID METAL SYSTEMS

Norbert F. Schwarz

Österreichisches Forschungszentrum Seibersdorf
Wien
Austria

## INTRODUCTION

One of the problems encountered in the liquid sodium corrosion of austenitic steels and other materials is the dependence of the corrosion rate on the down stream position. This effect has been described in the literature extensively and explanations for this phenomenon have been presented. Accordingly the following causes can be considered to dominate the down stream effect

- saturation of metallic element dissolution
- consumption of a corrosion inducing agent, e.g. $O_2$
- poisoning of active corrosion sites by a corroding or dissolving species
- hydraulic phenomena, growth of a laminar sublayer.

There are two reasons why the down stream effect is of special interest; firstly knowledge of the down stream function allows to extrapolate measured corrosion rates to the maximum possible rate and secondly an interpretation of the down stream function will have implications on the general explanation of measured corrosion rates. Since the geometry of the test section used in our experiments has an unusually large $L/d_H$ ratio (length/hydraulic diameter) of 630 it will be the aim of this paper to give a presentation of corrosion runs performed with special emphasis on results being connected with the interpretation of down stream phenomena.

## EXPERIMENTAL

All tests were performed in the high temperature sodium loop system HT 1. In the maximum temperature test section, two specimen

49

holders were arranged in series, a low velocity holder with 32
corrosion samples before a high velocity holder with 63 corrosion
samples. When not otherwise stated, the informations given refer
to the high velocity specimen holder only. Runs 1.10 and 1.12 were
interrupted corrosion runs with weight change measurements and
SEM investigations of selected samples in consecutive time intervals,
so that the kinetic behaviour could be studied.

Table 1.   Test parameters of considered runs

| | exposure conditions | | | | measured sodium conditions in ppm | | | | |
|---|---|---|---|---|---|---|---|---|---|
| run | temp. °C | vel. m/s | time accumulated h | cold trap Temp. °C | $C_O$ | $C_O$ | $C_{Ni}$ | $C_{Cr}$ | $C_{Fe}$ |
| 1.06 | 734 | 4.2 | 1500 | 150 | 3.5 | 5 | | | |
| 1.10.1 | 733 | 4.0 | 117 | 180 | 8.0 | | | | |
| 2 | | | 342 | | | 15 | 3 | 5 | 15 |
| 3 | 738 | | | | | 10 | 2 | 3 | 11 |
| 4 | 1577 | | | | | 15 | 1 | 1 | 3 |
| 5 | 2637 | | | | | 10 | 1 | 1 | 3 |
| 1.12.1 | 730 | 2.5 | 117 | 170 | 6.0 | 50 | 4 | 7 | 19 |
| 2 | | | 327 | | | 100 | 4 | 4 | 23 |
| 3 | | | 726 | | | 70 | 4 | 3 | 50 |
| 4 | | | 1500 | | | 60 | 3 | 4 | 15 |

All corrosion specimens in the test sections of run 1.10 and
the greater part of those in 1.12 were fabricated of the standard
material SS 1.4981 with the following chemical analysis:

| C | Cr | Fe | Mo | Mn | Nb | Ni | Si |
|---|---|---|---|---|---|---|---|
| 0.062 | 15.88 | bal | 1.94 | 1.39 | 0.94 | 16.68 | 0.45 |

In run 1.06 and 1.10 two different surface conditions of the
standard material were exposed together:

Mat. 1:  heat treated with 1070 °C/$H_2O$ and passivated
Mat. 2:  only passivated after fabrication.

Run 1.12 had samples of Ni base alloys in addition to the standard material exposed in the test section. More detailed information about specimens and test section geometry for run 1.10 and 1.12 can be found in (1, 2).

The sodium conditions presented in Table 1 need some discussion Oxygen concentration in the system is maintained with a cold trap, routine measurements of saturation temperatures are made with a continuously operating pluggingmeter. During all runs plugging-meter readings could be correlated satisfactorily to the cold trap temperature. Sampling of sodium for analytical purposes is done with a sampling device followed by off line destillation. Although analytical techniques have reached a high degree of accuracy, the sampling procedure itself has severe drawbacks since the sampling device is located in a bypass at relatively low temperatures. This introduces uncertainties due to possible spalling of deposits and possible chemical reactions not representative for the main system. A critical survey of oxygen concentration measurements has been presented by Claxton (3) and it has been recommended to assume the cold trap temperature as an accurate representation of the actual oxygen concentration. This has been considered in Table 1. With the oxygen correlation given by Noden (4) the theoretical oxygen concentrations can be compared to the measured concentrations. For run 1.10 the difference between theory and experiment is not too great, contrary ro run 1.12, where the measured oxygen concentration is exceedingly high, although it should be less than in run 1.10. For the measured Fe-concentration a similar behaviour can be observed. This behaviour reflects perhaps essentially altered sodium conditions, which at the moment can not be explained.

Documentation of measured corrosion rates

To facilitate data handling, all generated weight loss informations together with the relevant test parameters were stored in a computer memory. Connected to a graphic terminal, calculations and plots could be performed readily. As an example the processed data of run 1.06 are given in Fig. 1, which shows the accumulated weight loss of standard material SS 1.4981 after 1500 hours with and without heat treatment after fabrication of the samples (mat. 1 and mat. 2).

For all runs, documented in Table 1 a best fit curve of the form

$$F = A \exp. - B (L/d_H) + C \tag{1}$$

has been calculated for accumulated weight losses and for corrosion rates in the intermittent corrosion intervals.

For further discussions only the corrosion rates will be used. These are given in Table 2.

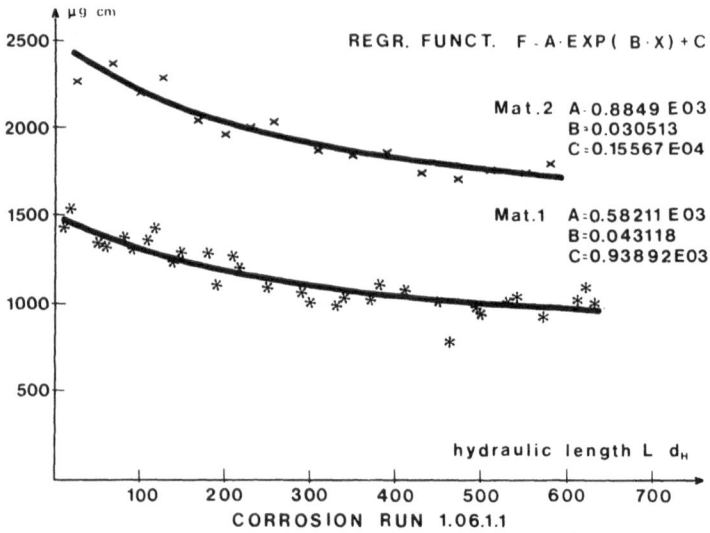

Fig. 1.   Accumulated weight loss of standard material SS 1.4981


Table 2.   Exponential fit for corrosion rates

| run | test interval $T \lfloor h \rfloor$ | mean exposure time $T_m \lfloor h \rfloor$ | A $\mu g/cm^2 h$ | B .10⁻³ | C $\mu g/cm^2 h$ | A+C $\mu g/cm^2 h$ | A/B $\mu g/cm^2 h$ | $\Delta$ C .10⁻³ $\mu g/cm^3$ |
|---|---|---|---|---|---|---|---|---|
| 1.1o.1 | 126 | 63 | 1,521 | 3,75 | 1,291 | 2,812 | 4o6 | 1,13 |
| 2 | 216 | 234 | | | | | | |
| 3 | 396 | 54o | o,352 | 6,34 | o,674 | 1,o26 | 56 | o,16 |
| 4 | 836 | 1157 | o,371 | 1o,34 | o,436 | o,8o7 | 36 | o,1o |
| 5 | 1o6o | 21o7 | o,736 | 5,97 | o,253 | o,989 | 123 | o,34 |
| 1.1o.1 | 126 | 63 | 1,398 | 6,5o | 1,662 | 3,o6o | 215 | o,59 |
| 2 | 216 | 234 | | | | | | |
| 3 | 396 | 54o | o,349 | 4,11 | o,791 | 1,14o | 85 | o,24 |
| 4 | 836 | 1157 | o,424 | 6,33 | o,578 | 1,oo2 | 67 | o,19 |
| 5 | 1o6o | 21o7 | o,663 | 3,77 | o,34o | 1,oo3 | 176 | o,49 |
| 1.12.1 | 117 | 59 | o,923 | 4,99 | o,151 | 1,o74 | 185 | o,82 |
| 2 | 21o | 222 | o,236 | 4,7o | o,26o | o,496 | 5o | o,22 |
| 3 | 399 | 526 | o,236 | 3,71 | o,231 | o,468 | 64 | o,28 |
| 4 | 774 | 1113 | o,375 | 4,27 | o,16o | o,536 | 88 | o,39 |

## DISCUSSION OF MEASURED CORROSION RATES

A detailed investigation of the down stream phenomenon after long corrosion times has been presented by Kolster (5). To facilitate comparison and discussion the same notations as used in (5) will be taken in the following. According to (5) the steady state corrosion rate $R_{St}$ up to at least 8 ppm oxygen in the sodium can be expressed as

$$R_{St} = K + C \, [O] \qquad (2)$$

With the rate constant K for the dissolution of metallic elements being dependent on down stream position according to the general relationship

$$K = \frac{c^o - c_{mo}}{R_d + R_t} \cdot \exp. - \frac{4L}{d_H v (R_d + R_t)} \qquad (3)$$

with

$$
\begin{array}{lll}
v & = & \text{sodium velocity} \\
c^o & = & \text{solubility of metallic elements in Na} \\
c_{mo} & = & \text{concentration of metallic elements in Na at the entrance} \\
& & \text{of the specimen holder (L=0)} \\
R_d & = & \text{reciprocal rate constant for deposition} \\
R_t & = & \text{diffusion resistance in the laminar sublayer}
\end{array}
$$

Comparing (3) with equation (1) the coefficients of this equation can be related to physical values:

$$A = \frac{c^o - c_{mo}}{(R_d + R_t)} \quad ; \quad B = \frac{4}{v(R_d + R_t)} \qquad (4)$$

From the properties of the exponential part of function (1) it follows, that the ration A/B is the integral of this function. Physically speaking, if saturation with corrosion products occurs, then

$$\frac{A}{B} = \frac{(c^o - c_{mo})v}{4} \quad [\,\mu g/cm^2 h\,] \qquad (5)$$

represents the release of corrosion products responsible for the down stream effect of an infinitly long test section. Since our test section shows strong saturation, this value will not deviate too much from the correct weight loss with finite test section length. With the known velocity in the test runs, the concentration difference at the entrance of the test section can be calculated.

$$\Delta c = (c^o - c_{mo}) = \frac{4 \cdot A}{v \cdot B} \quad \mu g/cm^3 \qquad\qquad (6)$$

This has been done for all runs. Results have been entered in Table 2. The results show the well known discrepancy of a concentration difference at the entrance of the test section in the order of $1.10^{-3}$ $\mu g/cm^3$ or approcimately $1.10^{-4}$ ppm which can not be reconciled with the experimentally determined solubility data of metallic elements which are in the order of ppm (6).

Run 1.10

In Table 2, C denotes the corrosion rate in the test interval, A represents the maximum value of the down stream dependent corrosion rate at the test section entrance. For run 1.10 these two values are presented over the exposure time in Fig. 2, annealed and cold worked, representing Mat. 1 and Mat. 2.

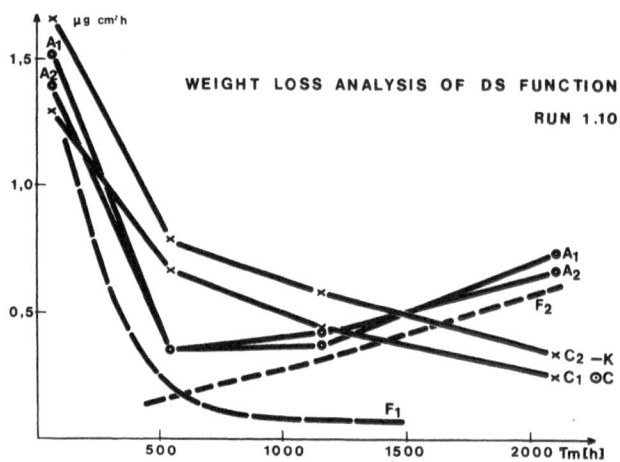

Fig. 2.   Corrosion rate C and maximum value A at test section
         entrance

From this graph it is obvious that the different metallurgical conditions of the standard material have practically no effect on A, but an influence on C, which additionally shows a strong dependency on exposure time. The results of earlier investigations (7) indicated that this second effect is due to the selective leaching process of metallic elements out of the steel matrix. This implies that in the total transport process of the metal atoms moving from the steel matrix into the liquid sodium, diffusion through the steel is the rate limiting step and that the mass loss will be proportional to the surface area as long as no saturation occurs. Fig. 3 shows the surface conditions of material 1 and 2 after 379 h of exposure.

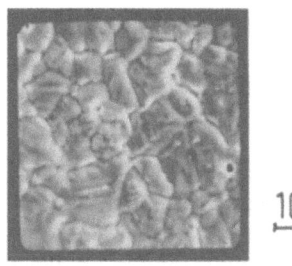 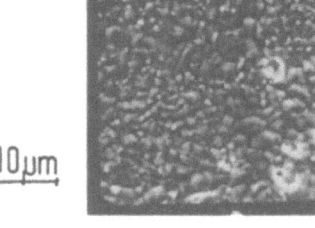

10µm

Mat. 1: 0.7 mg/cm$^2$          Mat. 2: 0.9 mg/cm$^2$

Fig. 3.   Corroded Surface after 397 h

It is clearly visible that the annealed material 1 has large grains at the surface whereas the material 2 has a damaged grain structure. Since corrosion, especially during the initial phase takes place preferentially by grain boundary grooving the higher effective surface of material 2 is evident. Calculations about the kinetic behaviour of A showed that an exponent of the exposure time of −0.43 fits both curves which is in agreement with an expected exponent between −0.50 for diffusion over a plane surface and −0.33 for the mass loss in grain boundary grooving (8). Since the down stream position is not affected by the build up of product C in the sodium, the solubility of the corrosion product has to be much greater than the concentration increase in the test section. During the first interval of run 1.10 an increase of $1.5 \cdot 10^{-4}$ ppm through the test section can be calculated necessitating a minimum solubility of corrosion products approximately two order of magnitude greater, that is $c^o_{min} > 10^{-2}$ ppm.

As a hypothesis it is assumed, that the corroding species is Ni which is being leached selectively out of the steel matrix. Strong evidence for this assumption can be found in the fact that calculation of the diffusion coefficient of Ni in austenite with the measured weight losses are in good agreement with the expected magnitude of this coefficient especially for short corrosion periods. More arguments in support of this hypothesis will be given later.

The down stream dependent part A in Fig. 2 shows, that there exists a minimum of the entrance corrosion rate with a relatively short exposure time of less than 500 hours. This could be explained by the assumption that two competing mechanisms of the corrosion process are operative. For short corrosion periods there is a strong down stream effect which goes to zero with increasing exposure time. In addition there may be a second contribution to the down stream effect which grows up to a limiting value. Both functions are tentatively drawn into Fig. 2 as F1 and F2. There is no pronounced

difference for the two surface conditions for this kind of corrosion.
For F2 a possible explanation could be that with the continuous
leaching of the alloying elements and redistribution of surface
concentrations the surface of the steel reaches equilibrium with
respect to the dissolved elements in the sodium. For long exposure
times surface concentrations of 5 % for Ni and Cr have been measured,
for the iron approximately 90 % of surface concentration has to be
expected. When this equilibrium situation is reached each concentration increase by the dissolution of these elements, especially of
iron will necessarily result in a down stream effect.

Run 1.12

The relevant values of Table 2 are graphically presented in
Fig. 4. For comparison the same data for Mat. 1 of run 10 are
included. It is obvious that in run 1.12 in which a series of Ni-
base alloys have been exposed together with the standard material,
the constant-corrosion part C is suppressed considerably. Calculating
the Ni release of the test section during this run, an increase in
the order of ppm in the sodium can be expected during moderate
corrosion times if deposition in the cooled parts of the circuit
remains low. This behaviour should be expected since only 30 % of
the total sodium volume has a temperature lower than 600 $^{\circ}$C.

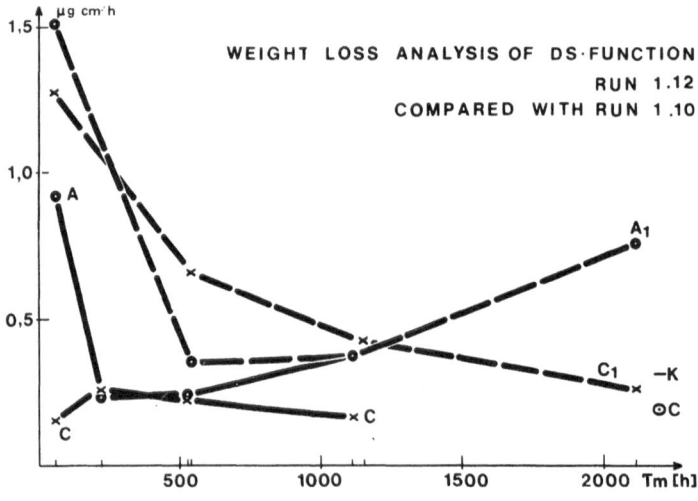

Fig. 4.   Weight loss analysis of DS function

Results from systematic investigations (9) disclosed, that in
low oxygen systems the deposition of Ni in the cold part is very
sluggish, so that in combination with our loop characteristics the
release of Ni in the test section should increase the concentration
considerably. Using the solubility data of Ni in sodium given by

Claar (6) an equilibrium concentration of 3.2 ppm should be expected at the maximum temperature and the measured Ni concentrations for run 1.12 of Table 1 signify, that the sodium is saturated with Ni. It can not be decided however, if the values C represent still a reduced Ni corrosion rate or if another corrosion process has to be considered.

The corrosion rate A differs in the two runs in so far as only the hypothetical function F 1 is reduced in 1.12. If F 2 shows the same tendency for longer corrosion periods it is unaltered with respect to run 1.10. Information from deposition studies show (10) that during the initial period at low oxygen levels the deposition products contain primarily Cr-carbides which deposit already in the high temperature region above 600 °C. A very high chemical activity should therefore be expected in the test section. An indication of this high activity of Cr-carbide has been found in a previous run 1.02 in which pure iron samples have been exposed simultaneously with standard material (11). A high portion of the weight gain could be attributed to a Cr carbide diffusion process. A strong indication of the down stream dependency of the Cr loss during run 1.10 is given in (12) where the down stream position shows approximately only half the selected Cr loss than the upstream position. In relation to run 1.10 this possibly selective Cr loss is lower which could be attributed to the lower theoretical oxygen content of run 1.12. The actual situation seems to be far more complicated, since contrary to the theoretical oxygen concentration, relatively high iron and oxygen concentrations have been measured which still wait for interpretation.

CONCLUSIONS

Experimental evidence and theory show that Ni is selectively leached out of the steel matrix. The rate dominating step is diffusion through the steel matrix to the surface. There is no down stream effect of this process as long as equilibrium conditions are not reached. The absolute amount of corrosion rate is proportional to the effective surface. That means that surface conditions respectively surface roughness is an important parameter. Theoretically for long corrosion times this rate should go to zero.

The down stream dependent part of the corrosion rate is influenced by two different processes. For short times a strong down stream effect results from Cr loss associated with oxygen and carbon in the sodium and probably of carbides in the grain boundaries. This down stream effect becomes small after an initial period of 400 hours which corresponds to the time needed to reach ferritisation of the surface.

After that period a down stream factor becomes dominante which is not influenced by the Ni content and should be associated with the atomic dissolution of iron and in the long run with the dissolution of the other alloying elements as well.

Since corrosion rates in the investigated intervals are still far from equilibrium conditions a comparison with the quantitative down stream model presented by Kolster (5) is not possible at the moment. According to this model the expected steady state corrosion rate of our experiments should be given by equation (2) with the rate constants

$$K = 1.2 \cdot 10^{11} \exp\frac{-48500}{RT} \left[ mg/cm^2 \ a \right] \ and$$

$$C = 3.8 \cdot 10^2 \exp\frac{-14000}{RT} \left[ mg/cm^2 \ a \ ppm \right]$$

The first term giving the down stream dependency which is thought to result from element dissolution and the second term the iron oxygen reaction which shall not be down stream dependent. These two terms are added on the right hand side of Fig. 2 and 4 for 730 °C and with the oxygen concentrations of 8 and 6 ppm respectively.

From this comparison it can only be suspected, that the oxygen reaction rate of equation (2) is included in the above defined Ni-solution term and that the increasing down stream effect reflects elemental dissolution from a surface which is near equilibrium with the contacting sodium.

For a more detailed understanding, longer corrosion periods with corrosion rate determinations during short intervals will be carried out.

REFERENCES

1.  N. F. Schwarz, G. E. Rajakovics: Int.Conf.on Liquid Metal Technology in Energy Production, Seven Springs, Champion Pa. May 3-6, 1976.
2.  H. Konvicka, N. F. Schwarz: 2nd Int. Conf., Richland, Wash., April 20-24, 1980.
3.  K. T. Claxton: Journal of Nucl. Enery, 21, (1967) 351-357.
4.  J. D. Noden, Journal.Brit.Nucl.En.Soc. 12, (1973), 329-331.
5.  B. H. Kolster: 2nd Int.Conf. on Liquid Metals in Energy Production, April 1980.
6.  T. D. Claar: Reactor Technology 13, (1970), 124-146.
7.  N. F. Schwarz, G. E. Rajakovics: ÖZE 29, 4 (1976), 135-142.
8.  W. W. Mullins: Trans.of the Metallurgical Soc. of AIME 354, Vol. 218, (1960).
9.  B. H. Kolster: Journal of Nucl.Mat. 55 (1975), 155-168.
10. B. H. Kolster, L. Bos: 1st Int.Conf.on Liquid Metal.Techn. in Energy Production, Champion, Pa, May 3-6, 1976.

11. G. Hofer, N. Wieling: Reaktortagung 1971 des DAF.
12. H. R. Konvicka et al.: 1st Int. Conf. on Liquid Metal Techn. in Energy Production, Champion, Pa, May 3-6, 1976.

# AUSTENITIC STAINLESS STEEL ALLOYS WITH HIGH NICKEL CONTENTS IN HIGH TEMPERATURE LIQUID METAL SYSTEMS

Heinz R. Konvicka, Norbert F. Schwarz

Österreichisches Forschungszentrum Seibersdorf
Wien
Austria

## INTRODUCTION

The use of liquid sodium as a coolant for fast breeder reactors has been successfully demonstrated. More recently, further applications as heat transfer fluid have been considered for solar power plants and topping cycles in fossil fuelled power plants. From an economic and safety point of view, numerous studies focussed on materials compatibility with liquid sodium at high temperatures, thus resulting in a large data base on corrosion behaviour, change of mechanical properties and sodium chemistry. However, due to the number of parameters involved in liquid metal systems, a large scattering of these data is observed.

At the Research Centre Seibersdorf a liquid metal technology programme was startet in the mid-1960s. So far, the basic interest focussed on the behaviour of austenitic stainless steel in flowing liquid sodium at 730 °C (1000 K). The test run Nr. 1.10 was conducted in five intervals up to a total exposure time of 2637 hours to collect information about the kinetics of bulk and selective corrosion phenomena in stainless steel (SS W. Nr. 1.4981). Results have been reported earlier (1, 2), but many questions especially with respect to the interpretation of downstream and velocity effects remained unanswered. Therefore, an analogous test run was planned using the same sodium conditions, but in addition to the standard test material (SS W. Nr. 1.4981) Fe-Cr-Ni base alloys with various Ni contents from 15 to 70 wt% were exposed in regularly spaced positions of the high and low sodium velocity pile of the test section. Preliminary results after three intervals (total exposure time of 728 hours) have been already reported (3). This paper deals with the findings after completion of the fourth interval (1498 hours).

EXPERIMENTAL TECHNIQUES

The sodium test loop HT 1 as well as the specimens and the test
section configuration have been described earlier in detail (4). The
sodium enters first the low velocity test pile (pile 2) and subsequently the high velocity test pile (pile 1). The sodium flow is
accelerated by a factor of about 10. The standard specimen geometry
can be described as a short tube ($\emptyset_o$ = 19 mm, $\emptyset_i$ = 15 mm, h = 10 mm).
Fe-Cr-Ni base alloy specimens have been machined from sheets ($\emptyset_o$ =
19 mm, $\emptyset_i$ = 15 mm, h = 4 mm). In order to be consistent with specimen
positions in test run 1.10, standard test material specimens of 6 mm
height have been added after each base alloy specimen. The chemical
analysis of the standard material is given in Table 1. All specimens
were heat treated after machining (1080 $^{o}$C/20 min). They were arranged in six equivalent stacks (Fig. 1), four of them exposed in
the high velocity pile and two in the low velocity pile.

Table 1.   Chemical analysis of standard test material

$$[ wt \% ]$$

|        | C     | Cr    | Fe   | Mo   | Mu   | Nb   | Ni    | Si   |
|--------|-------|-------|------|------|------|------|-------|------|
| 1.4981 | 0.062 | 15.88 | bal. | 1.94 | 1.39 | 0.94 | 16.68 | 0.45 |

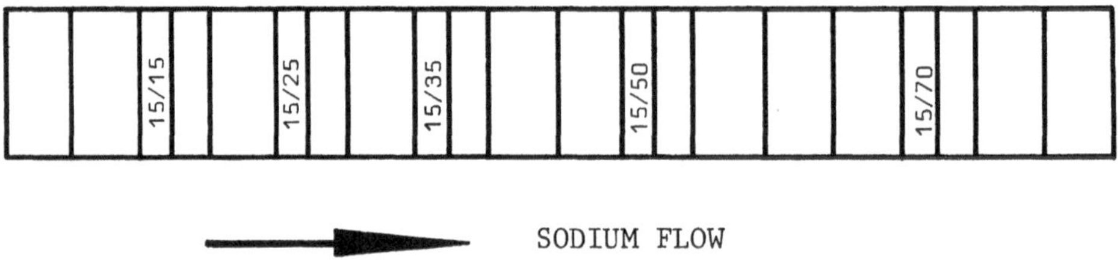

SODIUM FLOW

Fig. 1.   Specimen arrangement in one stack

Four test run intervals have been completed so far. The next
interval up to a cumulative exposure time of 2700 hours is planned.
Interval times and parameters are available from Table 2.

Table 2.   Test parameters

| run     | T $[^oC]$ | $V_{Na}$ $[msec^{-1}]$ | t $[h]$ cumulative |
|---------|-----------|------------------------|--------------------|
| 1.12 A  | 730       | 2.5/0.3                | 117                |
| 1.12 B  | 730       | 2.5/0.3                | 327                |
| 1.12 C  | 730       | 2.5/0.3                | 726                |
| 1.12 D  | 730       | 2.5/0.3                | 1500               |

RESULTS

Weight Change of Corrosion Specimens

After termination of the test interval D all procedures concerning loop shut-down, dismantling of the test section, cleaning
of samples and weight change measurements were accomplished in an
identical manner as for previous runs. The corrosion rates observed
during interval D are presented in a semilogarithmic plot of the
corrosion rate R versus the downstream-parameter $L/d_{Hydraulic}$ (Fig.2)
For clearness, only the data for the Fe-Cr-Ni base alloys are shown.
Corrosion rates of the standard test material are considerably
lower than in test run 1.10.

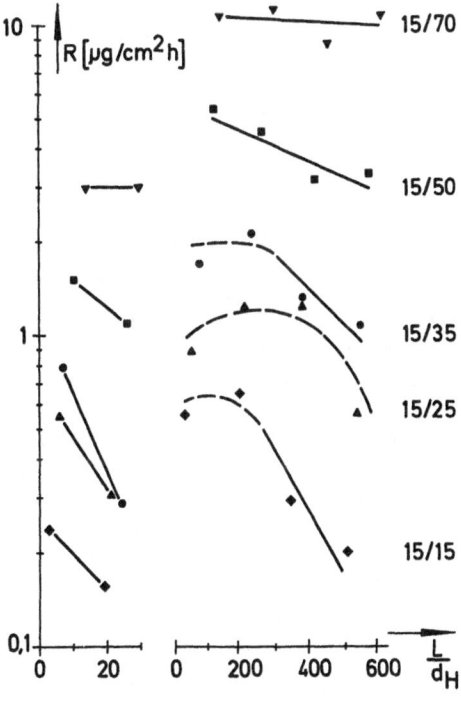

Fig. 2   Corrosion rate of base alloys during exposure D

Sodium Chemistry

The concentrations of oxygen, iron, chromium and nickel in
sodium were analyzed during each test interval (Table 3). The sodium
cold trap temperature was kept constant at 170 °C, what corresponds

to 6 ppm oxygen in sodium according to the Noden equation (6).

Table 3.   Sodium chemistry data

concentration [ppm]

| run | 0 | Fe | Cr | Ni |
|-----|-----|-----|-----|-----|
| 1.12.A | 50 | 19 | 7 | 4 |
| 1.12.B | 100 | 23 | 4 | 4 |
| 1.12.C | 70 | 50 | 3 | 4 |
| 1.12.D | 60 | 15 | 4 | 3 |

DATA ANALYSIS AND DISCUSSION

From Fig. 2 it is evident that the increasing nickel content
of the alloy results in an increase of the corrosion rate. The
downstream dependent corrosion rate is still very pronounced. Through
the data points straight lines are tentatively drawn which would
represent an exponential downstream function. If the downstream
effect arises from saturation of corrosion products in sodium, a
higher corrosion rate should lead to a higher downstream effect.
In a log R-vs.-$L/d_H$ graph this would mean, that the straight lines
connecting the points in the downstream position should be parallel
to each other. Obviously this cannot be observed. An explanation
can be given as follows: Since the test section is not composed of
one single material, the sodium will contain corrosion products
representative for an average of all corroding specimens. Those with
a high nickel content will be in contact with sodium containing
relatively low concentrations of corrosion products compared to
their own corrosion rate. For specimens with low nickel-contents the
situation is contrary. This means that the difference in the driving
force for dissolution (or corrosion) in the up- and downstream
position is altered in such a way that there results a lower downstream effect for the nickel rich specimens and a higher downstream
effect for the specimens with lower nickel contents as would be
expected from a test pile consisting only of one of these materials.
This observation is valid for both the high and low velocity test
sections. As a consequence of this discussion, the nickel content
actually plays a dominant role on the downstream effect for exposure
period D. This requires a nickel concentration in sodium which is
very close to saturation. Some experimental evidence exists supporting this assumption.

In Fig. 3 corrosion rates are represented as a function of the
average exposure period. The general behaviour of the corrosion
kinetics becomes now clearly indicated. For base alloys with up to
35 % nickel a normal kinetic behaviour with steadily decreasing

corrosion rates can be found. The 70 % nickel alloy shows a maximum
corrosion rate near 300 h with a steady decrease for longer corrosion periods. Investigation of surface conditions by SEM shows
that this increase in corrosion rate could be attributed to an
increase of the effective surface of the corroding samples during
the first few hundred hours. For the material with 50 % Ni this
behaviour is less pronounced, but still visible. An interesting
observation related to the downstream position which has been added
in Fig. 2 for material 15/70 pile 1 stack 4. It seems that corrosion
in the downstream position has a time lag, which would again point
to the fact that the sodium has a high concentration of corrosion
products (e.g. nickel). The same general behaviour can be observed
for the low velocity test section. For stainless steel with normal
alloy composition the kinetic behaviour can be understood as a
selective leaching process by which nickel and chromium concentrations are reduced in a surface layer until equilibrium corrosion
conditions are reached. For the nickel rich material only minor
selective effects can be found, although some redistribution of
alloying elements seems to take place.

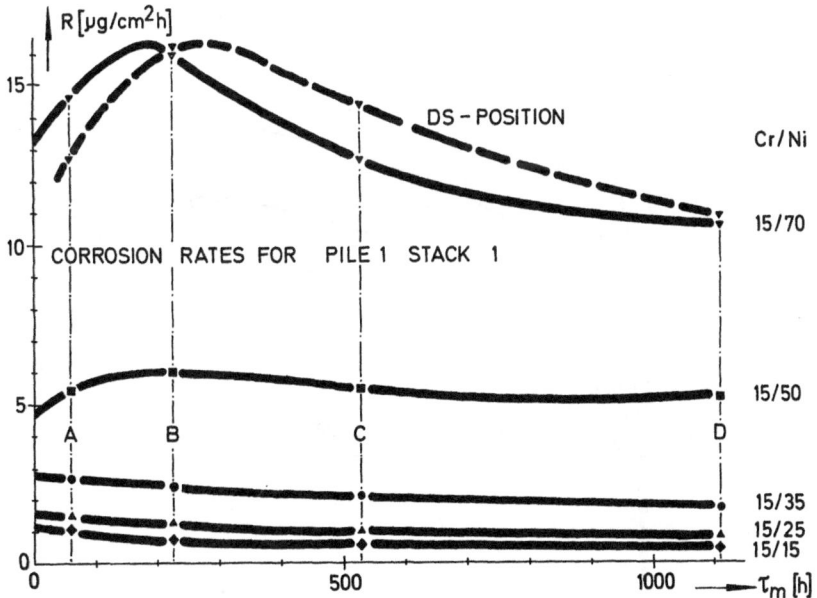

Fig. 3.   Corrosion rates for pile 1 stack 1

Another interesting phenomenon is observed when the weight
loss of the base alloy specimens of pile 1, stack 1 is plotted versus
the exposure time on a semilogarithmic diagram (Fig. 4). Connecting
the data points, at 725 and 1500 hours respectively, for each base
alloy by linear interpolation results in 5 almost parallel lines
intersecting the Y-axis of the diagram at certain values $\Delta g^{o}$. The

slope of these lines varies only from 3.3 x 10$^{-4}$ to 3.8 x 10$^{-4}$. This
indicates that the time dependence of the weight loss after an
initial corrosion period is mainly governed by a corrosion rate
limiting process independent of the initial nickel content of the
alloy. Thus, some correlation between the $\Delta g^o$-values and the initial
nickel content of the alloys has to exist (Fig. 5). To allow for the
different densities of the base alloys, the nickel concentration is
given in g/cm$^3$-units. Two regimes can be identified, where almost
linear relationships between $\Delta g^o$-values and the nickel content exist.
This supports earlier findings (3), which postulated similar corrosion behaviour of the base alloys up to 35 % nickel content on the
one hand side and of base alloys with more than 50 % nickel on the
other hand side.

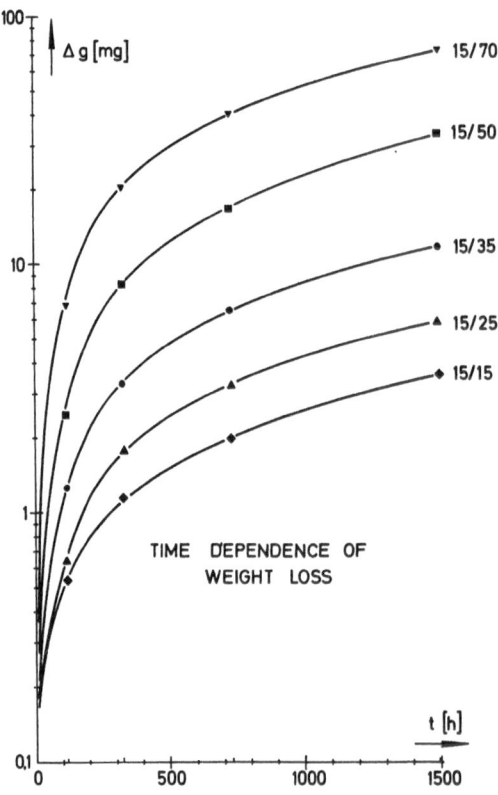

Fig. 4. Time dependence of weight loss

However, no satisfactory physical explanation for those findings
is available at present. Future work will be directed towards a
closer understanding of this correlations.

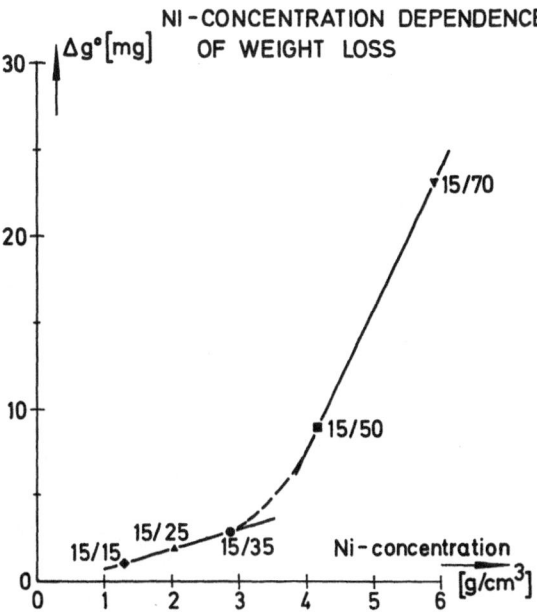

Fig. 5. Ni-concentration dependence of weight loss

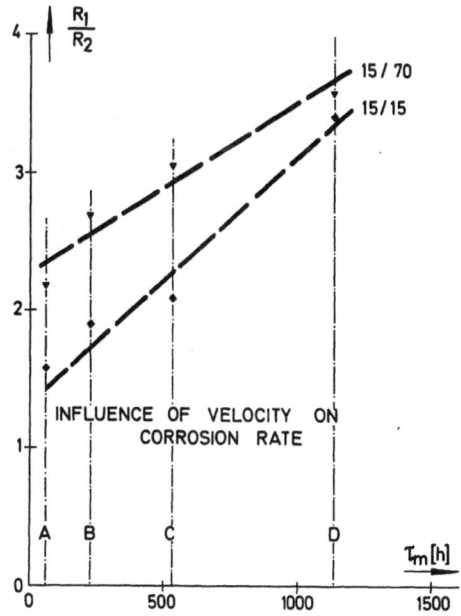

Fig. 6. Influence of velocity on corrosion rate

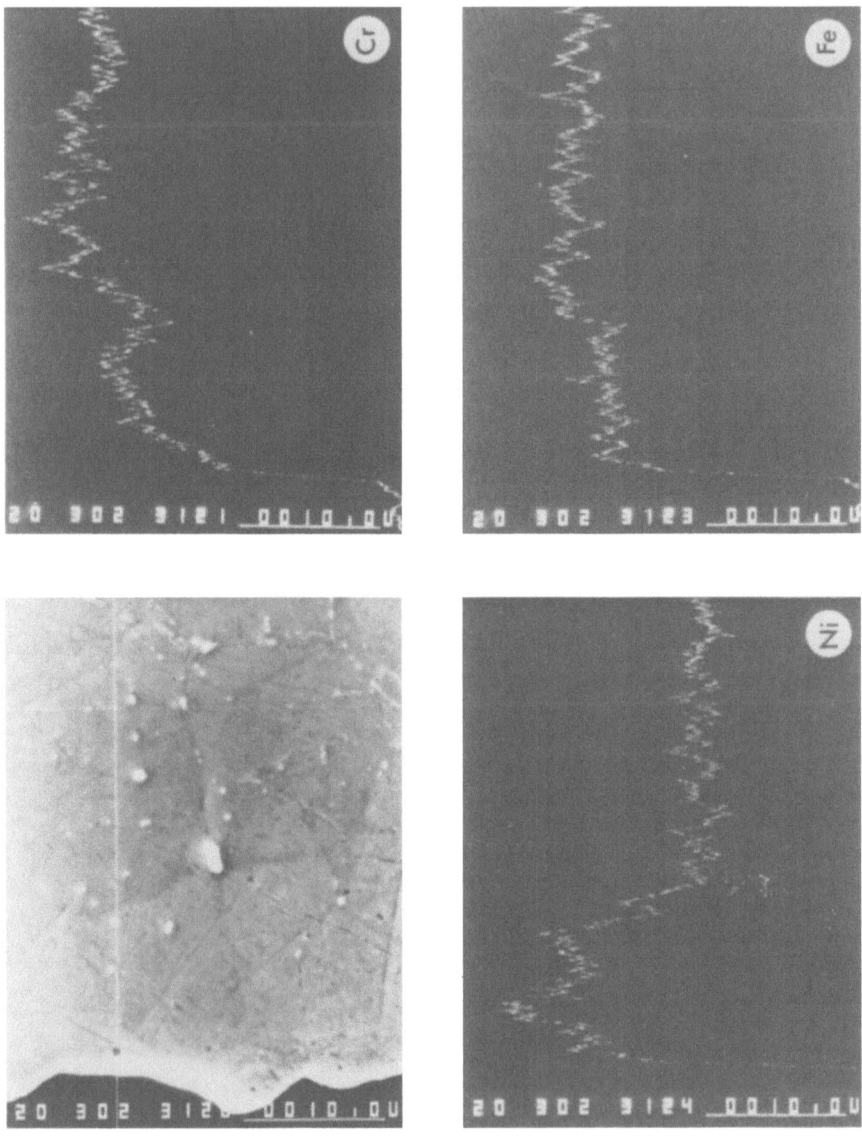

Fig. 7  Nickel enrichment on standard material

If one relates the corrosion rate of the last specimen of the low velocity test section to the corrosion rate of the first specimen in the high velocity test section, the effect of an order of magnitude change in velocity on corrosion rates can be estimated. The general situation for the two extremely different alloy compositions 15/15 and 15/70 are shown in Fig. 6. For short exposure times a velocity change has a small influence on the 15/15 alloy and a high one on the 15/70 alloy. This can be expected from literature, since the corrosion rate of nickel is strongly dependent on the velocity, whereas iron-base alloys show a smaller velocity dependence, with a limiting value at 3 - 6 m/sec. With increasing exposure time this ratio encreases for both materials, but the ratio for the 15/15 alloy grows stronger with time and comes close to the value for the 15/70 alloy after long exposure times. Since both materials show basically the same response to velocity changes as a possible explanation, the surface composition of the two materials seems to become similar, at least with respect to the corroding species. This is in accordance with results from the analysis of weight change data and corrosion kinetics.

RESULTS OF POST-EXPOSURE SPECIMEN ANALYSIS

The base alloy and standard material specimens from pile 1, stack 2, have been prepared metallographically and subjected to electron microprobe analysis. Chromium and nickel depletion are rather pronounced for the 15/15 and 15/25 base alloys. Some ferritic grain boundaries can be detected, but in general corrosion attack is limited to a depth of about 15 μm only. For all materials with higher nickel contents no pronounced concentration profiles of alloying elements are observed. Difficulties with the analytical procedures increase with increasing nickel contents, since the surface-near regions show increasing grain boundary attack resulting in formation of holes, what effects the quantitative analysis considerably.

The standard material specimen exposed immediately downstream of the 15/70 alloy shows an appreciable amount of nickel diffusion into the material. Up to a depth of ~ 10 μm the nickel content seems to be increased by a factor of 2 (Fig. 7). This could be interpreted as a supersaturation of sodium with respect to nickel within a very thin surface-near sodium flow area, which is even smaller than the laminar flow zone, and, consequently, leads to a nickel deposition on nickel poor material in downstream positions. There, support is given to the conclusions derived above.

CONCLUSIONS

After cumulative exposure time of 1500 hours, the corrosion rates of all five Fe-Cr-Ni base alloys tend to become constant. The corrosion rate of the standard material is considerably reduced in comparison to test run 1.10. This is due to the limited selective corrosion phenomena because of sodium chemistry conditions. Pronounced differences in the kinetics of 15/15 and 15/70 base alloys are observed during the initial corrosion period. Afterwards, the time dependence of the weight loss becomes similar and, thus, is independent of the initial nickel content of the alloy. Nickel transport from nickel rich alloys to nickel poor alloys is observed and may be attributed to supersaturation of sodium by nickel within very small surface-near sodium flow areas.

Several findings still lack an in-depth physical explanation. Future work will be directed towards an improved understanding of the results of this work.

ACKNOWLEDGEMENTS

The authors wish to thank Mr. G. Groboth for his efforts in performing the electron micro-probe analysis of all specimens.

REFERENCES

1. H. Konvicka, K. Komarek, I. Schreinlechner, International Conference on Liquid Metal Technology in Energy Production, Seven Springs, 1976, CONF-760503-P1, p.400.

2. N. F. Schwarz, G. E. Rajakovics, International Conference on Liquid Metal Technology in Energy Production, Seven Springs, 1976, CONF-760503-P2, P.623.

3. H. Konvicka, N. F. Schwarz, 2nd International Conference on Liquid Metal Technology in Energy Production, Richland, Washington, April 20-24, 1980.

4. N. Schwarz, G. Rajakovics, BNES International Conference on Liquid Alkali Metals, Nottingham, 1973, p.233.

5. N. F. Schwarz, "Corrosion Rate and Downstream Effect in Liquid Metal Systems", Intern.Seminar on Material Behaviour and Physical Chemistry in Liquid Metal Systems, Karlsruhe, FRG, March 24-26, 1981.

6. J. D. Noden, Journal.Brit.Nucl.En.Soc. 12, pp.329-331 (1973).

# INVESTIGATION OF IMPURITY DEPOSITION IN A COLD TRAP OF A SODIUM LOOP

Helga Schneider

Kernforschungszentrum Karlsruhe
Karlsruhe
Fed. Rep. of Germany

## INTRODUCTION

The main function of a cold trap is the purification of the loop sodium in respect to oxygen, but several other compounds or elements are deposited there, too.

To gain more informations upon these purification effects of a cold trap in an inactive sodium loop, the sodium at the coolest region of the cold trap of the test circuit KP 1 was analyzed. The cold trap was operated during 8150 hours at 120 and 145 $^{\circ}$C, respectively. A scheme of this cold trap is shown in Fig. 1.

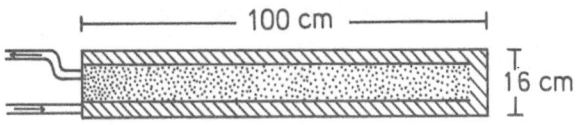

Fig. 1   Scheme of the cold trap KP 1

It consists of an outer tube closed on the bottom having a lenght of 100 cm and a diameter of 16 cm, and an inner tube filled with wire mesh. The sodium enters into the gap between the outer and the inner tube passing the lower end of the cold trap with the lowest temperature. There, the direction of the sodium flow is changed, the sodium is passing through the inner tube and returning into the loop.

EXPERIMENTAL PROCEDURE AND RESULTS

In a glove box the cold trap was cut in the region of the lowest operating temperature as shown in Fig. 2.

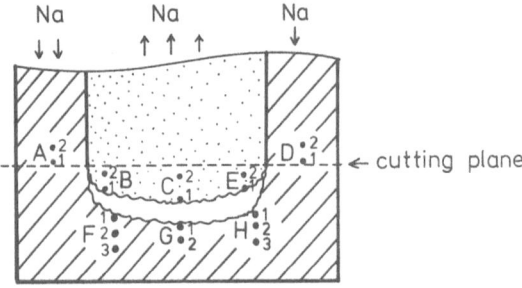

Fig. 2   Sectional drawing of the cold
trap

Cutting through the diameter a cavity was observed below the lower end of the inner tube, probably formed during cooling to room temperature causing a solidification of the sodium. Samples were taken from different places of both sides of the cutting plane characterized by letters and numbers in the diagram.

The remaining sodium from the bottom was used for particle analysis.

In Table 1 the results of estimations of carbon, present in different chemical states (2), are shown.

Table 1.   Contents of Carbon

ppm C

|                                      | A   | B   | D   | E   | F     | H  |
|--------------------------------------|-----|-----|-----|-----|-------|----|
| Carbide                              | 23  | 2   | 19  | 3   | 130   | 2  |
| Carbonate                            | 0.5 | 0.5 | 2   | 0.5 | 33    | 4  |
| Cyanide                              | 6   | 0.5 | 3   | 0.5 | 2/40  | 2  |
| elemental + acid-unsoluble carbides  | 255 | 46  | 709 | 75  | 50    | 65 |

In the first line the values of carbon contents present as water-soluble carbide are noticed. In the gap between the outer and the inner tube (position A) the content is about 20 ppm, below the filter of the inner tube (position B) the content is about one magnitude lower, and the highest value, more than 100 ppm, was detected in the surface layer of the cavity of the bottom on the side, where the sodium is entering into the trap (position F).

In comparison to the carbide contents the contents estimated as carbonate or cyanide carbon are low, the highest values being again at place F. The amounts of elemental carbon, including the carbon present as acid-insoluble carbides, are also significantly higher, the highest amount was at the location D, opposite to the entrance of the sodium. This may probably be caused by turbulences in the sodium flow on the bottom of the cold trap. Due to the carbide formation metallic elements were analyzed in the residue of combustion, and for example 20 ppm chromium were detected on the place H.

Table 2 shows some other nonmetallic impurities (3):

Table 2   Contents of several nonmetallic elements

| ppm | A | B | D | F | H |
|---|---|---|---|---|---|
| B | 1 | | | 12 | 4 |
| | 1 | | | 5300 | 6000 |
| CI | 2 | 500 | 110 | 1200 | 90 |
| | 3 | 180 | 34 | 350 | 60 |
| P | 1 | | | 37 | 9 |
| Si | 1 | | | 177 | 94 |
| N | 1 | 11 | 7 | 9 | 11 |

The contents of boron, phosphorus and nitrogen are in the ppm-level. The silicon contents are between 100 - 200 ppm. These elements are leached out of the structural material of the loop. Remarkable high contents of chloride were measured in the surface layer of the cavity. Chloride is an impurity of the sodium, in a sodium as received (DEGUSSA) we estimated about 20 ppm Cl (3). Obviously the chloride was collected in the cold trap sodium and during solidification of the sodium enriched in the surface layer of the cavity.

To demonstrate the degree of enrichment of other nonmetallic elements in the cold trap the results determined for carbon and nitrogen are compared with the contents in the loop sodium calculated from the determined activities using the foil equilibration method. For both elements contents were estimated lower than 0.1 ppm this means, that these elements were enriched in the cold trap at least for two orders of magnitude.

Table 3 shows the contents of several metallic elements (3):

Table 3.  Contents of several metallic elements

| ppm | | A | B | C | D | E | F | G | H |
|---|---|---|---|---|---|---|---|---|---|
| Fe | 1 | 6000 | | 60 | | 300 | 900 | | 100 |
|    | 2 | | 150 | | 130 | | 100 | 80 | 60 |
| Cr | 1 | 4000 | | 8 | | 30 | 200 | | 10 |
|    | 2 | | 20 | | 50 | | 65 | 50 | 17 |
| Ni | 1 | 900 | | 10 | | 40 | 80 | | 9 |
|    | 2 | | 10 | | 10 | | 20 | 20 | 10 |
| Mn | 1 | 120 | | 120 | | 300 | 50 | | 130 |
|    | 2 | | 120 | | 20 | | 75 | 100 | 17 |
| Cu | 1 | 400 | | 160 | | 470 | 250 | | 220 |
|    | 2 | | 400 | | 20 | | 200 | 150 | 17 |
| Zn | 1 | 300 | | 140 | | 250 | 60 | | 80 |
|    | 2 | | 90 | | 4 | | 70 | 40 | 9 |
| Bi | 1 | 4 | | 6 | | 12 | 40 | | 2 |
|    | 2 | | 20 | | 50 | | 60 | 12 | 12 |
| Sn | 1 | 2 | | 10 | | 10 | 3 | | 7 |
|    | 2 | | 5 | | 6 | | 8 | 6 | 7 |
| Pb | 1 | 11 | | 14 | | 13 | 12 | | 17 |
|    | 2 | | 18 | | 16 | | 17 | 13 | 17 |
| Co | 1 | 0.2 | | 0.2 | | 0.2 | | | |
| Ca | 1 | 10 | | 17 | | 30 | | | |

The contents of steel alloying elements determined at the sodium entrance are relatively high, they are considerably higher than in the center at the entrance of the inner tube. It can be concluded that these impurities are especially enriched at the location of the lowest Na temperature and direction change. As the values determined at the position F show, these elements are enriched in the surface layer of the cavity, too. The cobalt content is very low at all locations and amounts to less than 0.2 ppm. Zinc and copper are present in relatively high concentrations, whereas the elements tin, lead, bismuth and calcium are present only in a low concentration range.

To demonstrate again the degree of enrichment of these different elements in the cold trap sodium the contents of several elements determined in the loop sodium are shown in Table 4.

Table 4. Impurities determined in the loop sodium

| Element | Concentration (wppm) | |
|---------|---------------------------------------|------------------------------------------------|
|         | Sampling Nov. 78 unfiltered | Sampling Dez. 78 filtered (ca. 35 μm porosity of the filter) |
| Fe      | 8.0 ppm   | 4.0 ppm   |
| Cr      | 5.5 ppm   | 0.4 ppm   |
| Ni      | 1.0 ppm   | 0.25 ppm  |
| Mn      | 2.0 ppm   | 0.4 ppm   |
| Pb      | 10.0 ppm  | 22.0 ppm  |
| Sn      | 10.0 ppm  | 9.0 ppm   |
| Bi      | 10.0 ppm  | 1.0 ppm   |
| Zn      | 7.0 ppm   | 0.5 ppm   |

Two samples of this sodium were analyzed, the first sampling was made without filtering, in the second case a stainless steel filter having a porosity of about 35 μm was fitted into the sampling device. The results of these two samples are significantly different. The amounts of iron, chromium, nickel, manganese, bismuth and zinc are remarkably lower in the filtered sodium than in the unfiltered one. The content of tin is nearly the same and the content of lead is even higher in the filtered sodium sample.

Comparing the results of the analyses of the loop sodium with the results of the cold trap the following conclusions can be drawn:

The steel alloying elements are enriched in the cold trap up to some orders of magnitude.

The ratio iron to chromium and manganese, respectively, is similar in the cold trap and in the loop sodium.

The elements whose contents can be lowered by filtration are enriched in the cold trap, these are the steel alloying elements and also bismuth and zinc. Zinc is probably included in a Fe-O-Na-compound.

Lead and tin are not enriched in the cold trap and their contents were not lowered by filtration, too.

With respect to the concentration level of calcium it must be mentioned that this element has the tendency to depose on the surfaces of pipe walls and specimens, therefore the degree of enrichment in the cold trap is relatively low.

The residual content of sodium in the bottom of the cold trap, about 0.9 kg, was dissolved in iso-propanol. After addition of water to this solution having a greenish colour changing by reaction with

air in a brownish one, the unsoluble particles amounting to 0.07 %
were separated by filtration.

SEM-pictures of these particles showed their size varying
between 2 and 100 μm, as Fig. 3 shows; their composition is shown
in Table 5.

Table 5.  Particle composition

| Fe | = | 29.7 % | Mn | = | 2.9 % | Ni | = | 4.8 % | Cr | = | 8.7 % |
|----|---|--------|----|---|-------|----|---|-------|----|---|-------|
| Ti | = | 1.0 % | Si | = | 2.6 % | Al | = | 1.3 % | Co | = | < 0.1 % |
| Mo | = | < 0.5 % | Nb | = | < 0.5 % | Cu | = | 0.4 % | Zn | = | 2.2 % |
| Pb | = | < 0.3 % | Bi | = | < 0.3 % | Sn | = | < 0.5 % | C | = | 7.5 % |
| O | = | 15.7 % | Na | = | 1.2 % | $H_2O$ | = | 6.4 % | N | = | 0.8 % |

Fig. 3   SEM-picture of the particles

As expected the main component is iron. The ratio iron to
chromium and manganese, respectively, is similar to that in the
loop sodium and the cold trap sodium. The content of 2.2 % zinc
indicates the existence of solid compounds containing zinc.

Carbon seems to be present partly as carbide, since weak
chromium carbide reflexes were found by X-ray diffraction, and
partly as free carbon.

Additionally to these investigations the compositions of deposits on the wall of the cold trap were investigated. The inner
side of the outer tube had a nearly continuously dark surface.
This surface was analyzed using Scanning Auger electron spectroscopy (4).

Fig. 4 shows the Auger spectrum of the surface:

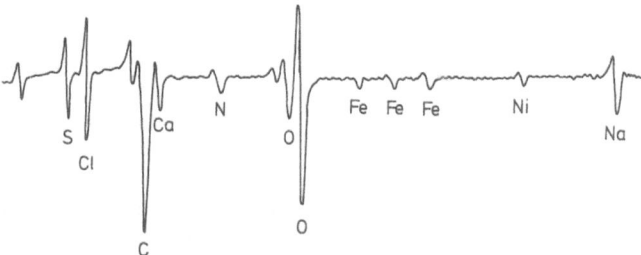

Fig. 4.   Auger spectrum of the surface of the
          cold trap wall

This indicates that the surface is covered with free carbon,
oxygen containing sodium compounds and with other nonmetallic impurities.

Fig. 5 presents the results of a profile analysis of the dark
surface layer:

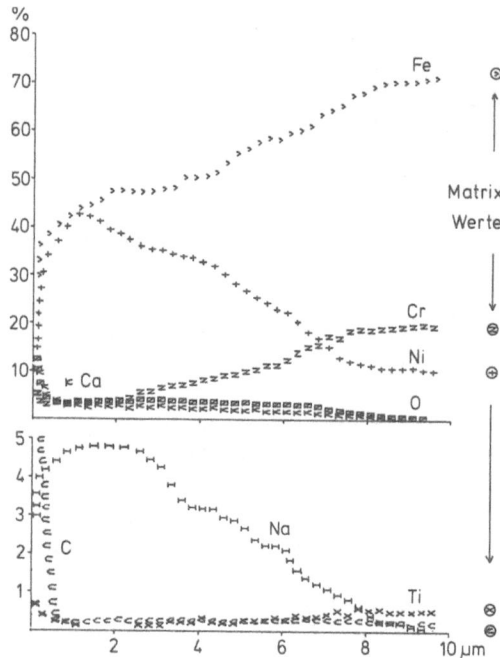

Fig. 5.   Composition of the surface layer of
          the cold trap wall

Nickel- and iron rich deposits were detected on the surface,
sodium is present in the outermost microns to about 5 %. The chromium content raises from about 3 % in 1 μm depth to about 20 % in
a depth of 10 microns. The oxygen content is below 5 % and the
presence of calcium indicates that this layer was formed by deposition. Manganese was not present in these deposits.

After removal of about 2.5 μm of the surface by argon ion
sputtering Auger distribution images were made. Fig. 6 shows the
Auger image for the distribution of nickel and the Auger spectrum
of a point analysis at the point y, which is indicating apart from
nickel a small amount of iron and the presence of sodium.

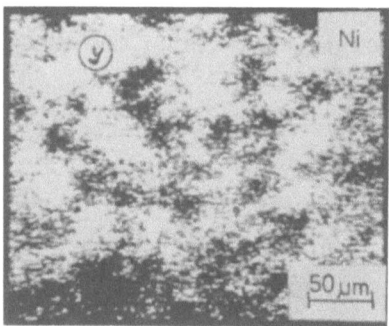

Fig. 6.   Ni-distribution image (bottom) and Auger-point analysis
          on the point y (top)

Fig. 7 shows the distribution images for iron and chromium
and the Auger spectrum of the point analysis at the point x. This
spectrum shows the presence of a small amount of chromium, but no
sodium peak is to be seen. These results indicate that iron is
accompanied by chromium and nickel by sodium, respectively.

Comparing the results of the investigations of the deposits
on the surface layer of the cold trap wall, exposed to sodium at
a temperature of max. 145 °C with results of surface analyses of
steel samples exposed to sodium at higher temperatures it can be
stated, that on steel specimens exposed in downstream position in
the spanish loop ML 1 at 590 and 500 °C (5) nickel is showing a
tendency for enrichment in direction to the surface, too.  But no
enrichment of manganese was detected on the cold trap wall as

it was determined on the surface of a steel sample exposed 6000 h
in downstream position of the Spanish loop at 500 $^\circ$C and also
detected on the surface of a steel sample 1.4919 exposed 10120 h at
350 $^\circ$C in a cooler region of the HT-loop (6).

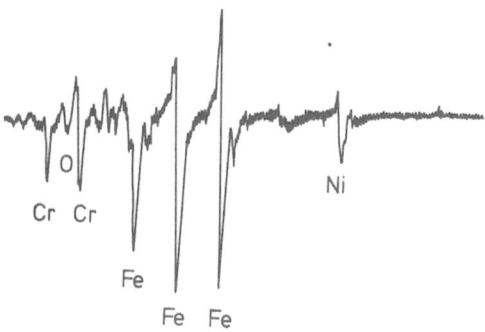

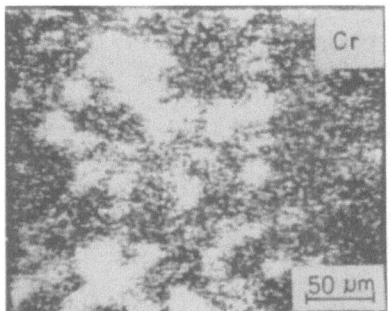
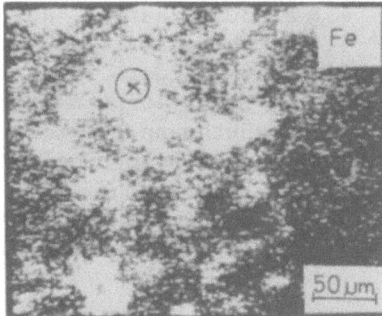

Fig. 7   Fe- and Cr-distribution images (bottom) and Auger point
         analysis on the point x (top)

CONCLUSIONS

     Summarizing it can be concluded, that the distribution of
impurities enriched in the cold trap is inhomogeneous. Therefore,
an estimation of the total amount of impurities deposited in the
cold trap is not possible. This inhomogeneous distribution is caused
by the very low solubility of most elements in sodium at low temperatures, therefore the impurity elements are more dispersed than
dissolved in the sodium. The enrichment of impurities in the cold
trap sodium compared to the loop sodium is in the range of some
orders of magnitude, apart from lead and tin being present in the
cold trap sodium in the same concentration range as in the loop
sodium. Therefore it can be concluded that a cold trap has no
purification effect in respect to these two elements lead and tin,
which is of interest in view of the possible influence of these
elements on the crack growing (7).

The particle size gives an explanation of the purification effect of the filter in the sampling device of the sodium loop.

Finally, it is of interest, too, that manganese is not deposited on the walls of the cold trap.

ACKNOWLEDGEMENT

I would like to thank Dr. H.U. Borgstedt for detailed discussions and Mrs. Ch. Gosgnach, Mrs. Z. Perić, Mrs. K. Schäfer, Mr. J. Biel and Mr. E. Nold for carrying out the special investigations.

REFERENCES

1. H. U. Borgstedt, G. Frees, KfK Nachrichten, 10, 3-4/78
      p. 41.
2. F. Lievens, Progress Report I, SCK-Mol, Sept. 1974.
3. H. Schneider et al., KFK 2267, 1976.
4. H. Schneider, E. Nold, KFK 2273, 1976.
5. H. Schneider, Werkstoffe und Korrosion, 30, 358, 1979.
6. H. Schneider, Proceedings of Int. Conf. on Liquid Metal
      Technology in Energy Prod., Champion 1976, p. 716.
7. H. Huthmann, G. Menken, H.U. Borgstedt, H. Tas, "Influence
      of flowing Sodium on the Creep Rupture and Fatigue
      Behaviour of Type 304 SS at 550 $^{o}$C", Second Intern.Conf.
      on Liquid Metal Technology in Energy Prod., April 20-24,
      1980, Richland, Washington/USA.

# EXPERIENCE WITH MULTIPLE IMPURITIES DETECTED BY PLUGGING INDICATORS IN SODIUM SYSTEMS

M. Rajan and R. D. Kale

Reactor Research Centre
Kalpakkam
India

## INTRODUCTION

Plugging indicators are rugged instruments which have been widely used to monitor the impurity levels of reactor sodium circuits and other sodium test loops. Their main advantage lies in that they give direct information concerning the plugging of flow passages in terms of what is called plugging temperature.

These instruments suffer from one draw back in that their response is non-specific. However, they are sufficiently sensitive to indicate presence of multiple impurities in sodium. In this paper, the authors have discussed their experience with multiple impurities detected by plugging indicators. Attempts have also been made to identify the nature of these impurities by studying their precipitation kinetics and its comparison with published literature.

## DESCRIPTION OF SODIUM SYSTEMS AND PLUGGING INDICATORS

The impurity behaviour discussed here was studied in two different sodium systems in which commercial grade sodium was charged through microfilters. Sodium charged in the two systems was obtained from two separate suppliers. Plugging indicator-A (PI-A) (Fig. 1) was installed on the system A containing 2.75 tonnes of sodium and indicator-B (PI-B) was installed on the system B containing 1.3 tonnes of metal. These systems were primarily meant for studies of performance of components such as sodium-sodium heat exchanger, sodium pump and control rod drive mechanisms. The flow restrictions in PI-A consists of 12 sq.holes of 1 x 1 mm and those in PI-B consists of 16 holes of 1 mm dia. Both plugging

indicators were operated in the manual mode, wherein, the cooling
rate was controlled manually by adjusting the air flow rate.

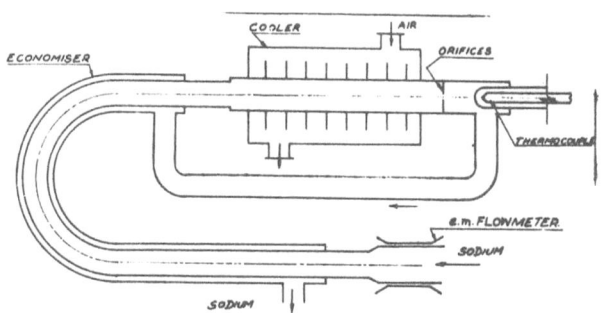

Fig. 1.   Plugging indicator

DETECTION OF MULTIPLE IMPURITIES

The behaviour of multiple impurities in a sodium system is
complex in nature. The effects are more predominant during the
initial purification of sodium before the system is in equilibrium
with cold trap. It is, however, quite difficult to know the exact
identity of impurities unless one is equipped with instrument for
continuous monitoring of specific impurity. Smith et al. have
recently studied impurity behaviour in the plugging indicator using
electrochemical monitors (1). In our sodium systems both plugging
indicators PI-A and PI-B have detected multiple impurities. However, the nature of impurities detected in the two systems appears
to be different. This is discussed below.

EXPERIENCE WITH PI-A

Plugging indicator-A detected two impurities having plugging
temperatures of 340 $^{\circ}$C and 165 $^{\circ}$C (2). The first impurity$^{x)}$ causing
plugging around 340 $^{\circ}$C showed a few important characteristics. The
impurity could be removed by gradually reducing the cold point
temperature of the cold trap. However, after a few hours of cold
trapping at 270 $^{\circ}$C, a new plugging run could not establish plugging
temperature in the expected temperature range close to the cold
point temperature. On the other hand plugging occured at a much
lower temperature of about 165 $^{\circ}$C, indicating that the impurity
plugging at higher temperature did not cause plugging below 270 $^{\circ}$C.
This indicated clearly that the first impurity has a very low
solubility below 270 $^{\circ}$C.

---

$^{x)}$Throughout the discussion, the first impurity refers to the one
  plugging at higher temperature and the second corresponding to
  lower temperature.

To reveal the presence of second impurity while the first
impurity existed in sodium it was decided to take plugging runs
at a faster cooling rate so that the first impurity could be
skipped. For this, the first impurity was redissolved in sodium
by heating the cold trap. The plugging indicator was cooled at a
faster rate (5 °C/min) from an initial temperature of 350 °C. A
partial plug of the first impurity was formed as evidenced by a
small decrease in flow (Fig. 2). However the second impurity did
not appear in the orifice on further cooling which finally resulted
in freezing of sodium. It is possible that the second impurity was
crystallizing elsewhere in the system say upstream the orifice, as
the presence of first impurity could have provided enough nucleation
sites at the lower temperature.

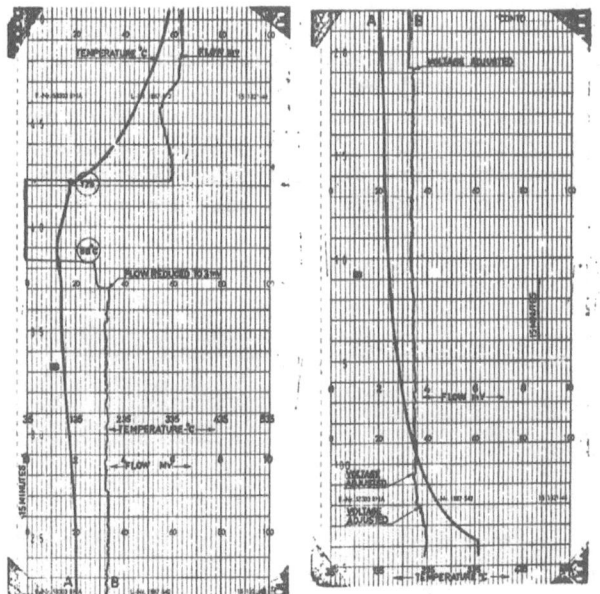

Fig. 2    Skipping of impurities by PI-A

EXPERIENCE WITH PI-B

PI-B detected two impurities in another sodium system containing
about 1.3 tonnes of sodium. The plugging temperatures of the two
impurities were 350 °C and 200 °C.

To detect the presence of second impurity with both impurities
existing in system, skipping of the first impurity was again
attempted in PI-B. The plugging indicator was cooled rapidly
(25 °C/min) for the purpose. Since sufficient time for crystallization was not available the first impurity plugged the orifice

partially. Such a flow tracing is given in Fig. 3. A decrease in
flow occured when the temperature was below 350 $^\circ$C. However, when
the orifice temperature was reduced below 200 $^\circ$C, the second
impurity must have prevented further growth of first impurity at
the orifice as shown by the flat portion of the flow curve. On
raising the temperature above saturation temperature of second
impurity, the flow continued to decrease due to build up of first
impurity.

The interference by second impurity was studied by cooling
the orifice at different rates. This becomes clear when we compare
flow traces of Fig. 3 and 4. The flat position of the flow trace
(Fig. 4) after the first flow decrease indicates the interference
by second impurity. As the temperature was reduced below the
saturation temperature of the second impurity, further flow decrease
was solely due to the second impurity crystallizing at the orifice.
The interference by second impurity at the orifice is very complex
in nature and is explained by Montevideo et al (3). As the temperature was reduced below the saturation temperature of the second
impurity, it started crystallizing out on the surface of the first
impurity. Since further sites were not available for nucleation and
crystal growth, the first impurity did not grow at the orifice.

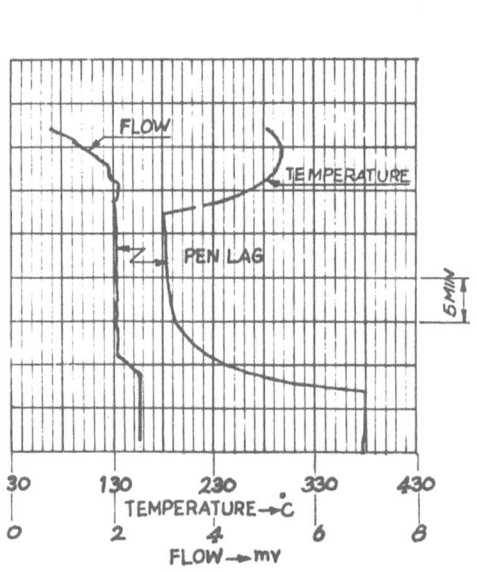

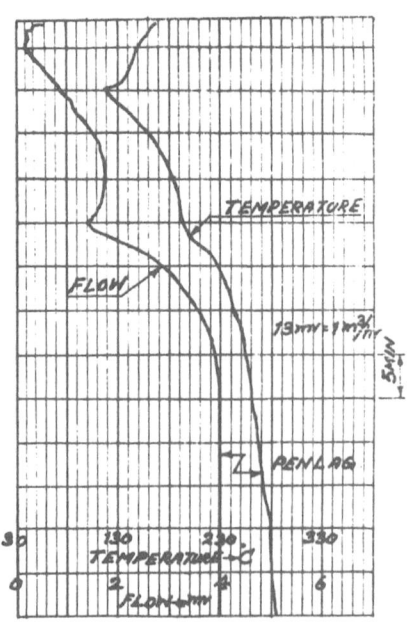

Fig. 3   Skipping of first                Fig. 4   Multiple impurity
         impurity in PI-B                          plugging in PI-B

IDENTIFICATION OF IMPURITIES

An attempt was made to identify the plugging species with the help of relations reported in the literature. Gwyther et al. have studied the plugging traces of their plugging indicators and they have observed that the plugging/unplugging rates have a direct dependence on clear orifice flowrates (4). The rate gradient values for plugging and unplugging were correlated using the equation

$$\frac{dFo}{dt} = AV^m \, \Delta C^n \, e^{-E/RTo}$$

where $dFo/dt$ = the gradient of the orifice flow rate ($m^3/sec^2$)

   $V$ = the velocity at the orifice m/sec

   $\Delta C$ = the super saturation, weight fraction of solute

   $To$ = orifice temperature in degress K

   $R$ = gas constant J/K mole

The quantities A, m, n and E are obtained by fitting techniques and reported (4).

Using above equation and constante (for oxygen) and the data for PI-B, gradient of the orifice flow rate was calculated in the case of first impurity. Since on line impurity monitors were not available the saturation temperature of the impurity was estimated from the plugging temperature. This was considered valid in view of slow cooling rate, $<2\ ^{\circ}C/min$ used during plugging runs. The calculated and experimentally observed gradients are compared below:

| Plugging run | Orifice velocity m/sec | log $\Delta C$ | log $\frac{dFo}{dt}$ calculated (oxygen) | log $\frac{dFo}{dt}$ experimental |
|---|---|---|---|---|
| 1 | 6.03 | −3.75 | −6.83 | −6.76 |
| 2 | 6.03 | −3.75 | −6.76 | −6.75 |
| 3 | 4.93 | −4.25 | −7.26 | −7.25 |
| 4 | 6.03 | −3.79 | −6.72 | −6.67 |
| 5 | 3.65 | −4.10 | −7.12 | −7.28 |
| 6 | 8.16 | −3.75 | −6.67 | −6.55 |
| 7 | 5.95 | −3.75 | −6.77 | −6.67 |

It can be observed that very good agreement exists between calculated and experimental values and thus it can be concluded that oxygen is the impurity causing plugging at $350\ ^{\circ}C$ in PI-B. A calculation of the oxygen impurity pick up from the entire loop surface based on a reported value of 2 g/m$^2$ (5) yielded a value of 140 ppm. This together with oxygen that would have passed in the loop from the initial charge would correspond to an oxygen saturation temperature nearly equal to the observed plugging temperature of 350 $^{\circ}$C. This also strongly suggests that the first impurity detected in PI-B is oxide.

Similar attempts were made to identify the second impurity in PI-B. The flow rate gradients of the second impurity when it deposits over the first impurity (Fig. 4) were found to be nearly same as calculated from the hydride data (4). This suggests that the impurity plugging at 200 $^{\circ}$C (second) is likely to be hydride.

Of the two impurities detected by PI-A, the first impurity has shown to be quite insoluble below 270 $^{\circ}$C. The exact nature of this impurity is not known. The second impurity appears to be oxide as evidenced by chemical analysis of sodium samples (2) taken when first impurity was completely eliminated by the cold trap.

CONCLUSIONS

An attempt has been made to understand behaviour of multiple impurities with the help of plugging indicators. It has been possible to identify impurities by studying the flow traces and comparing their characteristics with those available in the literature, thanks to detailed studies conducted by C.E.G.B., U.K.

ACKNOWLEDGEMENT

The authors thank the management of R.R.C. for permission to publish this paper. They also acknowledge with deep gratitude the contribution of their late colleague Mr. R. Selvaraj in performing the work reported here.

REFERENCES

1.  C. A. Smith, P. A. Simm and G. Hughes, Analysis of hydride and oxide deposition and resolution in sodium in relation to plugging meter behaviour - Nucl. Energy 1979, Vol. 18, No. 3.
2.  R. D. Kale, R. Selvaraj and M. Rajan, Experience in the operation of sodium purification loop in the Reactor Engineering Laboratory, RRC-11, Reactor Research Centre, Kalpakkam 603102, India 1976.
3.  D. A. Montevideo and C. Biery, Multicomponent impurity detection with a sodium plugging indicator, International Conference on liquid metal technology in energy production, Champion, Pennsylvania, May 1976.
4.  J. R. Gwyther et.al. Developments in Plugging Meters and cold traps for LMFBRs. Second International meeting on liquid metal technology in Energy Production, Richland, Washington, April 1980.

5.  R. Abramson et.al. Evolution of sodium purification,
    Rapsodie, Phenix, Superphenix, International Conference
    on liquid metal technology in Energy Production,
    Champion, Pennsylvania, May 1976.

PLUGGING INDICATOR - MEASUREMENT OF LOW IMPURITY CONCENTRATIONS

AT A CONSTANT ORIFICE TEMPERATURE

Damien Feron

CEN de Caderache

France

## INTRODUCTION

Plugging indicators have been used extensively to measure impurity concentrations in liquid sodium systems. Simplicity of design, installation and operation is the most commonly quoted advantage. If there is no problem when the plugging temperature is important, it is more difficult to measure a plugging temperature when the impurity concentration is low. Then, the plugging temperature is often reported as inferior to 110 $^{\circ}$C. This measurement is sufficient for the technicien. Sometimes, however, a more precise indication is required in cases where the implantation either an oxygenmeter or a hydrogenmeter is impossible. We propose a use for plugging indicator which satisfies this type of requirement.

## PRINCIPLE OF THE PLUGGING INDICATOR

The plugging indicator shown in Fig. 1 is the most commonly used in France. It is often the only on-line monitoring system. In this apparatus, the sodium flows in the outer annulus and cools the sodium flowing out the center tube. The plugging indicator is cooled by blowing gas (often air) and electrically heated. The orifice ring is situated at the end of the annulus, near a thermocouple.

In parallel with this apparatus, there is a special device, shown in Fig. 2, and called "constant pressure drop system". The pressure drop of this device is independent of the flow rate within a certain range. So this system is used to maintain the flow through the plugging indicator independent of the flow variations on the other parts of a sodium loop.

89

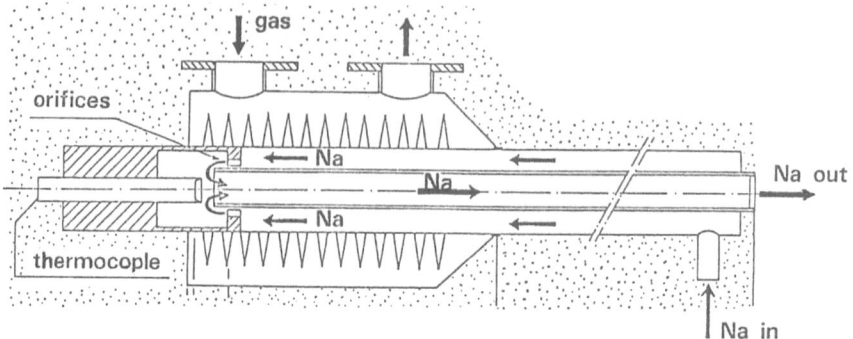

Fig. 1  Principle of Plugging Indicator

Fig. 2 Plugging Indicator and
       Constant Pressure Drop
       Device

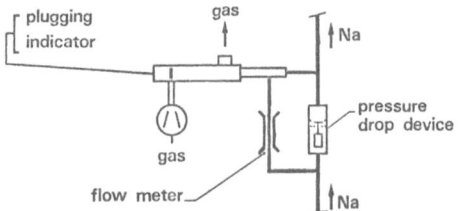

The plugging indicator is equiped with an automatic control
system. So, the temperature of the orifice ring decreases at a
given and constant rate until a restricted flow corresponding to
a given ratio of the initial flow. Reheating is then started. The
temperature when the sodium flow begins to decrease is the plugging
temperature. It is possible also to held the temperature constant
at a given value as shown on the Fig. 3.

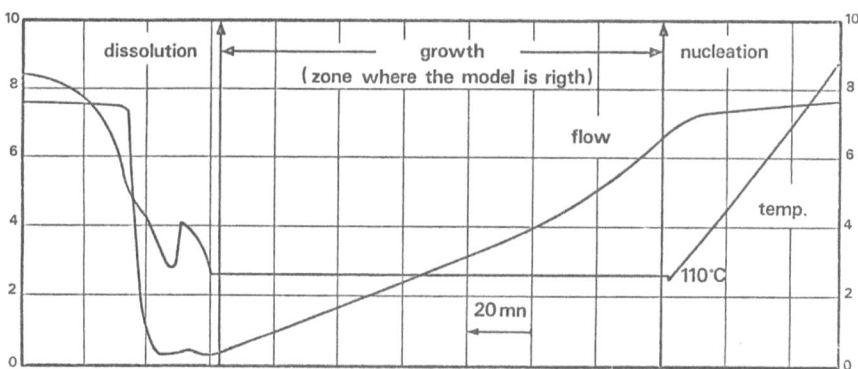

Fig. 3 Constant Temperature Plugging Run

MATHEMATICAL MODEL

The analytical model which describes the evolution of a plug,
has been explained in details (4). Let us remember the main
assumptions:

1. Plugs are assumed to grow only inside of the orifices and there is not any difference between the orifices.

2. The pressure drop during the plug is constant. This is a consequence of the constant pressure drop system. So, the variation of the sodium flow rate, F, can be expressed by the following equation:

$$F/Fi = 1 - f \tag{1}$$

where $Fi$ = initial flow rate before starting cooling operation

$f$ = fraction of an orifice plugged by the precipitate.

3. The evolution of crystals inside the orifices is divided in two steps:

- the first step is the formation of crystals (nucleation) on clean surfaces

- the second step is the growth of these crystals. The overall rate of this growth is

$$dm/dt = kA\ (C-C_s)^g \tag{2}$$

where $m$ = mass of impurity deposited
$t$ = time
$A$ = surface area for mass deposition
$C$ = impurity concentration in sodium
$C_s$ = saturation impurity concentration in sodium based on orifice temperature
$k$ = constant of the overall growth rate
$g$ = order of the process

The mathematical model does not describe the whole plugging curve, but only the portion of the curve where the growth is assumed to be the only mechanism of the plugging (Fig. 3).

The mass of impurity which is deposited at the time t, is given by:

$$m = \alpha\ \rho\ f\ ri^2\ 1 \tag{3}$$

where

$1$ = orifice length
$ri$ = initial width of on orifice
$\alpha$ = number of orifices
$\rho$ = density of the deposited impurity

The fraction of orifice plugged by the precipitate is given by the following equation (2):

$$f = (ri^2 1 - r^2 1)/ri^2 1 \qquad \qquad (4)$$

where

r = width of an orifice at the time t.

So, the surface area for mass deposition is:

$$A = \alpha 4 \ 1 \ r = \alpha 4 \ 1 \ ri \ (1 - f)^{1/2} \qquad \qquad (5)$$

Derivation, substitution and rearrangement of these equations to isolate F as function of the operating variables produce:

$$F^{-1/2} (dF/dt) = -4 \ k \ (C-C_s)^g \ \frac{Fi^{1/2}}{\rho ri} \qquad \qquad (6)$$

If the impurity concentration and orifice temperature are constant, the second number of equation (6) is constant. So the integration of (6) produces:

$$F^{1/2} = -\frac{2 \ Fi^{1/2}}{ri} \ k \ (C-C_s)^g \ t + K \qquad \qquad (7)$$

where

K  = constant.

This equation (7) can be used in two different ways:

a - The overall growth rate is unknown. Then the equation (7) is used to calculate the growth rate parameters (k and g) if the impurity concentration, C, and the orifice temperature are known. It is by this way that the overall rates given in the next paragraph were calculated.

b - The overall growth rate is known. The equation (7) can be used to calculate the impurity concentration C.


OVERALL CRYSTAL GROWTH RATES

Table 1 shows the overall crystal growth rates of sodium oxide, sodium hydride and the mixing oxide plus hydride at saturation equilibrium. The experimental method was described before (4). One may note that the overall rate orders are greater than 1. The growth is not under mass-transfer (diffusion)control. This result is wellknown in crystal growth from solution.

MEASUREMENT OF THE CONCENTRATIONS

Pollution assumptions

A limitation of the plugging indicator is its non specificity for any given impurity. However, for the calculation method reported below, three assumptions are possible:

- only one impurity precipitates: NaH
- only one impurity precipitates: $Na_2O$
- two impurities precipitate    : NaH and $Na_2O$
  at the saturation equilibrium (the two impurities have the same saturation temperature).

Calculation method

In order to know the concentration C of the impurity in liquid sodium, there are:

- a plugging curve as shown in Fig. 3. The temperature of the orifices is held constant at a value where the crystal growth rates where calculated.

- the equation (7) where all parameters are known.

Then, the method is the following:

1. to do an assumption about pollution

2. to choice the right rate

3. to calculate the slope, a, of $F^{1/2} = f(t)$
   The Flow F versus the time t is given by the portion of the plugging curve where the model is right.

4. to calculate the mass of crystals, $m_s$, which crystallize by unity of time and of area:

$$m_s = a\rho \; \frac{ri}{2 \; F^{1/2}} \tag{8}$$

5. to obtain the concentration C of the assumed impurity (s):

$$C = C_s + \frac{m_s}{k} \; 1/g \tag{9}$$

Application

In the previous paragraph, we have reported the general method. When the parameters of the plugging indicator are known, the method becomes easier. By instance, with the plugging indicator generally used on french reactors or sodium loops (Fig. 1), equation (9) is written:

$$[H] = [H]_s + \left(\frac{0,111}{k\tau}\right)^{1/g} \tag{10}$$

$$[0] = [0]_s + \left(\frac{0,274}{k\tau}\right)^{1/g} \tag{11}$$

where $\tau$ is the time required by the decrease of the flow rate F from 90 % of its initial value Fi, to 50 % of this initial value.

If we assume that the growth rates are right at the temperature of 110 °C, crystal growth parameters of the equations (10) and (11) are known and also the value of the concentrations at the orifice temperature saturation $[H]_s$ and $[0]_s$. So these equations are reported on Table 2. It is more convenient to use graphs. Therefore on Fig. 4, concentration of assumed impurity is reported versus time.

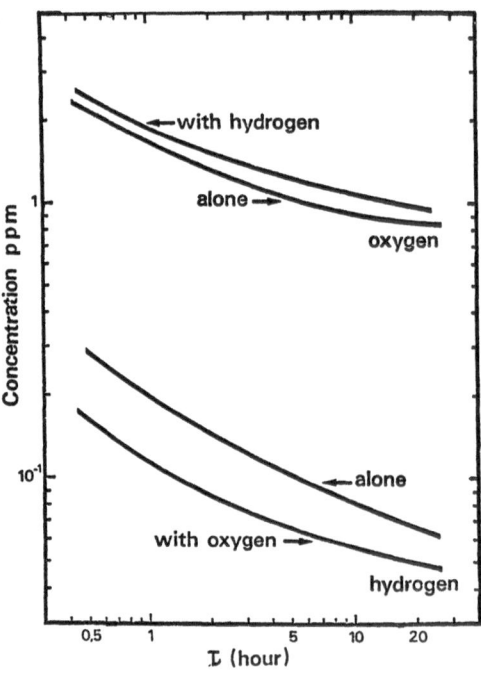

Fig. 4   Concentration of impurities as a function of time

If we want to use this method with our plugging indicators, we have only to do three things:

1. to obtain a plugging curve with a constant orifice temperature of 110 °C,

2. to read on this plugging curve the time $\tau$ required by the decrease of the flow rate from 90 % of its initial value to 50 %,

3. to report this time $\tau$ on the Fig. 4.

## Precision

If we assume mainly that there is no error due to solubility curves and that the temperature of orifices are known with a magnitude of $\pm$ 1 °C, the concentrations are calculated with an incertitude of about 20 %.

## CONCLUSION

The use that we propose for the plugging indicator is based on an analytical model which links the concentration of assumed impurity (oxygen and/or hydrogen) to the time required by a determined plug of the orifices. It is easy to use this method because its sums up by a reading on a graph.

Table 1.  Overall Crystal Growth Rates ($\Delta C$ ppm)

| Impurity | Temperature | Rate (kg/hm2) |
|---|---|---|
| NaH | 115 °C | $3.0 \, \Delta \, C^{1.8}$ |
| $Na_2O$ | 110 °C | $0.30\Delta \, C^{1.5}$ |
| $Na_2O$ saturation | 111 °C | $0.21\Delta \, C^{1.8}$ |
| NaH  equilibrium | | $6.1 \, \Delta \, C^{1.6}$ |

Table 2.  Relation Between Assumed Impurity Concentration and the Time $\tau$ (h) with an Orifice Temperature of 110 °C

| Assumed Impurity | Concentration (ppm) |
|---|---|
| NaH | $[H] = 0.037 + 0.160 \, (\tau)^{-1/1.8}$ |
| $Na_2O$ | $[O] = 0.74 + 0.94 \, (\tau)^{-1/1.5}$ |
| Saturation   $Na_2O$ | $[O] = 0.74 + 1.16 \, (\tau)^{-1/1.8}$ |
| Equilibrium  NaH | $[H] = 0.037 + 0.082 \, (\tau)^{-1/1.6}$ |

On large sodium loops, this method was found to be very sensitive and so it had been possible to detect very small leakage of pollution. But, at present time, a limitation of the method and of the plugging indicator generally is its nonspecificity for any given impurity.

REFERENCES

1.  R. F. Strickland-Constable, "Kinetics and Mechanisms of crystallization" Academic Press 1968.
2.  C. C. McPheeters, J. C. Biery, Nuclear Applications 6, 573 - 581, 1979.
3.  I. Yamamoto, S. Sumida, N. Izumi, Sagawia, J. of Nuclear Science & Technology 16, (6) 383 - 389, 1979.
4.  D. Feron "Mechanisms of sodium purification by cold traps" Thesis of Doctor-Ingineer - Institut du Génie Chimique Toulouse (France). 28th May 1979.
5.  J. D. Noden RD/B/N 2500, 1972.
6.  A. C. Whittingham, J. Nuclear Materials 60, p. 199, 191, 197.

INFLUENCE OF LITHIUM ON THE STRUCTURAL STABILITY OF TWO AUSTENITIC

STAINLESS STEELS OF THE TYPE 316 AND 18 Mn-10 Cr

Erich Ruedl, Vittorio Coen, Takao Sasaki, Horst Kolbe

Commission of the European Communities, Joint Research
Centre, Ispra Establishment
Ispra, Italy

INTRODUCTION

Liquid lithium is considered for tritium breeding and reactor coolant applications in fusion reactors. Many studies have been carried out on the corrosion behaviour of austenitic stainless steels of the type Ni-Cr in that liquid metal environment as these steels, especially 316, are envisaged as structural materials in the blanket of the first generation of reactors. Recently the study was extended to various commercial low Ni, Mn-Cr austenitic steels (1). Samples of these materials were exposed at 873 K to pure Li, in static tests, for times ranging from 500 to 1500 hours. The polished cross-sections were investigated by scanning electron microscopy and X-ray microanalysis. It was observed that for all the steels in contact with pure Li, a corrosion layer depleted in Mn and enriched in Ni is formed. In some of these materials, a grain boundary attack was also evidenced in the unterlying matrix.

In the present study, we have carried out a more detailed analysis on one of the materials previously investigated, and considered most promising, i.e. a very low Ni, Mn-Cr austenitic steel with trade name AMCR which did not show any grain boundary attack (1). We have investigated the surface corrosion layer and effected a phase and elemental study of the matrix beneath it using various electron microscopy techniques and quantitative X-ray microanalysis.

For comparison we have studied in the same experimental conditions, samples of AISI 316.

97

EXPERIMENTAL DETAILS

The two materials investigated were of the commercial type. The AMCR was supplied by Creusot-Loire in the form of a solution annealed, 5 mm thick plate. The AISI 316 steel, in the form of a 150 µm thick solution annealed sheet, was supplied by Goodfellow Metals. The chemical composition of the two materials is listed in Table 1.

Table 1.  Composition (wt%) of the examined alloys

| Material | Fe | Cr | Mn | Ni | Si | Mo | C | N | S | O |
|---|---|---|---|---|---|---|---|---|---|---|
| AMCR | 71.0 | 10.0 | 17.5 | 0.67 | 0.57 | – | 0.205 | 0.057 | 0.008 | 0.040 |
| AISI 316 (EN58J) | 64.3 | 18.3 | 1.5 | 12.90 | 0.60 | 2.4 | 0.080 | 0.043 | 0.007 | 0.010 |

From the Mn-Cr steel disc specimens of 3 mm diameter and 400 µm thickness were prepared by spark-machining. The discs were mechanically grinded to 100 - 200 µm thickness. From the 316 sheet discs 3 mm in diameter were prepared by mechanical punching.

Discs of both materials were heat treated at 873 K under vacuum ($10^{-5}$ Pa) for 1500 hours in electron beam weldel AISI 304L steel capsules filled with pure Li containing 17 ppm nitrogen in the case of AMCR and 60 ppm for AISI 316. More details about the experimental procedure are described in ref. 1. After exposure to Li and subsequent cooling to ambient temperature, the discs were removed from the Li and washed with water and alcohol. For comparison some discs were annealed in the same experimental conditions but without Li at 873 K for 1500 hours.

In order to carry out the structural and compositional examinations, the discs were treated as follows:

a) for the study of the surface layer formed on the samples exposed to Li, the discs were backthinned by electropolishing in a solution of 90 % acetic- 10 % perchloric acid at 288 K;

b) for the examination of the matrix of specimens exposed to Li or annealed in the absence of Li, the discs were electrolytically thinned from both sides in the same solution. In such a way, electron transparent regions were obtained at   50 to 100 µm depth depending on the thickness of the samples. The specimens were subsequently studied by scanning and transmission electron microscopy, by electron diffraction and quantitative, energy dispersive, X-ray microanalysis.

RESULTS

AMCR steel

Observation of solution annealed samples revealed that their matrix contained equiaxed grains of up to ~ 40 μm diameter. Within the grains planar defects in the form of microtwins and stacking faults were observed. The grains have an austenitic structure (γ-phase, fcc). After annealing in the absence of Li, a great number of particles precipitate both at grain boundaries and within the grains. These precipitates have a fcc structure with a lattice parameter close to $M_{23}C_6$ carbides so that it is reasonable to consider them as such. Their structural relationship with the matrix is the following:

$$(001)_{M_{23}C_6} \parallel (001)_{fcc} \text{ with } [100]_{M_{23}C_6} \parallel [100]_{fcc}$$

An X-ray microanalysis of the carbides indicated that they have a complex composition that varies slightly from particle to particle. Iron, manganese, chromium and nickel were identified as metallic constituents and their concentration range is given in Table 2.

Table 2.  Compositional range (wt%) of the main constituents of the $M_{23}C_6$ carbides identified in AMCR after annealing at 873 K for 1500 hrs with or without Li

| Condition | Cr | Fe | Mn | Ni |
|---|---|---|---|---|
| Only annealed | 35.5 - 38.5 | 30.7 - 33.7 | 24.5 - 27.0 | 3.0 - 7.5 |
| Annealed in the presence of Li | 67.0 - 75.0 | 16.0 - 21.0 | 7.0 - 12.0 | 1.0 - 2.5 |

The matrix of the samples annealed in the absence of Li is fully austenitic. Compositional changes that occurred due to carbide precipitation, were not sufficient to induce phase transformations. Different results were found on samples treated in Li. On account of corrosion, such specimens reveal a layer, 10 μm thick, depleted in Mn. The top of this layer is strongly enriched in Ni and has a fcc structure (γ-phase). The rest is mainly composed of Fe and Cr ( 84 % Fe  14 % Cr,  2 % Ni, Mn, Si), has a bcc structure (slightly deformed) and is magnetic (α-phase). No precipitates are revealed within this layer but pores are often observed. The region beneath is not face-centered cubic as in the samples annealed in the absence of Li but has a hexagonal close-packed structure. Some of the grains are almost completely split up in regions of hcp structure in twin relationship (Figs. 1 and 2). Within such grains and on their respective boundaries, no

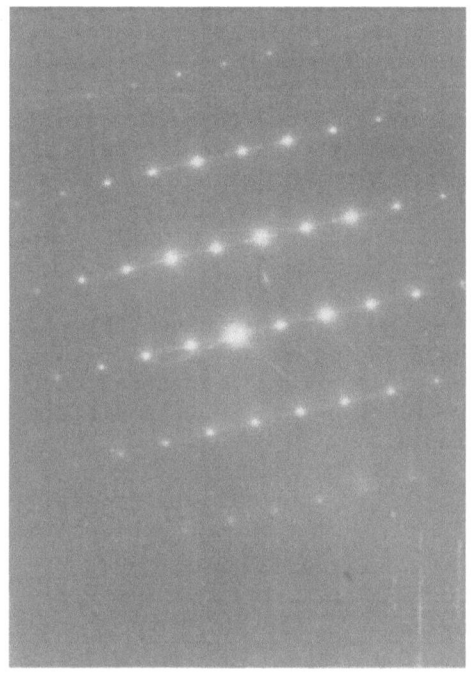

Fig.1 Transmission electron
micrograph of solution
annealed AMCR steel heat
treated in Li at 873 K for
1500 hrs. The austenitic
matrix ( 50 μm below the
surface of the disc) is
fully transformed to the
hcp ε-phase. Faults on
(0001) planes, no precipitation of carbides.

Fig.2 Electron diffraction pattern
of hcp lattice from region
selected in Fig. 1 at the
(2110) orientation. Streaks
due to faulting.

carbides were observed. Still beneath, other grains are observed
that retain some austenite (Fig. 3). In such regions, of duplex
structure, carbides are present along the grain boundaries as seen
in the samples annealed in the absence of Li. Annealing experiments
carried out in the transmission electron microscope have shown that
the regions of hcp structure transform to the fcc structure at
temperatures well below the test temperature.

The structural relationship observed for the fcc and hcp structure is the following:

$$(111)_{fcc} \parallel (0001)_{hcp} \text{ with } [\bar{1}01]_{fcc} \parallel [1\bar{2}10]_{hcp} \text{ and } [01\bar{1}]_{fcc} \parallel [\bar{2}110]_{hcp}$$

An analysis by electron diffraction showed that the observed
hcp structure corresponds to the non-magnetic $\varepsilon$-phase described by
Gordon-Parr (2) which forms martensitically in austenitic Fe-alloys
of low stacking fault energy by cooling below the $M_s(\varepsilon)$ temperature
(see for instance ref. 3). This temperature is usually below ambient
temperature for the more highly alloyed austenitic steels (4). X-ray
microanalysis of adjacent regions with hcp for fcc structure did not
show any difference in composition. Moreover, no change of composition in the hcp regions of Li-exposed specimens was found with
respect to the austenite matrix of solution annealed specimens. A
compositional analysis of the carbides formed in regions of duplex
structure, indicated that their compositions differed considerably
from that of the carbides precipitated during annealing in the
absence of Li, as can be seen in Table 2.

In order to ascertain whether the phase instability observed
in the matrix beneath the corrosion layer is peculiar to thin
specimens or occurs also in massive samples, rectangular specimens
measuring 10 x 5 x 2 mm that had been treated in Li together with
the discs, were also analyzed. A study of their cross section after
appropriate etching revealed that the hcp structure is also formed
in massive samples and extends to about 100 µm into the matrix.

AISI 316 steel

Examination of solution annealed 316 specimens showed that
their matrix is austenitic fcc. The included grains, of diameter
up to ~ 30 µm, are equiaxed. Within the grains, precipitates rich
in Mn and Si are sometimes observed. In the annealed material,
$M_{23}C_6$ carbides are also occasionally found on grain boundaries,
the matrix is austenitic as in the solution annealed material.

On the specimens heat treated in Li, a thin discontinuous
layer depleted in Ni ( 90 % Fe, 8 % Cr and 2 % Ni) and having
a deformed bcc structure, is present. The structure of the matrix
beneath this layer is austenitic, no grain boundary attack being
observed (Fig. 4).

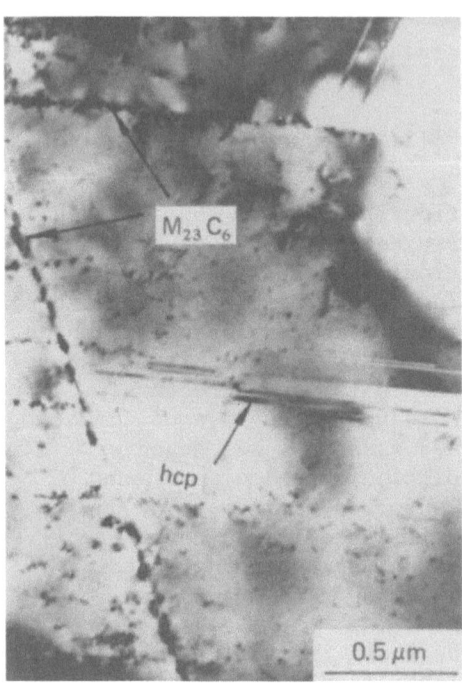 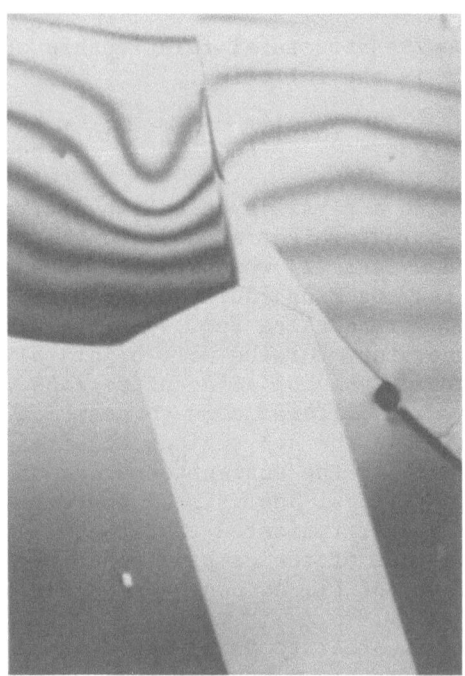

Fig.3 Transmission electron
      micrograph of solution
      annealed AMCR steel heat
      treated in Li at 873 K for
      1500 hrs. The austenitic
      matrix ( 80 μm below the
      surface of the disc) is
      only partially transformed
      into the hcp phase. Car-
      bides present on boun-
      daries.

Fig.4 Transmission electron micro-
      graph of solution annealed
      AISI 316 steel heat treated
      in Li at 873 K for 1500 hrs.
      The matrix is austenitic fcc.

DISCUSSION

The present results show that in the austenitic Mn-Cr steel AMCR heat treated in Li for 1500 hrs at 873 K, the corrosion effect is not confined to a surface layer but extends into the matrix to a considerable depth. The ε-phase is formed in a region of about 100 μm thickness beneath the corrosion layer. The $M_{23}C_6$ type carbides precipitated in this region, have a different composition with respect to those formed during annealing in the absence of Li.

We presume that the surface corrosion layer is formed at the test temperature. On its top, the structure remains fcc the austenicing element Mn being replaced by Ni. Beneath, on account of Mn depletion, the austenite structure becomes destabilized and a slightly deformed bcc structure is formed. The γ- to ε-phase transformation observed beneath the corrosion layer is diffusion-less martensitic. The martensitic type of transformation is con-firmed by the identical composition of the hcp and austenite phases in regions where they coexist and by the type of morphology developed by the newly formed phase. It is very probable that this transformation occurs during cooling from the test temperature as confirmed by the in-situ annealing experiments.

The formation of the ε-phase is strongly related to the stacking fault energy (3, 4) and may be induced therefore in AMCR exposed to Li by a reduction of this energy (and a respective increase of the $M_s(\varepsilon)$ temperature) due to compositional changes. Cr, Mn, Ni and Si can be excluded among the elements which determine the stacking fault energy in AMCR since no significant change in the concentration of these elements with resepct to the solution annealed matrix was noticed in regions of hcp structure. The observed transformation is therefore most probably related to a decrease in the carbon and/or nitrogen content. These elements which also strongly influence the stacking fault energy escape detection by EDS X-ray microanalysis. An experimental observation pointing to a decrease in the carbon content in the matrix region beneath the corrosion layer which is almost completely transformed to the hcp phase on cooling from the test temperature, consists in the absence of carbides in such region. A decrease in the nitrogen content of several Mn-Cr austenitic steels after heat treatment in Li for 1500 hrs at 873 K has been reported (1). For a confirma-tion of the above assumption, the determination of the C and N content in the regions of interest would be necessary. The different composition of the carbides formed along grain boundaries in the corrosion affected zone (where the γ- to ε-phase transformation is only partial) is another indication for compositional changes occurring in AMCR to a certain depth below the corrosion layer. We presume that the decrease in Mn content in these carbides is due to a local depletion of Mn along the grain boundaries. Evidence for the depletion of Mn along grain boundaries was recently obtained

in a parallel study on the commercial low Ni, Mn-Cr austenitic
steel Nitronic 32 corroded in Li under the same conditions as in
the present work (5).

The question arises why the Ni-Cr austenitic steel also
examined does not undergo a martensitic transformation as observed
in AMCR. It can be assumed that compositional changes also occur
in the matrix of the latter steel when heat treated in Li.
Beskorovainyi et al (6) have determined concentration profiles of
Ni, Cr and C near the surface of a Ni-Cr austenitic steel; considerable changes in the concentrations of these elements, to a
certain depth from the surface are reported. Most probably the
compositional changes occurring in the matrix of the 316 type
steel examined by us do not increase sufficiently the $M_s(\alpha')$ and/or
$M_s(\varepsilon)$ temperatures so as to induce a martensitic transformation on
the bcc $\alpha'$ and/or hcp $\varepsilon$-phase on cooling from the test to ambient
temperatures.

REFERENCES

1.  V. Coen, H. Kolbe, L. Orecchia and T. Sasaki, J. Nucl.
    Mat., 85-86, December II, Part A, p. 271 (1979).
2.  J. Gordon-Parr, Acta Cryst., 5 (1952), p. 842.
3.  D. Peckner and I. M. Bernstein, Handbook of Stainless
    Steel, (Mc Graw Hill Book Co., 1977), Chapter 4,
    p. 4-23.
4.  M. V. Bowkett, AERE-R 9361 (1979).
5.  To be published.
6.  N. M. Beskorovainyi, V. K. Ivanov and V. V. Petrashko,
    High Purity Metals and Alloys, Eds. V. S. Emel'yanov
    and A. I. Evstyukhin, translated from Russian
    (Consultants Bureau New York, 1967), p. 131.

# COMPATIBILITY OF CERTAIN CERAMICS WITH LIQUID LITHIUM

Inge Schreinlechner and Fritz Holub

Österreichisches Forschungsinstitut Seibersdorf
Wien
Austria

## INTRODUCTION

Low-atomic-number ceramics are considered to be candidates for the application in fusion reactors, as a review about the ceramic utilization in nuclear industry concerning current status and future trends revealed (1).

Thorough evaluations of the properties of several low-z-ceramics in respect to this application have been performed (2, 3).

A most recent review on selected problems of chemical aspects of controlled nuclear fusion states, as one of the high-priority items for near term chemical research, the development of ceramic blanket-container materials, compatible with coolant fluids to at least 800 $^{\circ}$C (4). Besides materials for first walls, there will also be the need for insulators (1). Some of the requirements for fusion reactors are also valid for other advanced energy technology areas, such as MHD storage batteries.

### Candidate Materials

Among the series of low- Z-ceramics some carbides, nitrides and oxides are considered to be the most promising materials for the application in fusion reactors (i.e. SiC, $Be_2C$, TiC, BN, $Si_3N_4$, $Al_2O_3$, BeO) (1).

### Compatibility with lithium

For some oxides, namely $ThO_2$, BeO, $Y_2O_3$ and HgO, as well as some nitrides, AlN, BN and $Si_3N_4$ thermodynamic assumptions about

the compatibility with liquid lithium have been performed and seemed
favourable (5). Experimental results however are very rare in open
literature. Of the most promising ceramic materials, as named above,
only results of a static capsule test of $Si_3N_4$ or $Si_3N_4 + Al_2O_3$ at
400 °C in liquid lithium between 10 and 300 h are available (6).

    In this paper preliminary results of some available ceramic
materials, namely BN, $Si_3N_4$, MgO and $ZrO_2$ (stabilized with CaO or
$Y_2O_3$ respectively) after static capsule tests in liquid lithium
at elevated temperatures are reported.

EXPERIMENTAL

    In order to compare results of different corrosion tests it
is essential to state the quality of the ceramic materials as well
as the purity of the lithium. The impurity content and the size
and distribution of the pores with respect to grain boundaries is
considered to be as important in the determination of corrosion
behaviour, as is the quality of the lithium in respect to nonmetallic impurities.

Ceramic materials

    1.  Materials pressed or sintered from powders, at the
Technical University Vienne (7).

    BN:              99.5 % ESK[x)], spec. weight 2.25; pressed.
    $Si_3N_4$:          99.5 % ESK[x)], spec. weight 3.2 + 5 % MgO binder,
                     hot pressed.
    $Si_3N_4$:          99.8 % ESK[x)], spec. weight 2.3 reaction sintered.
    MgO:             99.5 %, Veitscher Magnesit, Radenthein, sintered.

[x)]ESK: Elektroschmelzwerk Kempten

    2.  Origin of ceramic crucibles

    MgO:             Pure Magnesia, Morgan Refractories, GB
    $ZrO_2$:            + CaO stabilized, Comp. 1191, Corning Glass Works,
                     USA.
                     + $Y_2O_3$ stabilized, Comp. 1372, Corning Glass Works,
                     USA.

Lithium:  Granules, 1 - 6 mm, Ventron 99.9 % as received. Nitrogen
          content: 334 ppm (determined in our labs (8)).

Capsules:  Type AISI 316-Cb stainless steel capsules were after
           filling sealed by welding under argon atmosphere.

## Test conditions

In 6 experiments the ceramic samples were inserted in the lithium during the test. Namely BN and $Si_3N_4$ respectively were cut in cubes of the size of about 10 x 10 x 15 mm and fixed to a steel rod by two clamps. MgO was taken as delivered, in an egg-shape of the length of about 25 mm, and wrapped with a steel wire (see Fig. 1). In another experiment three crucibles, namely MgO, $ZrO_2$ (CaO or $Y_2O_3$ stabilized) were filled with lithium, placed in a SS-capsule under Ar-atmosphere (Fig. 2). The temperature was kept at 600 °C or 850 °C respectively. The time of corrosion varied from 50 hours to 500 hours.

## RESULTS

In Fig. 3 - 5 a MgO-pellet is shown before and after the corrosion test. Grain boundary attack has taken place together with a weight loss of about 5 %. The materials BN and $Si_3N_4$ were broken. The reaction product of BN was recovered by hydrolysis of Li in water and filtration of the slurry. From $Si_3N_4$ the Li was removed by distillation. With $ZrO_2$-crucibles no more metallic Li was left. In Fig. 2 the gravel of the reaction products are to be seen.

The results of X-ray powder diffraction analysis are given in detail in Table 1.

## DISCUSSION

BN: Since the X-ray powder diffraction analysis showed no more BN, Li is assumed to have reacted with BN to form intermediate products (i.e. $Li_3BN_2$), which were hydrolysed to the borates, when dissolved in water.

$Si_3N_4$: Of the several known compounds between Li-Si-N, only $Li_2SiN_2$ was found. Since the ratio of Si : N of this reaction product is on the side of higher N content in comparison to the starting product, nitrogen must have been withdrawn from the system. Indeed, the content of N in Li decreased from 334 ppm to 220 ppm N, which is however not sufficient to cover the nitrogen balance. The Si-content in Li was proved to have increased from 100 ppm to 2000 ppm Si. There are no data known about the solubility of Si in Li. From the mass balance yet it seems that some N out of the Li has reacted, and excess Si has entered the lithium phase. The system has never been in contact with the atmosphere.

$$Si_3N_4 + Li(N) \rightarrow Li_2SiN_2 + Li(Si).$$

MgO: An abrasive of the surface showed no Li-Mg-O reaction products.

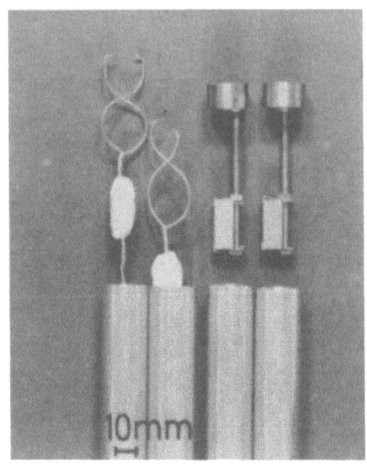

Fig. 1   Test materials
         MgO egg shaped
         BN  )
         Si$_3$N$_4$) cubes

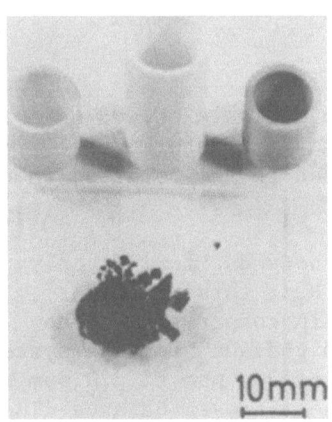

Fig. 2   Test materials
                ) left: ZrO$_2$+Y$_2$O$_3$
         crucible) middle: MgO
                ) right: ZrO$_2$+CaO
         gravel: reaction products
                 of ZrO$_2$ after
                 168 h at 850 $^{\circ}$C
                 in Li

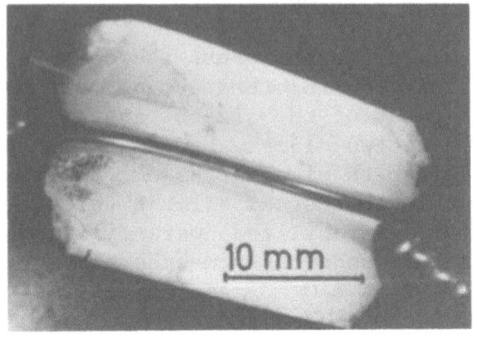

Fig. 3   MgO-surface before
         corrosion

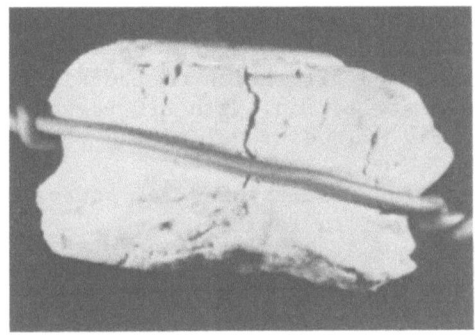

Fig. 4   MgO after 500 h at
         600 $^{\circ}$C in Li

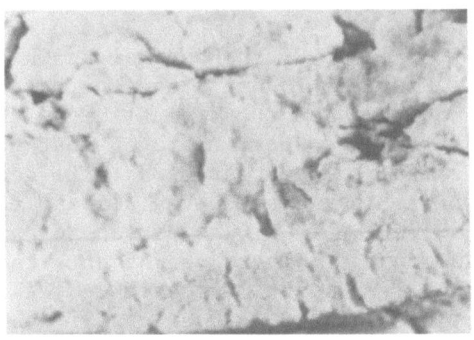

Fig. 5   Magnification of Fig. 4
         (3,2 x)

$ZrO_2$ stabilized with CaO or $Y_2O_3$: The entire lithium has reacted to form only one product, namely $Li_2ZrO_3$, of which the monoclinic and tetragonal phases are present in comparable amounts. Of the initial cubic modification of $ZrO_2$ with the stabilizing elements CaO or $Y_2O_3$, nothing was left. The reduction of $ZrO_2$ by lithium is not the favourable reaction for these conditions, since neither ZrO, nor $Zr_3O$ or Zr (which could have lateron be oxidized to $ZrO_2$) were detected. There are also no reaction products found between CaO or $Y_2O_3$ with lithium (i.e. $CaLi_2$, $LiYO_2$). Since however the contents of the stabilizing compounds of CaO or $Y_2O_3$ lie between 12 % or 6 % respectively, it must be assumed that these elements have again been built into the crystal lattices of $Li_2ZrO_3$.

CONCLUSION

According to the available data of free energy of formation (9), see Fig. 6, MgO is more stable then $Li_2O$ over the whole temperature range, whereas $ZrO_2$ seems more stable than $Li_2O$ only above 1000 K. Of the nitrides in question, BN seems more stable than $Li_3N$ above 500 K, $Si_3N_4$ only above around 1300 K. Except for BN the experimental results agree with the thermodynamic expectations.

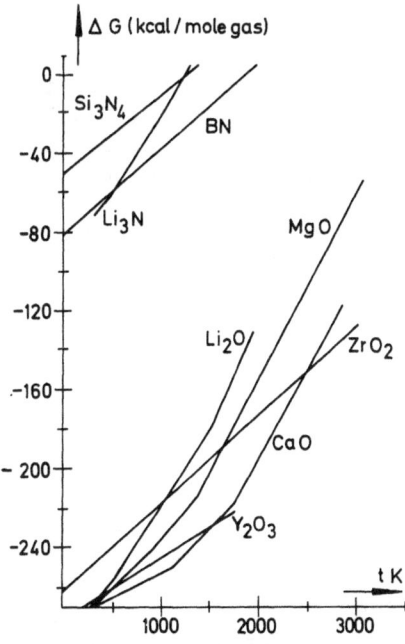

Fig. 6   Free energy of formation versus temperature

Table 1. Results of X-ray powder diffraction examination

| Material | product | found | missing |
|---|---|---|---|
| BN | residue after hydrolysis in $H_2O$ | borates, traces BN | $Li_3BN_2$ (presumably intermediate product) |
| $Si_3N_4$ | residue after distillation, no contact with atmosphere | $Li_2SiN_2$ | $Si_3N_4$, $Li_3N$, $Li_2O$, $Li_4Si$, $Li_8SiN_4$, $LiSi_2N_3$, $Li_5SiN_3$ |
| MgO (600 $^{\circ}$C) | abrasion from surface after contact with $H_2O$ | MgO | Li-Mg-O reaction products |
| $ZrO_2$(+CaO)  $ZrO_2$(+$Y_2O_3$) | )reaction product )after contact )with atmosphere | )$Li_2ZrO_3$ )monoclinic )   + )$Li_2ZrO_3$ )tetragonal | ) initial $ZrO_2$ cubic ) (CaO, $Y_2O_3$- ) stabilized) Zr, $ZrO_2$, CaO, $Y_2O_3$, $Li_2O$, $Li_2ZrN_2$, $Li_3N$, ZrN, $Zr_3O$, ZrO, $LiYO_2$, $CaLi_2$ |

The influence of nonmetallic impurities in lithium is considered to be of great significance (10). Unfortunately the oxygen-content of lithium was not known, but with a concentration of   300 ppm N in Li, BN and $Si_3N_4$ were not compatible with Li at 600 $^{\circ}$C; MgO is compatible to a certain extent with Li at 600 $^{\circ}$C. $ZrO_2$ stabilized with CaO or $Y_2O_3$ has completely reacted with Li at 850 $^{\circ}$C after 168 h.

ACKNOWLEDGEMENT

    The authors gratefully acknowledge the experimental support of Mr. J. Kozuh and Mr. N. Rinke.

REFERENCES

    1.  J. T. A. Roberts, Review 12, Ceramic utilization in the Nuclear Industry: Current Status and Future Trends (Part III) Powder Metallurgy International Vol. 11, No. 3, (1979) 125 - 129.
    2.  L. H. Rovnev, G. R. Hopkins, Nucl. Techn. 29, (1976) 274 - 302.

3.  J. Bressers, W. Van Witzenburg, High Temperature Materials
    Problems in Fusion Reactors, Rev. Int. Hautes Temp.
    refract., France, 13 (1976) 237 - 249.

4.  J. H. De Van et al., Chemical Aspects of Controlled
    Nuclear Fusion, Atomic Energy Review, Vol. 18, No. 2,
    IAEA, Vienne 1980, 553 - 574.

5.  R. N. Singh, Compatibility of Ceramics with Liquid Na
    and Li, J. Amer. Ceram. Soc., Vol. 59, No. 3 - 4,
    (1976), 112 - 115.

6.  R. N. Singh, W. D. Tuchig, Compatibility of $Si_3N_4$ and
    $Si_3N_4$ + $Al_2O_3$ with Liquid Na and Li, J. Amer. Ceram.
    Soc., Vol. 58, No. 1 - 2 (1975), 70 - 71.

7.  W. Wruss, Institut für chemische Technologie anorganischer
    Stoffe, Technische Universität, 1060 Wien, Getreide-
    markt 9.

8.  I. Schreinlechner, P. Sattler, J. Kozuh, Determination of
    Trace Impurities in Alkali Metals, Mikrochimica Acta,
    in press.

9.  T. B. Read, Free Energy of Formation of Binary Compounds,
    The MIT Press Cambridge, Massachusetts, London,
    England (1971).

10. M. G. Barker, I. C. Alexander, Influence of Dissolved
    Nitrogen and Carbon on Reactions of Cerium (III) and
    Cerium (IV) oxides with Liquid Lithium, J.C.S. Dalton,
    (1975) 1464 - 1466.

CHEMICAL ASPECTS OF THE CORROSION BEHAVIOUR OF TYPE 316 STAINLESS

STEEL IN LIQUID LITHIUM AND LIQUID SODIUM

Marten G. Barker, Stephen A. Frankham, Paul G. Gadd,
David R. Moore

The University of Nottingham
United Kingdom

INTRODUCTION

Stainless steel Type 316 is the alloy most favoured for use
in both fast reactors and fusion reactor systems. Although the
corrosion behaviour of 316 steel in sodium has been extensively
studied much less is known about its performance in liquid lithium.
This paper compares the interactions of the non-metals oxygen,
carbon, nitrogen and silicon dissolved in the liquid metals lithium
and sodium with the component metals in 316 steel. A particular
difference between sodium and lithium is the ease with which the
latter penetrates the grain boundaries of the steel, the influence
of this penetration on carburization and nitridation rates will be
discussed.

OXYGEN TRANSFER

A comparison of the thermodynamic data for the relevant
compounds of sodium and lithium is given in Table 1. From this
data it can be seen that oxygen dissolved in liquid lithium is
unlikely to interact with the components of stainless steels since
the formation of $Li_2O$ is energetically so much more favourable
than any binary or ternary oxides. The free energy of formation
of sodium oxide is much less negative and is comparable in value
to those of many ternary and binary oxides and hence the formation
of these compounds in liquid sodium is widespread. The oxides which
may be concerned with the corrosion of stainless steels in liquid
sodium are those of the elements Fe, Cr, Ni, Ti, Nb, Mn and Co.
Table 2  shows the oxide products which have been indentified as
beeing stable to liquid sodium.

113

Table 1. Free energy and solubility data for some alkali-metal compounds

| Compound | $Li_2O$ | $Na_2O$ | $Li_3N$ | $Na_3N$ | $Li_2C_2$ | $Na_2C_2$ |
|---|---|---|---|---|---|---|
| $-\Delta G$ at 900 K (kJ mol$^{-1}$) | 455 (1) | 289 (1) | 45 (2) | – | 89 (3) (at 873K) | – |
| Solubility wppm non metal | 12375 (4) | 1411 (5) | 160000 (6) | <<1 | 1100 (7) | 8.7 (8) |

Table 2. Oxides stable to liquid sodium

| | Fe | Cr | Ni | Ti | Nb | Mn | Co |
|---|---|---|---|---|---|---|---|
| Oxides stable to high oxygen Na | $Na_4FeO_3$ | $NaCrO_2$ | none | $Na_4TiO_4$ $NaTiO_2$ | $Na_3NbO_4$ | $Na_4Mn_2O_5$ | none |
| Oxides stable to low oxygen Na | none | $NaCrO_2$ | none | TiO | $Na_3NbO_4$ | $NaMnO_2$ MnO | none |

Space does not permit a detailed discussion of each of these systems but mention should be made of the lack of understanding in the Na-Fe-O system. The compound $Na_4FeO_3$ has only been observed at oxygen levels approaching saturation yet the corrosion rate of Fe in sodium shows an oxygen dependence down to very low oxygen levels. Not other oxide in the Na-Fe-O system has been shown to be stable towards liquid sodium. For example, the oxide $NaFeO_2$ – the iron analogue of $NaCrO_2$ is rapidly reduced by sodium to pure Fe and $Na_2O$. We have also examined the stability towards liquid sodium of mixed compounds of the type $Na(Cr_xFe_{1-x})O_2$, where x varies from 0 to 1, i.e., over the whole of the solid solution range and in all cases iron was removed as Fe metal and sodium oxide was formed together with $NaCrO_2$. These results clearly suggest that it is not possible to stabilise iron-oxygen bonds even by incorporation of iron into a stable lattice such as $NaCrO_2$.

NITROGEN TRANSFER

The behaviour of nitrogen is in direct contrast to that of oxygen in that transfer is most pronounced in liquid lithium systems and is only a minor effect in sodium. The free energy of formation of $Li_3N$ is much less than those of many binary and ternary nitrides and thus nitrogen transfer from liquid lithium to the steel surface will be predicted on thermodynamic grounds. Nitrogen transfer has been observed experimentally in reactions of metal oxides with lithium containing dissolved nitrogen (9). In these experiments the metal oxide is reduced directly to the metal (with $Li_2O$ being produced), and the finely divided metal then reacts rapidly with the dissolved nitrogen to produce both binary and ternary nitrides. The interaction of dissolved nitrogen in lithium with solid metal surfaces has been studied to a lesser degree but recent work at Nottingham has shown a good correlation between the products obtained in oxide reactions and those obtained as surface corrosion products on metal plates (Table 3). In Type 316 stainless steel the element most likely to react with nitrogen in lithium is chromium since the free energies of formation of the chromium nitrides CrN and $Cr_2N$ are far more favourable than for the nitrides of iron or nickel. A comparison of the free energies of formation of these nitrides with those of solutions of nitrogen in lithium shows that the formation of these nitrides will only reduce the nitrogen content of the lithium by a small degree (Fig. 1), equilibrium values being 60.000 wppm for CrN and 35.000 wppm for $Cr_2N$ at 900 K. X-ray diffraction of 316 steel surfaces exposed to lithium containing nitrogen does not show the presence of either nitride but does show the presence of an unknown corrosion product. An identical X-ray pattern is however given by the product of a solid-state reaction between Cr of $Cr_2N$ with $Li_3N$ at 700 $^\circ$C. This indicates the formation of a ternary nitride probably of the formula $Li_9CrN_5$. Although this compound has been reported by Juza (10) no X-ray pattern has been reported for the compound and some doubt still therefore exists as to the actual identity of the corrosion product.

Table 3.   Corrosion products observed in liquid lithium

| Metal or alloy | Impurity addition | Corrosion product | Metal or alloy | Impurity addition | Corrosion product |
|---|---|---|---|---|---|
| Fe | N | $Li_3FeN_2$ | Nb | N | $Nb_2N$ |
| Cr | N | $Li_9CrN_5$ | Nb | C | $Nb_2C$ |
| Cr | C | $Cr_{23}C_6 + Cr_7C_3$ | Ta | N | $TaN_{0.04} + Ta_2N$ |
| Ni | N or C or neither | Plate dissolves | Ta | C | $Ta_2C$ |
| | | | Mo | N | MoN phase |
| V | N | $V_3N$, VN; $Li_7VN_4$ | Mo | C | $Mo_2C$ |
| V | C | $V_2C$ | Zr | N | $Li_2ZrN_2 + ZrN$ |
| Ti | N | $Ti_2N$ | SS 316 | N | $Li_9CrN_5$ |
| Ti | C | $Ti_2C$ | SS 316 | C | $Cr_{23}C_6$ |

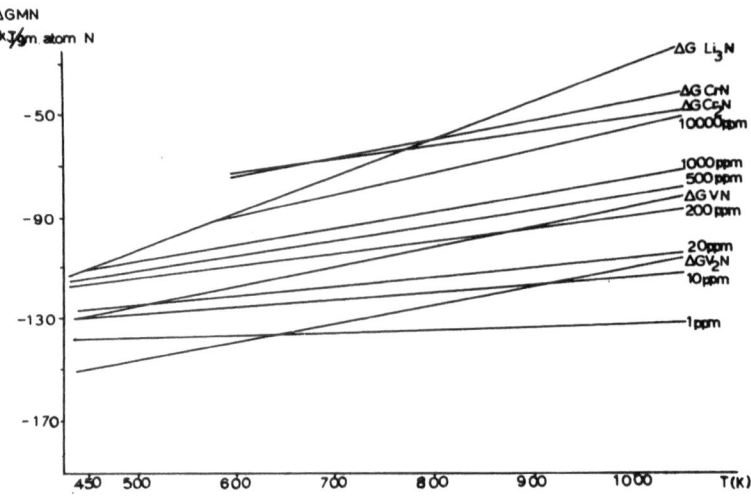

Fig. 1: Isopleth data for some nitrides in liquid lithium

CARBON TRANSFER

   Type 316 stainless steel shows almost identical behaviour
towards carbon in both sodium and lithium in that the carbide
$M_{23}C_6$ is formed in both systems. Extensive work in sodium systems
clearly shows that carbon transfer is both kinetically and thermodynamically controlled. At the solid liquid interface an equilibrium
carbide is formed but the diffusion of carbon into the matrix of
the steel may be affected by the microstructure of the steel.
Release of carbon into the liquid metal is also kinetically
controlled, different carbon sources e.g., graphite or $Fe_3C$ show
different release rates and removal of carbon from high carbon
steels is time dependet in that after the surface has been depleted
of carbon, the process is diffusion controlled. Despite these
problems it has been possible to measure the carburization of 316
steel at constants activities of carbon in both sodium and lithium.
In sodium it is possible to constantly measure carbon activity using
a carbon meter and we have shown that in small static systems
constant activity can be maintained over long time periods by using
powdered metal carbides as carbon sources. It was found that the
activity measured on a Harwell carbon meter agreed well with that
calculated from thermodynamic data and that the release rate from
the powdered carbide was greater than the combined removal rate of
the meter and the carburizing 316 steel specimens. Although it was
not possible to monitor the carbon activity in liquid lithium
equivalent experiments indicated that the same carbon sources gave
results which were consistent with a constant activity being
maintained. Metallographic examination of 316 steel specimens from

tests in lithium and sodium showed similar features, namely a bulk
carburized zone nearest to the surface and grain boundary carburization underneath. Calculation of the diffusion coefficients of
carbon in 316 steel from these experiments showed values in sodium
which were in good agreement with those previously obtained in
sodium (11, 12). Values obtained from specimens exposed to liquid
lithium were much higher than the sodium data (Table 4). A particular feature of the specimens exposed to lithium was the high rate
of lithium penetration. Polished specimens were found to self-
etch if left in air for a few days. The extent of penetration was
not found to be regular in depth but varied from 400 µm to 600 µm
for specimens exposed to lithium for 1 month. This compares with
sodium penetrations of about 20 µm over the same time period.
Examination of the specimens with a Cameca IMS3F ion microprobe
clearly showed the presence of lithium in the grain boundaries
to depths previously mentioned. It is therefore possible that the
larger diffusion coefficients measured in steels exposed to lithium
are the result of carbon being more readily moved into the steel
by the penetrating lithium rather than by diffusion in the steel
itself. Diffusion coefficients for nitrogen were also measured
from samples exposed to lithium containing high nitrogen levels
and these are included in the table for comparison.

Table 4.   Measured diffusion coefficients for carbon and nitrogen
           in Type 316 stainless steel exposed to liquid lithium
           at 650 $^{\circ}$C unit activity carbon or nitrogen

| Sample exposure time | Diffusion coeff. of carbon $(cm^2 s^{-1})$ | Diffusion coeff. of nitrogen $(cm^2 s^{-1})$ |
|---|---|---|
| 2 wks | $6.90 \times 10^{-11}$ <br> $6.92 \times 10^{-11}$ | $5.81 \times 10^{-11}$ |
| 3 wks | $6.95 \times 10^{-11}$ <br> $6.54 \times 10^{-11}$ | $5.65 \times 10^{-11}$ |
| 4 wks | $6.51 \times 10^{-11}$ <br> $6.67 \times 10^{-11}$ | $5.37 \times 10^{-11}$ |
| 4 wks | $6.41 \times 10^{-11}$ <br> $6.36 \times 10^{-11}$ | $5.27 \times 10^{-11}$ |
| Mean value | $6.65 \pm 0.22 \times 10^{-1}$ | $5.53 \pm 0.23 \times 10^{-11}$ |

SILICON TRANSFER

We have conducted a series of experiments which demonstrate that silicon reacts rapidly with nickel in both sodium and lithium. Nickel powder placed in sodium containing various amounts of silicon was found to react to form the silicides $NiSi$, $Ni_3Si$ and $Ni_3Si_2$, after heating at 600 °C for 2 days and distillation of the excess sodium at 350 °C. Similar reactions in liquid lithium gave the products $Ni_3Si$ and $Ni_2Si$ but not $Ni_3Si_2$. It would appear from these tests that silicon is being removed from the system possibly as $Li_{22}Si_5$ or even as a ternary nitride. The ternary nitrides $Li_2SiN_2$ and $Li_8SiN_4$ have both been observed as products in the reaction of silicon powder with solutions of nitrogen in lithium. It should also be pointed out that the ternary oxides $Na_2SiO_3$ and $Na_4SiO_4$ are stable towards liquid sodium at 600 °C.

Samples of Type 316 stainless steel exposed to high concentrations of silicon in both sodium and lithium showed different behaviour in the two metals. In sodium the steel surface reacted to form $FeSi$ as a surface product. Excess silicon was observed as a residue (on distillation of the excess sodium) in the form of $NaSi$. In lithium no product could be identified with certainty. The steel surface was covered with a product which flaked off when touched: X-ray diffraction showed a lattice constant almost exactly that of pure silicon but the phase was strongly magnetic. No residue was observed from the vacuum distillation of the excess lithium. Clearly more work is required in this area before any firm conclusions can be reached.

REFERENCES

1.  C. E. Wicks & F. E. Block, U.S. Bur. Mines Bull., 605, 1963.
2.  D. W. Osborne & H. E. Flotow, J. Chem. Thermodynamics, 1978, 10, 675.
3.  R. J. Pulham, Paper presented at this Conference.
4.  P. F. Adams, P. Hubberstey & R. J. Pulham, J. Less Common Met., 1975, 42, 1.
5.  A. W. Thorley, As quoted in J. Nucl. Energy, 1965, 9, 849.
6.  P. F. Adams, M. G. Down, P. Hubberstey & R. J. Pulham, J. Less Common Met., 1975, 42, 325.
7.  R. M. Yonco & M. I. Homa, Trans. Amer. Nucl. Soc., 1979, 32, 270.
8.  R. Thompson, Paper presented at IAEA Specialist Meeting on Carbon in Sodium, Harwell, England. November 1979.
9.  M. G. Barker & J. Bentham, J. Less Common Met., 1975, 40, 1.
10. V. R. Juza & J. Haug, A. Anorg. Allg. Chem., 1961, 309, 276.
11. A. W. Thorley & M. R. Hobdell, Paper presented at IAEA, Specialist Meeting on Carbon in Sodium, Harwell, England, November 1979.

12.  J. J. McGown & C. Bagnall, Paper presented at IAEA
     Specialist Meeting on Carbon in Sodium, Harwell,
     England, November 1979.

# COMPATIBILITY OF HIGH Ni, Fe-Cr ALLOYS IN Li AND IN Li CONTAINING TRACES OF LiH

Vittorio Coen, Horst Kolbe, Luigi Orecchia, Takao Sasaki

Commission of the European Communities, Joint Research
Centre, Ispra Establishment
Ispra, Italy

## INTRODUCTION

Liquid lithium is considered as a candidate material for
tritium breeding and reactor coolant applications in many magnetic
fusion reactor concepts. Austenitic stainless steels of the type
AISI 316 are envisaged on a priority base as containment materials
in the first generation of reactors and for this reason their
compatibility in static and flowing lithium has been extensively
investigated. In a previous work, we have studied the behaviour
of samples of AISI 316 and 304 in the presence of pure lithium and
in lithium containing traces of LiH (1). The mechanism of corrosion
is mostly influenced by the preferential dissolution of Ni with
subsequent destabilization of the austenitic structure and the
presence of 40 ppm H in Li seemed to exert a beneficial effect
on the extent of the corrosion layer. Following this study we have
examined the behaviour of Cr-Mn austenitic stainless steels (in
the same experimental conditions) which may be an interesting
alternative to the classical Ni-Cr steels (2). Also in this case,
the presence of traces of hydrogen seemed beneficial. We thought
that it would be of interest to extend the study to high Ni, Fe-Cr
alloys such as Incoloy 800, Incoloy 825, Uranus B6 and to Ni-Cr
alloys such as Inconel 600 and 601. These materials are engineering
alloys extensively studied, which on account of their good mechanical
properties have occasionally been considered for possible use as
containment materials for Li in fusion reactors. Their corrosion
resistance in liquid Li is not foreseen to be excellent on the
basis of what was found for lower Ni containing stainless steels.
This is confirmed by some Russian works on nickel and cobalt base
alloys mentioned by Bates (3). The same author has studied, in
static tests, alloys containing 45 - 47 % Ni as RA-333 and

121

Hastelloy X at 1163 K working in an argon atmosphere containing
0.15 % $N_2$ and with 1200 ppm of $N_2$ in Li. He found a severe corrosion.
Other authors have studied alloy 800 and alloy 600 in pure static
Li (4,5) stating that between 773 and 973 K they corroded significantly after 1000 hrs, but that the corrosion of alloy 600 could
be inhibited when 5 w% Al was added to Li. We have carried out
our tests at 873 K.

EXPERIMENT

Tests have been carried out for all the materials considered
with pure Li and with pure Li in the presence of a hydrogen partial
pressure of $2.13 \cdot 10^{-3}$ Pa corresponding to a concentration of
40 ppm H in Li at 873 K (1). The lithium in the form of ingots was
supplied by Koch-Light Laboratories Ltd. Coinbrook, Bucks. England.
It has a purity of 99.98 %. The chemical composition and origin of
the alloys studied are reported in Table 1. As can be seen, for
Incoloy 800 and Inconel 600, samples of different batches have
been investigated. Samples of approximately 10 x 10 x 3 mm were
out from the available sheets or rods and metallographically
polished. They were then degreased with acetone and heat treated
for 6 - 7 hrs at the test temperature under a vacuum of $1.3 \cdot 10^{-5}$
Pa. The same heat treatment was performed on all the containers
described below. All loading operations were carried out in a
stainless steel glove box which can be maintained at a vacuum of
$1.3 \cdot 10^{-4}$ Pa, and under pure argon. The glove box is equipped
with an electron beam welding device.

Table 1.    Chemical composition and origin of materials

| Material | Manufacturer | C%* | Mn% | Si% | S% | Cr% | Ni% | Mo% | Cu% | Fe% | Ti% | Al% |
|---|---|---|---|---|---|---|---|---|---|---|---|---|
| Incoloy 825 | Henry Wiggin & Co. Ltd. | 0.036 | 0.18 | 0.32 | 0.003 | 20.90 | 40.00 | 2.76 | 2.02 | 32.73 | 0.92 | 0.10 |
| Incoloy 800 Nr.1 rod φ 20 mm | " | 0.065 | 0.64 | 0.46 | 0.004 | 20.05 | 30.80 | | 0.21 | 46.85 | 0.56 | 0.36 |
| Incoloy 800 Nr. 2 rod φ 12.7 mm | " | 0.05 | 0.87 | 0.37 | 0.006 | 20.76 | 32.79 | | 0.48 | 44.11 | 0.36 | 0.18 |
| Incoloy 800 Nr. 3 sheet h = 3 mm | " | 0.05 | 0.95 | 0.46 | 0.007 | 20.65 | 31.68 | | 0.24 | 45.09 | 0.46 | 0.39 |
| Inconel 600 Nr. 1 rod φ 12.7 mm | " | 0.038 | 0.18 | 0.24 | 0.004 | 15.45 | 76.16 | | 0.13 | 7.70 | 0.10 | |
| Inconel 600 Nr. 2 sheet h = 3 mm | " | 0.072 | 1.30 | 0.36 | 0.004 | 17.3 | 70.86 | | 0.12 | 9.9 | 0.08 | |
| Inconel 601 sheet h = 1.5 mm | " | 0.055 | 0.19 | 0.37 | 0.003 | 22.8 | 60.5 | | 0.06 | 14.47 | 0.35 | 1.20 |
| Uranus B6 sheet h = 2 mm | Creusot - Loire | 0.012 | 1.33 | 0.68 | | 20.05 | 24.55 | 4.47 | 1.35 | 47.56 | | |

* Composition in weight percent

## Tests with pure Li

The samples to be studied are introduced in small 304 L stainless steel containers together with solid lithium (one container
for each family of alloys). The containers are electron beam welded
and heat treated at the desired temperature in resistance furnace
under vacuum:

a)  in a tubular furnace in static conditions,
b)  in a furnace in which an attachment that holds the containers
    is situated in a vacuum tight tube and rotates perpendicularly
    to the longitudinal axis of the tube at a speed of one run per
    minute.

## Tests with Li + LiH

The experimental set up for conducting these tests has been
described in detail elsewhere (1,2). It permits to work with pure
Li in the presence of a hydrogen partial pressure corresponding
to a defined concentration of hydrogen in Li. The reaction chamber
made of AISI 304L is heated under U.H.V. and a "quadrupole" gas
analyzer measures continuously the hydrogen partial pressure. All
the loading operations are carried out in the glove box described
above.

After the compatibility tests, both the reaction chamber for
the Li + LiH tests and the containers for the pure Li tests are
opened in the glove box, some Li is removed for analysis of the
nitrogen content and the metal samples are washed with water and
alcohol to remove the lithium. They are then metallographically
polished and examined by scanning electron microscopy, wavelength
dispersive, X-ray elemental distribution imaging and quantitative
energy dispersive X-ray microanalysis. Some X-ray diffraction
analyses have also been carried out on the surfaces of the corroded
samples.

## RESULTS

The tests have been carried out at 873 K for times up to
2150 hrs in pure Li and for 1000 hrs with Li + 40 ppm H in Li.
A detailed analysis of the behaviour of the different alloys
studied is given below:

## Incoloy 825

After 1000 hrs at 873 K in Li with a nitrogen content of 10 ppm
in static conditions, the corrosion layer has a thickness of 15 μm
as can be seen in Fig. 1. The same material treated in Li + LiH at
the same temperature and for the same time, although the nitrogen
content is higher: 300 ppm, shows a corrosion layer of only 3 μm
(Fig. 2). In a sample heat treated in Li ($N_2$ = 77 ppm) for 2150 hrs

at 873 K in the tubular furnace in static conditions, the corrosion
layer is   20 μm. Fig. 3 shows in detail what happens to the same
material after 2000 hrs: at 873 K in the rotating system, the nitrogen content of the Li is 20 ppm. As can be clearly seen in the
corroded zone, there is a substantial depletion in Ni and Cu and
an increase in the content of Fe and Cr. A quantitative analysis
from EDAX of the phases labelled shows that, while the composition
of the base material (M) is 40.3 % Ni, 33.8 % Fe, 21.5 % Cr in the
Ni depleted  A region Ni drops to 4.6 % while Fe increases to 51.9 %
and Cr to 36.7 %. X-ray mapping shows also a dark phase very rich
in Ti.

## Incoloy 800

In all our experiments we have loaded in the same container
one sample of each of the three batches available. After 1000 hrs
at 873 K in pure Li with a nitrogen content of 40 ppm (static
conditions), the behaviour of the three materials is somewhat
different even if the mechanism of attack is always the same. The
same materials in the presence of Li + 40 ppm H in Li and with
a nitrogen content of 300 ppm are practically unattacked. Fig. 4
illustrates what is mentioned above. In Fig. 5 the behaviour of
the three materials in two different runs of 2000 hrs at 873 K, in
the rotating device, is shown. A quantitative analysis from EDAX
of the phases present in sample No. 2 shows that, while the base
material has 30.8 % Ni, 46.9 % Fe, 20.7 % Cr in A, the Ni drops
to 7.2 %, Fe increases to 50.8 % and Cr to 39.8 %. In B, Ni drops
to 13.3 %, Fe increases to 64.8 % and Cr remains practically unchanged: 19.9 %. The black phase seen in the micrograph is rich
in Ti and contains Fe and Cr.

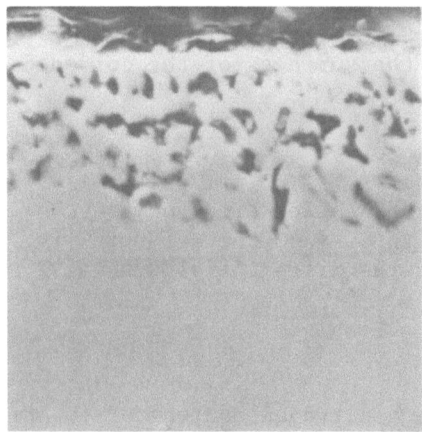 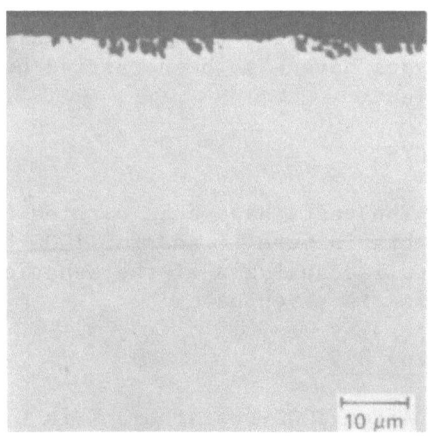

Fig. 1 Incoloy 825, 1000 hrs in   Fig. 2 Incoloy 825, 1000 hrs in Li
       Li 873 K, $N_2$=10 ppm.            + 40 ppm H in Li, 873 K,
       B.E.I. composition.                $N_2$=300 ppm. B.E.I.composition

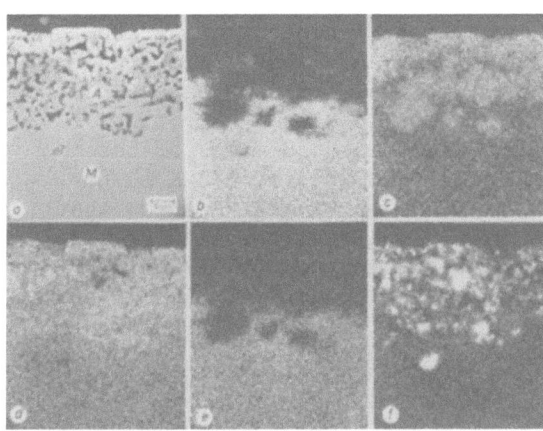

Fig. 3 Incoloy 825, 2000 hrs
        in Li, 873 K,
        $N_2$= 20 ppm,
        rotating device

a) B.E.I. composition,
b) Ni-Kα X-ray image,
c) Cr-Kα X-ray image,
d) Fe-Kα x-ray image,
e) Cu-Kα X-ray image,
f) Ti-Kα X-ray image.

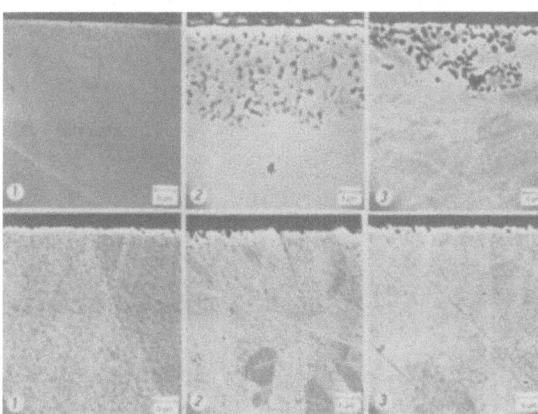

Fig. 4 Incoloy 800, batches
        1,2,3, 1000 hrs,
        873 K, B.E.I.
        composition.

upper row: Li $N_2$=40 ppm
lower ruw: Li+40 ppm H in
           Li
           $N_2$=300 ppm

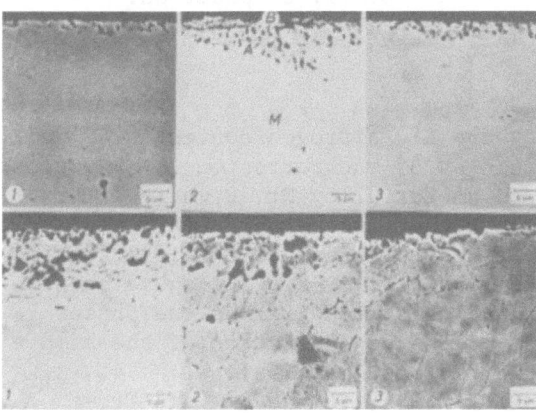

Fig. 5 Incoloy 800, batches
        1,2,3, 2000 hrs in
        Li, 873 K, rotating
        system.
        B.E.I. composition

upper row: $N_2$=15 ppm
lower row: $N_2$=40 ppm

Inconel 600

This material which has the highest Ni content among the
alloys studied, is highly corroded in Li and in Li containing
traces of hydrogen. The attack is identical for both batches
investigated but the thickness of the corrosion layer varies.
Often an intergranular penetration is noticed. Samples of batches
No. 1 and 2 are heavily corroded after 1000 hrs at 873 K in Li
+ 40 ppm H in Li (nitrogen content = 300 ppm). Fig. 6 shows what
happens to Inconel 600 of batch Nr. 1, while Fig. 7 illustrates
the behaviour of the same material after the same time and at
the same temperature in pure Li (nitrogen content = 10 ppm) in
static conditions.

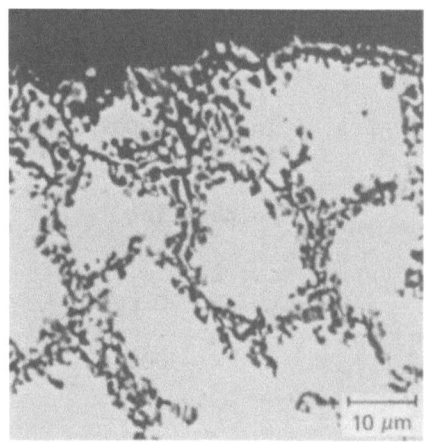
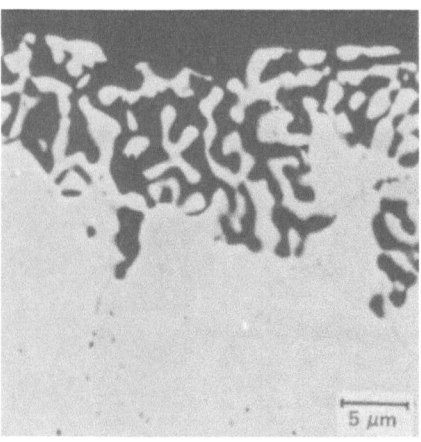

Fig. 6 Inconel 600, batch 1,          Fig. 7 Inconel 600, batch 1,
       1000 hrs, 873 K                        1000 hrs, 873 K
       Li + 40 ppm H in Li                    Li ($N_2$ = 10 ppm),
       ($N_2$ = 300 ppm),                     B.E.I. composition.
       B.E.I. composition.

After 2000 hrs at 873 K in pure Li (nitrogen content = 42 ppm),
in the rotating device, the thickness of the corrosion layer varies
from 25 μm for alloy Nr. 1 to 100 μm for alloy Nr. 2 while after
2150 hrs in the tubular furnace, at the same temperature, with pure
Li containing 24 ppm of nitrogen, the thickness of the layer varies
from 20 μm for alloy Nr. 1 to 57 μm for alloy Nr. 2. Fig. 8 shows
the latter case. A quantitative analysis from EDAX of the phases
labelled indicates the variation in the concentration of the main
elements present in the alloy. In the base material, Ni = 74.4 %,
Fe = 8.8 %, Cr = 15.1 %. In A, Ni = 7.9 %, Fe = 3.8 % and Cr =
88.3 %. In B, Ni = 60.9 %, Fe = 18.2 % and Cr = 18.4 %. In C,
Ni = 8.1 %, Fe = 9.7 % and Cr = 82.3 %.

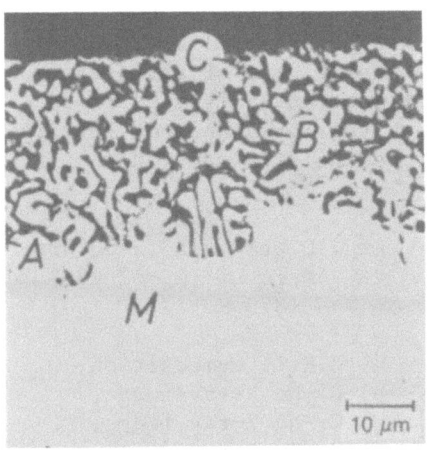

Fig. 8 Inconel 600, batch 2,
        2150 hrs, 873 K,
        Li (N$_2$ = 24 ppm),
        B.E.I. composition.

## Inconel 601

This material is severly attacked by pure Li. Fig. 9 is an
X-ray map of the main elements involved in the corrosion phenomenon
and refers to a sample heat treated in Li containing 10 ppm of
nitrogen for 2000 hrs at 873 K in the rotating device. The effect
of the presence of traces of hydrogen in Li is dubious in the case
of Inconel 601 although a slight improvement has been noticed.

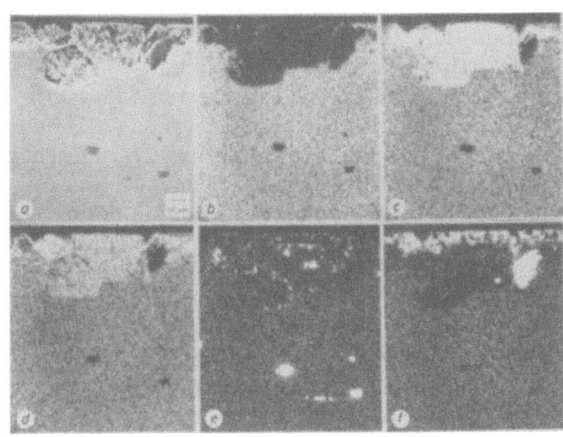

Fig. 9 Inconel 601,
        2000 hrs in Li
        (N$_2$=10 ppm),873 K,
        rotating device.

a) B.E.I.composition,
b) Ni-Kα X-ray image,
c) Cr-Kα X-ray image,
d) Fe-Kα X-ray image,
e) Ti-Kα X-ray image,
f) Al-Kα X-ray image.

## Uranus B6

This material behaves quite well in our experimental conditions.
We have not revealed any corrosion phenomenon in samples heat
treated at 873 K for 1000 hrs either in pure Li or in Li containing
traces of hydrogen. No reaction zone appeared in samples that were
heated at the same temperature for 2000 hrs, in the rotating device,

in Li with 50 ppm of nitrogen. A very slight reaction has been
noticed in a test in which the material has been for 2150 hrs at
873 K in the tubular furnace in Li containing 87 ppm of nitrogen
as can be seen from Fig. 10. An X-ray diffraction analysis carried
out on the surface of the sample has shown the appearance of ferrite
and carbides of the $M_{23}C_6$ type while before the treatment only
austenite was present.

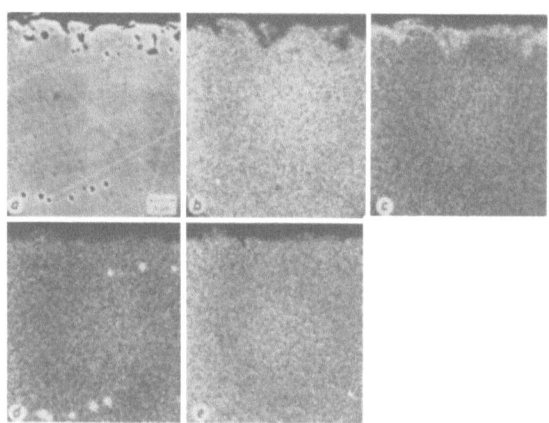

Fig.10 Uranus B6,2150 hrs
        in Li ($N_2$=87 ppm),
        873 K.

a) B.E.I. composition,
b) Ni-Kα X-ray image,
c) Cr-Kα X-ray image,
d) Mn-Kα X-ray image,
e) Fe-Kα X-ray image.

DISCUSSION

     For all alloys, in most cases, the corrosion layer was slightly
more relevant when the samples were heat treated in the rotating
system. The mechanism of corrosion is always the same: strong dissolution of Ni with the appearance of phases much richer in Cr and Fe.
In some cases, there is also a grain boundary penetration. On the
surface of all the alloys of the Ni, Fe, Cr type, only the austenite
phase was revealed by X-ray diffraction analysis before the tests,
while afterwards, ferrite and carbides were always evidenced together with the prevalent γ structure.

     Incoloy 825, Inconel 600 and Inconel 601 are severely attacked
by pure Li while in the case of Incoloy 800, there is quite a scatter
in the results, i.e. in most cases the material is corroded but the
entity of the corrosion layer varies even for samples that have
been subjected to an identical corrosion test.

     Uranus B6 which contains 25 % of Ni and 4.5 % of Mo, is practi-
cally unattacked by pure Li. Incoloy 825, which also contains Mo
(3 %), on the contrary, behaves very differently but if we consider the ratio Mo/Ni, it appears that the former material has a
Mo content three times higher. The carbon content is also three
times lower.

The very marked difference in behaviour of the examined materials in the presence or in the absence of traces of hydrogen observed in our previous works on AISI 316 and 304 and on the low Ni, Cr-Mn austenitic stainless steels, is not so evident in the case of the Ni rich alloys. Some results, however, seem to indicate that for Incoloy 825 and Incoloy 800, a marked improvement in the corrosion behaviour could be attributed to the presence of traces of hydrogen in the Li even if no explanation of the phenomenon can be given for the moment.

In the frame of our study of protective coatings against Li corrosion of AISI 316 and of steels or alloys with high Ni content (6) and following reports that additions of 5 wt% Al to Li "reduced drastically" the corrosion in the case of AISI 316 and Inconel 600 (7), we have carried out, in our experimental conditions, static tests in which we have studied the influence of the addition of Raffinal (Al 99.99 %) to pure Li on the corrosion resistance of Incoloy 825. Fig. 11 shows the cross section of a sample heat treated for 7000 hrs in a welded 304L container in which 5 wt% of Al was added to the Li. The nitrogen content at the end of the test was 35 ppm. A diffusion layer is evident. It consists of two intermetallic compounds: the first labelled B is rich in Fe and Al and has the following composition: Fe: 53.9 %, Al: 23.3 %, Ni: 15.2%, Cr: 6 %, Mo: 1.7 %. The second labelled A is rich in Ni and Al and has the following composition: Ni: 47.7 %, Al: 30.8 %, Fe: 17.6 %, Cr: 2.5 %, Mo: 0.7 %. Ni depletion is not avoided beyond the diffusion barrier: in the region labelled D we still find 17.3 % Al, 57.2 % Fe, 10.6 % Cr, 5.4 % Mo and only 7.1 % Ni. While in C the Al has dropped to 0.6 %, Cr has increased to 36.6 %, Fe: 46.6 %, Mo: 5.2 % and Ni: 9.1 %.

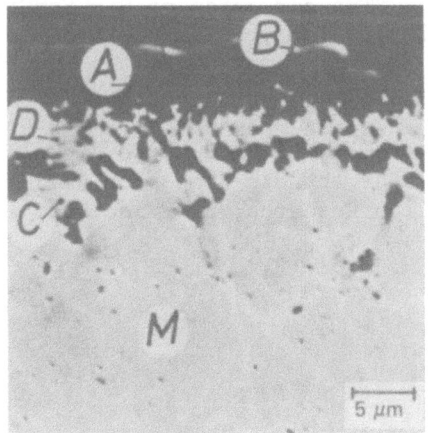

Fig. 11 Incoloy 825, 7000 hrs
at 873 K in Li+5wt% Al.
B.E.I. composition

The composition of this inner region is very similar to that
of the corroded layer reported in the case of the Incoloy 825
sample heat treated for 2000 hrs in the rotating device. A 15 μm
perturbated zone (between diffusion band and Ni deplation) after
7000 hrs is quite a progress with respect to the 25 μm after
2000 hrs at the same temperature in pure Li with only 20 ppm of
nitrogen.

It appears from this study that the use of bare high Ni alloys
in a liquid lithium environment seems problematic from a corrosion
point of view. The same can be said for the high Ni, Fe-Cr austenitic
stainless steels with the exception of a material such as Uranus B6
that seems to have a good Mo/Ni ratio. On the other hand, helium
build up under irradiation and activation considerations seems
also to discourage the use of these materials in a fusion environment.

REFERENCES

1.  V. Coen, H. Kolbe, L. Orecchia, T. Sasaki, "Influence of
    hydrogen on the lithium compatibility of AISI 316 and
    304 type stainless steels", Proceedings of the 7th
    Symposium of Engineering Problems of Fusion Research,
    Knoxville, October 25-28, 1977, Vol. 2, p. 1501.
2.  V. Coen, H. Kolbe, L. Orecchia, T. Sasaki, "Compatibility
    in Li and Li containing traces of LiH of low Ni, Cr-Mn
    austenitic stainless steels", Journal of Nuclear
    Materials, Vol. 85 and 86, 1979, p. 271.
3.  D. A. Bates, G. R. Edwards and D. L. Olson, "A evaluation
    of engineering alloys for high temperature lithium
    containment", Materials Performance, March 1980, p. 41.
4.  P. F. Tortorelli, J. H. De Van, "Corrosion tests of higher
    nickel alloys in static lithium", Summary in ORNL 5645,
    August 1980.
5.  P. F. Tortorelli, J. H. De Van, "The corrosion of Ni- Fe-Cr
    and Co-V-Fe alloys in static lithium", DOE/ET-0058/5,
    p. 141.
6.  to be published.
7.  P. F. Tortorelli, J. H. De Van, J. E. Selle and H. D. Upton,
    "Corrosion inhibition in systems of lithium with nickel
    bearing alloys", Proceedings of the 8th Symposium on
    Engineering Problems of Fusion Research, S. Francisco,
    November 13-16, 1979.

# CORROSION OF REFRACTORY METALS IN LIQUID LITHIUM

Norbert Rumbaut, Florint Casteels, Martin Brabers,
Michel Soenen, Hugo Tas, Jacky De Keyser

Studiecentrum voor Kernenergie
Mol Belgium

## INTRODUCTION

Since in fusion reactor concepts liquid lithium is considered
as a candidate for both the coolant and the breeder blanket material,
its action on refractory metals, as well as ferritic and austenitic
steels in chemically well-defined lithium is of major importance.

Non metals, such as nitrogen, carbon, oxygen and hydrogen are
corrosion initiators in Li and provoke mass transfer in cooling
circuits (1).

The measurement of the chemical activity is of importance
since the above-mentioned effects are governed by this thermodynamic property, rather than by the chemical concentration of
the non metals.

A thin foil technique for the determination of the nitrogen
activity has therefore been studied, since this element has a
considerable detrimental effect on the corrosion of potential
structural materials (2).

The "screening" tests performed on refractory metals in order
to select appropriate alloys for monitor foils offered at the same
time the opportunity of gaining information on their corrosion
behaviour in liquid lithium.

## EXPERIMENTAL MATERIALS AND CONDITIONS

The corrosion experiments and the nitrogen activity measurements have been carried out in a natural convection loop, the

characteristics of which have been published earlier (3). The
refractory metals studied were: Ti, Nb, Ta, Mo and W and the
monitor foils were Nb-Mo and V-Cr alloys. AISI 304 stainless steel
was also studied merely as a reference material for comparison.
The interstitial content of the "as received" metals is given
in Table 1. Experimental conditions during exposure to liquid Li
are summarized in Table 2.

Table 1.  Analysis of the refractory metals and alloys before
          corrosion experiments

|                     | ppm O    | ppm N | ppm C |
|---------------------|----------|-------|-------|
| W                   | 380      | 10    | 40    |
| Ta                  | 110      | 20    | 110   |
| Mo                  | 100      | 20    | 40    |
| Nb                  | 160      | 70    | 30    |
| Ti                  | 1 430    | 30    | 230   |
| DEW 202             | -        | -     | 460   |
| 50 - 50 at% Nb-Mo   | 190      | 160   | 41    |
| 30 - 70 at% V-Cr    | 2 520    | 330   | -     |

Table 2.  Experimental conditions during the different series of
          experiments[x)]

| Series of Experiments | Temperature ($^\circ$C) | Time (h)            | Nitrogen content (ppm) |
|-----------------------|-------------------------|---------------------|------------------------|
| I                     |                         |                     |                        |
| (W,Ta,Mo,Nb,Ti,       | 700                     | 168-336-504-672     | 175 $+$ 20 ppm         |
| AISI 304)             | 650                     | 168-336-504-672     | 175 $\pm$ 20 ppm       |
| II                    |                         |                     |                        |
| (Nb-Mo and V-Cr       | 700                     | 240                 | 150-175-640            |
| monitor foils)        | 700                     | 931                 | 330                    |
|                       | 700                     | 240                 | 5 000                  |

[x)]Temperature gradient in the loop during different tests is 250 $^\circ$C.
    The lithium velocity is 0.033 ms$^{-1}$.

EXPERIMENTAL RESULTS

Weight changes

    The changes in weight or thickness of the materials tested
in the first series of experiments are presented in Tables 3 and 4.

Table 3.   Decrease of foil thickness (expressed in micrometer) as a function of exposure time (h) and temperature ($^{\circ}$C)

| | 650 $^{\circ}$C | | | | 700 $^{\circ}$C | | | |
|---|---|---|---|---|---|---|---|---|
| Time | 168 | 336 | 504 | 672 | 168 | 336 | 504 | 672 |
| Alloy | | | | | | | | |
| Ta | | 0.18 | | 0.20 | 0.60 | 0.75 | 1.0 | 1.75 |
| Nb | 0.04 | 0.20 | 0.39 | 0.53 | - | 0.047 | 1.03 | 2.55 |
| W | 1.19 | 1.48 | 1.55 | 1.60 | - | 0.051 | 0.26 | 1.29 |
| DEW 202 (AISI 304) | 0.12 | 0.19 | 0.55 | 0.98 | 0.59 | 0.90 | 1.20 | 1.33 |

Table 4.   Weight increase expressed in mg.cm$^{-2}$ of Ti and Mo as a function of exposure time (h) and temperature ($^{\circ}$C)

| | 650 $^{\circ}$C | | | | 700 $^{\circ}$C | | | |
|---|---|---|---|---|---|---|---|---|
| Time | 168 | 336 | 504 | 672 | 168 | 336 | 504 | 672 |
| Alloy | | | | | | | | |
| Ti | 0.36 | 0.32 | 0.44 | 0.56 | 0.96 | 1.02 | 1.16 | 1.16 |
| Mo | 0 | 0.1 | 0.3 | 0.46 | 0.7 | 0.5 | 0.74 | 1.02 |

Whereas Ta, Nb, W and DEW 202 (AISI 304) showed a decrease of the foil thickness, Ti and Mo on the other hand gained weight to a considerable extent. A temperature increase from 650 to 700 $^{\circ}$C increases the corrosion rate by a factor five for Nb and Ta and a factor two for DEW 202. W corrodes initially at 700 $^{\circ}$C at a lower rate than at 650 $^{\circ}$C, due to a much longer induction period.

The pick-up of interstitial and substitutional elements is increased for Mo and Ti by a factor two by increasing the exposure temperature from 650 $^{\circ}$C to 700 $^{\circ}$C. The changes in thickness of the Nb-Mo foils used for the determination of the nitrogen activity in dynamic Li are very low and depend on the lithium quality and the niobium content of the foil. The corrosion attack increases with increasing nitrogen content of the lithium and an increasing Nb content in the foil. The corrosion attack remains limited to a few micrometers. Cr-V foils are very soluble in lithium at 700 $^{\circ}$C. The loss in thickness of the Cr-V foils increases with increasing nitrogen content and increasing vanadium content. The same trend has been noticed for AISI 304 namely an increase in corrosion rate with increasing nitrogen content of the Li.

## Microhardness changes

     Knoop microhardness measurements have been carried out in
order to study the result of the exchange mechanisms between
substitutional and interstitial elements present in the lithium
and in the foils of the unalloyed refractory metals. A considerable
hardening has been noticed for titanium. Its hardness increases
as well in the outer reaction layer as in the bulk material with
increasing exposure time. The thickness of the hard layer increases
with exposure time. The opposite effect has been observed for Mo,
where a gradual decrease of the microhardness in the bulk material
and the subsurface zone was measured. The bulk hardness values for
Ta are independent of exposure time and remain unchanged throughout
the foil. The observations are in accordance with calculated
distribution coefficients ($K_D$-values) for the metalloids between
Li and these metals (5).

## Chemical and structural changes

     Chemical and structural changes were examined with metallo-
graphy and microprobe, microscan and Auger Spectroscopy. The
unalloyed refractory metals present a pronounced difference in
their exchange mechanisms with non metals dissolved in liquid Li.

     In general Ta, Nb and Ti react to a considerable extent with
nitrogen whereas Mo, W and AISI 304 present a more neutral behaviour.
The exposed metals showed, exception made for Ta, an analogous
behaviour towards carbon. The oxygen content of Nb, Ta and W,
observed after exposure during 672 h at 700 $^\circ$C, is believed to be
due to the post-corrosion treatment with alcohol and water which
was necessary to eliminate the Li residue. Indeed, binary oxides
of these metals are unstable, i.e. have a much more positive Gibbs
free energy of formation than $Li_2O$ and ternary oxides like niobates
and tantalates of Li have hitherto not been observed in the presence
of Li (9).

     For Ti, the formation of an oxygen bearing corrosion product
is thermodynamically possible. There is also a considerable amount
of oxygen diffused into the bulk material of this metal, which is
in accordance with $K_D$ calculations (5).

     Most of the metals examined showed an induction period before
the onset of corrosion, indicating that a corrosion product has to
be nucleated. For all the metals a "levelling" effect was observed,
i.e. the corrosion rate decreased with time.

     Titanium developed a hard and coherent layer on its surface.
It is composed of nitrides and carbides and is about $3.10^{-6}$ m thick
after 168 h and $5.10^{-6}$ m after 672 h exposure at 700 $^\circ$C. The bulk
material picked up considerable amounts of nickel, iron and chromium.

Post-corrosion analysis revealed appreciable amounts of these steel components in the lithium: 460, 103 and 161 ppm resp. for Fe, Ni and Cr.

Niobium was considerably attacked by liquid Li with 175 ppm N at 650 $^o$C and especially at 700 $^o$C. The reaction layer is mainly composed of nitrogen, oxygen and carbon. The oxygen is believed to originate from the post-exposure treatment with alcohol and water to eliminate the lithium residue. The corrosion attack caused by penetrating and reacting lithium (with oxides present in Nb) has not been observed. This fact is in correspondence with earlier published data (4) reporting a threshold value for the initial oxygen content in Nb and Ta of respectively 400 ppm and 150 ppm. Under present experimental conditions of low oxygen content in the niobium, penetration reactions will not occur. The high distribution coefficient (5) for carbon and nitrogen in the Nb-C-Li and Nb-N-Li systems is conformable to the observed attack. The corrosion products are largely soluble in liquid lithium. This causes the Nb-foil to loose weight to a considerable extent (14 wt% for a foil of 50 μm thickness after 4 weeks.

Metallographic examination did not reveal any localized or preferential attack of tantalum. The very thin reaction layer ($\sim 10^{-6}$ m) contains the substitutional elements Ta and Ca and also carbon, nitrogen and oxygen. A concentration gradient for carbon and nitrogen exists in the first $10^{-6}$ m of the subsurface layer. The observed oxygen concentration is probably due to the post-exposure treatment. We believe that the corrosion attack proceeds much in the same way as for niobium, inter alia, because both metals are isoelectronical.

The low distribution coefficient for non metals in the Mo-Li system provokes a driving-out of these metalloids from the bulk of the foil towards the outer surface. Nitrides and oxides of molybdenum have indeed a high dissociation pressure (6). Microprobe analysis enabled the detection of a fine reaction zone ($2.10^{-6}$ m) alloyed with Fe, Ca, Cr, and Ni. Auger spectroscopy revealed the presence of oxygen, carbon and magnesium in the outer reaction layer and the scanning electron microscope showed evidence for the presence of a high density of symmetrical (fourfold symmetry) crystallites. These crystals contain oxygen, Fe, Cr and Ni and are believed to consists of a Fe-(Cr,Ni) molybdate.

Metallographic and the above-mentioned analytical techniques were not able to detect any attack or exchange mechanism between tungsten and substitutional and interstitial elements present in the dynamic lithium. The surface of the foils retained its metallic luster after 672 h exposure at 700 $^o$C, although the weight losses were considerable. This is an indication that the attack of this metal could be of a physical nature i.e. a mere dissolving in the liquid metal.

Selective leaching of Cr, Ni, Mn and Si out of the steel in
lithium is observed. This preferential dissolution process results
in the formation of a porous ferritic subsurface layer. The thickness of this layer increases from $7.10^{-6}$ m after 168 h exposure to
$1.10^{-5}$ m for an exposure of 672 h at 700 °C.

In a second series of experiments (conditions in Table 2),
Nb-Mo and V-Cr foils were equilibrated with Li, systematically
contaminated with N. Metallography and microprobe analysis were
not able to detect a pronounced preferential attack on the different
Mo-Nb and Cr-V specimens exposed during 240 h at 700 °C in Li
contaminated with 175 ppm, 330 ppm and 640 ppm of nitrogen.

The reaction layer remains limited to a few of a micrometer.

Auger spectroscopy revealed the preferential leaching from
niobium out of the Mo-Nb foils. The niobium concentration shows
a minimum at a few tenths of a micrometer below the surface of the
foil and reaches the original concentration at a distance of 1
micrometer from the lithium-foil interface. The general shape of
the niobium concentration-penetration profile is not dependent on
the nitrogen content, but the minimum value for niobium shifts
with increasing nitrogen content to lower values.

A typical curve is presented in Fig. 1.

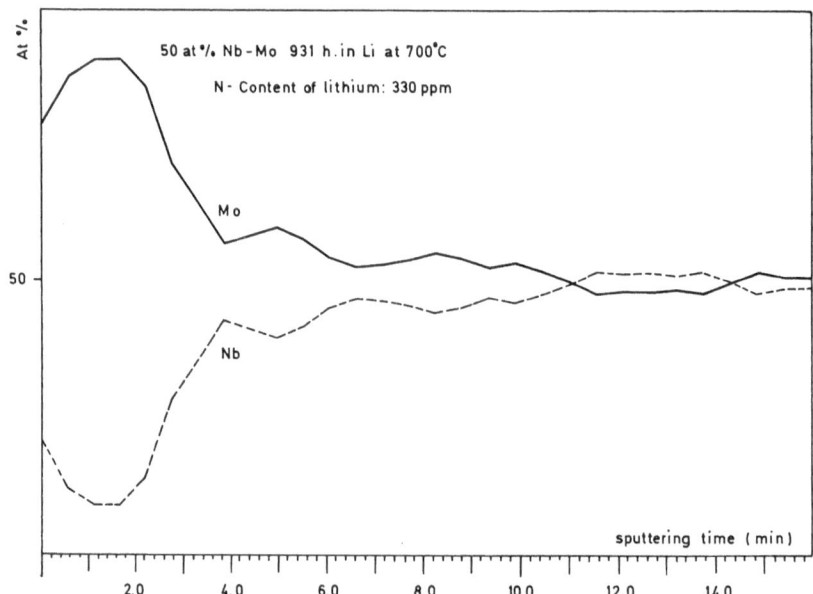

Fig. 1. Mo and Nb Concentration Profile in a 50 at% Nb-Mo Alloy
Tested for 931 h at 700 °C.

At 700 °C the nitrogen activities were as follows: 2.03 10$^{-5}$ for 175 ppm N, 3.84 10$^{-5}$ for 330 ppm N and 7.45 10$^{-5}$ for 640 ppm N.

They were calculated from an equation deduced elsewhere (7) and correspond also to partial nitrogen pressures of 4.10$^{-10}$, 1.5 10$^{-9}$ and 5.6 10$^{-9}$ atm respectively. The thermodynamic data for nitrogen in the equilibrated Nb-Mo foils were obtained from the work of Hörz and Steinheil between 2,100 and 1,600 °C (8). From their data, Rumbaut (7) obtained Fig. 2, which represents this activity as a function of the mole fraction of Mo in the reaction sublayer. The area between the horizontal dotted lines on the figure represents the region of activities which is most useful for the study of lithium corrosion, and which corresponds to a N-concentration range of 50 to 1 000 ppm in liquid Li.

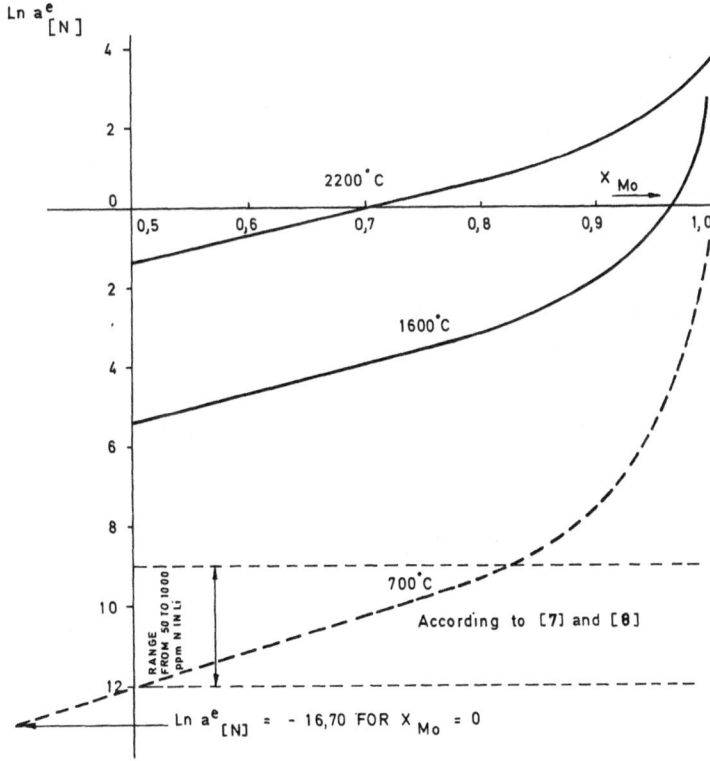

Fig. 2. Nitrogen Activity in a N-Saturated Nb-Mo Solid Solution as a Function of Temperature and Composition

The results of the experimental and theoretical calculated Nb/Mo ratio as a function of the nitrogen contamination level in

Table 5.   Experimental and calculated Nb to Mo ratio in nitrogen
           saturated solid solution in equilibrium with nitrogen
           contamined lithium

| Contamination of lithium with nitrogen (ppm) | Nitrogen partial pressure (atm) | Nitrogen activity | Ratio Nb to Mo | |
|---|---|---|---|---|
| | | | Experimental value | Theoretical (dotted line in Fig. 2) |
| 175 | $4.10^{-10}$ | $2.03 \ 10^{-5}$ | 0.50 | 0.57 |
| 330 | $1.5 \ 10^{-9}$ | $3.84 \ 10^{-5}$ | 0.38 | 0.42 |
| 640 | $5.6 \ 10^{-9}$ | $7.45 \ 10^{-5}$ | 0.29 | 0.32 |

lithium are presented in Table 5 for a 50-50 at% Nb-Mo alloy. The
accordance between the theoretically calculated and experimentally
obtained values may be considered satisfactory as the difference
between them is at most $\pm$ 15 %.

Post-exposure analysis of Cr-V alloys revealed the preferential
leaching of V out of the Cr-V foils. However, the chemical changes
could not be shown to be directly related to the nitrogen content
of the lithium.

CONCLUSIONS

The corrosion resistance of the refractory metals Ta, Nb and
W in dynamic Li is lower than expected, based on their solubility
in lithium and comparable with the resistance of AISI 304.

The corrosion mechanisms is governed by exchange mechanisms
of interstitial and substitutional elements present in the lithium
with the refractory metals. Mo behaves neutrally towards carbon
and nitrogen present in the lithium. The weight increase is
associated with the pick-up of Cr and Fe and the probable formation
of molybdates on the surface.

The behaviour of titanium is determined by its getter function:
the pick-up of carbon, nitrogen and oxygen and to a lesser extent
of the substitutional elements Ni, Cr and Fe.

A relation was observed between the minimum niobium content
in the reaction layer induced in Mo-Nb foils and the nitrogen
activity of the lithium.

ACKNOWLEDGEMENT

The authors are indebted to Euratom for the financial support.

REFERENCES

1.  J. H. Devan, J. E. Selle, A. E. Morris, Review of lithium
    iron-base alloy corrosion studies ORNL-TM-4927 (1976).
2.  P. F. Tortorelli, J. H. Devan, J. E. Selle, Effects of
    nitrogen and nitrogen getters on the corrosion of type
    316 stainless steel, CONF. 790313-3, CORROSION-79,
    March 1979, Atlanta (USA).
3.  M. Soenen, J. De Keyser, Experimental facilities at
    S.C.K.-C.E.N. in lithium technology, 2nd International
    Conference on Liquid Metals Technology in Energy
    Production, Richland, April 20, 1980.
4.  R. L. Klueh, Oxygen Effects on the Corrosion of Niobium
    and Tantalum by Liquid Lithium, Met.Trans. 5 (1974)
    p. 875.
5.  D. L. Smith, K. Natesan, Influence of Nonmetallic Impurity
    Elements on the Compatibility of Liquid Lithium with
    Potential CTR Containment Materials, Nuclear Technology
    22, (1974), p. 392.
6.  E. Fromm, E. Gebhardt, Gase und Kohlenstoff in Metallen,
    Springer Verlag, Berlin 1976.
7.  N. Rumbaut, Chemisch-thermodynamisch Gedrag van Niet-
    metalen in Dynamisch Lithium Thesis, University of
    Leuven (in preparation).
8.  G. Hörz and E. Steinheil, Gleichgewichtsuntersuchungen im
    System Niob-Molybdän-Stickstoff, Zeitschrift für
    Metallkunde 62 (1971) (849-900).
9.  M. G. Barker, "The Reaction of the Liquid Alkali Metals
    with Zr, Nb, Ta, Mo and W and their Oxides", The
    University of Nottingham 1976.

# DEPOSITS OF METALLIC IMPURITIES IN CRACKS OF CREEP-RUPTURE

# SPECIMENS OF TYPE 304 ss TESTED IN FLOWING SODIUM

Horst Huthmann        Hans U. Borgstedt        Hugo Tas

INTERATOM             Kernforschungszentrum    Studiecentrum voor
Berg. Gladbach        Karlsruhe                Kernenergie SCK/CEN
FRG                   FRG                      Mol, Belgium

## INTRODUCTION

Numerous structural and containment components of Liquid Metal Fast Breeder Reactors will be exposed to a flowing sodium environment at elevated temperatures. Therefore a joint program between INTERATOM, the Nuclear Research Center, KFK, and the Studiecentrum voor Kernenergie, SCK/CEN, was conducted to determine quantitatively the influence of a flowing sodium environment upon the creep-rupture behavior of Type 304 ss at 550°C. The sodium tests were performed in the three non-isothermal dynamic loops of these companies and for a valid comparison parallel tests in air were carried out with the same heat.

## MATERIAL AND TEST EQUIPMENT

All creep-rupture data presented were obtained using the same heat of X 6 CrNi 18 11-steel (DIN Type 1.4948), which is similar to the AISI Type 304 ss. The characteristic data of the material are summarized in Table 1. The creep-rupture tests were carried out on material in solution-annealed condition.

The specimens for the creep-rupture tests have a gauge length of 20 mm and a diameter of 4 mm. In order to avoid differences by fabrication all specimens were produced by INTERATOM.

The creep tests were conducted on conventional constant-load creep-rupture machines at INTERATOM, KFK, and SCK/CEN.

Figure 1 gives a schematic drawing of the used test facilities. These test machines are associated with sodium loops (1, 2, 3)

Table 1.  Characterization of Test Material

Type:              X 6 CrNi 18 11 (DIN 1.4948), similar Type 304 ss
Heat No.:          227766 (IA-No. 325)
Size:              plate, 40 mm thick
Heat treatment: solution annealed, 1060 $^\circ$C/water

Chemical composition (wt %):

| C | N | Cr | Ni | Mn | Mo | Co | P |
|------|------|-------|-------|------|------|-----|-----|
| .051 | .058 | 18.55 | 10.89 | 1.37 | <.01 | .03 | .01 |

| Si | Nb/Ta | Ti | S | Zn | Cu | Pb | Sn |
|------|-------|------|------|------|------|-------|--------|
| .37 | <.01 | <.01 | .007 | .024 | .047 | .0003 | <.000 |

Table 2.  Sodium Conditions

| Condition | INTERATOM | KFK | SCK/CEN |
|-----------|-----------|-----|---------|
| Test temperature ($^\circ$C) | 550 | 550 | 550 |
| Test position | at $T_{max}$ | at $T_{max}$ | at $T_{max}$ |
| $\Delta T$ in the loop ($^\circ$C) | 170 | 240 | 150 |
| Na velocity (m/s) at creep specimens | 4.3 | 3.0 | 3.6 |
| Oxygen saturation temperature ($^\circ$C) | <120 PL, 175 PL | 120 CT, 145 CT | 110 CT, 160 CT |
| Carbon activity | slight carburizing | not carburizing | slight carburizing |

PL: Plugging temperature,   CT: Cold trap temperature

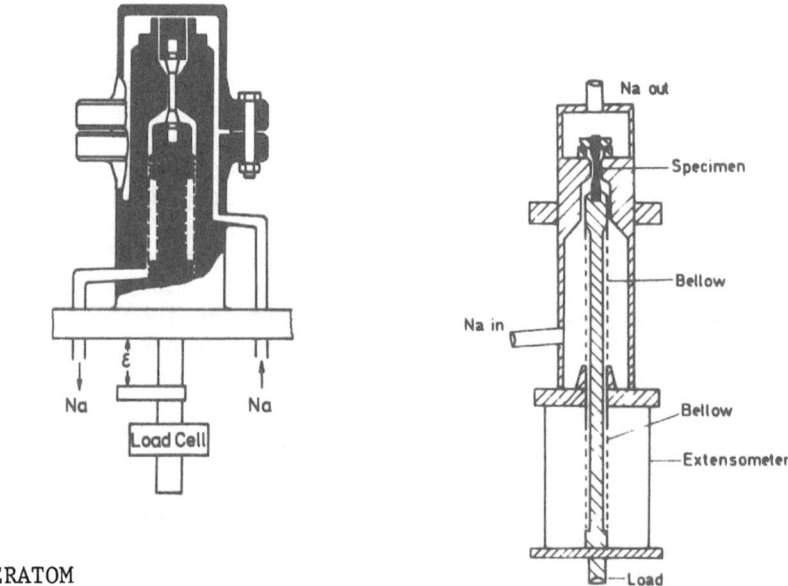

INTERATOM

KFK

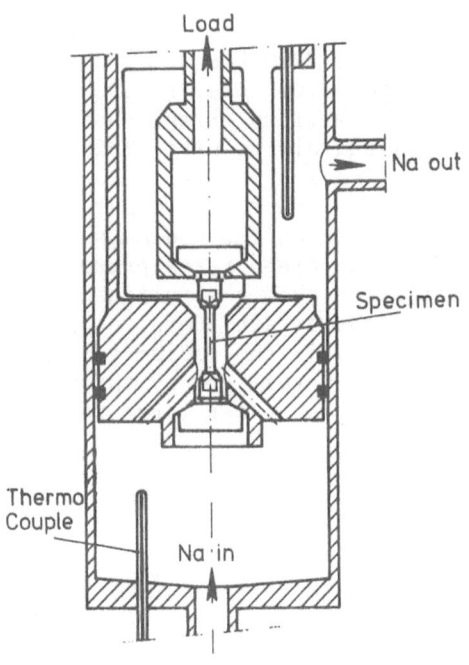

SCK

Fig. 1: Test equipments for creep testing in flowing sodium

providing a well-characterized sodium environment, under the conditions summarized in Table 2. The tests at INTERATOM, KFK, and SCK/CEN were carried out in comparable conditions at 550 °C, which was the maximum temperature in the loops. Oxygen content was controlled by cold traps and the oxygen saturation temperatures were varied from < 120 to 175 °C corresponding to < 1 to 6 ppm oxygen in sodium (4). The carbon contents of monitor foils (Type 304 ss), exposed to sodium inside the test positions, showed a final equilibrium value of about 800 ppm C in the loops at INTERATOM and SCK/CEN, and a value of about 300 ppm C in the KFK loop. Starting from an initial carbon content of 460 ppm C in the foil, this exhibits carburizing conditions in the loops at INTER-ATOM and SCK/CEN and a slight decarburizing environment in the KFK loop.

RESULTS

The uniaxial constant-load creep-rupture results carried out with X 6 CrNi 18 11 steel at 550 °C in flowing sodium of the INTERATOM loop are presented in Figures 2 to 4 together with the parallel test results obtained in air on the same heat. A comparison of the times to rupture attained in sodium and in air (Fig. 2) shows that below 1000 hrs rupture time, the in-sodium results are within the scatter of $\pm$ 1 $\sigma$ ($\sigma$ = standard deviation) of the air results. Above 1000 hrs, the times to rupture reached in sodium are below the scatter of the air-results. At present, only 3 results in the region from 1600 to 2700 hrs are available, and require confirmation by tests up to 10.000 hrs. Figure 3, which presents the time up to onset of tertiary creep, shows that there is no difference between sodium and air results. This reveals that the slight reduction in time to rupture is caused by a reduction of the tertiary creep time. Up to the onset of tertiary creep there is no difference between the sodium and air results. This is confirmed by the minimum creep rates presented in Figure 6. But the measured ductilities show a significant difference: Figure 4 shows that the creep strain (total elongation to rupture less the initial creep strain) in the sodium tests is lower than in the air tests. As shown by metallography this is a sodium effect, which occurs in the region of tertiary creep.

Contrary to these results obtained in the sodium loop at INTERATOM, is the creep rupture behaviour of the X 6 CrNi 18 11 steel measured in the loops at KFK and SCK/CEN. Figure 5 shows that the times to rupture, measured in these loops, are significantly reduced, depending upon the particular loop and the oxygen saturation (cold trap) temperature. The largest reduction time to rupture was measured in the loop of KFK. This reduction of time to rupture cannot be attributed to a reduction of the time of tertiary creep and in the sodium in the KFK loop even the minimum creep rate is higher than in air (Fig. 6).

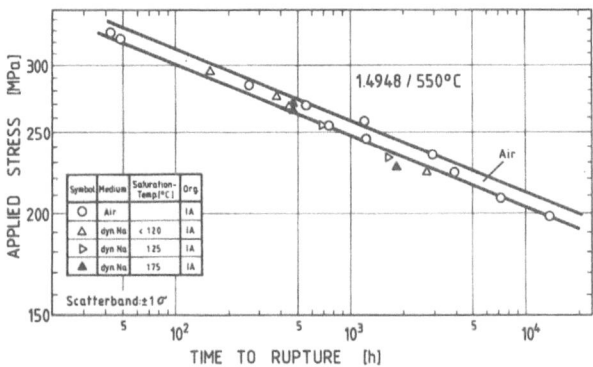

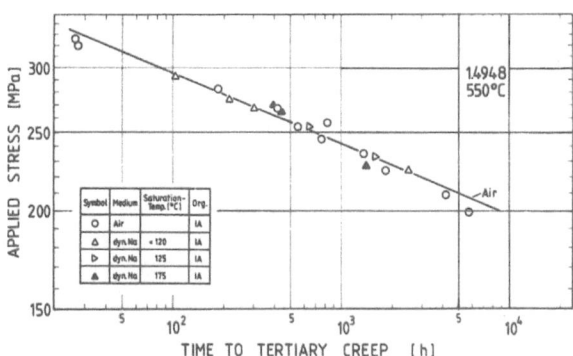

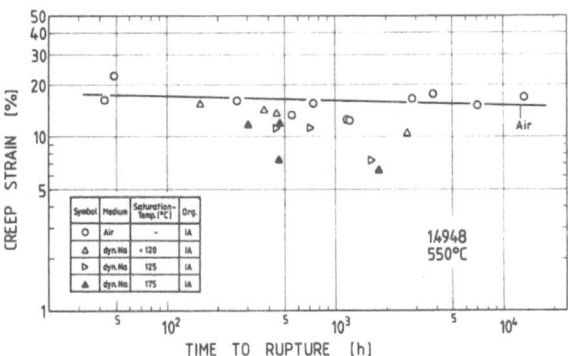

Figs. 2 - 4: Creep rupture behaviour of X6 CrNi 18 11 steel
(1.4948) at 550 °C in air and flowing sodium
(INTERATOM loop)

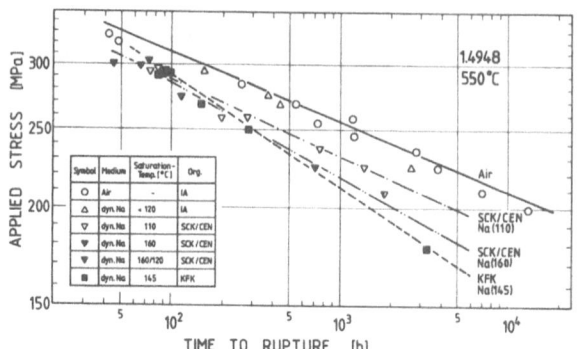

Fig. 5: Time to Rupture of X 6 CrNi 18 11 Steel (1.4948) at
550 °C in Air and in Flowing Sodium of Different Loops

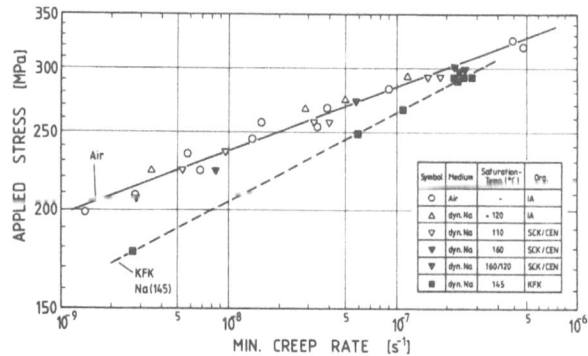

Fig. 6: Min. Creep Rate of X 6 CrNi 18 11 Steel (1.4948) at
550 °C in Air and in Flowing Sodium of Different Loops

Further creep rupture tests in the KFK sodium loop at 145 °C
cold trap temperature have shown that this harmful effect of sodium
does not occur with heat-treated (700 °C, 30 hrs) or cold worked
(10 %) specimens. Even the material AISI Type 316 ss did not show
a reduction of time in tests up to 2000 hrs in the SCK/CEN loop
at a cold trap temperature of 160 °C.

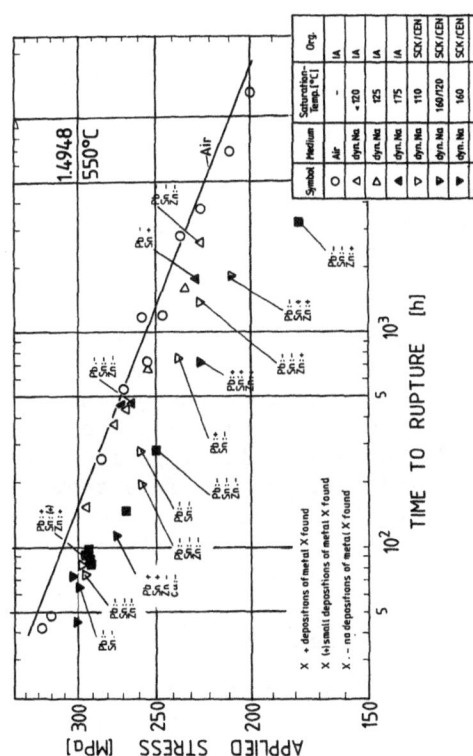

Fig. 7: Metallic depositions in cracks of creep-rupture specimens tested in flowing sodium in the loops at INTERATOM, KFK and SCK/CEN

But the benefical effect of a lower oxygen saturation temperature which was shown from the sodium tests of SCK/CEN with longer
times to rupture (see Fig. 5) was not examined in long term tests
in the KFK sodium loop. At the stress level of 249 MPa a cold trap
temperature of 120 °C does not increase the time to rupture
compared with 160 °C cold trap temperature in this loop. A high
oxygen saturation temperature cannot be the only reason for the
harmful sodium effect in the KFK and SCK/CEN loops, because the
in-sodium tests at INTERATOM at 175 °C oxygen saturation temperature did not show a significant reduction in time to rupture
compared to the times to rupture obtained in air.

POST-TEST INVESTIGATIONS

Comparison of the Sodium Qualities

The search for differences between the three sodium loops led
to a comparison of the level of metallic impurities which is given
in Table 3. The analysis of the sodium samples taken with Zr-
crucibles show that the sodium of the KFK loop contains 22 ppm Pb
and 9 ppm Sn, whereas the sodium of the INTERATOM loop contains
less than 4 ppm Pb and less than 0.5 ppm Sn. Comparison of the
samples taken with Ta crucibles shows that the SCK/CEN sodium also
contains more Pb and Sn impurities than the sodium of the INTER-
ATOM loop. A further analysis of sodium taken from the current
production of the DEGUSSA Company yields impurity levels of
0.35 ppm Pb, 0.88 ppm Sn and 0.20 ppm Zn. This shows that the
impurities due to these metallic elements in the sodium which
will be used for the SNR 300 are significantly lower than in the
sodium of the KFK and SCK/CEN loops.

EDX, Microprobe and ESCA Investigations

If the different impurities have any causal connection with
the reduction of time to rupture, then it should be possible to
find these elements in the cracks of the specimens. The results
of our search for depositions of metallic impurities by EDX
methods (energiedispersive X-ray) and microprobe are shown in
Fig. 7. It can be seen, that depositions of Pb, Sn and Zn are
found in specimens which were tested in sodium. In all specimens
tested in the sodium of KFK and SCK/CEN with times to rupture
greater than 500 hrs at least one element of these impurities was
found.

From the specimens tested in the sodium with higher purity
at INTERATOM, Sn was found in one specimen ($\sigma$ = 227 MPa, $t_R$ = 1821 h)
which was tested at an oxygen saturation temperature <120 °C with
2684 rupture time ($\sigma$ = 224 MPa) exhibits none of these impurities.
An example for the deposition of lead is given in Figure 8,
presenting the crack tip of a specimen tested at SCK/CEN.

Table 3.  Metallic Impurities in the Sodium of the Three Loops
          in ppm

| Element | SCK/CEN Ta-Crucible | INTERATOM Ta-crucible | Zr-crucible | Zr-crucible |
|---|---|---|---|---|
| Fe | 7.6 – 36.5 | 3.5 – 15.9 | – | 4.3;  4.2 |
| Cr | 4.5 – 23.5 | 0.9 – 5.6 | – | 0.4;  0.45 |
| Ni | 8  – 13 | – | – | 0.25; 0.25 |
| Mn | 2.1 – 5 | <0.1 – 0.2 | – | 0.4;  0.3 |
| Pb | 18  – 27 | <0.1 – 1.1 | 3.8  2.9 | 21;  22 |
| Sn | 7.2 – 15.4 | <5 | 0.3; 0.45 | 8;  9 |
| Bi | – | <5 | – | 0.8;  1 |
| Zn | – | <0.1 | 0.2; 0.2 | 0.3;  0.55 |
| Cu | 0.02– 0.2 | <0.5 – 2.4 | – | – |

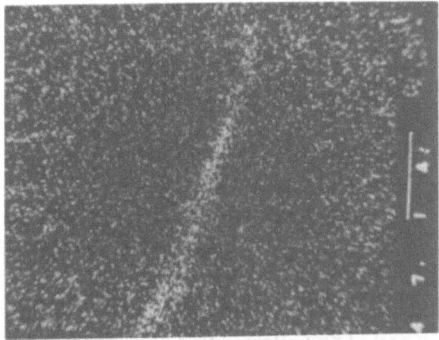

**X-Ray Image of Lead Distribution**

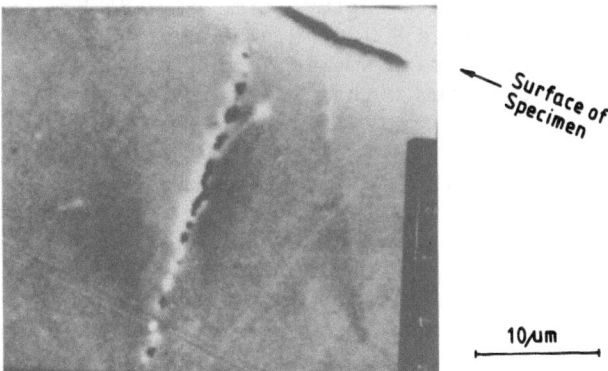

**Secondary Electron Image of Crack Tip**

Fig. 8: Demonstration of deposition of lead in a crack of a creep-
        rupture specimen tested in flowing sodium at SCK/CEN
        ($\sigma$ = 224 MPa, $t_R$ = 727 h)

Applied quantitative analyses by electron spectroscopy for chemical analyses (ESCA) demonstrate that the fracture surfaces of the specimens tested in sodium at SCK/CEN contain up to 2 % Pb, 1 % Sn, 8 % Zn and 8 % Cu. A parallel specimen tested in air exhibits 1.8 % Zn but no detectable quantities of Pb, Sn and Cu.

DISCUSSION

The different results of the creep rupture tests obtained in the loop at INTERATOM and, on the other hand, in the KFK and SCK/CEN loops led to the conclusion that the harmful sodium effects are due to the different sodium conditions. Differences in the carburizing or oxidizing conditions are not apparent nor explain the effects, but there are differences in levels of impurities of lead and tin. Because depositions of these elements were found in cracks of those specimens which were tested in the sodium loops with the higher impurities, the elements lead and tin are assumed to cause the harmful sodium effect on the creep rupture behaviour. The role of Zn ist not yet clear, because the Zn content was relatively high (240 ppm) in the as-received material.

In addition to an earlier paper (5) Sn was found in a specimen tested in the sodium loop at INTERATOM ($\sigma$ = 227 MPa, $t_R$ = 1821 h), too. This is not in contrary to the above assumption. If small impurities of sodium can reduce the rupture time, it is to be expected that these impurities can be active even in a smaller concentration.

This test at INTERATOM with 1821 hrs rupture time ($\sigma$ = 227 MPa) gives a sign of the influence of the oxygen saturation temperature upon the effect of the impurities, because a specimen with comparabl stress ($\sigma$ = 224 MPa) reached a rupture time of 2684 hrs in sodium with a saturation temperature < 120 $^\circ$C. This is in agreement with the results of SCK/CEN, where a smaller reduction of time to rupture was found in a sodium with a lower oxygen saturation temperature (see Fig. 5).

The assumption that the metallic impurities in sodium are responsible for the harmful effect on time to rupture is conformed by literature data. It is known that Zn, Pb and Sn act as severe embrittlers on steel (6 - 9). Even small quantities of these elements, for example, present in the steel as trace elements, can reduce the stress rupture properties. It was shown that a concentration of only 5 ppm Pb reduced the life-to-rupture of Nimonic 105 by half (10). It is assumed that the liquid sodium acts as an "inert carrier" (11) for the embrittling elements.

The expected mechanism is that these elements diffuse along grain boundaries into the material, this mechanism being promoted by an applied stress and by the sensitization, which took place in this material at 550 $^\circ$C.

The creep rupture tests at INTERATOM showed that in a sodium with low oxygen saturation temperature and high purity the only effect of sodium is a reduction of time within the period of tertiary creep, which depends on a reduction of ductility in this region. This effect, which is known from literature (12) as "Tertiary Creep Embrittlement" does not limit the use of the X 6 CrNi 18 11 steel as structural material.

ACKNOWLEDGEMENT

The authors gratefully acknowledge the staffs of KFK, SCK/CEN and INTERATOM in performing the tests in sodium and the contribution of Frau Dr. H. Schneider, KFK, Karlsruhe, in performing chemical analyses.

REFERENCES

1. H. Huthmann, E. D. Grosser, H. Tas, H. U. Borgstedt, in IAEA/IWGFR Specialists" Meeting on "Properties of Primary Circuit Structural Materials Including Environmental Effects", Bergisch Gladbach, FRG, Oct. 17 - 21, 1977

2. H. U. Borgstedt, G. Frees, KFK-Nachrichten 10, p. 41

3. M. Soenen, in "Alkali Metal Coolants", IAEA, Vienna 1967

4. R. L. Eichelberger Report AI-AEC-12685, 1968

5. H. Huthmann, G. Menken, H. U. Borgstedt, H. Tas, in Second International Conference on Liquid Metal Technology in Energy Production, April 20 - 24, 1980, Richland, Washington

6. A. H. Cottrell, P. R. Swann, The Chem. Eng., April 1976, p. 266

7. G. Herbsleb, W. Schwenk, Werkstoffe und Korrosion 28 (1977) p. 145

8. N. N. Breyer, Trans. ASM 61 (1968) pp. 219 - 232

9. J. C. Lynn, W. R. Work, P. Gordon, Mater. Sci. Eng. 18 (1975) pp. 51 - 62

10. G. B. Thomas, T. B. Gibbons, Metals Technology, March 1979, pp. 95 - 101

11. M. H. Kamdar, Prog. Mater. Sci. 15 (1973) pp. 289 - 374

12. R. S. Fidler, M. J. Collins, Atomic Energy Review 13, (1975) p. 3

# THE FATIGUE BEHAVIOUR OF TYPE 304 SS AT 550 °C IN FLOWING SODIUM

Horst Huthmann, Günther Menken

INTERATOM GmbH
Bergisch Gladbach 1
F.R.G.

## INTRODUCTION

The purpose of the present investigation is to determine the effects of a high temperature flowing sodium environment on the low cycle fatigue behaviour of the austenitic steel X 6 CrNi 18 11, (AISI Type 304 ss). The effects of the sodium on the structural material are influenced by the oxygen concentration of the sodium and by mass transfer phenomena, which occur in non-isothermal flowing sodium. In order to gain representative test results for the corrosive conditions of a LMFBR system it is necessary to perform the fatigue tests in a non-isothermal sodium loop.

## TEST EQUIPMENT AND MATERIAL

The facility for fatigue testing in sodium consists of a 100 kN MTS closed loop servohydraulic computer equipped fatigue machine, which is placed in one of six test sections of the non-isothermal sodium loop at INTERATOM. Fig. 1 gives a schematic diagram of this loop, which is used for creep strength and fatigue tests in flowing sodium.

The low cycle fatigue (LCF-) tests were performed under the following sodium conditions:

- test temperature and highest temperature in the loop: 550 °C,
- gradient of temperature in the loop: 170 °C,
- sodium velocity at the LCF-specimens: 2.5 m/s.

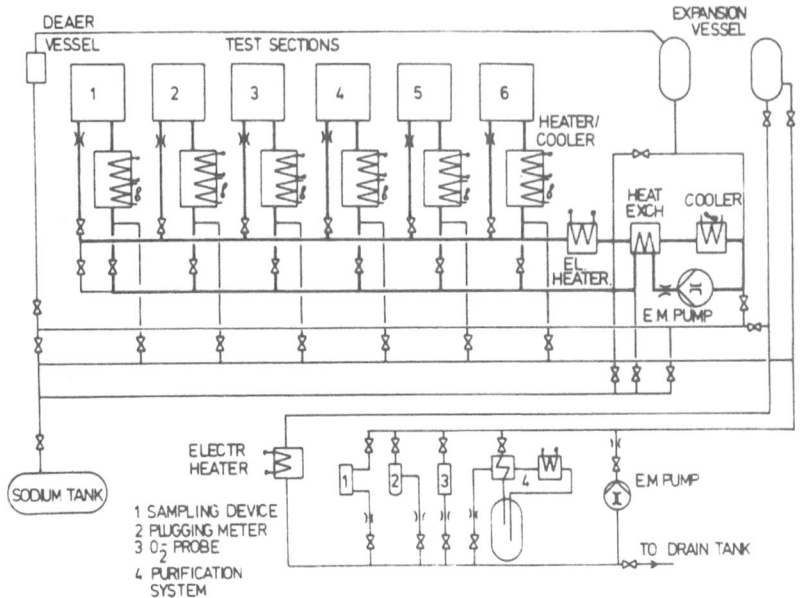

Fig. 1:  Creep strength and fatigue test loop of INTERATOM

The oxygen content was controlled by a cold trap and the plugging temperatures were varied from < 120 $^{\circ}$C to 175 $^{\circ}$C corresponding to < 1 to 6 ppm oxygen in sodium (1). The carbon content was controlled by monitor foils (Type AISI 304 ss) in the test position (550 $^{\circ}$C). Starting from an initial carbon content of 460 ppm the foils reached a final equilibrium value of about 800 ppm, this exhibits slight carburising conditions for the used test material.

Table 1.  Characterization of Test Material

Type:              X 6 CrNi 18 11 (DIN 1.4948), similar Type 304 ss
Heat No.:          227766 (IA-No. 325)
Seize:             plate, 40 mm thick
Heat treatment:    solution annealed, 1060 $^{\circ}$C/water

Chemical composition (wt %):

| C | N | Cr | Ni | Mn | Mo | Co | P |
|------|------|-------|-------|------|------|-----|-----|
| .051 | .058 | 18.55 | 10.89 | 1.37 | <.01 | .03 | .01 |

| Si | Nb/Ta | Ti | S | Zn | Cu | Pb | Sn |
|-----|-------|------|------|------|------|-------|--------|
| .37 | <.01 | <.01 | .007 | .024 | .047 | .0003 | <.000 |

The LCF-specimens were fabricated from the solution annealed X 6 CrNi 18 11-steel which is characterized in Table 1. The used hourglass specimen, shown in Fig. 2, with a minimum diameter of 8.8 mm and a gauge length of 21 mm were bounded by collars which enable the specimen to be welded into a bellow and to attach extensometers to the specimen. An elastic stress strain analysis by finite element calculation showed that the stress maxima, produced by the collars were limited to values smaller than the maximum stress in the middle of the hourglass specimen.

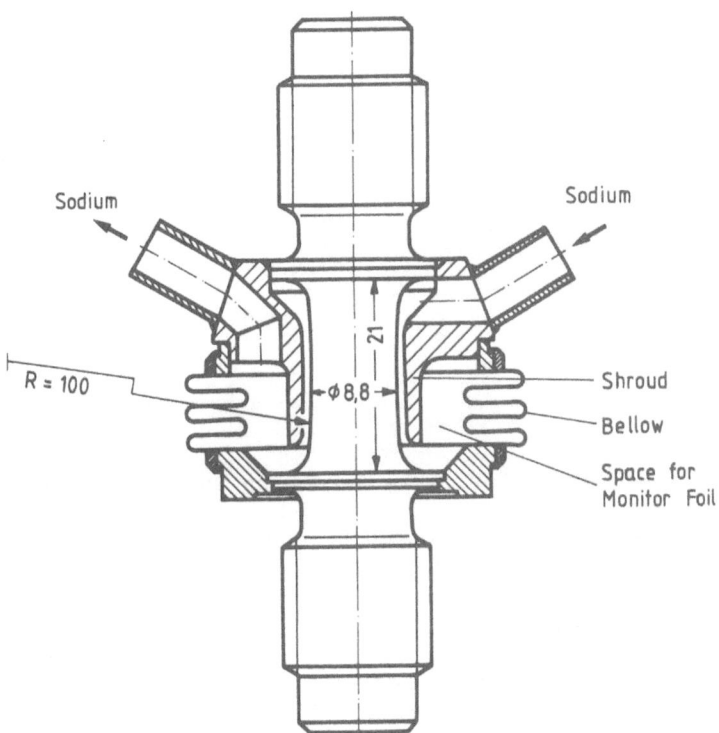

Fig. 2:  LCF-specimen for fatigue testing in flowing sodium

The shroud inside the bellow provides a sodium velocity of 2.5 m/s within the gauge length of the specimen and the free space between shroud and bellow was used for the exposure of a carbon monitor foil. Fig. 3 shows the equipment for the fatigue testing. The top end of the specimen is attached to the rigid portion of the fixture which is connected to the crosshead, while the bottom end of the specimen is attached to the drive rod that is connected to the actuator.

Load Cell

Extensometer
Argon

Sample Sodium
Inlet

Fixture of
Extensometer at
the Sample

Fig. 3:   Equipment for fatigue testing in flowing sodium

Specimen stress is determined by means of a load cell located between the rigid rod and the crosshead.

EXPERIMENTAL PROCEDURE

The low cycle fatigue tests were conducted in the axial strain-control mode at a strain rate of $3 . 10^{-3} s^{-1}$ and $3 . 10^{-5} s^{-1}$ with a fully reversed triangular wave form and zero mean strain. Prior to testing the specimens fabricated from the solution annealed material were pre-exposed to flowing sodium at 550 $^{o}$C for 500 h. This pre-exposure was done in the same sodium loop with identical sodium conditions as during the in-sodium tests. The purpose of this pre-exposure was to remove the chromoxide layers and to initate the selective corrosion process in the material surface.

In order to get a valid comparison parallel fatigue tests in air at 550 $^{o}$C were performed with the same test equipment. The specimens for these tests in air were thermally aged in argon at 550 $^{o}$C for 500 h. This was done in argon (quality: $O_2$ < 5 ppm, N < 10 ppm, $H_2O$ < 2 ppm) in order to avoid thick oxide layers.

Further tests in sodium and in air were done with a 0.5 h dwell period in tension. The cyclic parts of these tests were done in a strain-controlled mode with a strain rate of $3 \cdot 10^{-3}$ s$^{-1}$. At the maximum strain of $1/2 \ \Delta\varepsilon_t$ the stress reached was kept constant for 0.5 h, then the strain-controlled cycle with zero mean strain was continued.

RESULTS

The low cycle fatigue data for X 6 CrNi 18 11 steel tested in flowing sodium and in air at 550 °C are summarized in Table 2. The values for the total stress range $\Delta\sigma$ represent the data obtained at half of the fatigue lifetime ($N_f/2$). The values for the number of cycles to failure $N_f$ are determined by a reduction of the tensile load to a value lower than 40 % of the maximum tensile load.

Fig. 4 gives a plot of fatigue lifetime as a function of total strain range for X 6 CrNi 18 11 material tested in sodium and in air. The air test results at INTERATOM produced with $\dot\varepsilon = 3.10^{-3}$ s$^{-1}$ are in good agreement with results of TNO (2) on the same heat. The comparison between sodium and air shows, that in a strain range from about 0.4 to 1.5 % the fatigue lifetime in sodium is higher than in air, whereas at higher and lower strain ranges no differences in the lifetime occur. It is also shown by the in-sodium tests, presented in Fig. 4, that the different plugging temperatures of < 120°C and 175 °C do not change the fatigue lifetime.

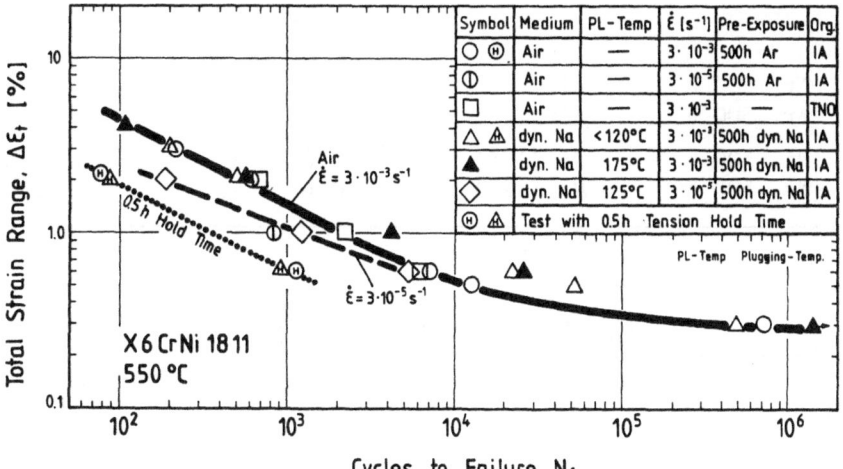

Fig. 4:   LCF data on X6 CrNi 18 11 steel

Table 2.   Low cycle Fatigue Data on X 6 CrNi 18 11 Steel (a)
Obtained in 550 $^{\circ}$C Sodium and Air (b)

| Test Number | Pre-Exposure at 550 $^{\circ}$C | Test Medium | $\dot{\varepsilon}$ $s^{-1}$ | $\Delta\varepsilon_t$ % | $\Delta\sigma_{1/2}$ MPa | $N_f$ |
|---|---|---|---|---|---|---|
| **1.) low cycle fatigue** | | | | | | |
| 58 | Ar, 500 h | Air | | 3.0. | 908 | 212 |
| 50 | Ar, 500 h | Air | | 2.0 | 801 | 608 |
| 65 | Ar, 500 h | Air | | 0.6 | 468 | 7114 |
| 45 | Ar, 100 h | Air | | 0.5 | 427 | 12976 |
| 57 | Ar, 500 h | Air | | 0.3 | 424 | 728991 |
| 68 | Na, 500 h | Na ( 175)(c) | $3.10^{-3}$ | 4.0 | 1004 | 104 |
| 54 | Na, 500 h | Na (<120) | | 3.0 | 940 | 210 |
| 53 | Na, 500 h | Na (<120) | | 2.0 | 822 | 536 |
| 73 | Na, 500 h | Na ( 175) | | 1.0 | 610 | 4104 |
| 56 | Na, 500 h | Na (<120) | | 0.6 | 475 | 22954 |
| 66 | Na, 500 h | Na ( 175) | | 0.6 | 471 | 26353 |
| 44 | Na, 100 h | Na (<120) | | 0.5 | 451 | 41160 |
| 51 | Na, 500 h | Na (<120) | | 0.3 | 415 | 493325 |
| 74 | Na, 500 h | Na ( 175) | | 0.3 | 446 | >1456300 |
| 105 | Ar, 500 h | Air | | 1.0 | 688 | 836 |
| 84 | Na, 500 h | Na (125) | $3.10^{-5}$ | 2.0 | 849 | 188 |
| 83 | Na, 500 h | Na (125) | | 1.0 | 692 | 1220 |
| 82 | Na, 500 h | Na (125) | | 0.6 | 579 | 5363 |

2.) low cycle fatigue with 0.5 h dwell period in tension (d)

$\Delta\varepsilon_c$ (e)

| Test Number | Pre-Exposure at 550 $^{\circ}$C | Test Medium | $\dot{\varepsilon}$ $s^{-1}$ | $\Delta\varepsilon_c$ % | $\Delta\sigma_{1/2}$ MPa | $N_f$ |
|---|---|---|---|---|---|---|
| 63 | Ar, 500 h | Air | | 2.0 | 805 | 79 |
| 81 | Ar, 500 h | Air | $3.10^{-3}$ | 0.6 | 568 | 1063 |
| 60 | Na, 500 h | Na (125) | | 2.0 | 829 | 88 |
| 78 | Na, 500 h | Na (125) | | 0.6 | 542 | 1037 |

(a)   Heat 227768 (IA-No. 325)
(b)   Mode: Axial Strain Control; Wave Form: Triangle with Zero Mean Strain
(c)   Na (175): Dynamic Sodium with 175 $^{\circ}$C Plugging Temperature
(d)   Dwell Periods in Tension with Constant Stress
(e)   $\Delta\varepsilon_c$: Total Cyclic Strain

The test with 0.5 h tension hold times show no influence of the sodium environment. For the cyclic strain range of 0.6 % this was not expected because the tests without hold times show a more than three times greater fatigue lifetime in sodium than in air at this strain range.

The low cycle fatigue tests in sodium with a strain rate of $3.10^{-5}$ $s^{-1}$ show a lower lifetime than the tests with the higher strain rate in sodium. But at 0.6 % strain range the lifetime in sodium with the lower strain rate is nearly the same as in air with the higher strain rate.

At 1.0 % strain range the parallel specimen tested in air with this low strain rate reached a lower lifetime than the specimen tested in sodium. This gives an indication that the beneficial sodium effect observed at the higher strain rate occurs at this low strain rate, too.

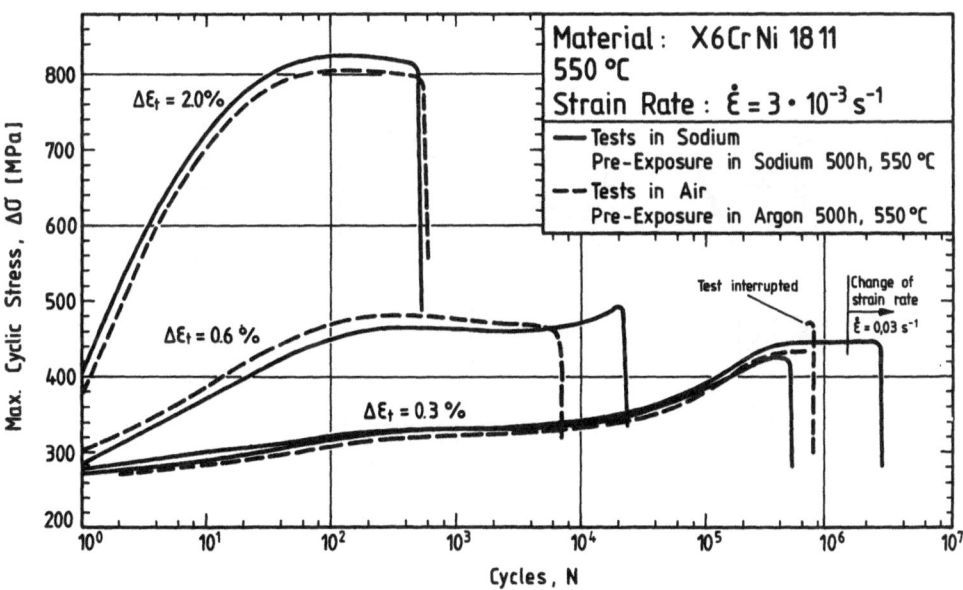

Fig. 5:   Cyclic strain hardening of X6 CrNi 18 11

Fig. 5 gives the cyclic strain hardening of the X 6 CrNi 18 11 steel measured in the fatigue tests in sodium and air. There is no marked difference caused by the environment. At 0.3 % total strain range a large amount of secondary hardening takes place as well in the sodium as in the air test. This secondary hardening even occurs in the sodium test with 0.6 % strain range, whereas in the air test at this strain range the specimen failed before.

The observed secondary hardening starts after about 3000 cycles and reaches its saturation after about $5.10^5$ cycles which corresponds in time from 1.7 h to 280 h for the test with 0.3 % strain range.

The cyclic strain hardening of the fatigue tests with 0.5 h tension hold periods are shown in Fig. 6. From the comparison of the tests in sodium and in air it follows no significant difference. The comparison with the fatigue tests without dwell periods demonstrates for the test with 0.6 % cyclic strain range an about 20 % higher amount of hardening for both test media. In these tests no secondary hardening appears explicitely.

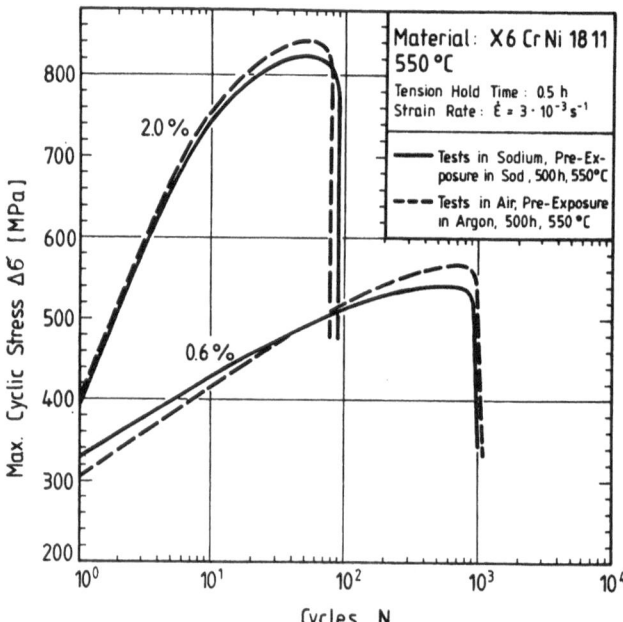

Fig. 6: Cyclic strain hardening in LCF test with hold time

RESULTS OF SCANNING ELECTRON MICROSCOPY

Scanning electron microscopy (SEM) was used to examine the low cycle fatigue specimens. The micrographs of the fracture surface of the specimens tested in air and in sodium give no indication for a different behaviour. This was demonstrated previously (3) for the specimens with 0.5 % strain range. But the side surface of the specimens tested at 0.5 % strain range show interesting differences between the air and sodium specimen (Fig. 7. Whereas the sodium specimen exhibits a considerable amount of plastic deformation in the side surface the air specimen shows an unchanged surface.

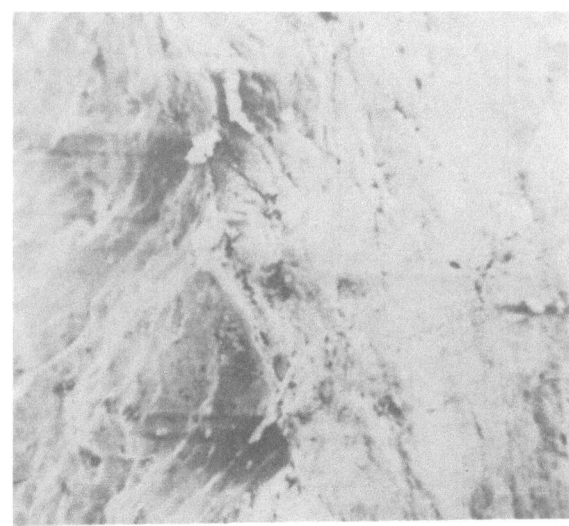

Sodium test, 550°C
Na 500 h; 3m/s
$\Delta\varepsilon_t$=0.5 %
$N_f$ =52466
    50μm

Air test, 550°C
Ar 500h
$\Delta\varepsilon_t$=0.5 %
$N_f$ =12976
    50μm

Fig. 7:  Transition between fatigue crack and specimen surface of
         X6 CrNi 18 11 steel tested in sodium and in air

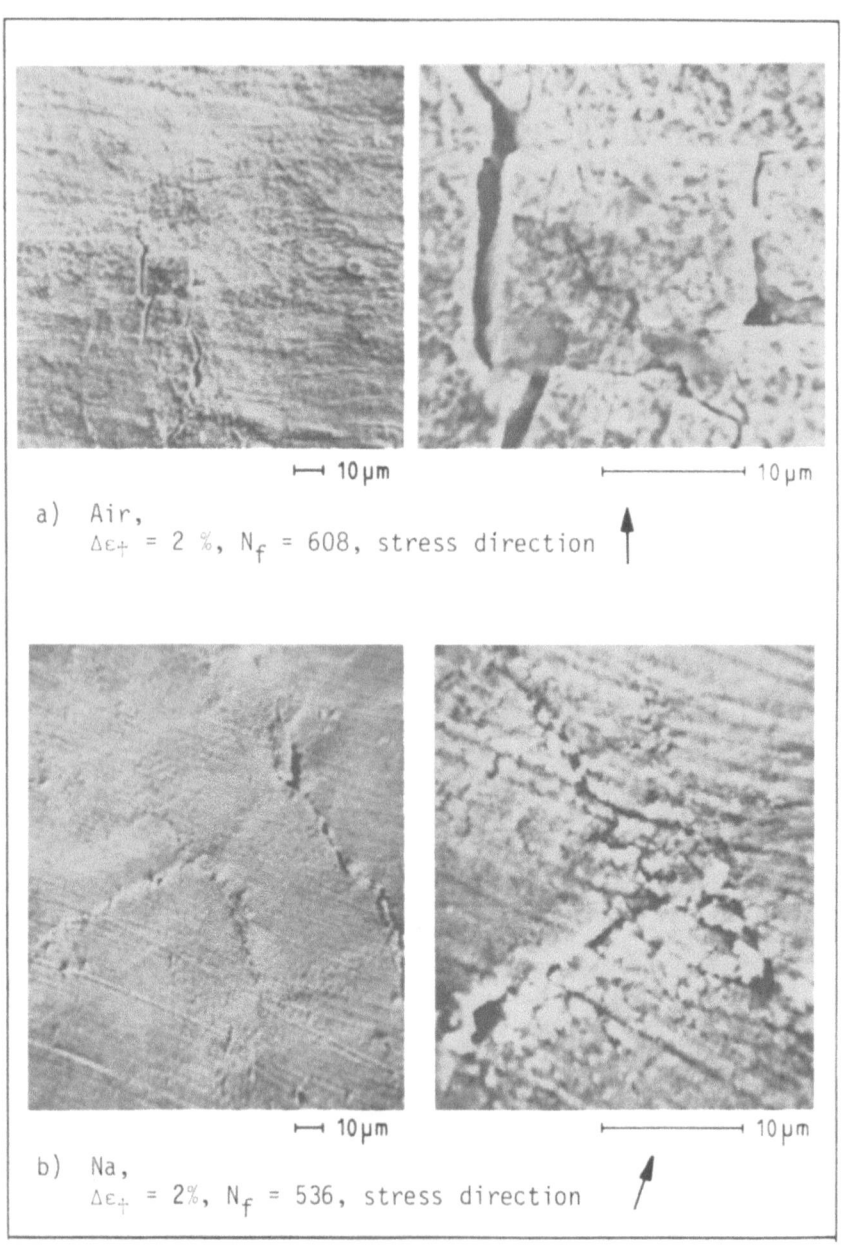

Fig. 8: Crack initiation on the fatigue specimens of X6 CrNi 18 11
       steel tested in air and in sodium

## DISCUSSION

This plastic deformation of the side surface of the sodium
specimen gives an indication for the longer lifetime in sodium than
in air. The reason for this could be that the sodium removes the
chromium oxyde layer of the stainless steel. As a result the formation of extrusions and intrusion (4) by sliding processes at the
free surface of the edge grains is not impeded. Further the new
slide planes at the surface are very reactive and the formation
of oxide layers can impede the further sliding process, too. Both
may lead to a lower plastic deformation at the surface of the air
specimen and the formation of high local stresses which give rise
to an earlier crack initiation in the specimens tested in air.

The crack initiation in sodium may be caused by cavitation
of the material along the slip-bands. Fig. 8b shows these cavitation
and cracks along slip-bands which take its course under an angle of
$45^{o}$ to the stress direction. On the parallel specimen tested in air
no slip-bands are to be seen and at this strain range of 2.0 % the
crack initiation occurs along the grain boundaries as it is shown
in Fig. 8a. Despite of this different mechanism of crack initiation
at 2.0 % strain range the lifetimes in sodium and in air do not
differ. Even at the low strain range of 0.3 % the fatigue lifetimes
in air are as high as in sodium. At this strain range the plastic
strain per cycle is very low (< 0.03 %) and no plastic deformation
occurs at the side surface of the specimen tested in air and in
sodium. Therefore sliding processes are not influenced by oxidation
at this strain range.

In addition, the secondary hardening which occurs at this strain
range reduces the plastic strain. The reason for this secondary
hardening may be the effect of precipitation of carbides induced
by cyclic stresses (3).

## ACKNOWLEDGEMENT

The authors gratefully acknowledge experimental work of
K. Kramer, H. Franken, P. Quintus and H. Eberhardt.

## REFERENCES

1. R. L. Eichelberger, Report AI-AEC-12685, 1968.
2. W. Husslage, B. R. Drenth; Second International Conference
    on Liquid Metal Technology in Energy Production, April
    20-24, 1980, Richland, Washington, USA.
3. G. Menken, W. Husslage, H. Huthmann; IAEA/IWGFR Specialists'
    Meeting on "Properties of Primary Circuit Structural
    Materials Including Environmental Effects", Bergisch
    Gladbach, FRG, Oct. 17-21, 1977.
4. Kh.G. Schmitt-Thomas, Z.Werkstofftech. 8 (1977) 250-258.

# FATIGUE AND CREEP-FATIGUE IN SODIUM OF 316 L STAINLESS STEEL

André Ardellier

CEN de Cadarache

France

## INTRODUCTION

Mainly due to thermal variations, structure components of sodium fast cooled breeder reactors are submitted to strains in the plastic strain range.

Present paper describes equipment and results obtained on type 316 L stainless steel at 450 °C and 600 °C with low-cycle fatigue and creep fatigue tests.

## EQUIPMENT AND EXPERIMENTAL PROCEDURE

The facility for testing in sodium consists of a mechanical actuated machine with associated sodium loop.

### Sodium loop

The sodium loop was constructed from 316 L stainless steel. It is possible to recirculate approximately 15 kg of sodium up to temperature of 650 °C.

Starting at the electromagnetic pump which is located in the cold leg of the main loop system, the sodium flows to the main heater. Sodium leaving the heater enters the test chamber; the desired sodium temperature for testing is automatically controlled by a thermocouple located at the inlet of the test chamber.

All tests where performed with a velocity of about 1 meter per second. After leaving the test chamber, the sodium flows through a finned tube air cooled heat exchanger, and back to the electro-magnetic pump.

A side loop is used to continuously cold trap the main loop sodium. About 10 % to 15 % of the main flow is circulated through the cold trap maintained at 110 $^{\circ}$C by an air blower operated by a temperature controller.

The 110 $^{\circ}$C cold trap temperature yields an oxygen concentration less than 1 ppm  according to Eichelberger standard.

In place of LCF test specimens it is also possible to locate foils of type 316 L - 304 L stainless steel and Fe-Mn alloy in order to determine carbon activity in sodium.

The foils are exposed at 600 $^{\circ}$C for 700 hours and then chemically analysed for equilibrated carbon in the foils. So, sodium loop provides a well definite sodium environment.

Test specimen

The test specimen shown in Fig. 1a is a cylindrical one. Its effective length was determined by means of an extentometer in order to take into account of the amount of elongation in the fixture and the transition region of the specimens.

The test specimens with a 4.5 mm diameter by 8 mm gauge length were fabricated from rods that had been annealed in argon at 1100 $^{\circ}$C and air cooled. The composition is given below:

C < 0.03    Si < 1.0    Mn < 2.0    Ni 10.9    Cr 16.8    Mo 2.0.

Test chamber and fixture

A mechanically push pull testing machine fitted with the test chamber was used for this fatigue testing.

The test chamber is welded to the specimen and connected by two pipes to the sodium loop (Fig. 1b). At the lower part of the test chamber a bellow constructed from 316 L stainless steel enables an axial elongation of ± 1.7 mm.

Test specimen heads are tightly held in two grips, which are screwed into the chucks of the machine.

The load is applied to the specimen test section through the test fixture assembly.

MEASUREMENT AND CONTROL

Strain amplitude is maintained constant during tests after stabilization.

a

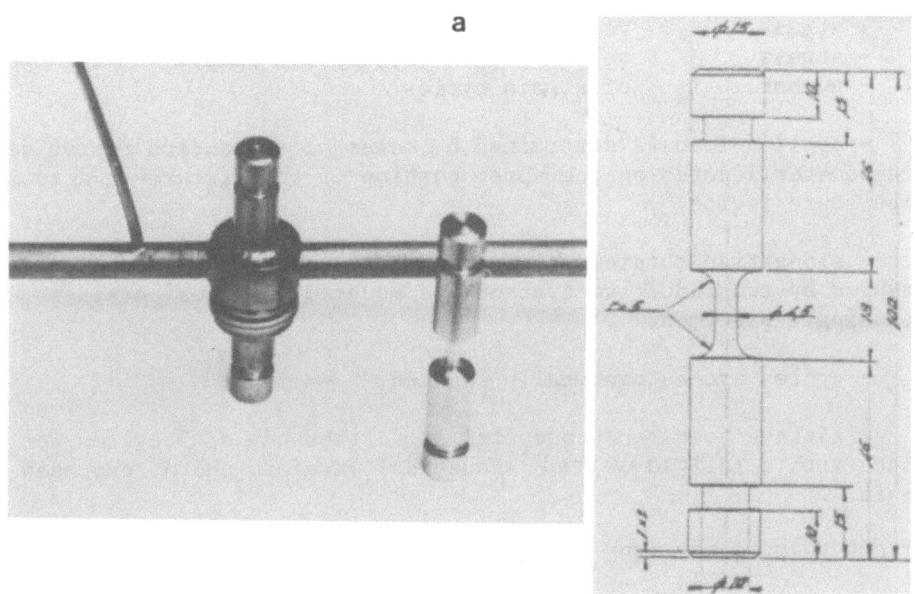

b

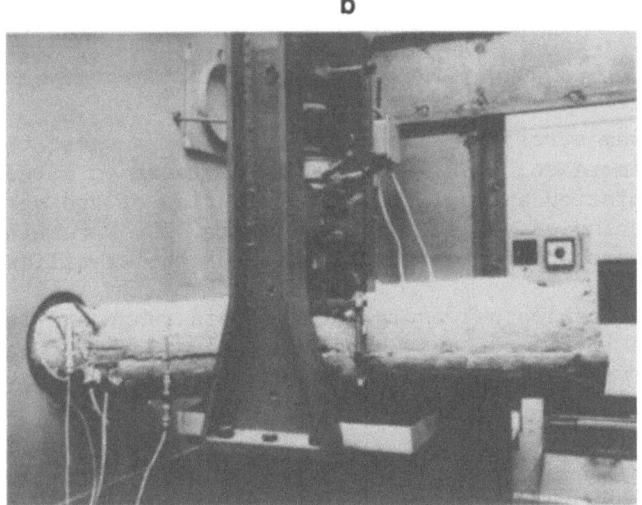

Fig. 1:   a)   Test Specimen
          b)   Fitting of Fatigue Machine on Sodium Loop

The important specimen parameters measured during tests are:

- temperature      versus time
- stress           versus time
- stress           versus strain (hysteretic loop)
- number           of strain cycles

- applied load is determined by means of a standard resistive extensometer located on the upper portion of the fixture in a room temperature region.

- elongation obtained by means of an eccentric driving rod is measured by two inductive transducers mounted on opposite sides of a single specimen.

- cycles are automatically counted.

- fatigue test was conducted at a frequency of 3 cycles per minute with a fully reversed sinusoidal waveform and a zero mean strain.

FATIGUE RESULTS FOR TYPE 316 L STAINLESS-STEEL

Tests were performed at temperatures of 450 $^\circ$C and 600 $^\circ$C with a sodium velocity of about 1 meter per second past the specimen gauge section. Due to continuously cold trapping the oxygen content was maintained less than 1 ppm.

Carbon activity in sodium was of the order of magnitude of $10^{-2}$, based on exposure of foils to the sodium.

The diagram stress vs. time given in Fig. 2a shows three phases during test: increase, stabilization and decrease of stress connected with fissuration and failure. The evolution of the stress-strain cycle at beginning of a test is given in Fig. 2b. Parameters of stable cycle corresponds to the beginning of the stabilized phase.

The relationships between plastic strain range and fatigue life are given in Fig. 3a and 3b respectively for 600 $^\circ$C and 450 $^\circ$C. The dashed curves in both figures represent comparativ data obtained in air.

At 600 $^\circ$C, the ratio of fatigue life in sodium to that in air increases by factors of 2 to 6 as the plastic strain range decreases from 4 % to 0.7 % which indicates that the environmental effect is more pronounced at longer lifetimes. At 450 $^\circ$C, this effect is more pronounced and reaches about one order of magnitude.

The cyclic stress-strain response for the two temperatures is shown in Fig. 3c and 3d. The cyclic stresses for the annealed

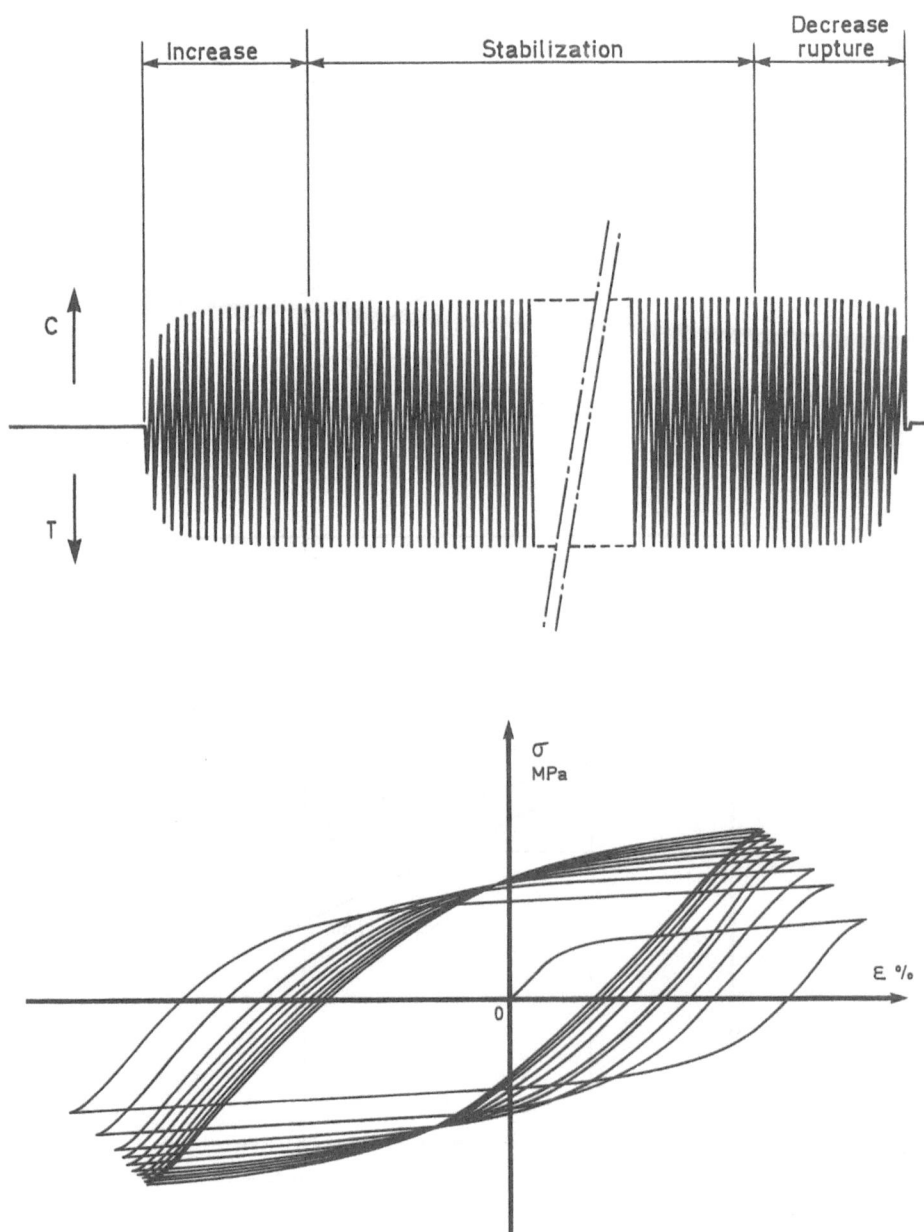

Fig. 2:  a)  Evolution of Stress during Fatigue-Test
         b)  Evolution of First Stress-Strain Cycles

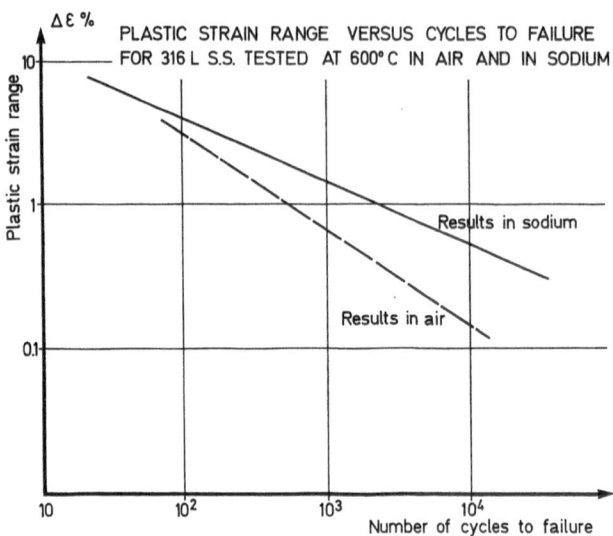

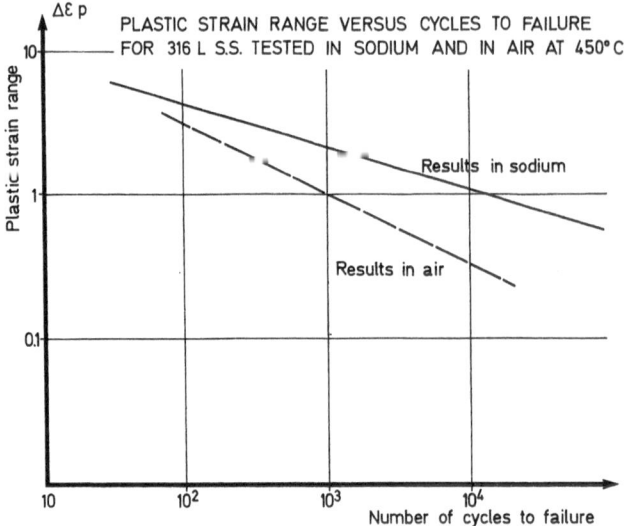

Fig. 3:   Plastic strain range versus cycles to failure
         a)   600 $^{\circ}$C
         b)   450 $^{\circ}$C

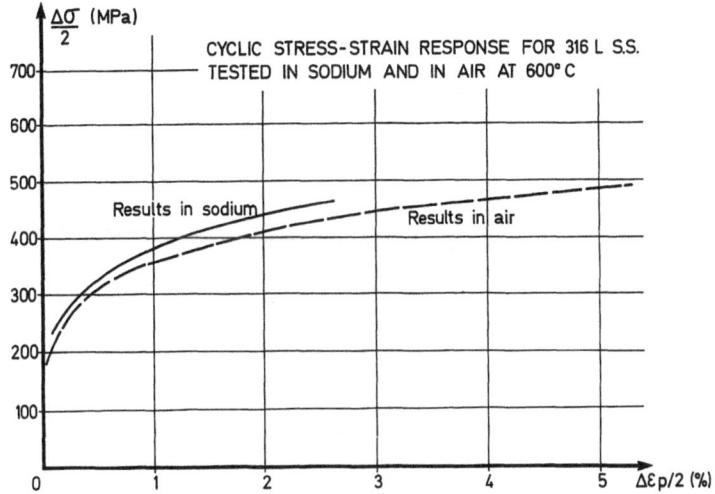

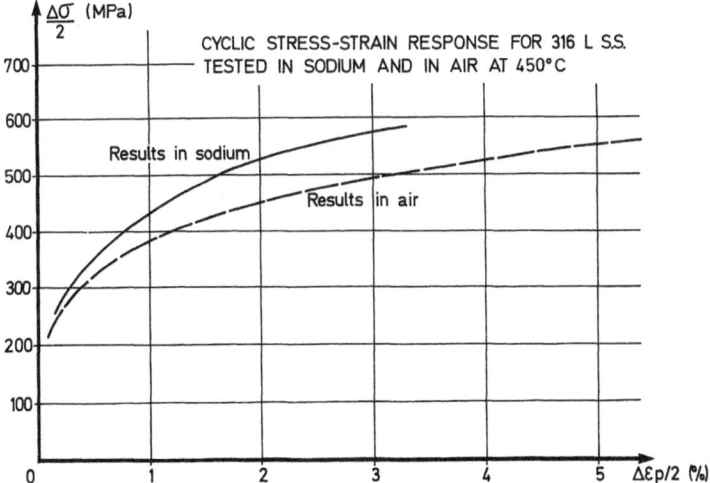

Fig. 3:  Cyclic stress-strain response
         c) 600 °C
         d) 450 °C

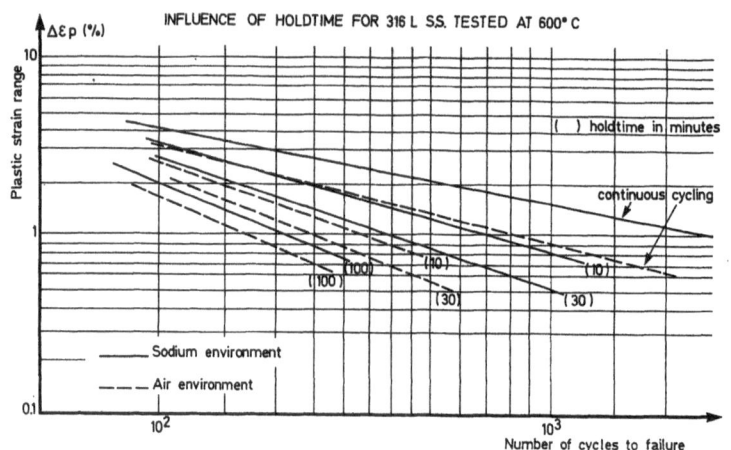

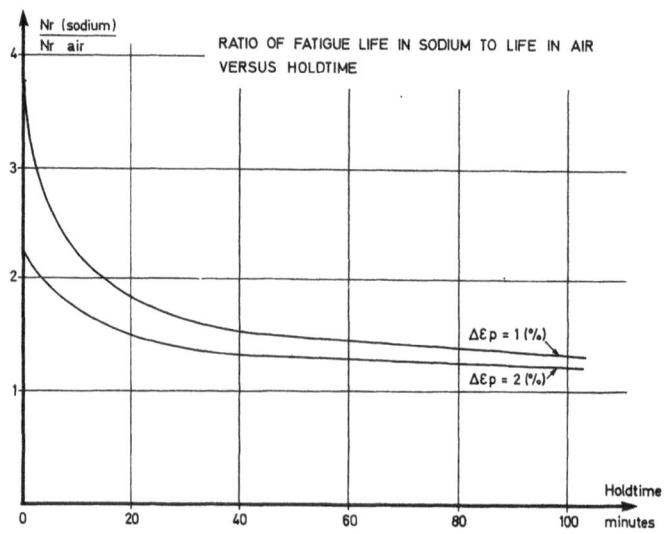

Fig. 4:   Creep-Fatigue Results
          a)   Test Temperature 600 °C
          b)   Influence of Holdtime

materials in sodium seem in agreement with the stresses obtained in air but slightly upper.

CREEP-FATIGUE INTERACTION RESULTS

The control waveform was modified to provide with holdtime in maximal tension.

Tests were run with holdtimes of 10, 30 and 100 minutes.

The realtionship between plastic strain range and cycles to failure is shown in Fig. 4a for 316 L S.S. tested at 600 $^{\circ}$C. The dottet lines represent comparativ data obtained in air.

These results (Fig. 4b) show that the beneficial effect observed with zero holdtime remains but decreases quickly up to 30 minutes holdtime. Then the evolution is not very pronounced.

CONCLUSIONS

Comparison with runs in air on type 316 L stainless steel shows a better low-cycle fatigue behaviour in a sodium environment. This beneficial effect can be attributed to the low oxygen content which limits the surface oxidization.

In the case of holdtimes in tension, it has been observed a decrease of the beneficial effect of the sodium. Notwithstanding it is to be noted that fatigue life in sodium remains about at less 25 % upper that in air (at less up to holdtimes of 100 minutes).

ACKNOWLEDGEMENT

The author wish to thank Mr. Petrequin and Mr. Motto for results in air. Mr. P. Baque for his comments on carbon activity in sodium, and Mr. C. Dumeyniou for his participation to the tests.

INFLUENCE OF FLOWING SODIUM ON THE CREEP PROPERTIES OF FERRITIC

STEEL R 8 (EM 12)

Hugo Tas, Florint Casteels

S.C.K. Mol

Belgium

INTRODUCTION

A sufficiently high creep strength is one of the requirements for candidate wrapper materials for fast sodium cooled reactors. The metallurgical stability and mechanical properties of ferritic stainless steels can be altered in sodium by carburization - decarburization phenomena (1) and physicochemical processes (2). The behaviour of ferritic stainless steels in sodium depends not only on exposure temperature and time but also on the quality of the sodium.

Since only limited information is available related to the possible deterioration of the creep properties of R 8, a number of creep tests both in vacuum and in dynamic sodium have been carried out allowing the determination of the influence of a well qualified sodium on secondary creep rate, time-to-rupture and fracture mechanism. These results will be correlated to chemical and metallurgical changes appearing in the steel during sodium exposure.

EXPERIMENTAL TECHNIQUES

Uniaxial creep tests on R 8 (Belgian name for EM 12) (Fe 9Cr 2Mo 1.5V 0.5Nb Si Mn 0.1 % C) have been carried out on two test apparatus mounted in parallel on the high temperature test section (650 $^{o}$C) of the Na-2 loop at Mol. Identical specimens have been tested under vacuum (1.10$^{-5}$ torr).

The stresses at which the tests have been performed both in sodium and in vacuum are given in Table 2. The specimens, fabricated

175

Table 1.  Chemical composition and metallurgical conditions of
          alloy tested

| Element wt % | C | Mn | Si | Cr | Mo |
|---|---|---|---|---|---|
| | 0.08 | 0.93 | 0.265 | 8.88 | 1.91 |

| Element wt % | V | Nb | Co | B |
|---|---|---|---|---|
| | 0.53 | 0.42 | 0.014 | $0.5 \times 10^{-3}$ |

Swaged at 1100 $^\circ$C, followed by a thermal treatment of 30' at
1100 $^\circ$C and air cooling. Final heat treatment: 2 h at 750 $^\circ$C
under argon.

Table 2.  Survey of creep tests

| Test | Medium | Creep stress $N/mm^2$ | $t_R$ | $\Delta\varepsilon_{t(\%)}$ | Secondary creep rate %/h |
|---|---|---|---|---|---|
| 1 | vac | 88.29 | 110 | 62.4 | 0.11885 |
| 2 | vac | 88.29 | 94 | 59.2 | 0.16154 |
| 3 | vac | 49.05 | 4042 | 85.2 | 0.00086 |
| 4 | vac | 49.05 | 3258 | 57.3 | 0.00162 |
| 5 | Na | 88.29 | 162 | 66.5 | 0.098 |
| 6 | Na | 88.29 | 165 | 62.5 | 0.043 |
| 7 | Na | 49.05 | 5313 | 67 | 0.00127 |
| 8 | Na | 49.05 | 6298 | 64 | 0.00228 |

Table 3.  Chemical analysis of sodium impurities (wppm)

| Oxygen | Total carbon | Al | Ag | Bi | Cr |
|---|---|---|---|---|---|
| 2.8 - 12.4 | 2.9 - 4.4 | 0.047 | 1.00 | 2.56 | 0.137 |

| | | Cu | Fe | Mn | Ni |
|---|---|---|---|---|---|
| | | 0.16 | 0.88 | 0.044 | 0.16 |

| | | Si | Pb | Sn | |
|---|---|---|---|---|---|
| | | 0.092 | 110 | 16 | |

from rod material, have a gauge length of 20 mm and a diameter of
4 mm. The chemical composition and metallurgical conditions are
given in Table 1. The maximum exposure time in sodium reached
6298 hours with a sodium velocity of 3.6 $m.s^{-1}$. Periodical analyses
of the sodium have been performed. The results for total carbon,
oxygen and metallic impurity content are given in Table 3. The
cold trap was operated at 160 $^\circ$C with the cold leg at a temperature
of 380 $^\circ$C.

MICROSTRUCTURAL CHANGES

The microstructure of the material corresponds with a duplex
structure with aligned delta ferrite grains and bainitic grains.
A large number of carbide precipitates (Fe, Cr, Nb)$_x$C$_y$ appears
after short time creep testing particularly in the delta ferrite
(Fig. 1a). Prolonged exposure at 650 $^\circ$C results in a coarsening of
the carbide phase. A large number of carbides is found to be present
at the original ferrite – bainite interface (Fig. 1b). A Laves
phase of the (Fe, Cr, Si)$_2$(Mo, Nb) type appears after long term
annealing at 650 $^\circ$C (Fig. 2). This phenomenon fits earlier published
data (3). Sodium exposure results in the development of a sodium
affected zone. This layer is characterized by the following compositional changes of substitutional elements: selective leaching of
chromium, manganese and silicon and a pick-up of nickel. The composition of the surface layer as a function of exposure time is given
in Table 4. The nickel pick-up extends to depths of respectively
10 μm and 20 μm for the two exposure times.

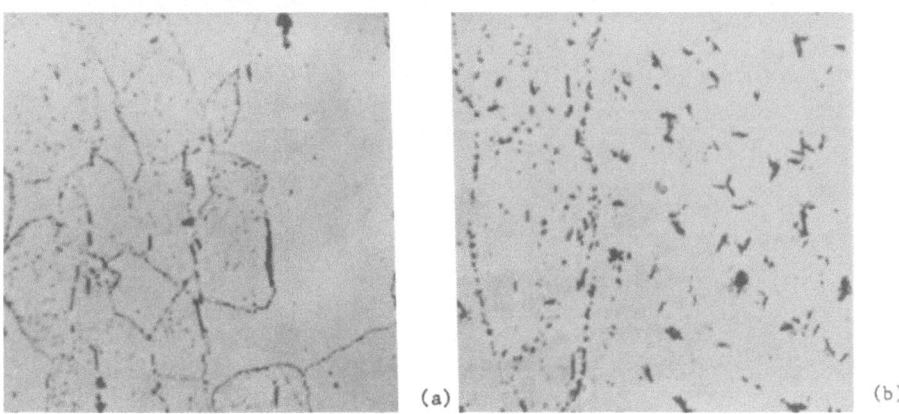

(a)                                          (b)

Fig. 1.   Longitudinal sections of sodium tested creep specimens
          10 μm

          a: after 165 hours at 650 $^\circ$C: Nb carbides
          b: after 6298 hours at 650 $^\circ$C: Nb carbides and plate-like
                                            Laves phase

No gradients of substitutional elements are present in the
surface layers of the specimen creep tested for 110 hours in vacuum.

Metallographic and microprobe analyses carried out in the
surface and subsurface layers of the sodium exposed specimens did
not reveal any change in the density of the (Fe, Mo, Nb) carbides
(Fig. 2). The earlier statement related to the growth of the
carbide phase with increasing exposure time is confirmed by the
microprobe images.

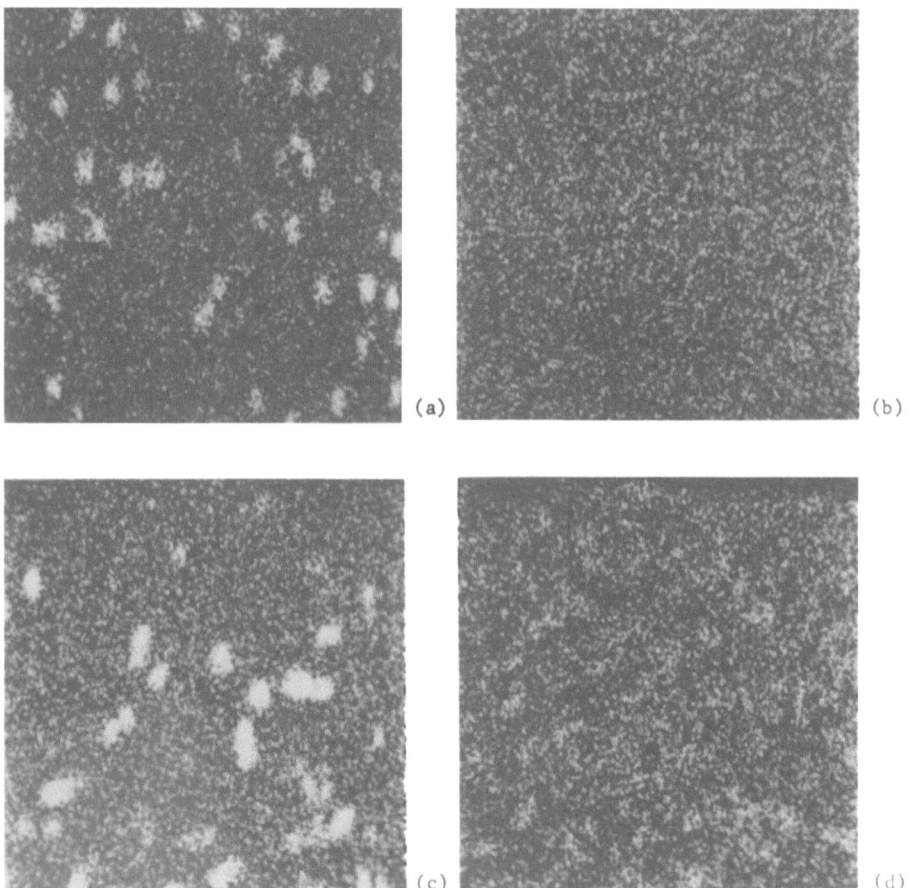

Fig. 2.   Growth of Nb-rich carbides and Mo-rich Laves phase during
          creep test at 650 °C in sodium (top edge corresponds with
          outer surface specimen)
          a: Nb-X ray image,  165 hours at 650 °C          ‚20 μm‚
          b: Mo-X ray image,  165 hours at 650 °C
          c: Nb-X ray image, 6298 hours at 650 °C
          d: Mo-X ray image, 6298 hours at 650 °C

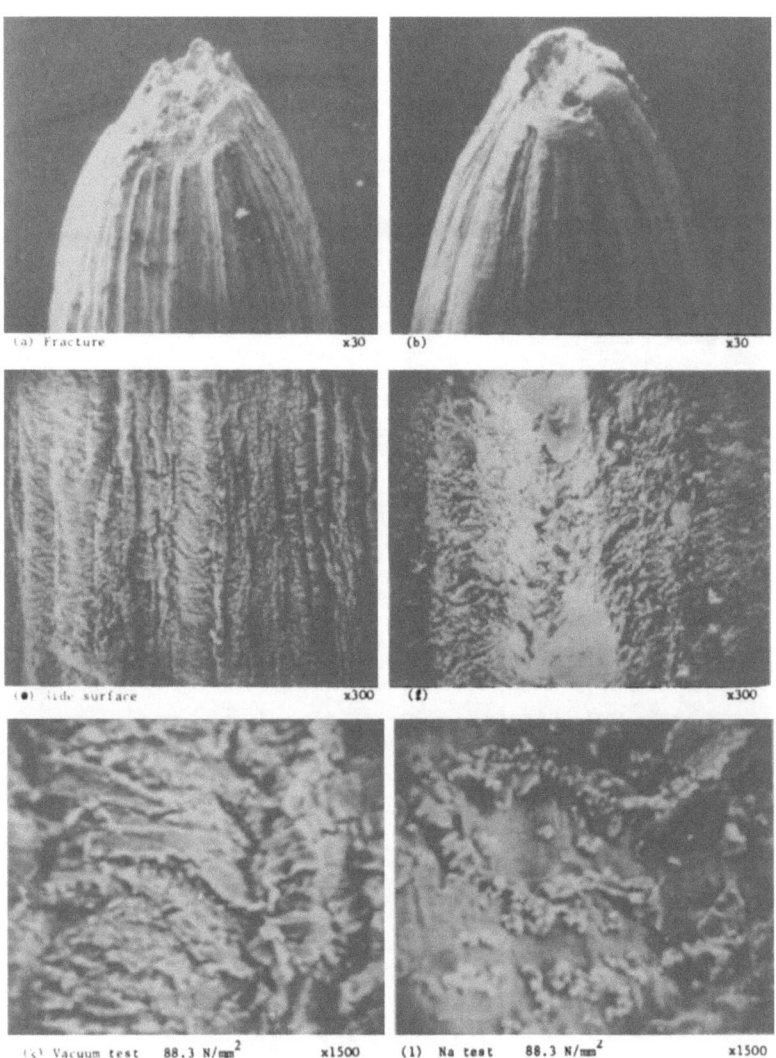

(a) Fracture                    x30        (b)                          x30

(e) Side surface                x300       (f)                          x300

(g) Vacuum test    88.3 N/mm$^2$    x1500  (l)  Na test    88.3 N/mm$^2$    x1500

Fig. 3: Fracture and surface morphologies
Part 1

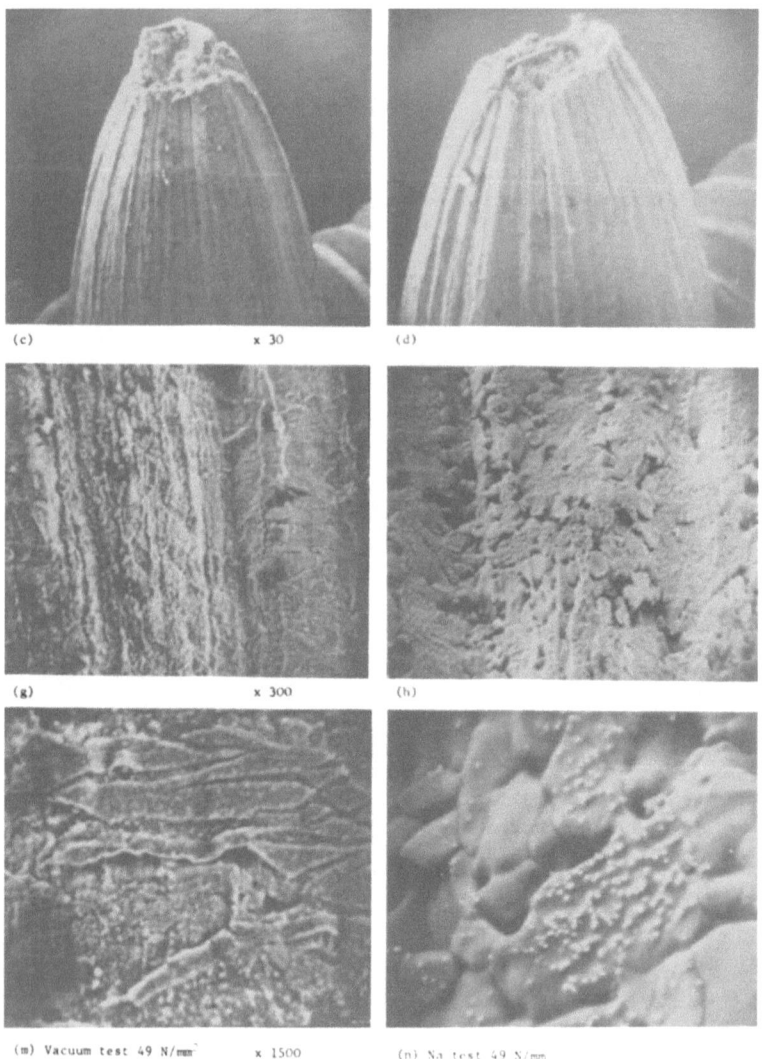

(c)                    x 30          (d)

(g)                    x 300         (h)

(m) Vacuum test 49 N/mm²      x 1500      (n) Na test 49 N/mm

Fig. 3:   Fracture and surface morphologies
          Part 2

MECHANICAL TESTS

The comparison of the values of secondary creep rate, rupture time and total elongation at fracture obtained in vacuum and sodium result in the following conclusions (see Table 2):

1.  The secondary creep rate is not affected.
2.  A slight increase of time-to-rupture is found for the sodium exposed specimens.
3.  The total elongation at fracture is not influenced.

SURFACE MORPHOLOGY AND FRACTURE

Microscan analyses of the specimens creep tested in sodium and in vacuum revealed a change in their surface morphology.

The fracture mode in both environments is of the ductile type with pronounced necking (see Fig. 3a, b, c, d). The duplex nature of the microstructure (aligned delta ferrite grains and bainitic grains) results in the formation of longitudinal ribbons on the side surface of the specimens.

The oxide layer originally present on the surface of the specimen breaks down during vacuum creep testing (Fig. 3k, m).

This oxide layer is partially dissolved in the sodium during the 110 h creep test experiment (Fig. 3f, l). Longer creep testing results in the formation of a surface morphology typical for sodium corrosion: small holes develop at isolated places and combine at grain boundaries and interphase boundaries resulting in the formation of cracks (Fig. 3h).

The white dots present on the surface of long duration experiments have been identified as Mo-rich corrosion products and Pb-rich of Pb-Ni deposits (Fig. 3n). The presence of Pb-rich and Pb-Ni deposits can be attributed to the high Pb impurity level of the sodium.

DISCUSSION AND CONCLUSIONS

Mechanisms influencing the creep properties of structural materials in a liquid environment are diffusion controlled processes such as leaching or pick-up of substitutional and interstitial elements (1) and physicochemical processes (2) such as general corrosion, surface pitting, wedging induced crack formation, grain boundary softening and absorption and reaction of surface active impurities promoting early fracture (3).

Thermal ageing of the EM 12 alloy results in a pronounced coarsening of the carbide phase and the nucleation and growth of a Laves phase of the $(Fe, Cr, Si)_2$ $(Mo, Nb)$ type (4).

This ageing effect is responsible for the deviation of linearity of the creep curve represented in the $\log \sigma$ - $\log t$ diagram (5).

The observed preferential leaching of manganese, chromium and silicon and the pick-up of nickel result in the formation of a sodium affected zone with a thickness of 10 μm and 20 μm for exposure times of respectively 165 and 6298 h. The compositional change in the outer layer of the affected zone, as measured by microprobe analyses, cannot have led to the formation of austenite. Possible structural changes induced by interstitial element transfer could not be detected by metallographic and microprobe analyses.

Parameters describing the creep properties of the R 8 (EM 12) steel, tested both in vacuum and sodium such as secondary creep rate, total elongation at rupture and the onset of tertiary creep are in each others scatter band.

The slight increase of time-to-rupture for the sodium exposed specimens compared to values obtained in vacuum can be attributed to carburization and the formation of very fine carbides not detectable by conventional metallographic techniques. A similar phenomenon has been found during creep testing of a dispersion strengthened ferritic steel in sodium (6). Although the creep tests in sodium have been performed in sodium loaded with metallic impurities such as Pb and Sn, no promotion of crack growth and loss of ductility have been observed. This phenomenon may be due to the transgranular fracture mode of this steel and the relatively high testing temperature (2), the embrittling affect of these impurities on the 304 type stainless steel observed in the same loop (4) being associated with an intergranular fracture mode and occurring at 550 °C (4).

Table 4.  Composition of surface layer as a function of exposure time in sodium

| Exposure time (h) | Elements (wt.%) | | | | | |
|---|---|---|---|---|---|---|
| | Fe | Ni | Cr | Mn | Mo | Si |
| 165 | 92.97 | 1.16 | 3.90 | 0.21 | 1.60 | 0.15 |
| 6 298 | 90.91 | 1.26 | 6.26 | 0.08 | 1.47 | 0.02 |

REFERENCES

1. S. Yuhara, H. Atsumo, International Conference on Ferritic
   Steels for Fast Reactor Steam Generation - London
   30/5/77 - 2/6/77 - paper 47.
2. M. H. Kamdar, Progress in Materials Science 15 (287- )(1973).
3. F. L. Versnyder, H. J. Beattie Jr., Transactions ASM 47
   (1955) (214 - ).
4. H. Huthmann, G. Menken, H. U. Borgstedt, H. Tas, Second
   International Conference on liquid metal Technology
   in Energy Production - Hanford, April 1980. paper 41.
5. M. Caubo, J. Mathonet, Revue de Métallurgie (1969)
   (345 - ).
6. H. Tas, F. Casteels, Paper presented at the Symposium
   "Sodium Coolant Technology" Dimitrovgrad 15-22/10/1977.

# SODIUM LOOPS FOR MATERIAL BEHAVIOR TESTING IN FLOWING SODIUM IN KFK

Hans Ulrich Borgstedt, Günther Drechsler,
Günter Frees and Egon Wollensack

Kernforschungszentrum Karlsruhe
7500 Karlsruhe
Fed. Rep. of Germany

## INTRODUCTION

Alkali metal corrosion can influence the mechanical high temperature properties of structural materials by changing their contents of interstitial elements (C, N, O). On the other hand liquid metals can also cause brittle behaviour of materials. The preceding tests (1) have demonstrated that sodium accelerates the crack propagation rate in creep rupture tests. Though sodium changes mainly surface near areas of the pipes the tests made before do not indicate a significant sodium influence on the fatigue life of stainless steel spacimens (2). To study the influence of proper liquid sodium environment on creep crack growth and on fatigue cracking behaviour, two new loops have been constructed, one of them connected to two fatigue testing systems, the other containing four creep rupture test sections.

## SODIUM LOOP FOR FATIGUE TESTS FARINA

The sodium loop for testing the fatigue behaviour and fatigue crack growth rates of structural materials has been designed to expose specimens connected to material testing machines at the maximum of the loop temperature. Additionally the sodium is forced to pass the specimens with a flow velocity of realistic value of 3 m/s. Since the purity of the sodium seems to be of major influence on the test results, purification and purity measuring devices are foreseen.

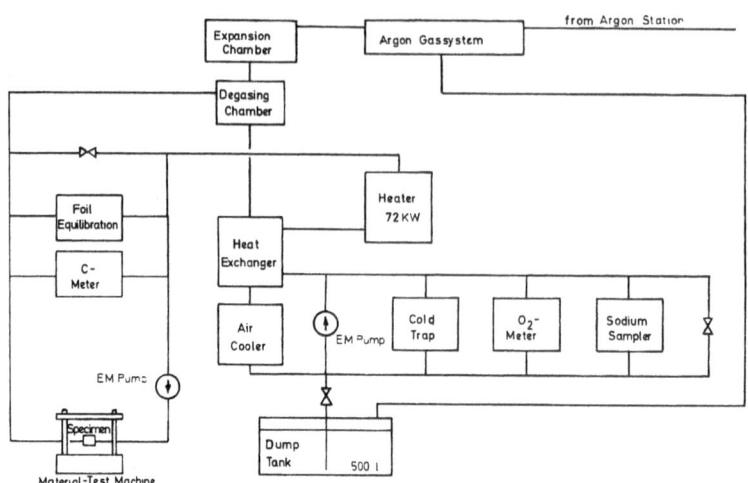

Fig. 1: Flow diagram of the sodium fatigue and crack growth loop
FARINA

The main loop is a figure eight loop with an electomagnetic
pump of a capacity of 5 m$^3$/h at 4.5 bar and a sodium/air cooler in
the cold part, a recuperative heat exchanger and a heater of 72 kW
to supply the test sections with sodium of the desired temperature
and flow velocity. Fig. 1 shows the flow diagram of this loop.

The hot part of the circuit contains a foil equilibration
chamber to estimate carbon activities at up to 973 K in a by-pass
line and another by-pass for the introduction of a carbon meter
probe. The cold part of the loop is by-passed by a cold trap, an
oxygen meter module and the sodium sampler. The two parallel test
by-pass lines are equipped with their own electromagnetic pump.
The two testing machines (MTS Systems 100 kN servohydraulic machines)
can be applied for lcf tests as well as for crack propagation tests.

The encapsuled lcf Na specimens made of the SNR structural
material, stainless steel X6 CrNi 18 11, have been pre-corroded
within their capsules (see Fig. 2). After the 3000 h exposure to
flowing sodium at the test temperature (823 K) we have stored them
under frozen sodium to prevent any contacts to the atmosphere. Due
to dificulties in the external strain control the in-sodium tests
are delayed by some months. The design of a test chamber for crack
propagation test using CT specimens is under the way.

Fig. 2: Encapsuled lcf Na specimen to be mounted to the test sections


SODIUM LOOP FOR CREEP RUPTURE TESTS CREVONA

     The second loop designed for creep rupture tests, CREVONA, has
principally the same design as the loop FARINA, as can be seen from
the Fig. 3 showing the flow diagram of CREVONA. The high temperature
circuit of the system contains four parallel test sections equipped
with loading and strain measurement devices for creep rupture tests.
The design of this device is shown by Fig. 4. The load is measured
by a calibrated load cell, the strain by two inductive extensometers.

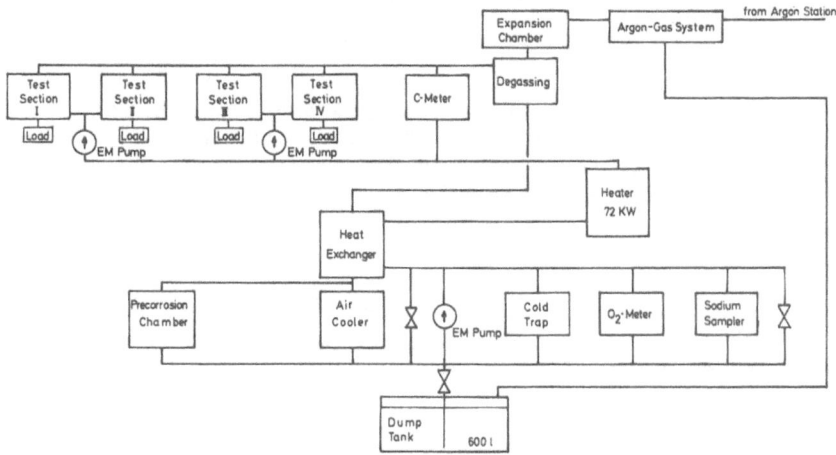

Fig. 3: Flow diagram of the sodium loop for creep rupture tests
        CREVONA

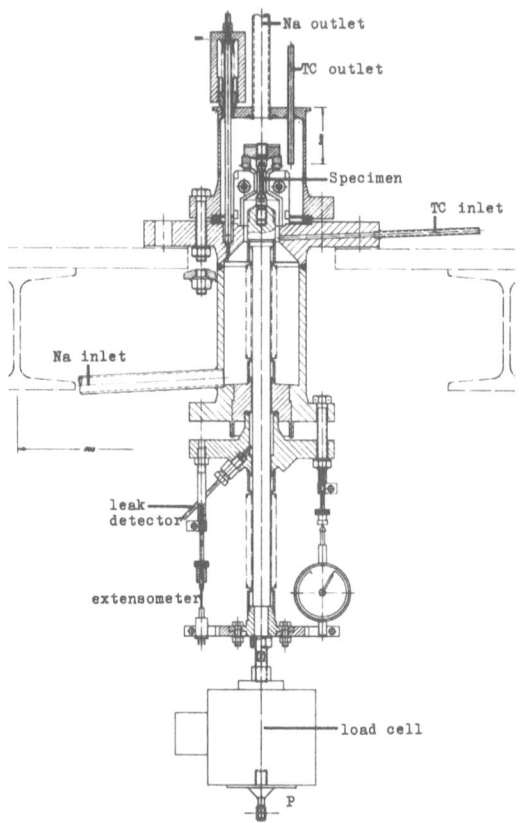

Fig. 4: Design of the creep rupture test section of CREVONA


The loop CREVONA has additionally a test section for pre-corrosion of fatigue specimens within their capsules. This test section is fitted to the loop in a manner that it can easily be replaced by another test section.

DATA  EVALUATION DEVICE

The thermocouples, flow meters and oxygen monitors of both loops and the load cells and extensometers of CREVONA are connected to a 128 kBytes AEG 80-20/4 processor unit, the block diagram of which is shown by Fig. 5.

The equipment consists of a graphic display, a point display, a line printer, a plotter, and two teletype printers. The processor unit prints out every hour an operation report and gives read out or plots of actual states of the creep rupture curves.

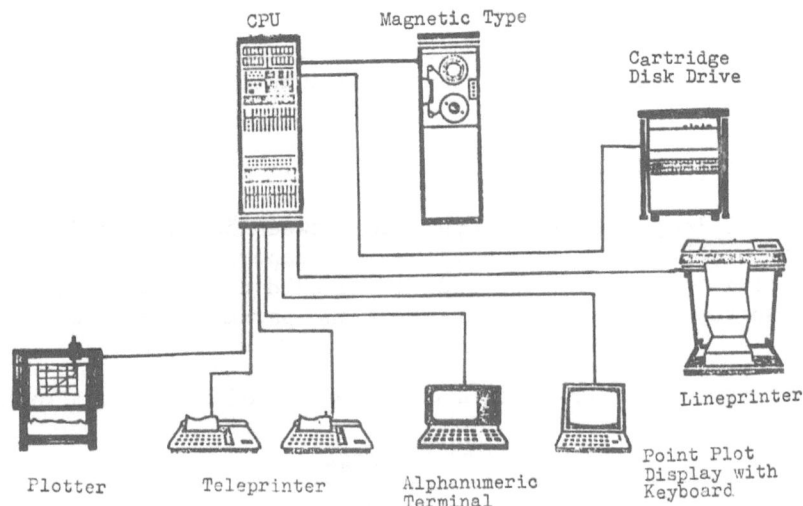

Fig. 5: Block diagram of the processor unit of both Na loop

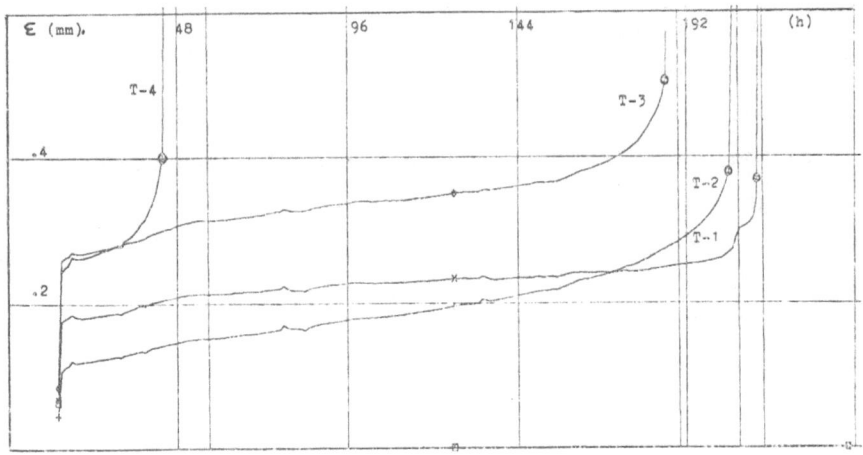

Fig. 6: Plots of the creep rupture curves of all four test sections
in the first creep rupture test

After the creep rupture life of the specimens, the whole creep curves can be printed out as is shown for the results of the first test run of CREVONA in Fig. 6.

The processor control of operation of both loops enables us to let them run without operational personnel by night and during the weekends.

OPERATION EXPERIENCE

The two loops have been operated between 500 and 1500 hours. They have reached the desired values of temperatures and flow velocities in the test areas and in the cold traps. The Table 1 gives a survey on the characteristics and parameters in both loops.

Table 1. Characteristics and parameters of the sodium loops

|  | FARINA | CREVONA |
|---|---|---|
| Sodium volume | 400 l | 500 l |
| Capacity of main heater | 72 kW | 72 kW |
| Number of test sections | 2 | 4 + 1 |
| Maximum temperature of test sections | 550 $^{o}$C (600 $^{o}$C) | 550 $^{o}$C (600 $^{o}$C) |
| Flow velocity in test sections[+)] | 3 m/s | 3 m/s |
| Cold leg temperature | 400 $^{o}$C | 400 $^{o}$C |
| Purification system | cold trap | cold trap |
| Purity measurement |  |  |
|     carbon | foil equilibration at 700 $^{o}$C | |
|     oxygen | emf cell | emf cell |
|     dissolved metals | sampling device | sampling device |
| Operation experience | 500 h | 1500 h |

[+)] in the test sections designed for the present tests

In FARINA, a General Electric oxygen meter cell is working satisfactorily, in CREVONA it worked only about 100 h and must be replaced due to a leakage. Several sodium samples have been taken from both loops. The Table 2. gives a survey on the impurity levels measured in the sodium of FARINA, in CREVONA the impurities are on the same low level.

Table 2.   Sodium Purity in FARINA

| Cold trap temperature | | 398 K (125 $^{\circ}$C) | |
|---|---|---|---|
| Oxygen | chemical analysis | 6.3/7.3 | ppm |
| | oxygen meter | 5 $\pm$ 2 | ppm |
| Carbon | foil equilibration | 0.1 | ppm |
| Nitrogen | foil equilibration (chemical activity) | 9 x 10$^{-4}$ | |
| Iron | | 3.7/2.2 | ppm |
| Chromium | | 0.8/0.2 | ppm |
| Manganese | | 0.2/0.2 | ppm |
| Nickel | | 0.2 | ppm |
| Tin | | 0.9 | ppm |
| Bismuth | | 0.75 | ppm |
| Lead | | 0.35 | ppm |
| Zinc | | 0.20 | ppm |

ACKNOWLEDGEMENTS

We gratefully acknoledge the aid in analytical chemistry given by Dr. H. Schneider and the assistance we got from Mrs. Z. Perić and Mr. G. Wittig.

REFERENCES

1.   H. Huthmann, G. Menken, H.U. Borgstedt, H. Tas, Second Internat. Conf. on Liquid Metal Technology in Energy Production, Richland, WA, USA, April 20-24, 1980
2.   G. J. Zemann, D. L. Smith, Nuclear Technology 42 (1979) 82-89

CORROSION PRODUCT RELEASE INTO SODIUM FROM AUSTENITIC STAINLESS

STEEL

William F. Brehm and R.P. Anantatmula

Hanford Engineering Development Laboratory
Richland
USA

## INTRODUCTION

During operation of a sodium-cooled nuclear reactor, the fuel cladding and in-core materials will undergo material loss (corrosion) by the flowing sodium. An experimental program to characterize the release and transport of corrosion products has been active at our laboratory since 1966. The purpose of this paper is to describe the results of radioactive and non-radioactive corrosion product release from austenitic stainless steel obtained during the study and to identify some release mechanisms. The previous excellent reviews of material corrosion in sodium (1, 2) did not discuss release of manganese and cobalt, the two most significant radioactive species. Discussion of control measures against radioactive material transport, and general summaries of that problem have also been presented elsewhere (3, 4, 5).

## EXPERIMENT

Specimens used for radionuclide transport studies were 20 % cold worked AISI 316 stainless steel fuel cladding tubes, irradiated in EBR-II. Typical specimen composition is 65 % Fe, 17 % Cr, 13 % Ni, 2.5 % Mo, 1.5 % Mn, 0.5 % Si, 0.05 % C, minor amounts of N, P, S. Specimens were cut from parts of fuel assembly where no composition gradients were produced during irradiation and sodium exposure (temperature about 370 $^{\circ}$C).

Sodium exposure was done in two Source Term Control Loops; a schematic of the loop is shown in Fig. 1. The loop sodium inventory was about 18 liters. Test loops were run at specimen temperatures of 604 $^{\circ}$ or 538 $^{\circ}$C ($\pm$ 3 $^{\circ}$C), cold leg temperatures of 427 $^{\circ}$ or

360 $^\circ$C$_4$(+ 5 $^\circ$C), with a sodium velocity of 6.7 m/s, Reynolds number
4.4x10$^4$ at the specimen surface. Oxygen level was controlled by
cold traps and hot traps and monitored by electrolytic oxygen
meters calibrated by vanadium wire analysis (6). Excellent
reproducible results were obtained throughout the test program
using these techniques over a range of three orders of magnitude
oxygen activity. Nominal oxygen activity levels were 2.5+0.2 ppm,
0.5+0.1 ppm, and < 0.01 ppm. Post specimen examination included
weight change analysis, sequential incremental analysis to determine
subsurface composition and radiochemical gradients. In addition to
weight loss data, specific release rates of Fe, Cr, Ni, Mn, Mo,
54Mn and 60Co for each specimen were obtained from the sequential
incremental analysis data.

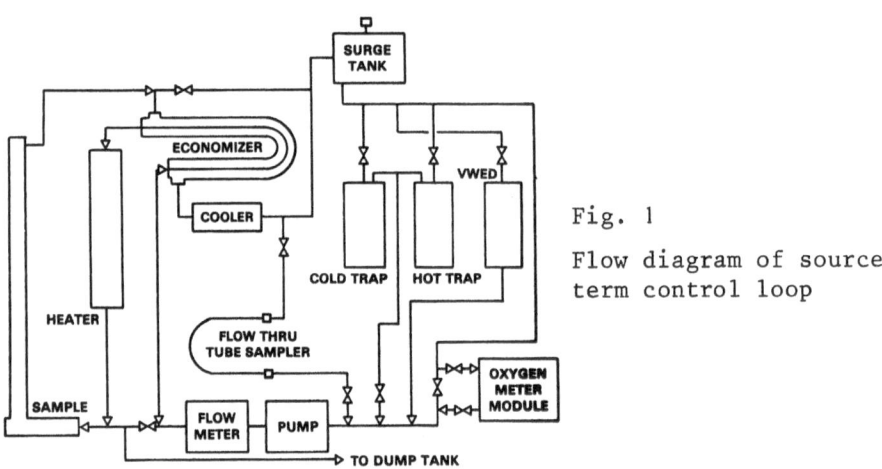

Fig. 1

Flow diagram of source
term control loop

Further details of the experiments and system operation are
given in Ref. 7 and 8.

RESULTS

Weight Loss Data

The weight loss data were fitted to a linear equation of the
type R = at + b; where R = weight loss in $\mu$g/mm$^2$, t is in 1000 hour
units, and a and b are constants. The weight loss data at 538 $^\circ$C
and 604 $^\circ$C can be represented by the equations:

R = (0.23 + 0.015) t + (0.10 + 0.11) (538 $^\circ$C, 0.5 ppm oxygen)
R = (1.15 + 0.02)  t + (1.26 + 0.14) (604 $^\circ$C, 0.5 ppm oxygen)
R = (2.86 + 0.11)  t + (0.27 + 0.34) (604 $^\circ$C, 2.5 ppm oxygen)

The + values in the parentheses are 90 % confidence limits obtained from the least squares analysis. The experiment in STCL-2 at 538 °C and 2.5 ppm oxygen was terminated at 4,000 hours, and only one set of weight change data was available. The average weight loss was 2.0+0.22 μg/mm². The equivalent corrosion rates were calculated from the time coefficient to be 0.26 μm/yr (0.5 ppm oxygen), 0.55 μm/yr (2.5 ppm oxygen) at 538 °C and 1.27 μm/yr (0.5 ppm oxygen), 3.15 μm/yr at 406 °C (2.5 ppm oxygen) respectively. The weight loss increased by a factor of 2.1 at the higher oxygen level at 538 °C, and by a factor of 2.5 at 604 °C. These data are shown graphically in Fig. 2 and 3. The weight loss data at 604 °C, < 0.01 ppm oxygen are shown in Fig. 2. It can be seen that there was no decrease in weight loss at 604 °C, < 0.01 ppm oxygen. This latter result trends to support the hypothesis of Weeks and Isaacs (1) that at very low oxygen activities, a solubility-driven corrosion mechanism dominates. It is seen in the next section that iron is the only corroding species whose release was a strong function of oxygen level.

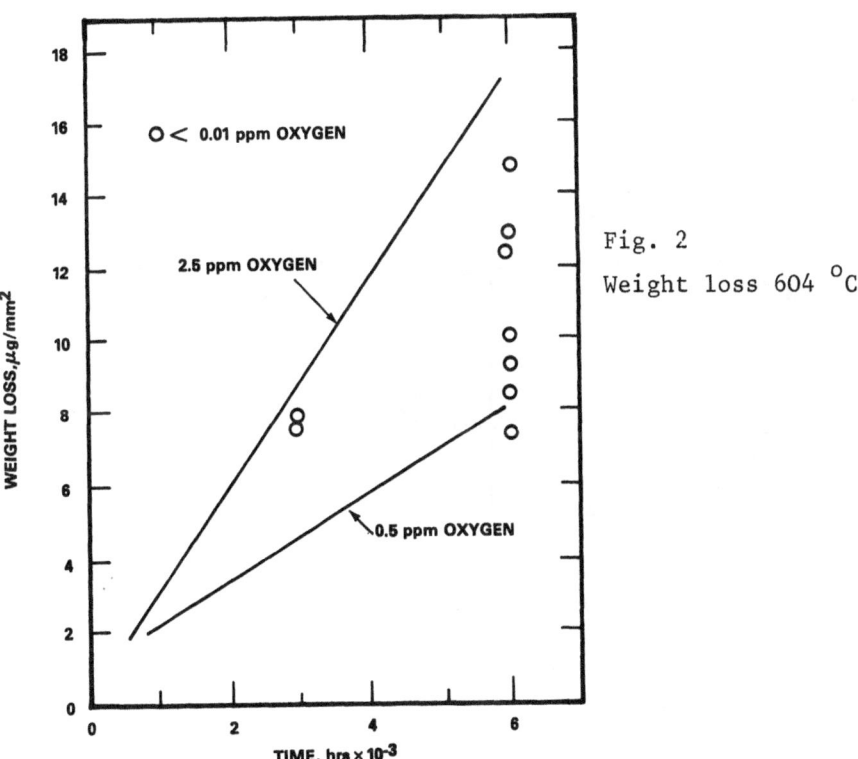

Fig. 2

Weight loss 604 °C

The temperature coefficient of the weight loss, as seen in Fig. 4, can be extrapolated to an activation energy of 150 kJ/mol or 36 kCal/mol in good agreement with a published correlation (9).

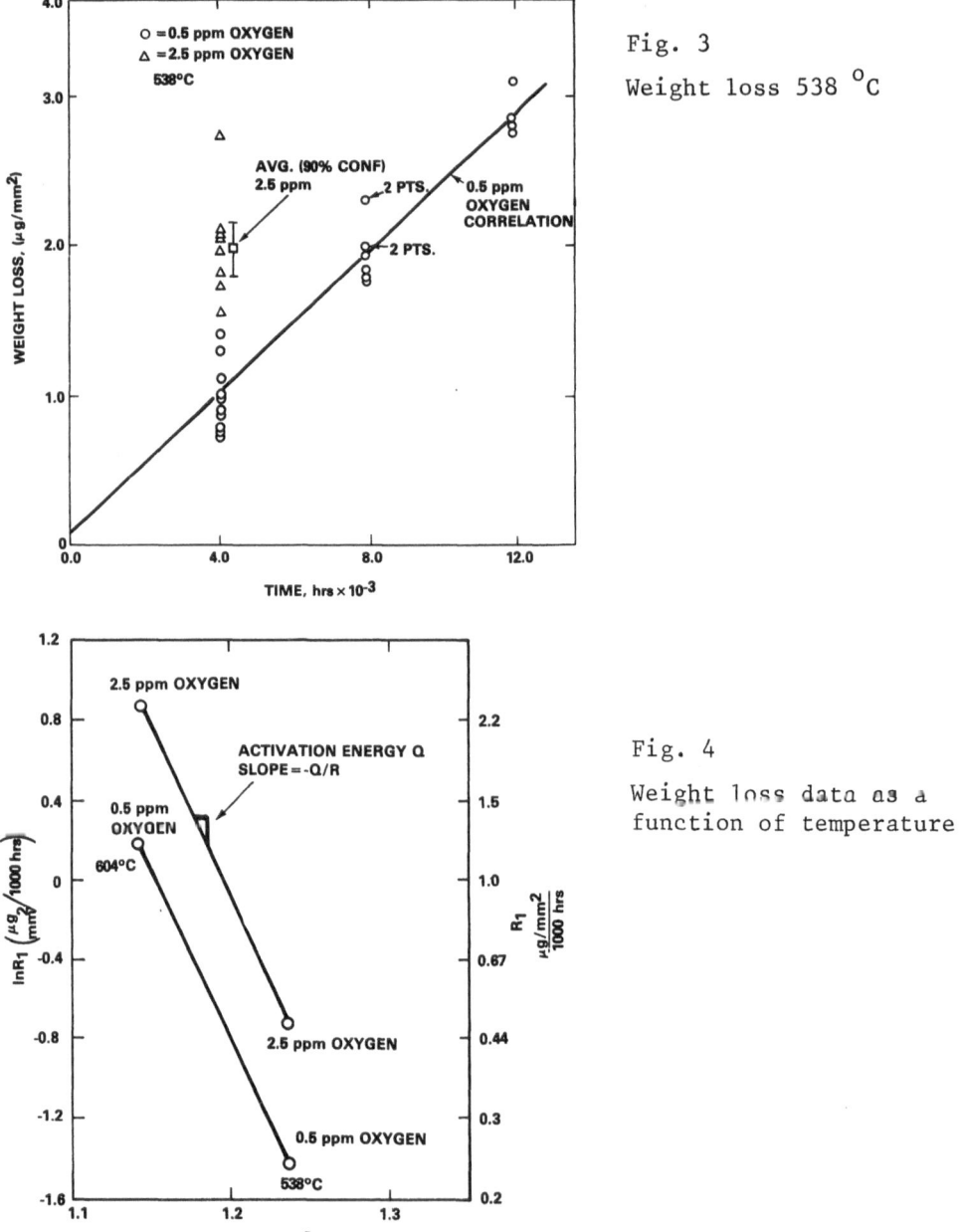

Fig. 3

Weight loss 538 $^{o}$C

Fig. 4

Weight loss data as a function of temperature

Weight Loss of Alloy Constituents

    Least-squares fits for linear and logarithmic correlations for individual element loss (same units as before) have been developed and are given below for 604 $^{o}$C.

$$604\ ^\circ C$$

|  | 0.5 ppm oxygen | 2.5 ppm oxygen |
|---|---|---|

Fe $R=(0.14 \underline{+}0.03)t + (0.19 \underline{+}0.11)$     $R=(1.28 \underline{+}0.12)t + (-0.87 \underline{+}0.43)$

   $R=(0.25^{+0.04}_{-0.03})t$ $^{(0.71 \underline{+}0.11)}$     $R=(1.30^{+0.33}_{-0.26})t$ $^{(0.59 \underline{+}0.19)}$

Cr $R=(0.48 \underline{+}0.03)t + (0.61 \underline{+}0.09)$     $R=(0.74 \underline{+}0.03)t + (0.39 \underline{+}0.10)$

   $R=(1.12 \underline{+}0.05)t$ $^{(0.60 \underline{+}0.03)}$     $R=(1.14 \underline{+}0.06)t$ $^{(0.78 \underline{+}0.05)}$

Ni $R=(0.46 \underline{+}0.03)t + (0.47 \underline{+}0.12)$     $R=(0.63 \underline{+}0.03)t + (0.37 \underline{+}0.11)$

   $R=(0.88 \underline{+}0.04)t$ $^{(0.71 \underline{+}0.03)}$     $R=(1.01 \underline{+}0.04)t$ $^{(0.73 \underline{+}0.04)}$

Mn $R=(0.061 \underline{+}0.005)t + (0.07 \underline{+}0.02)$   $R=(0.10 \underline{+}0.01)t + (0.04 \underline{+}0.03)$

   $R=(0.11 \underline{+}0.004)t$ $^{(0.78 \underline{+}0.03)}$     $R=(0.13 \underline{+}0.01)t$ $^{(0.86 \underline{+}0.05)}$

The four columns do not sum exactly to the total weight loss because we have not computed Mo, Si or C loss. It can be seen, however, that the only element with a significant increase in release rate at a fivefold increase in oxygen content is iron. This result is consistent with previous hypotheses of Weeks and Isaacs (1). It will also be noted that more chromium and nickel than iron are released at 0.5 ppm oxygen; this result is consistent with the observed preferential leaching, a solubility-driven corrosion mechanism and the lack of sensitivity of weight loss to oxygen level at < 1 ppm.

## 54Mn Release

The 54Mn was preferentially (superstoichiometrically) released from the 316 SS at all temperatures and oxygen conditions. The time rate of release is shown in Fig. 5. Logarithmic correlations for 54Mn release, normalized to percent of specimen inventory are as follows:

$604^\circ C$, 0.5 ppm oxygen - $\log_{10}R=0.703 \underline{+}0.02 \log_{10}t-(0.556 \underline{+}0.012)$
$604^\circ C$, 2.5 ppm oxygen - $\log_{10}R=0.792 \underline{+}0.04 \log_{10}t-(0.482 \underline{+}0.02)$
$538^\circ C$, 2.5 ppm oxygen - $\log_{10}R=0.67 \underline{+}0.12 \log_{10}t-(1.04 \underline{+}0.11)$

There was no significant difference in release of 54Mn at different oxygen levels at either temperature. The superstoichoimetric release (about a factor of 3:1) was caused by attaining a very low concentration of 54Mn at the surface of the material and outward diffusion of 54Mn into the sodium. Normalized diffusion profiles for 54Mn are shown in Fig. 6 and 7. The inflection in the curve near the edge is the ferrite-austenite boundary. Note that the ferrite layer thickness is <2 μm at 604 $^\circ C$ and <1 μm at 538 $^\circ C$. The difference in 54Mn total release between 538 $^\circ C$ and 604 $^\circ C$, at equivalent times and 0.5 ppm oxygen is about a factor of three (3), a less steep temperature dependence than for weight loss.

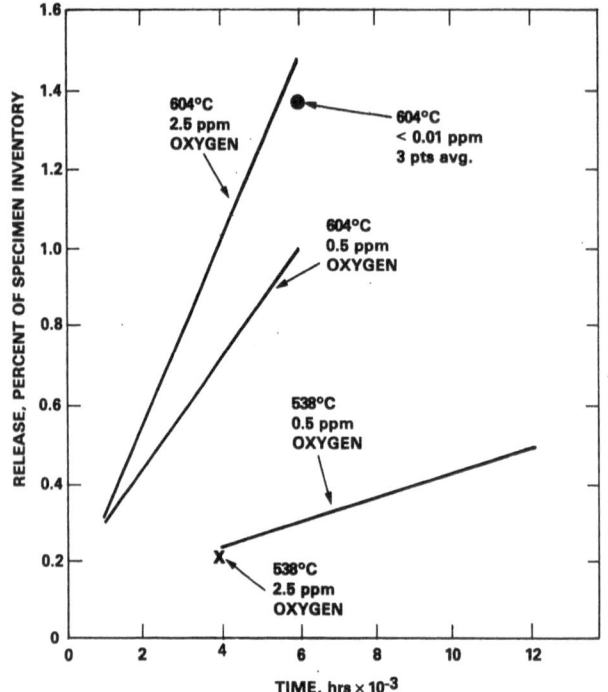

Fig. 5

$^{54}$Mn Release

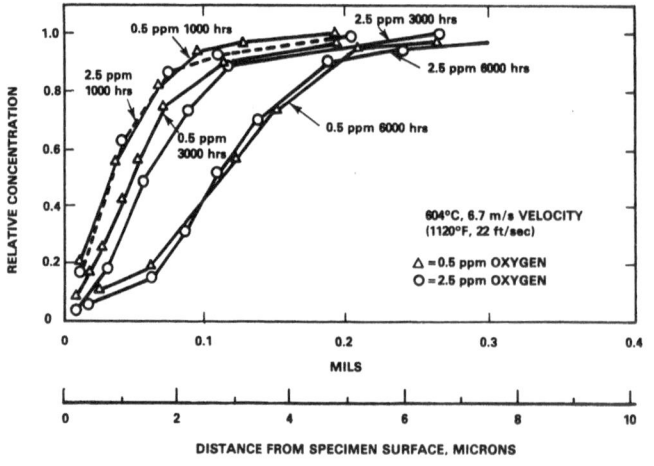

Fig. 6

Growth of
depleted zone

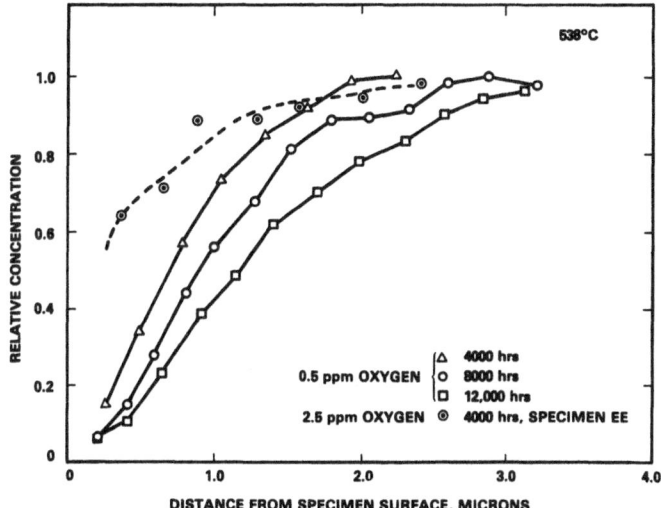

Fig. 7

Growth of depleted
zone for 54Mn
depletion at 538 $^{\circ}$C

The time exponent of 0.67 to 0.79 on the 54Mn release curves
suggest a mixture of surface reaction and grain boundary diffusion.
The factor of three difference in 54Mn release corresponds to an
activation energy of 196 KJ/mol[x)] or 47 Kcal/mol[x)]; in good agreement with Smith and Hales data (10) for 54Mn grain boundary
diffusion, which should be the dominant diffusion mechanism at
these temperatures. The diffusion coefficient was higher than the
Smith and Hales value, however. The reason for this result is
discussed in the section following 60Co release.

60Co Release

The 60Co release data, expressed in linear form, are as
follows (R in percent of specimen inventory, t in 1000 hour units):

604$^{\circ}$C, 0.5 ppm oxygen R= (0.007 $\pm$ 0.002)t + (0.038 $\pm$ 0.010)
604$^{\circ}$C, 2.5 ppm oxygen R =(0.045 $\pm$ 0.009)t + (0.01 $\pm$ 0.03)
538$^{\circ}$C, 2.5 ppm oxygen R =(0.0076$\pm$ 0.002)t + (0.005 $\pm$ 0.014)

These data are plotted in Fig. 8. At both temperatures, a
significant effect of oxygen on 60Co release was observed between
0.5 and 2.5 ppm oxygen, but 60Co release was not reduced at <0.01 ppm

[x)]The diffusional release flux is proportional to $D^{1/2}$

oxygen; in fact, it was increased substantially. The release of 60Co was substoichoimetric, that is, 60Co was preferentially retained in the steel, with less preferential retention at 538 $^{\circ}$C.

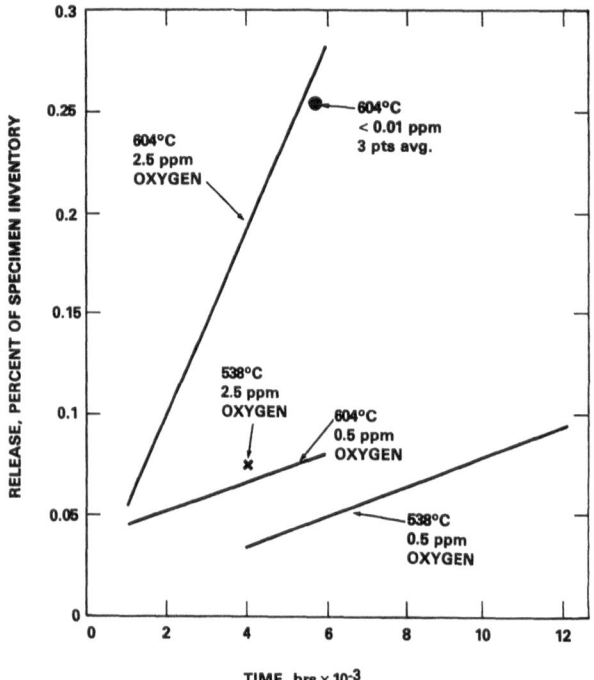

Fig. 8

60Co Release

This result can be seen in Fig. 9. However, there were large uncertainty bands in the 60Co release data. As a result of the lower amount of preferential retention, the reduction in 60Co release was only a factor of 1.7 between 604 $^{\circ}$C and 538 $^{\circ}$C. Apparently the solubility of cobalt or activity of cobalt in steel changes in this temperature range such that the cobalt prefers to remain in the steel at higher temperature, and would show more nearly stoichiometric corrosion at lower temperature. This fact needs to be kept in mind when estimating radioactive material transport in a reactor. The normalized ratio of Co/Mn release was about ten (10) at 604 $^{\circ}$C and about five (5) at 538 $^{\circ}$C. The result that 60Co release was sensitive to oxygen level cannot be explained readily since the oxide of cobalt is thermodynamically less stable than that of sodium and iron. However, since cobalt, iron and molybdenum are all preferentially retained in the steel, an Fe-Co-Mo rich surface develops, and one hypothesizes that the cobalt and iron are released together in particulate or complex form. This hypothesis could explain the sensitivity of 60Co release to oxygen (except at < 0.01 ppm oxygen) and the tendency of 60Co to be picked up in magnetic fields around pumps and flowmeters.

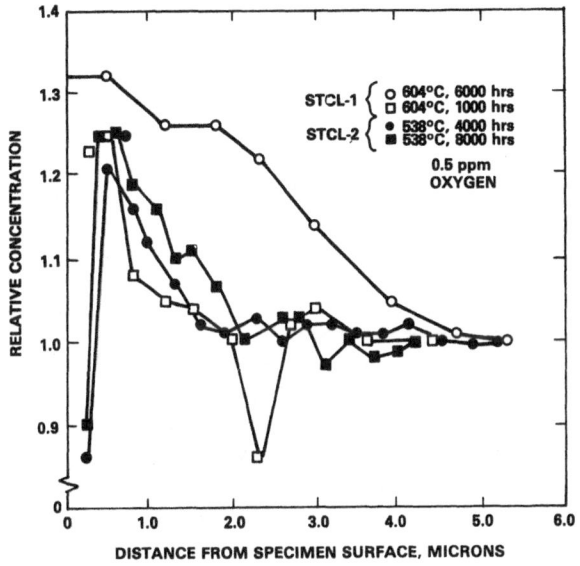

Fig. 9

Comparison of 60Co
accumulation at 604 °C
and 538 °C

## Low Velocity Data/Evidence of Enhanced 54Mn Diffusion

Leakage of sodium around a regenerative heat exchanger tube-
to-shell junction in STCL-2 diverted the sodium flow from the test
section, causing the sodium velocity near the specimens to drop
to approximately 0.6 m/sec. The test parameters were 604 °C hot
leg temperature, 0.1 - 0.3 ppm oxygen in sodium, and 2,450 hours
test time. The defective weld junction was discovered during the
test and was repaired prior to the tests at 538 °C in STCL-2. The
data obtained in the low velocity test are compared below with the
usual high velocity corrosion correlations from STCL-1. It is
apparent from the data that the material release rates were much
lower at the low velocity. This is especially true for the weight
loss, which is a factor of 15 lower at the low velocity, whereas
the 54Mn release decreased only by a factor of 4.

## Comparison of Low- and High-Velocity Corrosion and Nuclide Release Rates

| Low-Velocity (<1 m/sec) 2450 hs,604°C,0.5-0.3ppm oxygen | | High-Velocity (6.7 m/sec) 3000 hs,604°C,0.5ppm oxygen |
|---|---|---|
| R | | |
| (4 irradiated specimens) | | |
| $\mu g/mm^2$ | 0.26 | 4.7 |
| 54Mn release % | 0.15 % | 0.60 % |
| estimated 54Mn diffusion coefficient $cm^2/sec$ | $3 \times 10^{-16}$ one specimen | $1-5 \times 10^{-15}$ |

In Fig. 10 are plotted the 54Mn diffusion profiles at the two
velocities. The "depleted zone" thickness is smaller in the low
velocity specimen, which accounts for the observed difference in
54Mn release. The diffusivities of 54Mn from specimens at the two
velocities also differ by a factor of 10. The 54 Mn diffusivity in
the low velocity specimen agrees well with that reported by Smith
and Hales (10). Fig. 11 shows a comparison of subsurface concentrations of Cr and Ni at the two velocities. The depletion of Ni and
Cr near the surface is obviously much larger at the high velocity.
The depletion of Ni and Cr not only lead to a ferrite layer
formation near the surface, but also excess vacancies within the
austenite adjacent to the ferrite layer. Although austenite
vacancies and ferrite layer formation occur at the low velocity,
the thickness of ferrite layer and the number of vacancies in
austenite are considerably smaller at the low velocity. This can
account for enhanced 54Mn diffusivity at the high velocity,
resulting in the observed greater 54Mn release. It is to be
remembered that the "high" velocity of 6.7 m/s is typical of sodium
in reactor fuel pin bundles.

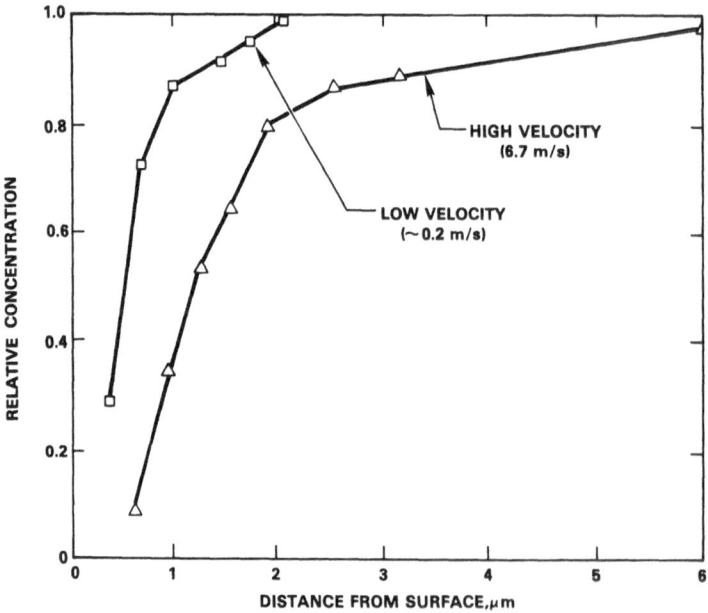

Fig. 10

54Mn Diffusion
Comparison
− 604 °C

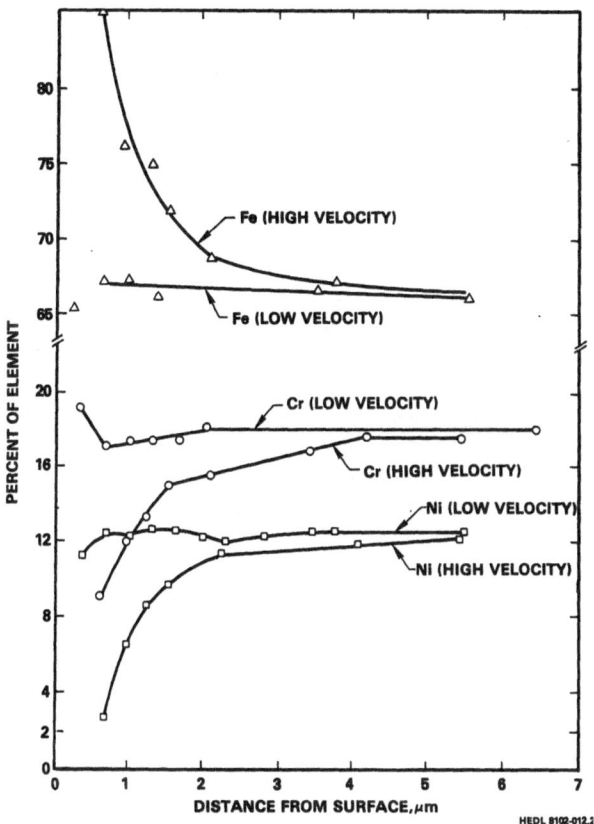

Fig. 11

Elemental Diffusion Profiles

HEDL 8102-012.2

## CONCLUSIONS

The results of this study show that corrosion of austenitic stainless steel in sodium is a complex process involving several mechanisms for the different alloy constituents. The selective and enhanced release of 54Mn can be explained by accelerated diffusion in the outer layer of corroding steel; the enhanced diffusion is in turn caused by vacancies left behind by preferential release of Cr and Ni. It is very likely that 60Co is released with iron during the corrosion process. The results of the study confirmed many of the hypotheses of Weeks and Isaacs (1) concerning the role of oxygen in corrosion of steels by sodium.

## ACKNOWLEDGEMENT

This work was sponsored by the United States Department of Energy under Contract Number DE-AD14-76FF02170.

REFERENCES

1.  J. R. Weeks and H. S. Isaacs, Advances in Corrosion
    Science and Technology, Vol. 3, pp 1-66 (1973).
2.  R. S. Fidler, M. J. Collins, Atomic Energy Review, Vol.3,
    pp 3-50.
3.  W. F. Brehm, "Radioactive Corrosion Product and Control",
    paper IVB-1, Proc.Int.Conf.on Liquid Metal Technology
    in Energy Production, Champion, PA, USA (1976) (CONF-
    760503.
4.  W. F. Brehm, et al, "Radioactive Mass Transport in Sodium-
    Cooled Nuclear Reactors", HEDL-SA-1870, Second Int.
    Conf.on Liquid Metal Technology in Energy Production,
    Richland, WA, USA (1980).
5.  H. Feuerstein et al, Atomic Energy Review, Vol. 17, No. 3,
    pp 698-762, September, 1979.
6.  D. L. Smith and R. H. Lee, ANL-7891 (1972).
7.  W. F. Brehm, et al, "Techniques for Studying Corrosion and
    Deposition of Radioactive Materials in Sodium Loops",
    paper 18, Proc. 1 AEA, Specialists Meeting on Fission
    and Corrosion Product Behaviour in Primary Systems of
    LWFBR's, Dimitrovgrad USSR (1975) IWGFR-7.
8.  L. E.Chulos, "Operational Techniques Employed for the
    Liquid Sodium Source Term Control Loops", paper 11-A-8
    in Ref.4 proceedings.
9.  C. Bagnall, D. C. Jacobs, WARD-NA-3045-23 (1973).
10. A. F. Smith and R. Hales, Metal Science, Vol. 9, pp 181-84
    (1975).

# CORROSION AND MASS TRANSFER IN LIQUID SODIUM AS A FUNCTION OF THE CHEMICAL COMPOSITION OF THE SURFACE OF AUSTENITIC STEELS AND ALLOYS

Boleslav Eremiás, Miroslav Fresl and Frantisek Storek

G. V. Akimov State Research Institute of the Protection of Materials and State Research Institute of the Iron Metallurgy, Prague, CSSR

## INTRODUCTION

During the exposure of austenitic steels and alloys in high-temperature sodium, we can observe not only weight changes but also changes in the composition of the surface concentration of main and minor alloying elements. These changes result from the selective extraction of elements, from the diffusion and mass transfer of alloying elements, or from the production of scales and deposits. The knowledge of these changes allows a better understanding of the corrosion mechanism and of the mass transfer in austenitic steels and alloys in liquid sodium.

Therefore, the Sodium Laboratory of the State Research Institute for the Protection of Materials has reserved a part of the experimental programme for the study of these problems. For this study two types of materials were chosen: unstabilized 18Cr14Ni3Mo steels with different surface preparation and six variants of austenitic alloys corresponding approximately to the composition of Alloy 800.

## EXPERIMENTAL

Austenitic steel specimens (see Table 1), used for a 6000 h exposure to sodium at 700 °C, were 60 mm long tube specimens (6 mm od by 0.3 mm wall) with a different state of the surface:

- as received, but without final annealing and with 25 to 30 % cold deformation (designated "Z")

- as received (with final annealing at 1050 °C in split ammonia in a continuous furnace (designated "P")

- after final annealing at 1050 °C in split ammonia, however with a surface treatment by electrochemical polishing of the internal surface and mechanical polishing of the external surface (designated "M")

Austenitic alloys corresponding approximately to the composition of Alloy 800 (see Table 1), used for a 1000 h exposure to sodium at 600 °C, were flat slabs 60 x 12 x 0.5 mm with gauge length 40 mm and total surface area about 9 cm². By contrast to tube specimens, they were abraded with metallographic papers down to 3/0.

Both types of specimens were degreased with gasoline and acetone in the soxhlet apparatus and weighed.

Table 1.   The compositions of the materials tested

| wt % | C | Mn | Si | Cr | Ni | Mo | Ti | N | Al |
|------|-----|-----|-----|------|------|------|------|------|------|
| Z,M | 0.063 | 1.03 | 0.22 | 17.58 | 13.83 | 2.56 | - | 0.087 | - |
| P | 0.063 | 1.03 | 0.22 | 17.58 | 13.83 | 2.56 | - | 0.097 | - |
| 13 | 0.082 | 0.78 | 0.62 | 20.70 | 31.56 | - | 0.63 | 0.061 | 0.24 |
| 13E a) | 0.091 | 0.76 | 0.65 | 20.85 | 32.68 | - | 0.64 | 0.055 | 0.23 |
| 13V b) | 0.082 | 0.74 | 0.67 | 20.04 | 32.35 | - | 0.63 | 0.059 | 0.24 |
| 14 | 0.029 | 1.36 | 0.35 | 20.30 | 32.53 | - | 0.56 | 0.074 | 0.25 |
| 14E | 0.031 | 1.36 | 0.39 | 20.61 | 31.84 | - | 0.59 | 0.075 | 0.21 |
| 14V | 0.047 | 1.15 | 0.34 | 19.77 | 32.85 | - | 0.56 | 0.055 | 0.22 |

a) Symbol E for electro-slag remelted heats
b) Symbol V for vacuum arc remelted heats

TESTING EQUIPMENT

Tube specimens were placed in special cassettes allowing the exposure from the sodium side of external as well as internal wall the tubes. The cassettes with samples were placed in the hot leg of a convection sodium loop (1) made of CSN 17246 (0.1C/19Cr/9Ni/Ti) steel. In individual sections of the loop there were the following temperatures in the course of the corrosion tests:

a) 6000 h experiment
   in hot leg:  700 °C
   in cold leg: 600 °C
   in dilatation vessel: 400 °C
   in cold trap: 110 °C

b) 1000 h experiment
   in hot leg:  600 °C
   in cold leg: 500 °C
   in dilatation vessel: 400 °C
   in cold trap: 110 °C

In view of the determination of the temperature-dependence of the solubility of oxygen in liquid sodium in the temperature range of 110 - 400 $^{\circ}$C by means of vacuum distillation method developed in our Institute (2) the concentration of oxygen $C_{o/Na}$ in our sodium convection loops is regulated by the temperature of the cold traps T, using the relationship

$$\log C_{o/Na}/ppm/ - 6.318 - 2355/T \qquad (1)$$

For the said temperatures of the cold trap, therefore the content of oxygen in sodium was presumed to be about 1.5 ppm in both experiments. The sodium content in both experiments was about 6 1, the velocity of sodium circulation during the corrosion tests was about 0.1 m/s.

The 6000 h experiment was divided into two parts, i.e. the convection loop was opened after 2000 h of the exposure, a half portion of samples was taken out of the loop under an argon atmosphere for evaluation and the cassettes were completed by new samples. The remaining and new samples were then exposed to sodium for 4000 h so that evaluation of the exposure to sodium could be done after 2000 h as well as 4000 h and 6000 h exposures to sodium at 700 $^{\circ}$C. The specimens taken out of the loop were cleaned with ethyl alcohol and water, dried and weighed. Besides this the samples of each material placed at the entry to the hot leg were analyzed on the EMX-SM microprobe with the aid of crystal spectrometers. For the surface analysis, the accelerating voltage of 15 kV was chosen, the current to the sample of about 200 nA and the electron beam diameter, about 2 - 3 microns. In the course of the analysis, the rastered area of the surface was 100 x 100 μm so that the average analysis of the surface layer up to the depth of 1 - 2 μm was obtained as a result. Maximal reduction of the roughness of the sample surface was achieved by pressing the samples surface between mirror-polished surfaces of steel hammers with a hardness of HRc 63 with the use of a pressure of 700 MPa. Thus adjusted surfaces were essentially planar and with the help of standards polished by usual procedures it was possible to obtain well applicable results of the analyses.

In the 1000 h experiment similarly as in the 6000 h experiment, the specimens were taken out after the exposure to sodium of the loop under an argon atmosphere, cleaned with ethyl alcohol and water, dried and weighed. Besides this the samples of each material placed at the entry to the hot leg were not only analysed by microprobe, but also checked for average carbon and nitrogen contents after exposure to sodium by coulometric determination (carbon) and by a distillation method (nitrogen). Prior to it mechanical properties were measured using an electronic tensile testing machine (Instron) and compared with the values obtained on the same samples after the same times of exposure in argon at 600 $^{\circ}$C.

RESULTS AND DISCUSSION

Table 2 presents a time dependence of weight changes for austenitic steel with three different states of the surface. Table 3 presents for comparison data of the initial surface composition and surface composition after 2000 h, 4000 h and 6000 h of the exposure to sodium at 700 $^{\circ}$C. These data represent mean values obtained from the microanalysis of the external and internal surfaces of tube specimens prior to the exposure and after it.

Table 2.   Time dependence of weight losses ($mg/dm^2$) of austenitic steels with different state of the surface in liquid sodium at 700 $^{\circ}$C

| State of the Surface | Time of Exposure | | |
|:---:|:---:|:---:|:---:|
| | 2000 h | 4000 h | 6000 h |
| Z | 47.86 | 79.32 | 110.53 |
| P | 54.96 | 76.31 | 92.90 |
| M | 50.94 | 71.40 | 86.04 |

Whereas the time dependence indicated these mean values to be different for different surface treatments of the same type of the steel, for the initial chemical composition the mentioned mean value obtained from the microanalysis of the external and internal surfaces was the same.

Table 3.   Comparison of the chemical composition of the austenitic steel with different state of the surface in the course of the exposure to sodium at 700 $^{\circ}$C

| Surface state | Concentration at surface (wt %) | | | | | | Exposure time (h) |
|:---:|---|---|---|---|---|---|:---:|
| | Fe | Cr | Ni | Mo | Mn | Si | |
| | 67.22 | 16.96 | 11.90 | 3.41 | 0.41 | 0.10 | 2000 |
| Z | 69.59 | 15.18 | 10.60 | 4.31 | 0.27 | 0.05 | 4000 |
| | 71.57 | 13.50 | 9.37 | 5.30 | 0.25 | 0.01 | 6000 |
| | 67.41 | 16.32 | 11.37 | 4.40 | 0.40 | 0.10 | 2000 |
| P | 68.89 | 15.00 | 10.37 | 5.39 | 0.30 | 0.05 | 4000 |
| | 70.00 | 14.00 | 9.62 | 6.12 | 0.25 | 0.01 | 6000 |
| | 67.27 | 16.80 | 11.37 | 4.03 | 0.43 | 0.10 | 2000 |
| M | 68.76 | 15.70 | 10.36 | 4.86 | 0.27 | 0.05 | 4000 |
| | 69.69 | 14.93 | 9.62 | 5.50 | 0.25 | 0.01 | 6000 |
| Z,P,M | 63.23 | 18.76 | 13.21 | 2.44 | 1.73 | 0.47 | 0 |

With respect to the fact that a "normal" corrosion process governed by the solubility will occur, the results presented in Tables 2 and 3 were interpreted on the basis of the known data concerning the effect of the content of corrosion products of steels in sodium on their corrosion behaviour. The attention was paid to Soviet works (3, 4) where it was found that the weight changes resulting from corrosion processes are different with different saturation of sodium by alloying elements removed from the surface layers of the materials exposed to the high-temperature sodium. On the basis of this fact and of a conclusion that the concentration of components in the steel surface for a given time of the exposure to sodium is different for different saturation of sodium in respect to these components, the correlation can be considered for observed weight changes of the steel according to the following relationship

$$\Delta m_t = k \sum_{i=1}^{n} (x_i^o - x_{i,t}) \, x_{Fe,t} \left[ 1/x_{Fe}^o \right] \qquad (2)$$

where $\Delta m_t$ = weight change of the steel sample after time t of the exposure to sodium $(mg/dm^2)$ ;
$x_i^o$ = mol fraction of the i-th component in the sample surface prior to the exposure to sodium (the same for iron, i.e. $x_{Fe}^o$);
$x_{i,t}$ = mol fraction of the i-th component in the sample surface after time t of the exposure to sodium (the same for iron, i.e. $x_{Fe,t}$);
k = ratio of the weight change of the sodium surface layer on the steel sample to mol fraction changes of individual components in the same layer $(mg/dm^2)$.

By treating the data from Tables 2 and 3 one could find a value k for all the three states of the surface of the austenitic steel after 2000 h, 4000 h and 6000 h of the exposure to sodium at 700 $^{\circ}$C. These values are listed in Table 4 and from them it is obvious that at given sodium parameters (temperature, oxygen content in the sodium and the sodium flow velocity) and given state of the steel surface k is a time-independent constant, that may be used for calculating weight changes from mol fraction changes of individual components in the steel surface after any time of the exposure to sodium.

Table 4. Values of k, of austenitic steel of Table 3

| Surface state | Time of exposure | | |
| --- | --- | --- | --- |
| | 2000 h | 4000 h | 6000 h |
| Z | 987.78 | 987.70 | 987.70 |
| P | 987.65 | 987.65 | 987.72 |
| M | 987.68 | 987.67 | 987.67 |

In our experiment the material specimens exposed 1000 h at
600 °C to liquid sodium with an oxygen content of 1.5 ppm and a
flow velocity of 0.1 m/s, indicated different weight gains in a
range of 10.8 to 16.2 mg/dm$^2$ (see Table 5) depending on the chemical
composition of the surface prior to the exposure. In Table 6 there
are data from the surface microanalysis prior to the exposure and
after it for investigated materials. It is obvious from Table 6, that
during the exposure to sodium the concentration of Ni, Mn, Fe, Si,
Fi and Al are decreased as a result of the selective removal of
these elements by sodium, whereas the surface is enriched in chromium
and oxygen. Even in this case, these data were interpreted according
to relationship (2), but considering oxygen as a component present
in the sample surface after the exposure to sodium and observed weight
changes as a negative ones. The value of k was found other than in
the case of stainless steels, nevertheless it was essentially independent of moderate deviations in the chemical composition of investigated materials (see Table 5). The treatment of data from Tables
5 and 6 can demonstrate the effect of the chemical composition of
the surface prior to the exposure to sodium when assuming the application of a concept of the effect of concentration of main
alloying elements (Ni, Fe, Mn, Si, Al, Ti) on the diffusion penetrability of scales on basis modified $Cr_2O_3$ - the assumed oxidation
product on the alloy 18Cr35Ni47Fe (5). This concept assumes that
the main and minor alloying elements which are present in the surface prior to the exposure can be incorporated into $Cr_2O_3$ and affect
its diffusion penetrability in either positive or negative sense,
but always in the ratio of their mol fractions in the surface prior
to the exposure.

Table 5.    Weight gains (mg/dm$^2$) and value of k for alloys corres-
            ponding to Alloy 800 after 1000 h exposure to sodium at
            600 °C

| Material | Δm | k | Material | Δm | k |
|----------|-------|--------|----------|-------|--------|
| 13 | 12.11 | 149.06 | 14 | 10.75 | 149.08 |
| 13E | 16.15 | 149.06 | 14E | 12.77 | 149.11 |
| 13V | 14.45 | 149.15 | 14V | 12.82 | 149.10 |

According to this concept the weight changes in its absolute
value can be expressed using the relationship

$$|\Delta m| = y_{Ni}x_{Ni}^o + y_{Mn}x_{Mn}^o + y_{Fe}x_{Fe}^o + y_{Si}x_{Si}^o + y_{Al}x_{Al}^o + y_{Ti}x_{Ti}^o \quad (3)$$

where $|\Delta m|$ = weight change in its absolute value after the exposure
of the sample to sodium ( mg/dm$^2$ );
$y_{Ni} \ldots y_{Ti}$ = relative contributions from alloying elements which
can affect change in weight due to their presence in the surface

prior to the exposure and their transfer to scales formed during the exposure $_o$ mg/dm$^2$;
$x_{Ni}^o$ ... $x_{Ti}^o$ = mol fractions of these alloying elements in the sample surface prior to the exposure to sodium.

Table 6.  Comparison of the chemical composition of the specimens of Table 5 prior to 1000 h exposure to sodium at 600 $^o$C and after it

| Material | | Concentration at surface (wt %) | | | | | | |
|---|---|---|---|---|---|---|---|---|
| | | Fe | Cr | Ni | Mn | Si | Ti | Al |
| 13 | before | 48.39 | 16.04 | 33.80 | 0.77 | 0.35 | 0.45 | 0.19 |
| | after | 23.80 | 23.03 | 18.00 | 0.48 | 0.23 | 0.16 | 0.15 |
| 13E | before | 49.34 | 16.57 | 33.38 | 0.73 | 0.32 | 0.46 | 0.20 |
| | after | 27.75 | 23.22 | 12.74 | 0.44 | 0.20 | 0.17 | 0.16 |
| 13V | before | 46.68 | 16.05 | 35.08 | 1.07 | 0.45 | 0.45 | 0.20 |
| | after | 25.10 | 23.14 | 17.70 | 0.40 | 0.51 | 0.16 | 0.17 |
| 14 | before | 48.85 | 15.26 | 34.26 | 0.86 | 0.22 | 0.34 | 0.20 |
| | after | 20.95 | 22.98 | 16.92 | 0.50 | 0.10 | 0.13 | 0.17 |
| 14E | before | 48.07 | 15.26 | 34.98 | 0.93 | 0.21 | 0.37 | 0.19 |
| | after | 22.45 | 23.26 | 17.65 | 0.50 | 0.10 | 0.14 | 0.15 |
| 14V | before | 49.18 | 15.29 | 33.59 | 0.98 | 0.33 | 0.35 | 0.26 |
| | after | 21.87 | 24.24 | 14.64 | 0.61 | 0.18 | 0.14 | 0.21 |

Table 7 shows the measured values of $y_{Ni}$, $y_{Mn}$, $y_{Si}$, $y_{Al}$ and $y_{Ti}$. Based on the sign and magnitude of these parameters, it is possible to draw a conclusion that under conditions of the experiment carried out, the diffusion penetrability of scales is increased by the presence of Ni, Ti and particularly Al, whereas Fe, Si and particularly Mn reduce this penetrability.

Table 7.  Relative contributions from alloying elements to scales formed during the exposure  (mg/dm$^2$)

| $y_{Ni}$ | $y_{Mn}$ | $y_{Fe}$ | $y_{Si}$ | $y_{Al}$ | $y_{Ti}$ |
|---|---|---|---|---|---|
| 36.863,5 | -324.879,1 | -25.256 | -45.814,5 | 730.930,7 | 73.496,5 |

There is ample evidence that tensile strength as well as yield strength of Alloy 800 can be increased and the ductility decreased after the long-term exposure to temperatures of 566 to 621 $^o$C (6). This effect of the temperature is explained by the precipitation of

the phase $Ni_3(Al,Ti)$, whose presence was demonstrated experimentally
(6). From the standpoint of prognoses of service life, for this type
of material so far practically no reliable information has been
available on the effect of the carbon and nitrogen transfer on the
values of the mechanical properties after periods of high-temperature
exposures in sodium in spite of a number of works relating to the
austenitic steels (7, 8). With respect to this fact an attempt was
made to correlate the values of the 0.2 % yield stress and ductility
with the carbon + nitrogen contents in samples measured after exposure in argon (unchanged carbon + nitrogen contents) and after
exposure in sodium (changed carbon + nitrogen contents) at 600 $^\circ$C
and for 1000 h. The established relationships are shown in Fig. 1
and 2. From both figures it is very clear that the mechanical properties after the exposure to sodium as well as argon can be considered as dependent on the chemical composition of the materials
being tested. In the six heats being studied, the heat 14 seems to
show the lowest change of mechanical properties after 1000 h exposure
to sodium at 600 $^\circ$C and simultaneously it also shows the lowest
carburization and denitrization as well as the lowest weight changes
(see Table 8).

Table 8.   Comparison of changes of the average content of carbon and
           nitrogen with weight gains of alloys corresponding to the
           composition of Alloy 800 after 1000 h of exposure to sodium
           at 600 $^\circ$C

| Material | $\Delta$C (wt %) | $\Delta$N (wt %) | $\|\Delta m\|$ (mg/dm$^2$) |
|----------|------------------|------------------|----------------------------|
| 13       | 0.007            | - 0.015          | 12.11                      |
| 13E      | 0.010            | - 0.020          | 16.15                      |
| 13V      | 0.009            | - 0.018          | 14.45                      |
| 14       | 0.005            | - 0.010          | 10.75                      |
| 14E      | 0.008            | - 0.016          | 12.77                      |
| 14V      | 0.008            | - 0.016          | 12.82                      |

CONCLUSION

    The results obtained have demonstrated that weight changes of
austenitic steels as well as alloys corresponding approximately to
Alloy 800 can be expressed as a function of the chemical composition
of the surface prior to the exposure to high-temperature sodium and
after it.

    From the results presented it can also be concluded that in the
case of Alloy 800 type a non-protective surface scales are probably
formed on these alloys during their 1000 h exposure to low velocity-
low oxygen sodium. In the presence of these scales changes in weight

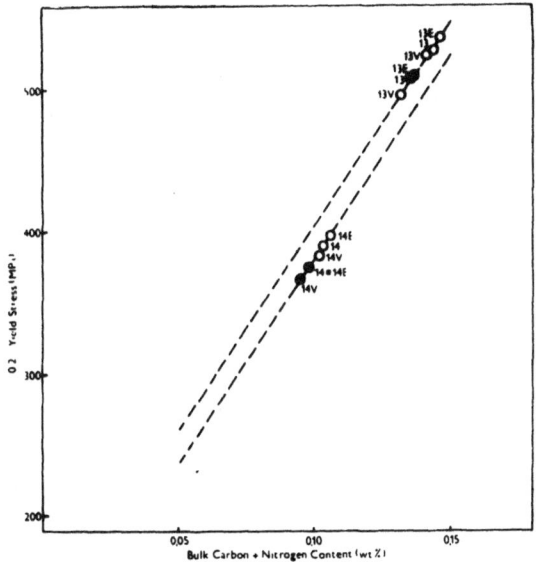

Fig. 1  Effect of the average C + N contents on the yield strength
        of the material tested after 1000 h in Ar (open circles)
        and after 1000 h in Na (full circles) at 600 °C

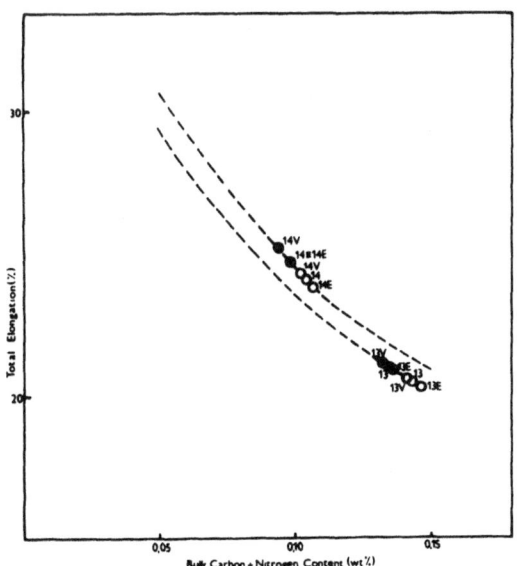

Fig. 2  Effect of the average C + N contents on ductility of the
        materials tested after 1000 h in Ar (open circles) and
        after 1000 h in Na (full circles) at 600 °C

depend on their diffusion penetrability, which is determined by the chemical composition of the surface prior to the exposure to sodium.

At the same time the results allow to conclude that the changes in the mechanical properties of these alloys after the exposure to sodium and argon can be evaluated in relation to the carburization and denitrization process occurring in sodium.

REFERENCES

1. G. Ilincev: Alkali Metal Coolants, Vienna, 1966 (IAEA, Vienna 1967) 131.
2. G. Ilincev: Kernenergie 18 (1975) 69.
3. V. N. Bykov, V. V. Zotov, N. D. Kraev: State and perspect trends of works on constructing Atomic Power Station with fast reactor (in Russian), Vol. 1, Obninsk 1968, 244.
4. N. D. Kraev, V. V. Zotov, O. V. Starkov: Kernenergie 21 (1978) 244.
5. W. B. A. Sharp: Corrosion Science 10 (1970) 283.
6. M. A. Cordovi: BNES Conference on Status Review of Alloy 800 &Reading, 1974) 1.
7. K. Natesan, T. F. Kassner, Che Yu Li: Reactor Technology 15, No. 4 (1972-73) 244.
8. B. Eremias, M. Fresl: J. Nucl. Mater 87 (1979) 345.

# PARTICULAR OBSERVATIONS ON THE BEHAVIOUR OF AUSTENITIC STAINLESS STEELS IN SODIUM CORROSION LOOP ML-1

Antonio Marin, Roberto Solano,
Segundo Barroso and Manolo de la Torre

Junta de Energia Nuclear
Madrid, Spain

## INTRODUCTION

The present corrosion program in the sodium loop ML-1 covered up to 9.000 hours with interruptions after each 1.000 hours in order to control the sodium corrosion effects on the tested materials. Shortly before reaching the programmed 9.000 hours a leak was detected in the high temperature test section (700 $^{o}$C) for which reason it was decided to interrupt the program.

In the evaluation of the specimens tested the effect of the sodium contaminated by the leak was considered of special concern in the corrosion processes.

## MATERIALS TESTED

The materials were selected in order to reproduce as closely as possible the materials found in the different components of the primary loop of a fast breeder reactor. Two austenitic stainless steels of German origin (1.4970 and 1.4981) and the nickel-base alloy Incoloy-800 were selected. The austenitic stainless steels 1.4970 and 1.4981 are considered as candidates for cladding of the fuel elements of the SNR-300 fast breeder reactor. The Incoloy-800 may be used as a structural material in the sodium cooled fast breeder reactors, because its good behaviour in water at high temperatures allows its use in the construction of water/sodium heat exchangers.

Table 1 shows the tested materials, their composition, test temperatures and exposure times.

Table 1.   Materials Tested in ML-1

| Material | Composition | Temperature ($^{\circ}$C) | Time (hours) |
|----------|-------------|---------------------------|--------------|
| 1.4970 | 15,15,Cr,Ni,Mo,Ti,B | 700$^{\circ}$C and 590 $^{\circ}$C | 9.000 hours |
| 1.4881 | 16,16,Cr,Ni,Mo,Nb | 700$^{\circ}$C and 590 $^{\circ}$C | 9.000 hours |
| Incoloy-800 | 18,32,Cr,Ni,Al,Ti | 570$^{\circ}$C and 500 $^{\circ}$C | 9.000 hours |

In the present work sheet specimens were studied according to the dimensions of the specimen holders of the ML-1. The sodium-exposed surface was 40 x 20 mm$^2$. The specimen thicknesses were between 5 and 3 mm.

EXPERIMENTAL CONDITIONS

The corrosion tests were carried out in the ML-1 Loop (1). This loop contains three independent circuits (Fig. 1), namely a main loop, a purification and corrosion loop. The main loop is a facility designed to provide the sodium velocity, temperature and impurity level conditions required in the testing of equipments and components exposed to dynamic sodium. It is fitted with piping, attachment points, electromagnetic pump, heaters, coolers and expansion tank. The purification loop is fittet with an electro-magnetic pump, heat exchanger, cold trap and plugging meter in order to maintain the required levels of purity. Finally, the corrosion loop is intended to provide the appropriate liquid-sodium environment necessary in the corrosion and mass transfer tests. It is fitted with an electromagnetic pump, heaters, coolers, sodium/sodium recuperative exchangers, expansion tank and an independent purification system of cold and hot traps. There are four test sections assembled in series with capacity for testing different materials, allowing the operation at different temperatures and velocities up to maximum values of 730 $^{\circ}$C (with a $\Delta t$ of 240 $^{\circ}$C) and 10 m/sec.

The particular design of this loop with four test sections assembled in series allows a correct selection of temperatures, in these sections, to simulate the primary system thermal gradients of a sodium cooled fast breeder reactor (Fig. 2).

EXPERIMENTAL WORK

The specimens were taken from the test sections at every interruption, cleaned with alcohol and running water until the water showed no alkalinity. Then they were dried in a vacuum oven and evaluated according to the following techniques:

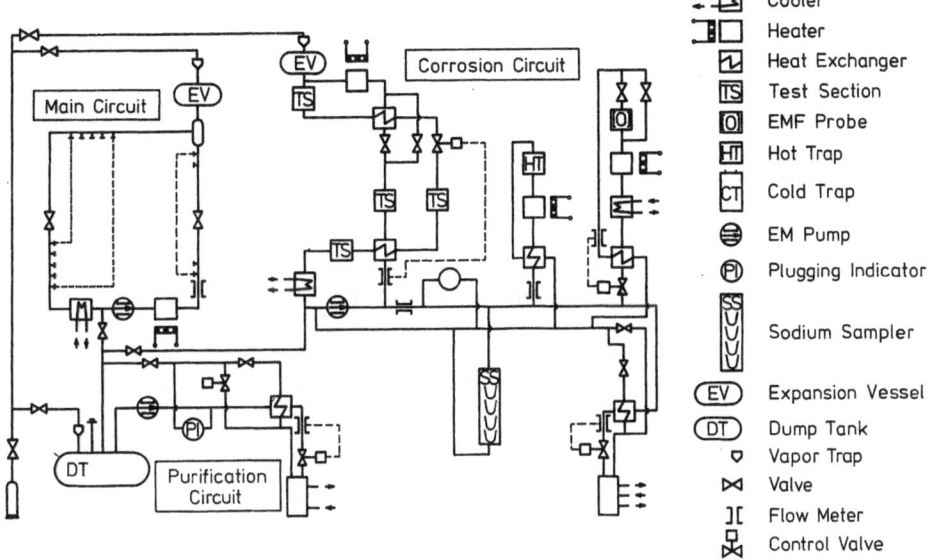

Cooler

Heater

Heat Exchanger

TS    Test Section

EMF Probe

HT    Hot Trap

CT    Cold Trap

EM Pump

PI    Plugging Indicator

Sodium Sampler

EV    Expansion Vessel

DT    Dump Tank

Vapor Trap

Valve

Flow Meter

Control Valve

Fig. 1.   Flow Diagram of ML-1

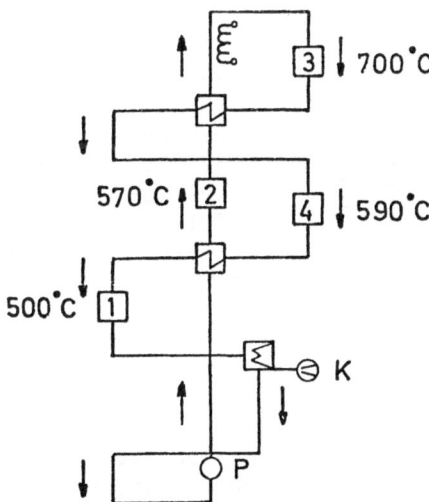

Fig. 2.   Test Sections of the
Corrosion Loop of ML-1

- Weight changes after each 1.000 hours of sodium exposure.
- Metallography.
- Scanning electron microscopy and energy dispersive X-ray analysis.
- Chemical analysis including surface layer analysis by glow discharge optical spectroscopy (GDOS) and Auger electron spectrometry (AES) (2, 3).

RESULTS

The specimen weight changes, increases in the deposition zones and decreases in the dissolution ones, allows "best fit" curves to be derived for the corrosion kinetics for the different test materials. The experimental weight change as a function of the exposure time gave the following parabolic equation (1):

$$\Delta W = A_o + A_1\left(\frac{t}{1000}\right) + A_2\left(\frac{t}{100}\right)^2 \tag{1}$$

where

$\Delta W$ = weight change $(mg/cm^2)$
$t$ = exposure time (hours)
$A_o$, $A$ and $A_2$ = empirical coefficients

In Table 2 the values of the empirical coefficients are given.

Table 2.  Coefficients of equation (1)

| Specimen | $A_o$ | $A_1$ | $A_2$ |
|----------|-------|-------|-------|
| A - 6 | - 0.125455 E$^{-1}$ | 0.92763 | - 0.282576 E$^{-1}$ |
| A - 14 | - 0.0625 | 0.058369 | - 0.463095 E$^{-1}$ |
| B - 4 | - 0.510909 E$^{-1}$ | 0.628106 | - 0.244697 E$^{-1}$ |
| B - 16 | - 0.641667 E$^{-1}$ | 0.499048 | - 0.264286 E$^{-1}$ |
| I - 5 | 0.657143 E$^{-1}$ | - 0.221429 E$^{-1}$ | 0.39285 E$^{-1}$ |
| I - 14 | - 0.838182 E$^{-1}$ | 0.1495 | - 0.901515 E$^{-1}$ |

From equation (1) the curves representating the corrosion kinetics of the materials were derived. Fig. 3 shows the curves corresponding to equation (1) as well as the experimental ones.

It can be seen that in the deposition zone, 590 $^o$C, the weight increase is continuous up to approximately 5.000 hours. After this time and up to 7.000 hours there are no significant changes and in the last part of the curve a significant decrease is observed. This anomalous behaviour during the last period of the test is interpreted as a consequence of contamination of sodium by the leak, which altered the established corrosion kinetics.

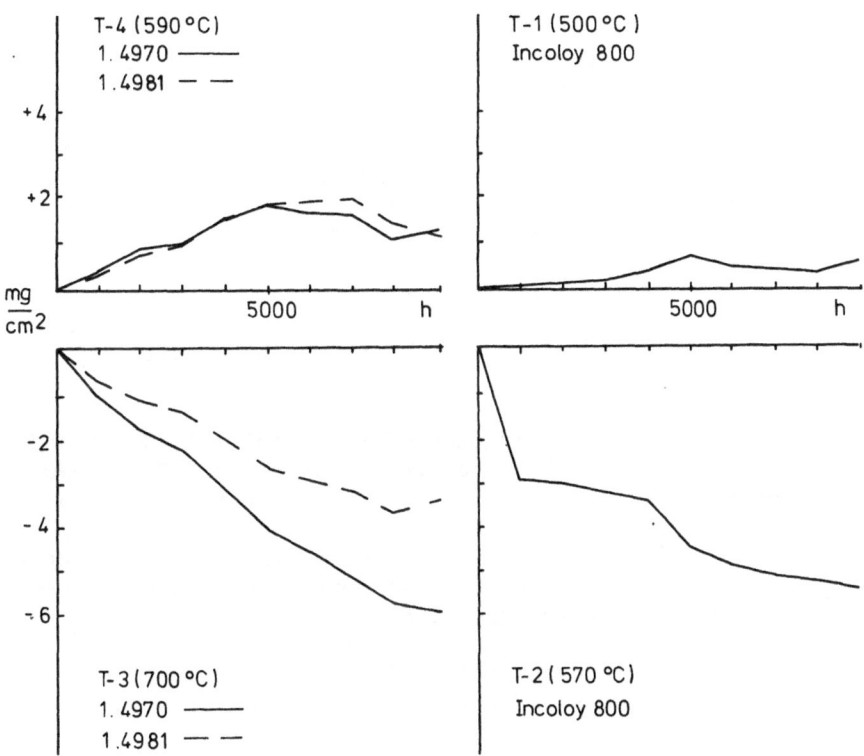

Fig. 3.   Kinetics of corrosion and deposition

In the test section corresponding to the dissolution process
a uniform loss of weight was observed for both materials, up to
8.000 hours, with the slope of the curve corresponding to the
material 1.4970 being higher than the corresponding to 1.4981.
Between 8.000 and 9.000 hours a decrease in the slope is observed
in the material 1.4981 which was much more significant than for
the material 1.4970. This can be explained by the formation of a
surface reaction layer which protects, in some way, the material.
This hypothesis will be reinforced later on with the X-ray
diffraction data.

In Fig. 3 are also shown the kinetics of the Incoloy 800
corrosion. It can be seen that the weight increase product at
500 $^{\circ}$C is very small and after 6.000 hours actually non-existent.
The curves corresponding the tests carried out at 570 $^{\circ}$C show strong
variations in the weight decrease especially during the first
1.000 hours of the test. A comparison between both curves indicates
that the specific corrosion velocity in the dissolution zone, 570 $^{\circ}$C,
is much higher than that observed in the deposition zone, 500 $^{\circ}$C.

The leak effect is less pronounced due to the lower test
temperature as well as the larger distance from the test section
T-3 (700 $^{\circ}$C) and the fact that the sodium has been partially
purified in the cold trap.

The metallographic examination of the steel 1.4970 exposed
to the sodium up to 9.000 hours at temperatures of 700 and 590 $^{\circ}$C
shows an intergranular corrosion process at high test temperature
(700 $^{\circ}$C) and a layer of crystals deposited when the test temperature
was 590 $^{\circ}$C. Figs. 4a and 4b show the micrographs corresponding to
the above mentioned phenomena. Fig. 4a corresponds to the dissolution process and Fig. 4b to deposition.

The metallographic examination of the steel 1.4981, tested at
700 $^{\circ}$C during 9.000 hours indicates the formation of a surface
layer where the grain boundaries are ferritic (see Fig. 4c). The
same material exposed to sodium at 590 $^{\circ}$C shows an austenitic
structure with intragranular precipitations and a deposition layer
(see Fig. 4d).

The Incoloy 800 exposed to sodium at 570 $^{\circ}$C during 9.000 hours
shows a rough surface in which many intergranular channels can be
observed (Fig. 5a). In Fig. 5b, corresponding to the same material
after electrochemical etching, wide grain boundaries can be seen,
a thin layer where the intragranular precipitation is rather strong
and some depleted zones in the vicinity of the grain boundaries.

In Figs. 5c and 5d the profiles and microstructure of Incoloy
800 exposed to sodium at 500 $^{\circ}$C are presented. The formation of a
very well defined deposition layer is shown.

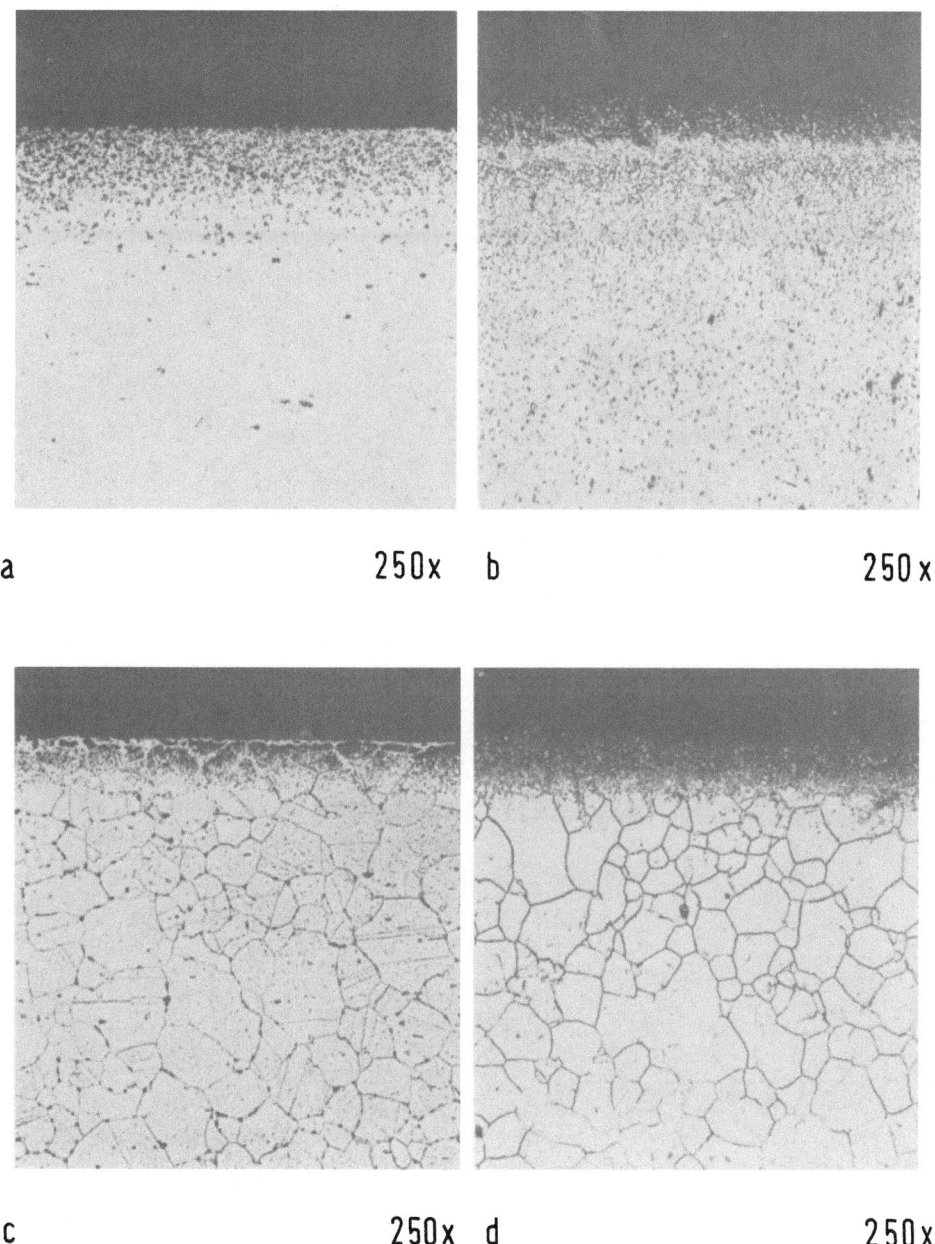

a                    250x    b                    250x

c                    250x    d                    250x

Fig. 4:   Metallographic examination of steel no. 1.4970
          a, b dissolution zone          c, d deposition zone

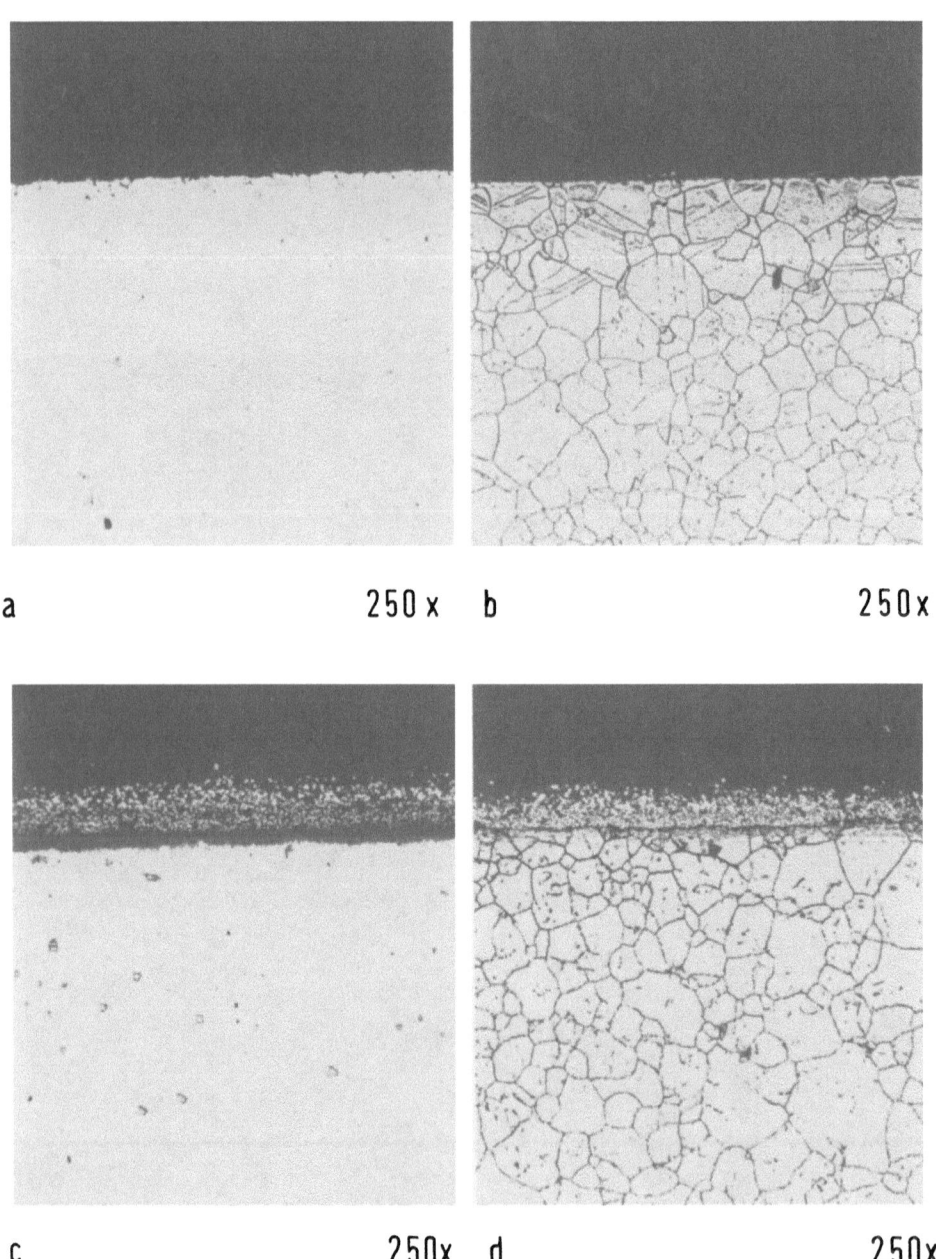

Fig. 5: Metallographic examination of Incoloy 800
        a, b dissolution zone          c, d deposition zone

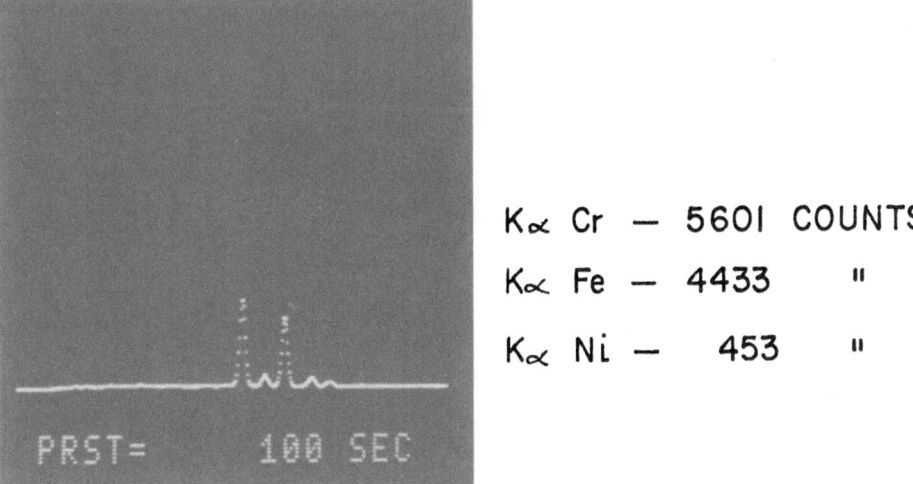

$K_\alpha$ Cr — 5601 COUNTS

$K_\alpha$ Fe — 4433      "

$K_\alpha$ Ni —   453      "

Fig. 6:  SEM examination of steel no. 1.4970 surface after 9000 h
         in ML-1

The SEM examination of the steel 1.4970 exposed to sodium at
700 °C during 9.000 hours shows a surface covered by small crystals
in the form of a lancet with well-defined contours. The presence
of this layer is contradictory with that of the dissolution process observed at 6.000 hours. The energy dispersive X-ray analysis
of this layer is also different to that observed for the 6.000 hours
test. A high increase is found in the chromium concentration as
well as a strong decrease in the nickel and iron concentrations.
Fig. 6 shows a S.E.M. micrography of the surface at 9.000 hours and
the corresponding spectrum analysis. Table 3 contains the counts
obtained by energy dispersive X-ray analysis which correspond to
the relative variation of the alloying elements concentration. The
anomalous behaviour is interpreted as a surface reaction process
between the specimens and the contaminated sodium during the last
part of the test. In order to identify the superficial reaction
products X-ray diffraction analysis and AES were carried out. The
AES show the presence of $N_2$ in the surface in a very high concentration as well as a big increase in the chromium concentration. The
X-ray diffraction allowed to identify the compound CrN (Chromium
nitride). Fig. 7 shows the X-ray diffraction record where can be
observed the peaks corresponding to the CrN compound.

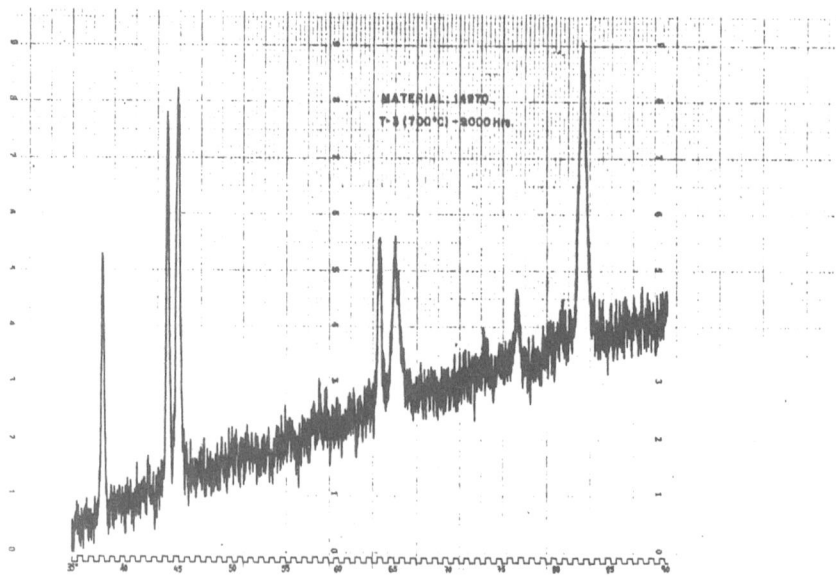

Fig. 7: X-ray diffraction of CrN

The steel 1.4970 exposed to sodium at 590 °C during 3.000 hours
presents many small crystals with well-defined contours typical of
a deposition process. The energy dispersive X-ray analysis shows,
in contrast to the base material a very high increase in the
chromium concentration and a depletion in the nickel and iron
content.

Table 3.  X-Ray Energy Dispersive Analysis of the Main Alloy
          Elements

| Material | | $T(^{\circ}C)$ | t (h) | Cr | Fe | Ni |
|---|---|---|---|---|---|---|
| | not exposed to sodium | | | 2.231 | 6.283 | 1.100 |
| 1.4970 | | 700 | 6.000 | 1.844 | 7.665 | 861 |
| | | | 9.000 | 5.601 | 4.433 | 453 |
| | exposed to sodium . | 590 | 3.000 | 4.345 | 2.255 | 729 |
| | | | 6.000 | 5.194 | 2.075 | 760 |
| | | | 9.000 | 4.659 | 2.571 | 866 |
| | not exposed to sodium | | | 2.624 | 5.740 | 945 |
| 1.4981 | | 700 | 3.000 | 2.593 | 8.113 | 1.226 |
| | exposed to sodium | | 9.000 | 7.313 | 3.916 | 507 |
| | | 590 | 3.000 | 4.965 | 1.988 | 807 |
| | | | 9.000 | 4.151 | 2.117 | 803 |
| | not exposed to sodium | | | 2.936 | 4.696 | 2.144 |
| Incoloy-800 | | 570 | 3.000 | 2.493 | 5.305 | 2.089 |
| | | | 6.000 | 3.254 | 3.914 | 1.907 |
| | exposed to sodium | | 9.000 | 2.645 | 4.499 | 2.450 |
| | | 500 | 6.000 | 2.963 | 4.041 | 1.471 |
| | | | 9.000 | 3.035 | 4.212 | 1.260 |

Note:  The figures correspond to the counts number obtained in the
       multichannel analysator in 100 seconds.

    At 6.000 hours of test both the morphology and the energy
dispersive X-ray analysis give some information on the nature of
the deposition process. In contrast at 9.000 hours exposure an
anomalous behaviour revealed by a surface morphology which seems
to show that part of the deposited material has dissolved and the
energy dispersive X-ray analysis shows, in comparing with the
results obtained at 6.000 hours, a depletion in the chromium
concentration and an increase in the iron and nickel content. The
energy dispersive X-ray analysis carried out in all the tested
specimens, at 590 $^{\circ}$C and exposure times of 3.000, 6.000 and 9.000
hours show the presence of Si, Ca and Na. In the deposited layer
Ca appears and an enrichment in Cr, $O_2$, Na is found by AES. It is
supposed that the most of the deposited layer is formed by a mixed

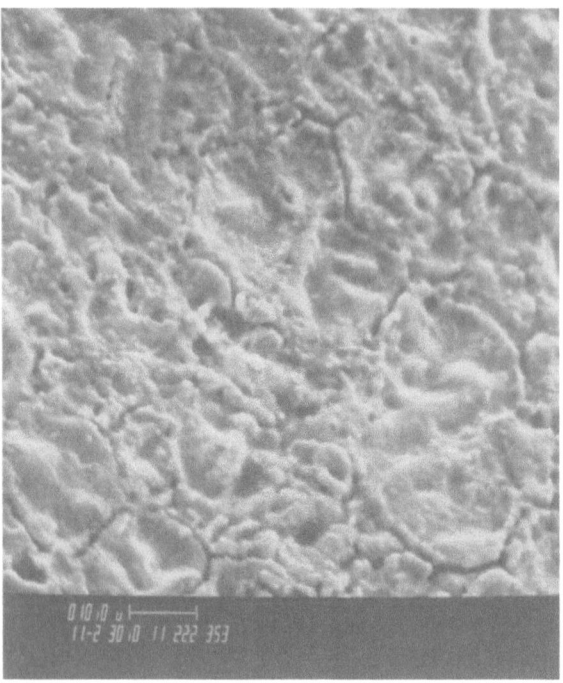

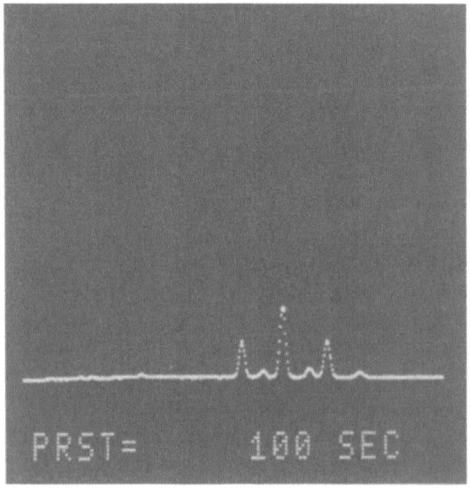

$K\alpha$ Cr — 2645 COUNTS
$K\alpha$ Fe — 4499      "
$K\alpha$ Ni — 2450      "

Fig. 8:   SEM examination of Incoloy 800 surface after 9000 h
          in ML-1

salt of the type $(Na, Ca)_x$ $(Cr, Fe, O)_y$ although it has not been possible to identify this compound by X-ray diffraction.

The behaviour of the steel 1.4981 exposed to sodium shows similar characteristics to the 1.4970 material discussed above.

The study of the Ni-rich alloy Incoloy 800 by S.E.M. shows that before exposure to sodium the material was strongly cold-worked indicated by the morphology. The X-ray energy dispersive analysis shows a high nickel content in this alloy and the presence of chromium and iron.

The material exposed to sodium at 570 $^\circ$C presents on its surface a progressive dissolution process with increasing time in the loop. Fig. 8 presents the morphology and analysis of a specimen exposed to sodium during 9.000 hours. One can see many cavities on the surface and an intergranular corrosion. The energy dispersive X-ray analysis shows alterations in the concentration of the main alloy elements; decreases in the chromium content at 9.000 hours and increases in the nickel content in contrast to the material not-exposed to sodium.

After exposure to sodium at 500 $^\circ$C it is observed in the same material the formation of a deposition layer consisting of rounded particles for which size and density are a function of the exposure time. The analysis point out a significant decrease in the Ni content on the surfaces as well as no significant variations in the Cr content.

ACKNOWLEDGEMENT

Authors are indebted to the contributions of Mrs. Dr. H. Schneider, J.L. Rodriguez Almazán, C. Martinez Pérez and Dr. H.U. Borgstedt to this work.

REFERENCES

1. F. Oltra Otra, E. Bojarsky, H. Leising, Energia Nuclear (Madrid) 19 (1975) 161.
2. M. de la Torre, F. J. Martinez Domenech, M. V. Garcia, B. de Celis, Energia Nuclear (Madrid) 19 (1975) 171.
3. H. Schneider, H. Schumann, Report KfK-2009 (1974).
4. H. Schneider, E. Nold, Report KfK-2273 (1976).
5. A. Marin, M. de la Torre, H.U. Borgstedt, Corrosion and mass transfer in ML-1 Sodium Loop, Conf.on Liquid Metal Technology in Energy Production, Richland, Washington, April 1980.

6.  H. U. Borgstedt, G. Frees and A. Marin, Corrosion and
    Carburization of Incoloy 800 in Liquid Sodium up to
    973 °K, Petten International Conference Alloy 800,
    March, 1978.

7.  A. Marin, R. Rodriguez Solano, H. U. Borgstedt, Energia
    Nuclear 19 (1975) 191-200.

8.  A. Marin, H. U. Borgstedt, Natriumkorrosionsversuche am
    Stahl x 10  Cr Ni Mo Ti 1515 (Werkstoff Nr. 1.4970),
    KfK-1574 (1972).

COMPATIBILITY STUDIES ON STAINLESS STEEL TYPE 347 IN LIQUID

SODIUM USING A THERMAL CONVECTION LOOP

H. S. Khatak and J. B. Gnanamoorthy

Reactor Research Centre
Kalpakkam
India

INTRODUCTION

    Thermal convection loops which are relatively inexpensive and
easy to fabricate have been found to be useful for screening alloys
prior to their full-fledged tests for compatibility studies in
liquid sodium using forced circulation loops. Type 347 stainless
steel was selected for studies in a thermal convection loop in
view of the possibility of improving its high temperature mechanical
properties by thermo-mechanical treatments (1). This paper describes
the construction and operation of a thermal convection loop made out
of type 347 SS and presents the results obtained after operating
the loop for 8061 h.

DESIGN AND CONSTRUCTION OF THE LOOP

    The main features of the loop are shown in Fig. 1. The loop
has been fabricated from 12.7 mm i.d., 20.6 mm o.d. seamless pipe
of type 347 SS. After locating the specimens as shown in Fig. 1
(Sectional view at A) the pipes were welded by TIG process. One
vertical section of the loop ("Hot leg") was heated by a resistance
furnace to a maximum temperature of 400 °C. The other parts of the
loop were naturally air-cooled to give a temperature gradient (ΔT)
of 100 °C. Thermocouples were located on the outer surface of the
piping at the entrance and exit sides of the hot leg as well as in
the middle of the cold leg to monitor the surface temperatures of
the piping. Under steady state conditions, the loop parameters were
as follows:

| | |
|---|---|
| Maximum temperature: | 400 °C |
| Minimum temperature: | 300 °C |
| Weight of Na in the loop: | 650 gms |
| Flow-head: | 720 mm |
| Velocity of sodium: | 19 cm/sec |

The method suggested by Epstein (2) was utilized for calculating the velocity of sodium.

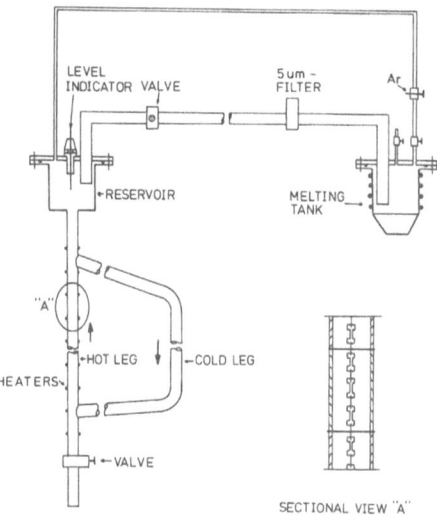

Fig. 1.   Thermal Convection Loop for Liquid Sodium

THE SPECIMENS AND THEIR LOCATION IN THE LOOP

6 mm dia rods of AISI type 347 SS (nominal chemical composition, Cr: 18.07 %, Ni: 10.69 %, Nb: 0.49 %, C: 0.05 %, Si: 0.53 %, Mn: 0.81 %) were cold rolled to strips of about 1 mm thickness. These strips were then solution annealed and water-quenched. Subsize tensile specimens of 12.7 x 4 x 1 mm gauge dimensions were obtained from the annealed strips. The specimens were mechanically polished, degreased, weighed and inserted in different sections of the loop except the bends. Two specimens of the thermomechanically treated 347 SS were located in the loop at the L/D ratio of 28 to 35. The thermomechanical treatment given was: rolling of annealed strips at 400 °C to reduction in thickness of 80 % and 84 % followed by cooling in air.

SODIUM CHARGING AND OPERATION OF THE LOOP

The loop was repeatedly evacuated and purged with argon gas
and then the solid sodium pieces charged into the melting tank were
melted and aged at 150 $^\circ$C for two days to precipitate the oxides
into the cooler bottom portion. From the melting tank, sodium was
transferred to the loop section through a 5 µm stainless steel
filter under argon pressure and at a temperature of 150 $^\circ$C. The
filling of the loop was guided by a spark-plug type level indicator.
The loop was then heated to 400 $^\circ$C and maintained at this temperature
for 3 days. After draining off the sodium, to remove the oxides from
the pipe's inner surface, the loop was refilled with sodium. Thereafter, while the hot leg was continued to be heated to the maximum
temperature, the cold leg heaters were switched off. The steady
state conditions were reached after a few hours.

The tests were terminated after a total exposure period of
8061 h and the specimens were taken out for examination after
draining the sodium from the loop. The adhering sodium from the
specimens were removed with alcohol. The specimens were finally
cleaned ultrasonically in acetone. After recording the weight
changes, visual, light optical and SEM examinations and tensile
tests were carried out on these specimens to evaluate the effects
of sodium exposure.

RESULTS AND DISCUSSION

Visual examinations revealed that the specimens from the hot
leg had a green deposit on the surface while those from the cold
leg had a black deposit. Earlier studies (3, 4) have indicated that
the green deposit observed in the hot leg is mainly due to sodium
chromite.

The weight changes in the specimens have been plotted as a
function of the L/D ratio in Fig. 2. Generally the specimens have
suffered a loss in weight in the hot leg. The black deposits
observed on the cold leg specimens are expected to contain elements
leached out from the hot leg as well as the impurities present in
the sodium. However, the weight changes observed on the specimens
in the cold leg were neither regular nor significant. Since the
temperature of the pipe surface is less than that of the specimens
in the cold leg, it is presumed that most of the deposition has
taken place on the pipe surface.

The results of the tensile tests at room temperature have been
shown in Fig. 3. The horizontal lines at 585 N/mm$^2$, 70 percent and
240 N/mm$^2$ are respectively the UTS, percentage elongation and YS of
the unexposed material. A strain rate of $6.56 \times 10^{-4}$ sec$^{-1}$ was used
for the tensile tests. Considerable decrease in ductility was observed in the specimens from the hot leg, the lowest being at

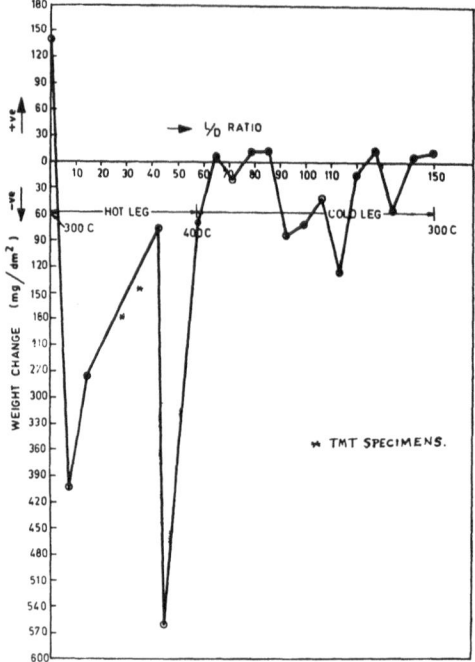

Fig. 2.   Weight changes in Sodium-exposed samples.

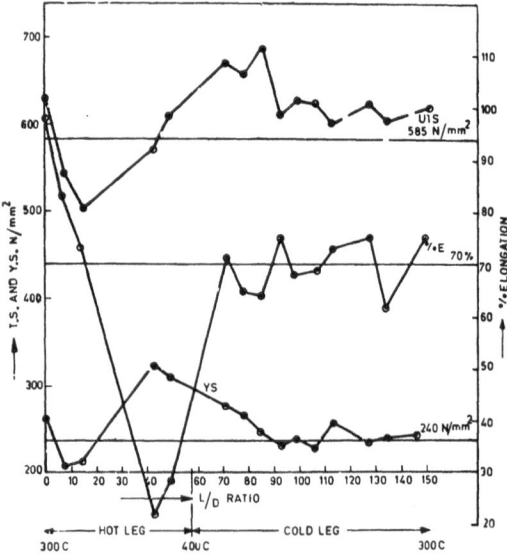

Fig. 3.   Results of tensile tests at room temperature.

L/D = 42 (22 % elongation). There was decrease in UTS initially up
to L/D = 40 in the hot leg and thereafter the UTS value started
increasing. A peak value was obtained between L/D = 60 and 90 in
the cold leg. Yield strength values followed almost the same pattern
as that for UTS except that the increase started even earlier than
L/D = 40. The changes in the YS, UTS and percentage elongation
could be attributed to the microstructural changes resulting from
the intergranular selective leaching and/or carbon pick-up by/from
sodium. Selective leaching of elements such as Ni, Mn, C results
in the transformation of the austenite to ferrite. The microstructure
of a specimen at L/D = 42 (Fig. 4) indicated the formation of a
uniform ferrite layer of 8 µm thickness and an intergranular attack
of up to 20 µm. Examination of the pipe surface at the corresponding
position showed an intergranular attack up to 30 µm depth.

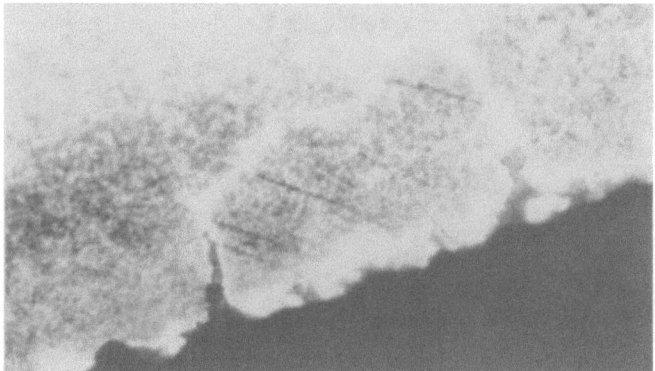

Fig. 4   Photomicrograph of hot leg specimen (L/D = 42) showing
         uniform ferrite layer and IG attack (750 X)

     The effect of carburization upon low and high temperature
mechanical properties of austenitic stainless steels is fairly
well established. As the carbon content increases tensile strength
increases and ductility decreases. With carbon concentration in
the alloys greater than 0.5 Wt % fracture occurs with no measurable
ductility. The photomicrograph of the specimen from the cold leg
at L/D = 71 (Fig. 5a) shows carburized grain boundaries. A peak
in the UTS value was obtained at this point. No significant change
in microstructure was seen in the specimen at L/D = 141 (Fig. 5b).
This is in agreement with the absence of any difference in the
values of UTS and YS of the material before and after sodium exposure at this location. Examination of the specimen from the hottest
point in the loop (L/D = 57) also revealed some carburization. It
therefore appears that the changes in the mechanical properties
were mainly due to carburization.

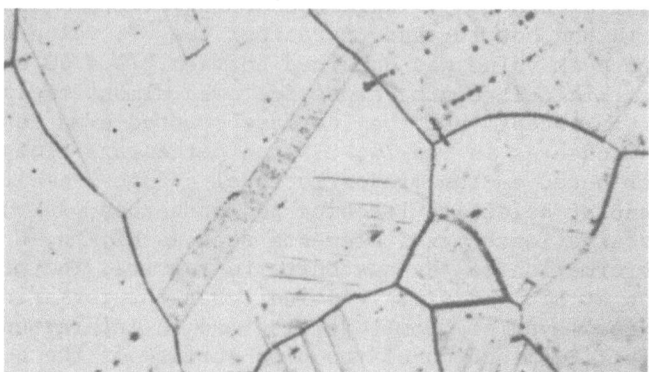

Fig. 5a   Photomicrograph of cold leg specimen (L/D = 71) showing
          carbide formation at the grain boundaries (192 X)

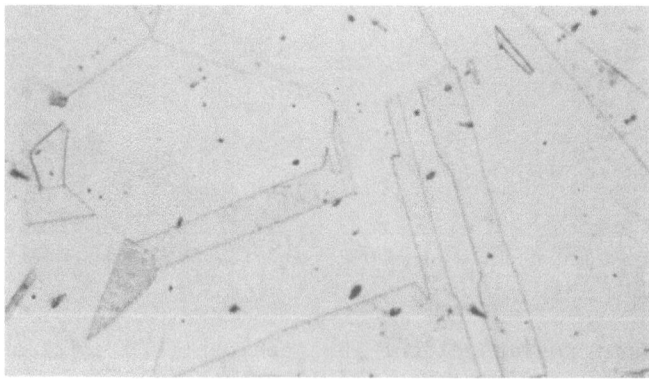

Fig. 5b   Photomicrograph of cold leg specimen (L/D = 141) showing
          no microstructural change (192 X)

        Fig. 6 shows the SEM photomicrographs of surfaces of exposed
specimens at L/D = 71 and 141. It is observed that the deposit at
L/D = 71 is greater, both in respect of quantity and particle
size, than that at L/D = 141. This agrees with the observation
of Roy et al (5) that the tendency of deposition of chromium and
carbon is more at a point where the temperature starts decreasing.
Thus there is an enhanced possibility of carburization at the
entrance to the cold leg.

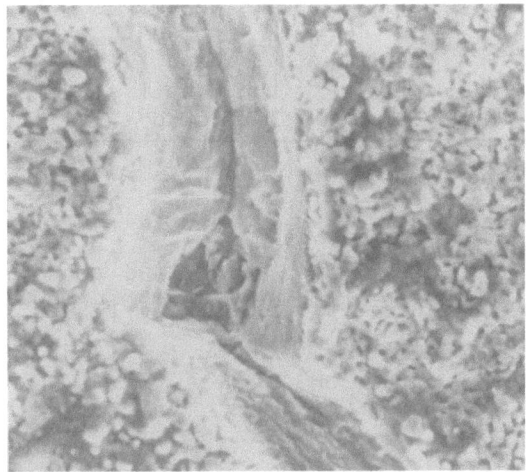

Fig. 6a   SEM Micrograph of specimen at L/D = 71 (2000 X)

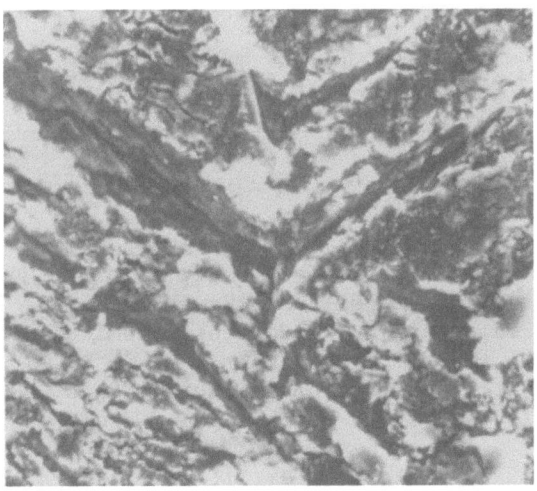

Fig. 6b   SEM Micrograph of specimen at L/D = 141 (2000 X)

Table 1. shows the results of the tensile tests at 600 °C. The thermomechanically treated samples showed a large reduction in UTS and YS values after sodium exposure. Heavy intergranular attack was observed on these specimens. Therefore this type of thermomechanical treatment on 347 SS appears to be unsuitable for sodium service at high temperatures.

Table 1.   Results of tensile test at 600 $^{o}$C

| Sl.no. | specimen heat treatment | position in the loop L/D ratio | Y.S.$_2$ N/mm$^2$ | T.S.$_2$ N/mm$^2$ | % Elongation |
|---|---|---|---|---|---|
| 1 | annealed and unexposed | - | 120 | 460 | 44 |
| 2 | annealed and exposed | 71(cold leg) | 200 | 475 | 47 |
| 3 | annealed and exposed | 92(cold leg) | 155 | - | - |
| 4 | annealed and exposed | 57(hot leg) | 210 | 440 | 17 |
| 5 | thermomechanically treated, 80 % reduction in thickness at 400 °C & cooled in air (unexposed) | - | 750 | 835 | 9 |
| 6 | -- do -- (exposed) | 35(hot leg) | 200 | 395 | 25 |
| 7 | thermomechanically treated, 84 % reduction in thickness at 400 °C & cooled in air (unexposed) | - | 850 | 950 | 6 |
| 8 | -- do -- (exposed) | 28 | 220 | 390 | 30 |

REFERENCES

1.  J. J. Laidler, Trans. Am. Nucl. Soc., 13, 2 (1970), 589.
2.  L. F. Epstein, Proc. Int. Conf. Peaceful Uses of Atomic Energy, United Nations, New York (1956) 9, 311.
3.  R. S. Fidler and M. J. Collins, Atomic Energy Review, 13, 1 (1975) 9.
4.  R. H. Hiltz, Corrosion by Liquid Metals, (Eds) J. E. Draley and J. R. Weeks, Plenum Press, New York (1970) 63.
5.  Prodyot Roy, et al, ibid., 1.

# SOLUBILITY OF IRON IN LIQUID LEAD

I. Ali-Khan

Kernforschungsanlage Jülich
Jülich
Fed. Rep. of Germany

## INTRODUCTION

The use of molten lead in metal and chemical industry is well known. It's further applications in nuclear and non-nuclear fields are under consideration. An INTOR (International Tokamak Reactor) study (1) proposes molten lead alloys with lithium and lithium compounds as breeding material for fusion reactors due to their excellent nuclear physical properties. Lead-Bismuth alloys and other metals are being considered as target materials for a spallation neutron source (2).

As new possibilities of utilization of lead are showing up, the problem of finding structural materials for containment remains hard pressing (1 - 7).

In addition to the studies of corrosion and mechanical properties of structural materials, the knowledge of their solubility behaviour (7) in molten metals is necessary prior to deciding for their use in a specific system.

At the nuclear research center of Jülich the corrosion behaviour of a large number of different iron base and iron containing alloys and refractory metals in molten lead - possibly to be used as a heat transport medium - was investigated. An evaluation of the literature showed that the solubility data of iron in molten lead were not conclusive.

O. Cutler Shepard and Ralph Parkman (8) determined the solubility of iron in molten lead and found a break at 904 °C in their solubility vs. temperature curve. They calculated ΔH solubility values for α- and γ-iron in molten lead to be as following:

ΔHs = 14000 cal/mole for α-phase in the temperature range of 538 - 816 °C; and 51000 cal/mole and 167000 cal/mole for 904 and 1010 °C respectively. According to D.A. Stevenson and J. Wulff (9) the ΔH and ΔS values for α-iron are: ΔH = 27900 cal/mole, ΔS = 7.0 cal/K · mole. For γ-iron they found: ΔH = 22000 cal/mole and ΔS = 1.9 cal/K · mole. Contrary to this Brasunas (10) found no reaction of molten lead with iron upto 1000 °C. K.O. Miller and J.F. Elliot (11) found no break in their solubility curve. The phase diagramm Fe - Pb (12) shows no solubility of iron in molten lead. For a comparison the solubility values of different authors are represented in Fig. 3.

EXPERIMENTAL

The apparatus used for determining the solubility of Fe in molten lead is shown in Fig. 1 - 2. The associated hydrogen and argon gas purification facilities are described elsewhere (1).

The apparatus consisted mainly of the following parts:

1. Equilibrium furnace
2. Furnace to purify lead before using for solubility tests
3. Device with vacuum lock to take out the lead samples for analysis

The equilibration part of the apparatus was provided with a furnace and a holder for an $Al_2O_3$-crucible, in which solid metal specimens could be equilibrated with molten lead under inert atmosphere.

A vacuum lock was constructed with a heatable quartz container to purify the lead in molten state. The quartz container was provided with a 30 μm filter and could be slided down to the crucible through a gate valve for injecting molten lead into the crucible. The purification of lead was carried out at 750 °C by purging purified hydrogen through the melt in the quartz container to reduce the metallic impurities. The temperature of the melt was then brought down to 350 °C and kept over night at this temperature so that the reduced metallic impurities could be crystallized out and after coagulation would not pass along with molten lead through the filter into the crucible. The tests showed that by this method Fe-content of the lead could be reduced from 50 ppm to 5 ppm. After filling the crucible with molten lead the temperature of the furnace was raised to 1000 - 1100 °C and the melt was stirred by passing purified Argon gas through the melt. The equilibrium was reached in 50 h. The first sample from the equilibrated lead was taken out at 1000 - 1100 °C. After that the temperature was reduced every 100 °C to take melt samples at different temperatures down to about 350 °C.

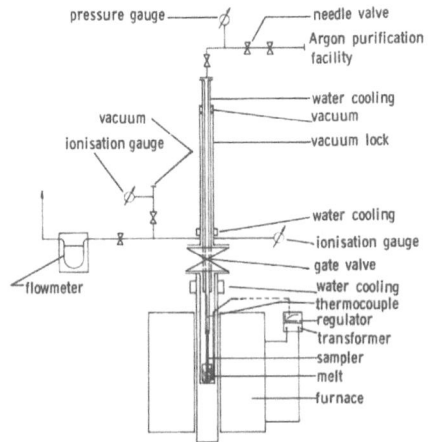

Fig. 1: Schematic diagram, apparatus as used for determining the solubility in molten lead

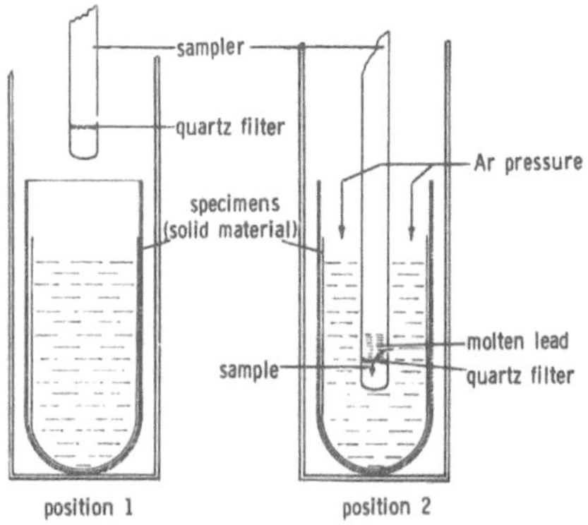

Fig. 2: Detail of the sampling procedure

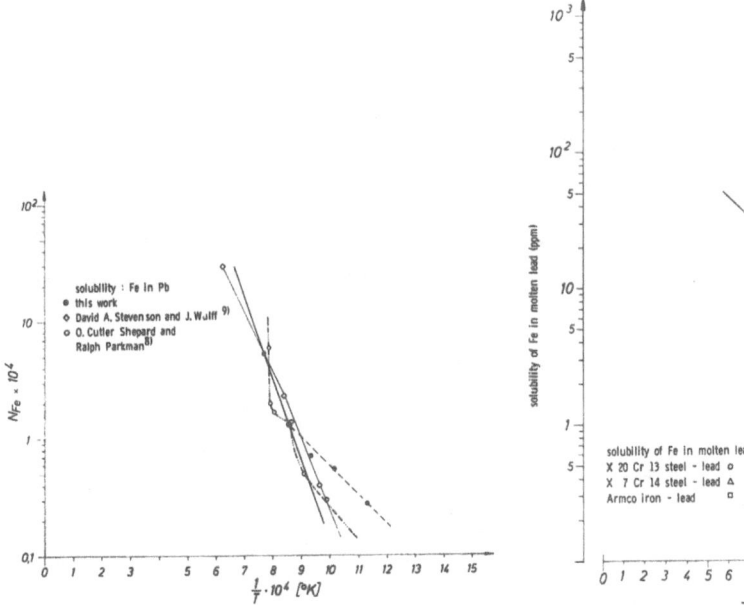

Fig. 3: Iron solubility in molten lead (ARMCO iron)

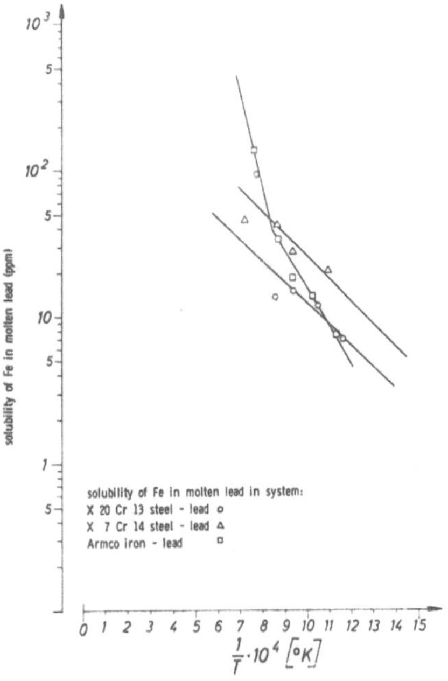

Fig. 4: Comparison with steels
X20 Cr 13
X 7 Cr 14

The specimens used in solubility tests were Armco Iron, and steels: X7 Cr 14, X20 Cr 14, X8 CrNb 17, and X8 CrTi 17.

RESULTS

The solubility data achieved in this work have been plotted as $N_{Fe}$ (N= mole fraction) against the reciprocal of temperature (Fig. 3). The curve shows a break at about 900 °C. This is related to the α-γ-phase transition in iron at 911 °C. ΔH and ΔS values have been calculated by inserting the equation (9)

$$\log N_{Fe} = - \frac{\Delta H}{2.303RT} + \frac{\Delta S}{2.303R}$$

and the results compared with the values given by other authors.

The equation is applied for evaluating the liquidus lines in an ideal eutectic system. The Fe – Pb system behaves hereby as an ideal system thermodynamically. Due to a break in the curve, two equations have been developed, and ΔH and ΔS values calculated for α- and γ-phase:

$$\log N_{Fe} = - \frac{7100}{T} + 6.23$$

for γ-phase

$$\log N_{Fe} = - \frac{2450}{T} + 2.24$$

for α-phase and ΔH and ΔS values have accordingly been calculated for the two temperature ranges.

ΔH = 32483 cal/mole; ΔS = 10.2 cal/K mole for γ-phase and ΔH = 11208 cal/mole; ΔS = -8.1 cal/K mole for the temperature range of 538 – 816 °C

A value of 57 283 cal/mole was obtained for ΔH at 904 °C.

This value lies 12 % higher than the one given by Cutler, Shephard and Parman (8) for this temperature. Solubility tests were also carried out by using the specimens of the chromium steels X7 Cr 14 and X20 Cr 14. The results are given in Fig. 4 as measured. The difference between the slope of the curves for α- and γ-iron and the chromium steels is considerable. The solubility of iron in molten lead was also determined by using the steel specimens of X8 CrTi 17 and X8 CrNb 17. In the case of these steels the iron solubility in molten lead was found to by 16 and 14 ppm respectively at 650 °C.

In Fig. 5 micrographs of Armco iron specimens after solubility tests are shown. Iron crystals, most probably at least partly precipitated on cooling, are observed. Another micrograph (Fig. 5b) shows the microstructure of an Armco iron specimen in the bent region with strong corrosion attack. The X 20 Cr 13 and X 8 CrTi 17 steel specimens show distinctly intercrystalline corrosion (Fig. 6 and 7) after higher temperature solubility tests.

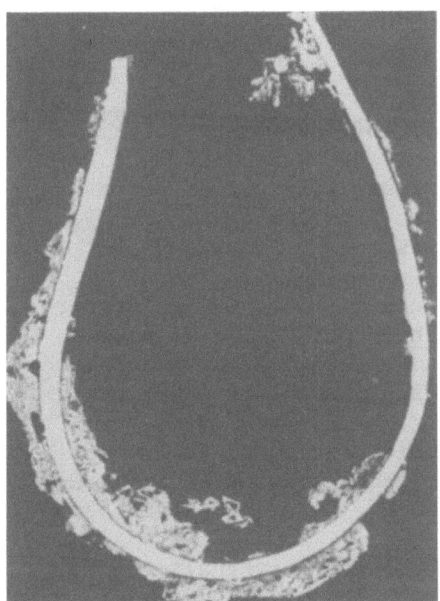
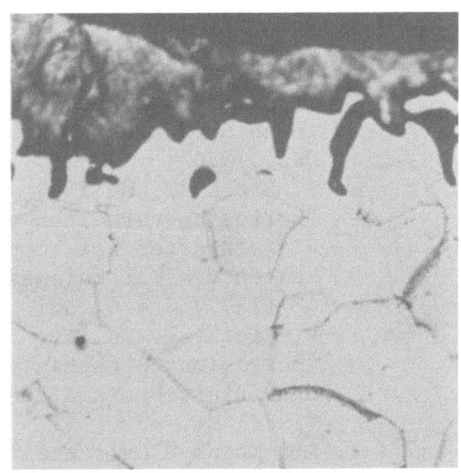

Fig. 5a and 5b

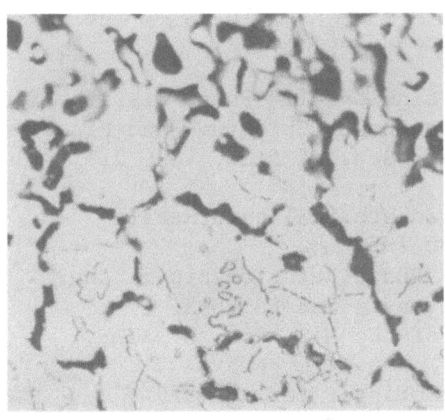
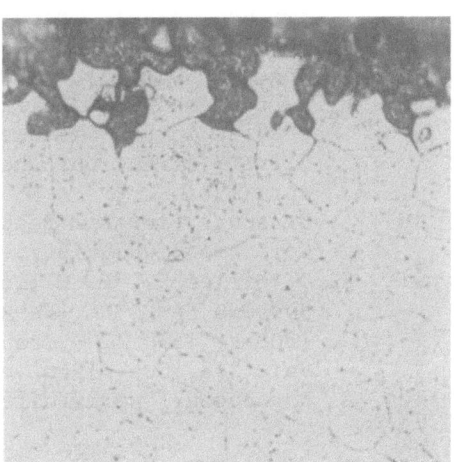

Fig. 6                      Fig. 7

(see text)

ACKNOWLEDGEMENT

Thanks are expressed to Prof. F. Waelbroeck, Director, Institute für Plasmaphysik, KFA-Jülich, for his support in preparing this paper

REFERENCES

1. INTOR (International Tokamak Reactor), USA-Contributions, the 3rd Meeting of the INTOR workshop 16-28 June, 1980, IAEA Vienna, EUR FU BR4/X 11 501/80 EDV 42
2. G. S. Bauer, Priv. Communication
3. I. Ali-Khan: "Korrosionsverhalten von Stählen in stationären Bleischmelzen", JÜL-694-RW, Sept. 1970
4. I. Ali-Khan, S. Krawczynski: "Korrosionsverhalten in Stählen und hochschmelzenden Metallen gegenüber Blei in anisothermen Kreislaufsystemen", JÜL-721-RW, Dez. 1970
5. F.-R. Block, W. Müller, J. Schneider and G. Stolzenberg: "Korrosionsverhalten verschiedener Stähle gegenüber Bleischmelzen bei Temperaturen bis 1400 K und Möglichkeiten des Korrosionsschutzes", Arch. Eisenhüttenw. 48 (1977) Nr. 6 Juni
6. F.-R. Block, W. Müller, J. Schneider and G. Stolzenberg: "Korrosionsverhalten verschiedener Werkstoffe gegenüber Bleischmelzen bis 1400 K", Stahl und Eisen 97 (1977) Nr. 15.
7. I. Ali-Khan: "Löslichkeit und Korrosionsverhalten an Eisen, Stahl, Molybdän, Niob, Tantal, Vanadium, Wolfram und Chrom in Bleischmelzen bei höheren Temperaturen", JÜL-661-RW, 1970
8. D. C. Shepard, R. Parkman: "Investigation of Material, for use in Heat Transfere System containing Molten Lead Alloys, ORO 38, Progress Rep. No. 2, Aug. 1 - Oct. 31, 1950
9. D. A. Stevenson, J. Wolff: "Liquid-Solid Phase Distribution Studies in the Systems Iron-Lead, Chromium- and Nickel-Silver", Trans. of the Met. Soc. of AIME, Vol. 221, Apr. 1961
10. A. Brasunas: Interim Report on static liquid-Metal Corrosion, ORNL-1647, June 1954
11. K. O. Miller, J. F. Elliot: "Phase Relationship in the Systems Fe-Pb-Ni, Fe-Ni-C(sat.) and Fe-Pb-Ni-C, 1300 to 1550 $^{\circ}$C", Trans. of the Met. Soc. of AIME, Vol. 218, Oct. 1960
12. E. Hornbogen: "Werkstoffe", Springer Verlag 1979

# CORROSION OF STEELS AND REFRACTORY METALS IN LIQUID LEAD

I. Ali-Khan

Kernforschungsanlage Jülich
Jülich
Fed. Rep. of Germany

## INTRODUCTION

Liquid metals are used in many technologies in the industry and are further interesting for their application in nuclear and non-nuclear fields (1, 2). According in recent proposals lead alloys can advantageously be used as breeding material in fusions reactor technology. They are also interesting as target materials for spallation neutron sources.

Recent efforts for determining the material behaviour in molten lead or its alloys have been concentrated mainly to explore the phenomenon of Liquid Metal Embrittlement (3, 4). The data on corrosion behaviour of lately investigated structural materials are generally in agreement with the results achieved previously (1-2). At the nuclear research center KFA-Jülich the corrosion behaviour of a large number of structural materials was investigated in liquid lead for its utilization as a heat transport medium. The tests were carried out with ferritic and austenitic steels, refractory metals and their alloys in stationary melt and in thermal convection loops. In view of the renewed interest in systems using molten lead these results as obtained by the author at the institute of reactor materials of KFA-Jülich are here presented and discussed in some more details than could previously be done.

## EXPERIMENTAL

The investigations in stationary molten lead were conducted with U-shaped specimen stripes of the size 10 x 100 mm clamped in

$Al_2O_3$-crucibles, which were, after filling with purified molten
lead, heated in a furnace as shown in Fig. 1 at 575, 650 and
750 °C over a period of 3250 h. For tests in circulating molten
lead thermal convection loops were used (Fig. 2). The lead used
for the tests was of 99,99 % purity (nominal analysis:
Bi < 0.005 %; Ag, As, Cu, Fe, Sb, Sn each < 0.001 %). It was,
however, further purified by purging hydrogen through the melt
before using for tests. The hydrogen and argon gas were purified
in gas purification facilities (1). After completing the runs
the specimens were investigated by using metallography and SEM
technique.

RESULTS

The tables 1 - 3 and the Fig. 3 - 14 show some of the results
of corrosion tests of chromium- (Tab. 1, Fig. 3 - 8) and chromium-
nickel steels (Tab. 2 and 3, Fig. 9 - 14) in stationary molten
lead and in circulating melt in thermal convection loops. In
stationary molten lead the corrosion attack was found to be more
severe in the bent region of the U-shaped specimens. With the
exception of the steels 1.4112 (X 90 CrMoV 18) and 1.4718
(X 45 CrSi 9) showing an attack depth of 150 and 300 µm respectively
the chromium steels were only little attacked at 575 °C. The
austenitic steels, however, showed in general higher corrosion
attack at 575 °C than at 750 °C (Fig. 9 - 10). The more severe
corrosion attack in austenitic steels at lower temperature is
mostly intercrystalline and is considered to be due to increased
precipitation of carbides and intermetallic phases in grain
boundaries in the lower temperature range rather than due to a
solubility effect of Ni in molten lead as suggested by other
authors (5, 6). The scanning electron micrography (SEM) shows
strong Cr-depleted zone (Fig. 14) in the corroded region of the
steel 1.4436 (X 5 CrNiMo 18 12). The results of thermal convection
loops are summarized in Tab. 3. The corrosion was found to be
initially localized and observed as corrosion patches along the
specimens length which after exposing to higher temperatures or
larger test periods united together to form a continuous corrosion
layer. The stabilized steels were strongly eroded in circulating
molten lead. The thickness of most of these steels was reduced
to 12 - 30 % of the original value as measured before the tests.

The specimens of nickel base alloys X 8 NiMoCr 60 18 (Mat. No.
2.4472) and X 8 NiMo 65 30 (Mat. No. 2.4482) and Hastelloy N were
completely destroyed at 750 °C in a thermal convection loop with
ΔT = 232 °C within 4 h and could not be further investigated. The
chromium steels 1.4021 (X 20 Cr 13), 1.4002 (X 7 CrAl 13), 1.4511
(X 8 CrNb 17) and 1.4510 (X 8 CrTi 17) were also tested at 750 °C
over 306, 575 and 811 h respectively.

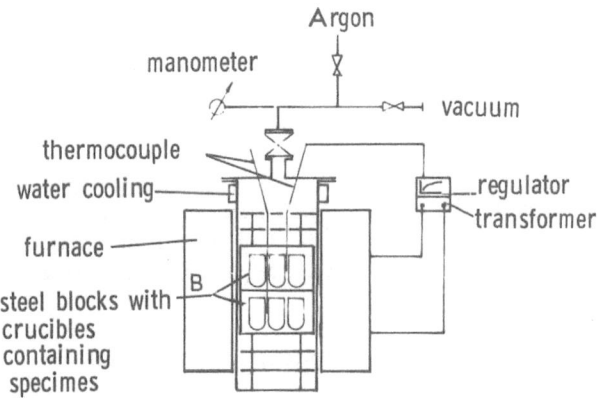

Fig. 1:   Tests in stationary melt

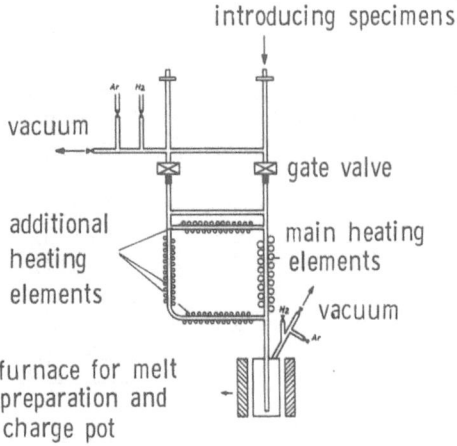

Fig. 2:   Thermal convection loop for testing corrosion behaviour
          of structural materials

Table 1.  Corrosion behaviour of chromium steel specimens in
          stationary molten lead and in thermal convection loop
          of the steel X 8 CrTi 17

| DIN Mat.-No. | Specimen test in stationary melt, 3250 h, corrosion attack, depth µm | | | Specimen tests in loops, hot leg max.Temp. 600 °C, $\Delta T$ in loops 232 °C, 1002 h – run | | | |
| --- | --- | --- | --- | --- | --- | --- | --- |
| | | | | specimen thickness | | | observed attack depth |
| | | | | before test | after test | | |
| | 575°C | 650°C | 750°C | mm | mm | % | µm |
| 1.4006 | 10 | 15 | 100 | 0.99 | 0.72 | 73 | 300–273 |
| 1.4024 | 15 | 20 | 50 | 1.77 | 1.46–1.48 | 82 | 220–200 |
| 1.4021 | k | 30 | 80–150 | 2.10 | 1.70–1.76 | 83 | 320–305 |
| 1.4034 | 15–70$^x$ | 70° | 90$^x$ | 1.48 | 1.12–1.15 | 76 | 260 |
| 1.4002 | 15–70$^x$ | 20 DZ | 20$^x$ DZ | 1.79 | 1.46–1.58 | 82 | 260–220 |
| 1.4713 | 25 | k | 30 DZ | 1.01 | 0.78–0.90 | 77 | 180–130 |
| 1.4724 | 10 | 15 | – | 2.08 | 1.78–2.02 | 86 | 200–187 |
| 1.4119 | 10 | 30 | 150 | 1.51 | 1.27 | 84 | 240–143 |
| 1.4122 | k | 20E | 175 | 3.91 | 2.72–2.80 | 70 | 240–227 |
| 1.4016 | k | k | 35 | 1.26 | 1.00–1.02 | 79 | 140–130 |
| 1.4511 | k | 30–60 | 10–35 | 0.94 | 0.68–0.72 | 72 | 170 |
| 1.4510 | k | 30 | 50 | 1.62 | 1.45 | 90 | 155–150 |
| 1.4922 | 10 | 25 | 100 | 1.23 | 0.80–1.00 | 70 | 250–155 |
| 1.4112 | 150 | 75 | k | 1.94 | 1.62–1.66 | 84 | 320–260 |

1)  =  U-shaped specimens, corrosion attack as observed in the
       bent region
k   =  no or only little corrosion
x   =  large material loss
DZ  =  diffusion zone
E   =  decarburisation

Table 2.  Corrosion behaviour of austenitic steel specimens in
          stationary molten lead at different temperatures and in
          thermal convection loop of steel X 10 CrNiMoNb 18 10,
          600 $^{\circ}$C,  T = 151 $^{\circ}$C, 1008 h - run

| DIN Mat. No. | Specimen test in stationary melt, 3250 h, corrosion attack, depth μm | | | Specimen test in loops, hot leg max. Temp. 600 $^{\circ}$C, $\Delta$T in loops 151 $^{\circ}$C, 1008 h - run | | | |
|---|---|---|---|---|---|---|---|
| | | | | specimen thickness | | | observed attack depth |
| | | | | before test | after test | | |
| | 575$^{\circ}$C | 650$^{\circ}$C | 750$^{\circ}$C | mm | mm | % | μm |
| 1.4301 | 40 | 10 | 25 | 1.64 | 1.28-1.52 | 78 | 140-80 |
| 1.4310 | 60-90 | 15 | 10 | 1.03 | 0.88-1.02 | 85 | 80-53 |
| 1.4300 | 10 | 20 | k | 0.86 | 0.74-0.78 | 86 | 80-53 |
| 1.4307 | 10 | 25 | 20 | 3.09 | 2.54-2.72 | 82 | 140-87 |
| 1.4543 | 40 | 25 | 150 | 2.50 | 2.33-2.36 | 93 | 100-62 |
| 1.4401 | 55 | 15 | 10 | 1.93 | 1.78-1.88 | 91 | 140-80 |
| 1.4449 | 40 DZ | 20 | 10 | 2.01 | 1.82-1.86 | 91 | 60-53 |
| 1.4436 | 25 | k | 70 | 1.11 | 1.00-1.02 | 90 | 80 |
| 1.4404 | 40 | 10 | 15 | 2.60 | 2.40-2.60 | 92 | 200-130 |
| 1.4438 | 80 | 35 | 20 E | 2.01 | 1.80-1.98 | 90 | 140-120 |
| 1.4550 | 110 | 35 | 20 | 0.83 | 0.68-0.76 | 82 | 70-55 |
| 1.4580 | 30-100 | 20 | 50-70 | 1.56 | 1.24-1.46 | 79 | 200-95 |
| 1.4541 | 15 | 20 | 30 | 1.62 | 1.44-1.52 | 89 | 100-70 |
| 1.4571 | 25-50 | 15 | 25 | 2.05 | 1.92-1.99 | 94 | 80-60 |
| 1.4981 | 20-40 | 30-75 | 100-50 | 0.90 | 0.74-0.86 | 82 | 120-90 |

1)  =  U-shaped specimens, corrosion attack as observed in the
       bent region
k   =  no or only little corrosion
x   =  large material loss
DZ  =  diffusion zone
E   =  decarburisation

Table 3.   Corrosion behaviour of austenitic steels in thermal
           convection loops at 600 °C, ΔT = 151 °C for 1008 h-
           and 2016 h-run

| DIN Mat.- No. | Spec. thickness before tests | Specimen tests in thermal convection loops, hot leg max. Temp. = 600 °C, ΔT in loops = 151 °C | | | | | |
|---|---|---|---|---|---|---|---|
| | | 1008 h-run | | | 2016 h-run | | |
| | | final spec. thickness | observed attack depth | | final spec. thickness | observed attack depth | |
| mm | mm | mm | % | μm | mm | % | μm |
| 1.4550 | 0.83 | 0.68–0.76 | 82 | 77–55 | 0.25–0.45 | 30 | 200–173 |
| 1.4543 | 2.50 | 2.33–2.36 | 93 | 100–62 | – | – | – |
| 1.4541 | 1.62 | 1.44–1.52 | 89 | 100–77 | 0.20–0.30 | 12 | 300–250 |
| 1.4961 | 0.95 | 0.79–0.80 | 83 | 140–89 | 0.22–0.70 | 23 | 300 |
| 1.4988 | 0.96 | 0.82–0.88 | 85 | 110–90 | 0.22–0.37 | 23 | 330–300 |
| 1.4981 | 0.90 | 0.74–0.86 | 82 | 120–90 | 0.30–0.33 | 33 | 200–163 |
| 1.4571 | 2.05 | 1.92–1.99 | 94 | 80–60 | – | – | – |
| 1.4580 | 1.56 | 1.24–1.46 | 79 | 200–95 | – | – | – |
| 1.4449 | 0.01 | 1.82–1.86 | 91 | 60–53 | 1.74–1.80 | 86 | 80–67 |
| 1.4306 | 0.96 | 0.88–1.02 | 92 | 100–60 | 0.65–0.93 | 68 | 120–60 |
| 1.4300 | 0.86 | 0.74–0.78 | 86 | 80–53 | 0.62–0.80 | 72 | 50–4? |
| 1.4305 | 2.91 | 2.58–2.64 | 89 | 100–70 | 2.63–2.74 | 90 | 150–73 |

     The corrosion attack was largely transcrystalline (Fig. 8).
The thickness of chromium steel specimens was not reduced in
750 °C tests as in the case of stabilized austenitic steels.   In
thermal convection loop tests the specimens V, Nb, Nb - 5-% V,
NB - 1-% Zr were not attacked at 625 °C over a period of 1658 h.
In another test a corrosion of max. 30 μm depth (Nb - 5-% V) was
observed in refractory metals specimens but only in the contact
region of specimens refractory metals/specimenholder (steel
X 8 CrTi 17). At 700 °C (ΔT = 152 °C) no corrosion attack was
observed in V and NB specimens in circulating molten lead over
a period of 306 h. In another investigation (1) conducted for
determining the solubility of refractory metals in molten lead,
the specimens of Mo, Nb, V, W, Ta showed no corrosion attack in
stationary lead up to about 950 °C. Ta specimens showed an attack
depth of max. 40 μm in bent region while the Cr specimen was
strongly attacked due to dissolution of Cr in molten lead when
exposed between 415 and 925 °C and between 405 and 985 °C
respectively. The results of corrosion tests can again be
summarized as follows:

1.  The corrosion tests in stationary liquid metals are important
    as they give information about the mode of corrosion in alloys.
    For a further evaluation of corrosion behaviour of alloys, the
    tests in circulating melt are, however, essential due to the
    phenomenon of erosion and mass transport.

2.  The tests showed that the corrosion attack was strongly
    increased due to stresses in material. The depth of corrosion
    attack was much higher in the bent region of U-shaped speci-
    mens than in the rest of the specimens.

3.  The larger corrosion attack in chrom-nickel steels at lower
    temperature is mainly due to the microstructural processes
    at grain boundaries causing a structural instability in material
    which makes them vulnerable to corrosion attack. A chemical
    reaction of molten lead with one or the other precipitated
    carbide or intermetallic phase is possible. An attack on the
    precipitates especially at the beginning, during the state
    of precipitation and in the grain boundary region seems to
    be more likely. The solubility effect of Ni in molten lead
    was apparently of minor consequence.

4.  Metallurgical factors increasing the stability of austenite
    in chromium - nickel steels tend to increase the corrosion
    resistance. It is suggested that carbon depletion in steels
    and corrosion attack in grain boundaries is responsible for
    material deterioration.

5.  Among chromium steels the steels with pure ferritic structure
    showed better corrosion resistance in molten lead.

6.  In thermal convection loops the erosion of the specimens and
    mass transport is strongly time and temperature dependent.
    Some of the austenitic steels containing Nb, Ti and Mo were
    reduced in their thickness to 12 - 33 %. The corrosion attack
    in chromium steels was found to be largely transcrystalline.
    Higher carbon content tends to intercrystalline attack in
    these steels. Chromium steels were not as much eroded in
    circulating molten lead as the stabilized austenitic steels.

7.  Refractory metals show excellent corrosion resistance in
    stationary as well as in circulating molten lead.


ACKNOWLEDGEMENT

    Thanks are expressed to Prof. F. Waelbroeck, Director of the
Inst. für Plasmaphysik, KFA Jülich, for his support in preparing
this paper.

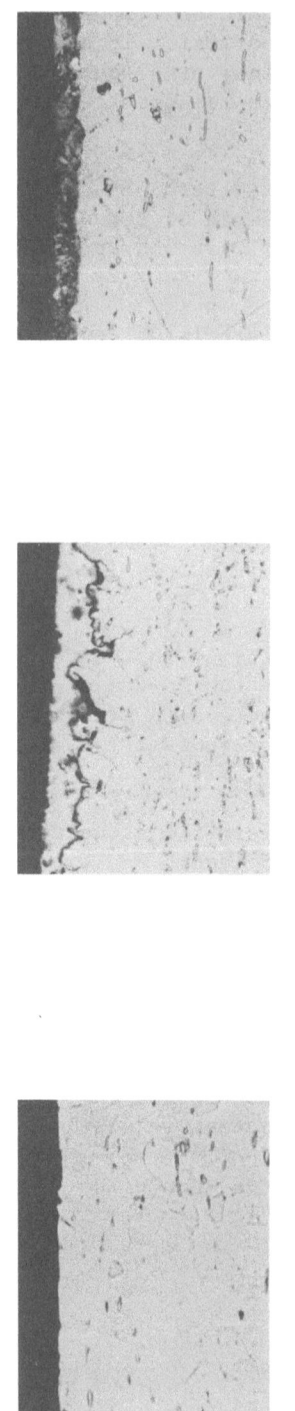

Fig. 3          200 X

Fig. 4          200 X

Fig. 5          200 X

Fig. 3 – 5: Steel X 8 CrNb 17, stationary molten lead tests at 575, 650 and 750 °C respectively, 3250 h-runs, Etch:oxalic acid

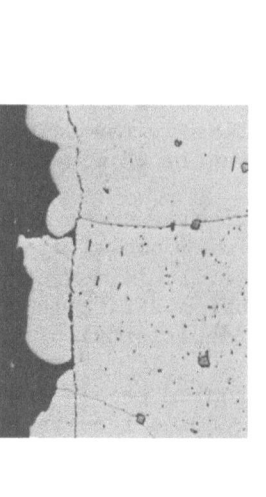

Fig. 6          200 X

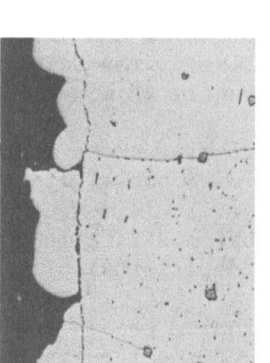

Fig. 7          200 X

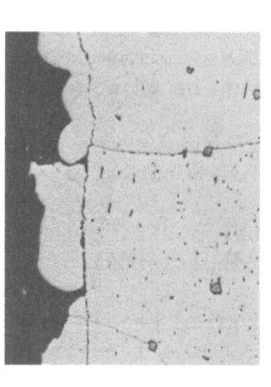

Fig. 8          200 X

Fig. 6: X 8 Cr 17, Fig. 7: X 8 CrTi 17: stationary molten lead tests at 750 °C, 3250 h-runs,
Fig. 8: X 8 CrTi 17, loop test: 750 °C, 811 h

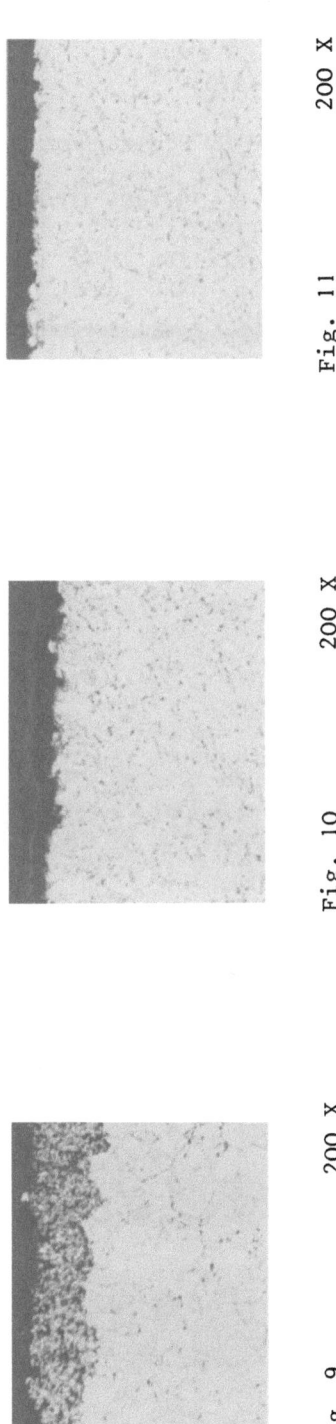

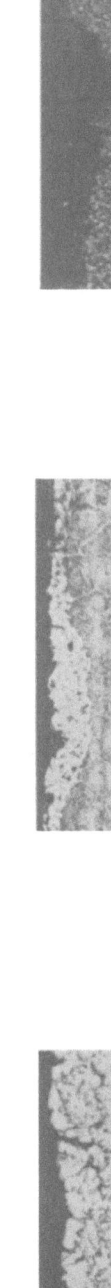

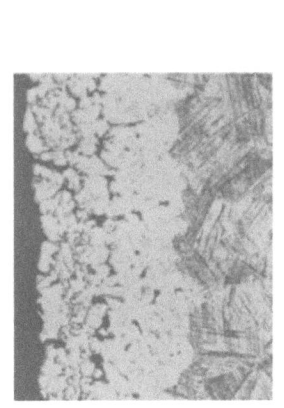

Fig. 9          200 X          Fig. 10          200 X          Fig. 11          200 X

Fig. 9 – 11:  steel X 2 CrNiMo 18 10, stationary molten lead tests at 575, 650 and 750 $^{\circ}$C respectively, 3250 h-runs, Etch: mixed acid

Fig. 12          200 X          Fig. 13          200 X          Fig. 14          200 X

Fig. 12: X 12 CrNi 17 7 loop test, 600 $^{\circ}$C, $\Delta T$ = 151 $^{\circ}$C, 2016 h-run,
Fig. 13 and 14: X 5 CrNiMo 18 12, loop test, 600 $^{\circ}$C  $\Delta T$ = 151 $^{\circ}$C, 1008 h-run, Etch: mixed acid

REFERENCES

1.  I. Ali-Khan; This meeting, JÜL-694-RW, 1970;
    JÜL-721-RW, 1970; JÜL-661-RW, 1970.
2.  F. R. Block, W. Müller, J. Schneider and G. Stolzenberg;
    Arch. Eisenhüttenwes., 48 (1977), Nr. 6.
3.  M. G. Nickolas, AERE-R 9199, Apr. 1979, HL 79/1404 (C 14).
4.  F. A. Shunk, W. R. Warke; Ser. Mat., 8 (1974).
5.  S. Banerjes; Proc. Int. Congress Met. Corrosion 1972, Tokio.
6.  J. V. Cathart and W. D. Manly; USAEC, Rep. ORNL-20008,
    Febr. 1956.

# USE OF COATINGS TO PROTECT STEELS AGAINST LEAD CORROSION AT HIGH TEMPERATURES

Franz-Rudolf Block and Volker Schwich

Technical University
Aachen
Fed. Rep. of Germany

## INTRODUCTION

New techniques in reactor technology favour the use of liquid lead on account of its advantageous physical and thermal qualities as heat carrier. Although the solubility of iron in pure liquid lead is low the corrosion action of lead in chemical processes limits the exposure time of the steels employed. Thus in lead bath patenting the containers must be repaired or even renewed every two years. Furthermore attack of the metal containers can be observed in lead ladles in which temperatures up to 1000 K are used for hardening and tempering. The use of other base metals is generally impossible because of prohibitive prices, low strength at elevated temperatures or poor workability. As an alternative it is possible to coat high temperature steels with suitable materials which are corrosion resistant against lead and adhere sufficiently to base metals.

## COATINGS

The formation of adequately adhering coatings is the basis for the use of different materials. Criteria for the evaluation of a coating method and its quality can only be given in connection with the field of application of the composite. Furthermore, the technical use of a coating method depends on the ability to coat large as well as intricate layers in a simple manner. Plasma-spraying permits the coating of large areas and also of components of every shape. In addition new combinations of spraying materials can improve corrosion protection.

STRUCTURE OF THERMAL SPRAYED LAYERS

    The structure of thermal sprayed layers differs considerably
from that of cast, shaped or galvanized layers. A characteristic of
this method is that slightly heated and highly superheated particles
are together propelled onto a relatively cold surface. There they
freeze in a very short time. This results in a laminar feather-like
layering of the coating material with incorporated unfused particles.
Depending on the spraying parameters all these layers are more or
less porous. This porosity is produced by bridging undercuts during
spraying, gas development and shrinking cracks. Fig. 1 shows the
structure of a typical plasma-sprayed molybdenum layer.

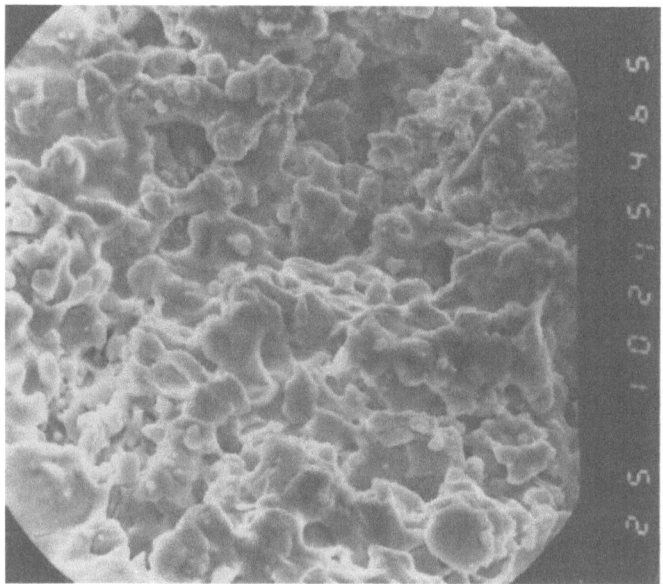

Fig. 1:   Structure of a typical plasma-sprayed Mo-layer; 1000 : 1;
          SEM

The porosity and structure depends on both the spraying method and
the spraying material.

    If spraying does not take place in vacuum a low oxide content
will be observed in some layers. While the particles are propelled
onto the surface oxides are formed by air and swept along with the
inert gas - the smaller the particles, the stronger the oxidation.
The amount of air in an argon-plasma flame reaches 80 % at a
distance of 50 mm from the nozzle. Oxidation can be diminished by
adding hydrogen to the argon gas, or boron and silicon to the
spraying powder. Molybdenum sprayed layers are free of oxides as the
molybdenum-trioxide ($MoO_3$) evaporates at 1066 K.

Another problem is that the composition of the coating does
not correspond to that of the spraying powder; for example, the
content of nitride in zirconium-nitride (ZrN) decreases from 14
weight-% to 8 - 9 weight-%.

ADHESION TO THE BASE METAL

The quality of a coating mainly depends on the adhesion to
the base metal. There are four binding mechanisms:

- mechanical interlocking
- physical adhesion
- chemisorption and
- metallurgical bonding.

Protection against corrosion attack can be lost by parts of
the layer chipping off or by cracks forming within the layer.

If hot and even partially fused particles are sprayed onto a
cold substrate a tensile stress is produced after cooling within
the layer. The stress can be reduced by yielding within the coating.
Thus a low E-modulus of the layer and a long cooling time are
favourable. On reheating the coated material stress is induced if
the thermal expansion coefficients of layer and base metal differ.
The formation of cracks and flakes also depends on the shape of the
working piece.

Layers with low thermal conductivity are susceptible to damage
by thermo-shock. A low thermal conductivity is produced by the
coating material itself or by a high porosity within the coating.
The adhesion of the coatings to the base metal can be increased by
intermediate layers having thermal expansion coefficients between
those of the base metal and coating. The tensile stress is then
distributed between the interfaces. In addition interfaces may be
expected to enhance corrosion resistance.

A further increase of the adhesion is achieved by roughening
and preheating the substrate. This results not only in a lowering
of the stresses but also in the formation of a metallurgical bond.
Finished sprayed composites can be subjected to heat treatment.
Pores are reduced or closed, oxides coagulate and elements of the
matrix material diffuse into the sprayed layer, usually improving
the adhesion. Fig. 2 shows the plasma-sprayed molybdenum layer of
Fig. 1 after high temperature annealing. Dense packed grains are
formed.

COATING MATERIALS

All metals and alloys, especially substances with a high
melting point can be applied by thermal spraying. However, the

mechanical and metallurgical properties of a sprayed material do
not necessarily correspond to the properties of the initial material.

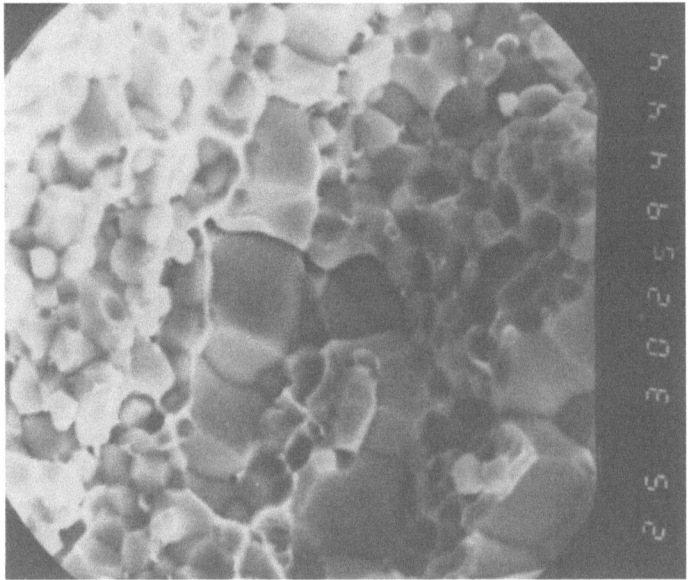

Fig. 2:   Plasma-sprayed molybdenum layer after a heat treatment;
          3000 : 1; SEM


     Most ceramic materials possess extreme hardness and a high
melting temperature due to the nature of their binding mechanism.
The plasma-spraying method is particularly adaptable for the
spraying of ceramic layers, as high temperatures are produced.

     After preliminary examination of a considerable number of
sprayed ceramic materials with regard to their behaviour in liquid
lead the following materials were chosen for their favourable properties:

- TiN, ZrN;
- TiB$_2$, ZrB;
- Wc, W$_2$C, Cr$_3$C$_2$ and
- alumina-magnesium spinel.

     Layers are sprayed using homogeneous powders and also powders
of continuously varying composition.

     Mixed layers with decreasing metal content towards the surface
have an excellent corrosion resistance against liquid lead and
adhere strongly to the base metal.

Several refractory metals are resistant against corrosion attack by liquid lead, principally tantalum, tungsten, niobium, zirconium and molybdenum.

Within this group molybdenum layers have shown the best adherence to preheated high temperature steels, in spite of their small thermal expansion coefficient as compared with steels. Table 1 shows the properties of the tested spraying-powders.

Table 1:  List of the examined spraying-powders

| Spraying-powder | particle size of the powder μm | thermal expansion coefficient (273 – 1273 K) $\times 1 . 10^{-6} K^{-1}$ |
|---|---|---|
| Mo | 5 – 45 | 5.8 |
| Mo | 5 – 22 | 6.2 |
| WC+Co-sublayer | 5 – 24/4.8-23 | 5.6/6.0 |
| AlMgO$_4$+Mo-sublayer | 5.6 – 22.4/5-22 | 10.9/6.2 |
| ZrN+Mo-sublayer | 10 – 38/5-22 | 7.2/6.2 |
| TiN+Mo-sublayer | 15 – 38/5-22 | 9.35/6.2 |
| mixed layer ZrN : Mo = 1 : 1 | 5 – 45 | 6.9 |

CORROSION TESTS

The corrosion behaviour of coated as well as of uncoated steels has been examined in lead melts at temperatures up to 1400 K. Table 2 gives a list of the tested steel qualities.

Table 2:  List of the examined steel qualities

| Standard No. (DIN) | Designation (DIN) | Standard No. (DIN) | Designation (DIN) |
|---|---|---|---|
| 1.4541 | X 10 CrNiTi 18 9 | 1.4864 | X 12 NiCrSi 36 16 |
| 1.4725 | CrAl 15 5 | 1.4876 | X 10 NiCrAlTi 32 20 |
| 1.4742 | X 10 CrAl 18 | 2.4778 | CoCr 28 Fe |
| 1.4762 | X 10 CrAl 24 | 2.4640 | NiCr 15 Fe |
| 1.4821 | X 20 CrNiSi 25 4 | 2.4816 | NiCr 15 Fe |
| 1.4828 | X 15 CrNiSi 20 12 | 2.4869 | NiCr 80 20 |
| 1.4841 | X 15 CrNiSi 25 20 | | |

In the experimental installation twelve sheets can be tested simultaneously. Each specimen is immersed in a separate lead bath.

The dependence of the corrosion rate on the flow velocity can be
ascertained by stirring the specimens with various rotating velocities in isothermal melts. Fig. 3 shows a section through the
installation.

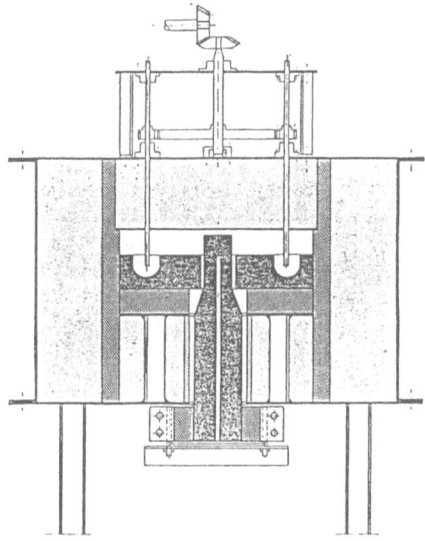

Fig. 3:   Section through the test installation

Chemical changes within the test sheets as well as in the
coatings have been followed with a Secondary Ion Mass Spectrometre
(SIMS). Changes in texture and surface morphology have been detected using both Scanning Electron Microscope (SEM) and X-ray
Microanalysis. Structure alterations have been determined by metallographic methods.

The experiments were conducted by immersing the specimens
coated on one side with molybdenum in lead melts containing 99.97
weight-% lead. Table 3 gives the analysis of the lead employed.

Table 3:   Analysis of the lead employed

| Pb weight-% | Bi -% | Ag -% | Cu -% | Fe -% | Sb -% | Zn -% | Sn -% |
|---|---|---|---|---|---|---|---|
| 99.9715 | 0.024 | 0.0010 | 0.0001 | 0.0001 | 0.0001 | 0.001 | 0.0001 |

The small proportion of lead oxide was not reduced. No tendency
for the molybdenum to escape as $MoO_3$ vapour could be detected, but
the formation of extremely stable lead oxide skins could be observed.

Fig. 4 shows a lead oxide skin on the plasma-sprayed molybdenum layer of the steel X 20 CrNiSi 25 4 (1.4821).

Fig. 4:   Lead oxide skin on the plasma-sprayed molybdenum layer of the steel X 20 CrNiSi 25 4; 1000 : 1; SEM

The formation of stable and protecting lead oxide skins on a molybdenum layer would appear to depend on the chemical composition of the base steel, because these skins have been only found in connection with silicon containing steel qualities e.g. X 20 CrNiSi 25 4 (1.4821), X 15 CrNiSi 20 12 (1.4828) and X 12 NiCrSi 36 16 (1.4864).

The diffusion of alloying elements through the layer is inhibited by the oxide skin. The diffusion rate depends on the quality of the base steel. Nickel especially tends to permeate the layer and to dissolve in the lead melt. Fig. 5 to 8 show the growth of nickel rich crystals through a molybdenum coating.

The molybdenum coated austenitic-ferritic steel X 20 CrNiSi 25 4 (1.4821) shows only slight nickel loss and a good resistance to corrosion attack by liquid lead up to 1350 K. A transverse section is shown in Fig. 9.

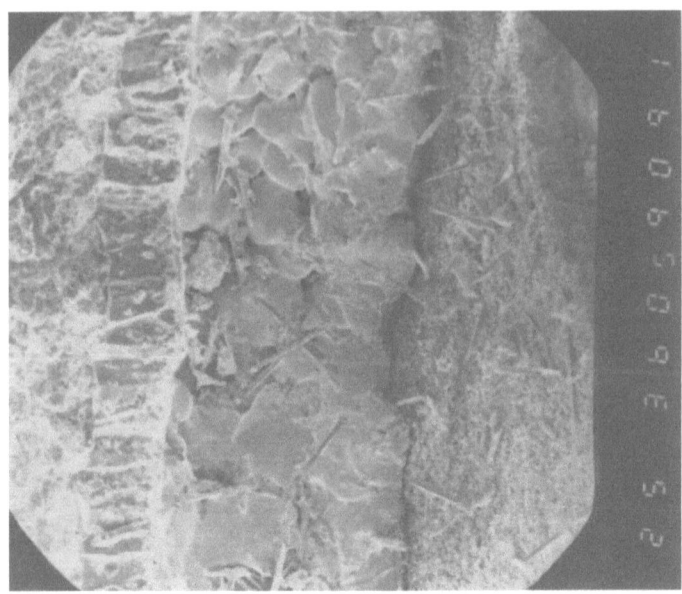

Fig. 5:   Nickel crystals in the molybdenum coating of the steel
          X 15 CrNiSi 25 20 (1.4841); 36 : 1; SEM

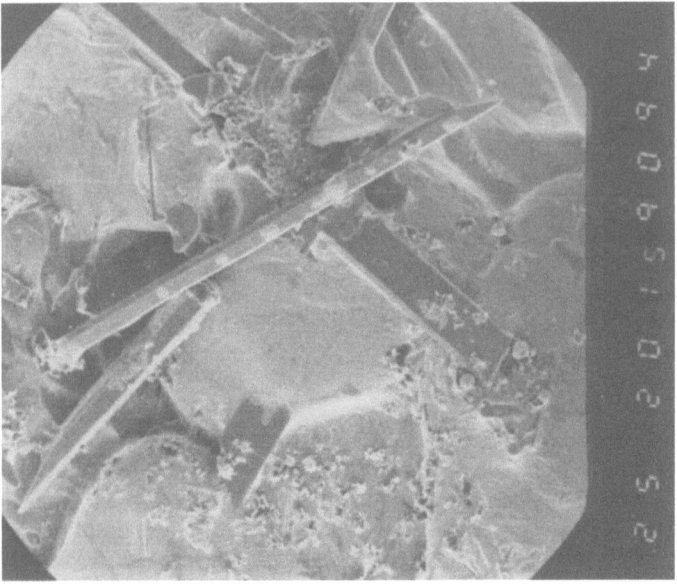

Fig. 6:   Nickel rich crystals within the molybdenum layer;
          200 : 1; SEM

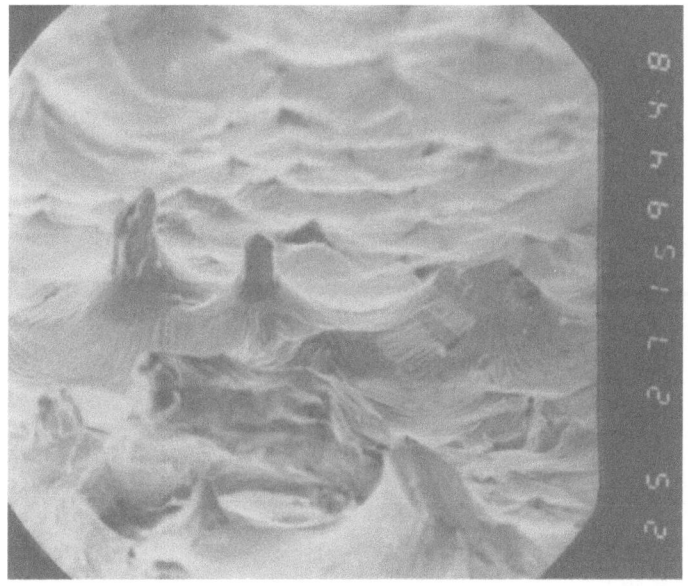

Fig. 7:   Nickel crystals grown through the surface of a molybdenum
          layer covered with a lead oxide skin; 300 : 1; SEM

Fig. 8:   Nickel crystals grown through the surface of a molybdenum
          layer; 1000 : 1; SEM

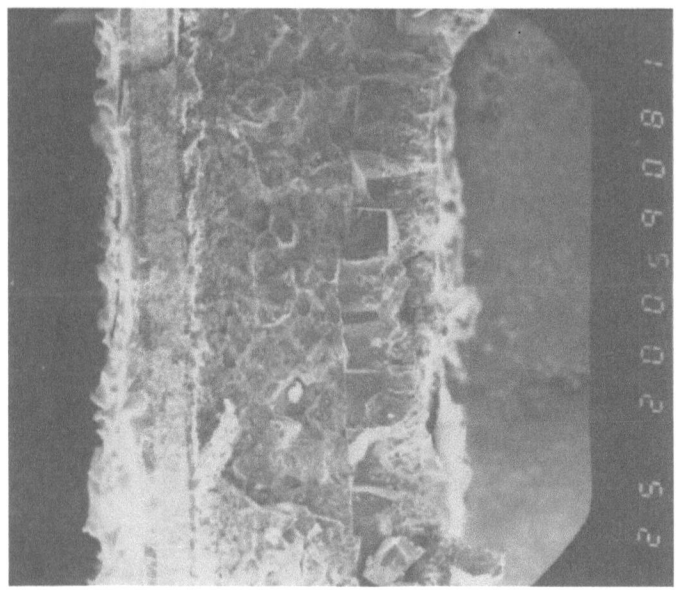

Fig. 9:   Transverse section through a one-sided molybdenum coated
          specimen of the steel X 20 CrNiSi 25 4; 20 : 1; SEM

        Considerable experimental work on a wide range of steel qua-
lities has confirmed that at temperatures higher than 1350 K and
with nickel contents higher than about ten percent corrosion attack
cannot be prevented even by coating. For example, austenitic steels
with more than 20 weight-% nickel suffer a disastrous damage after
100 hours in liquid lead at 1350 K. Fig. 10 shows lead penetration
on the grain boundaries of the special steel NiCr 15 Fe.

        Coating nickel rich steels with homogeneous layers of Mo, WC,
$Al_2O_3$ or $Cr_2O_2$ only retards destruction. Strongly adhering dense
layers which can inhibit the nickel diffusion have not yet been
sufficiently investigated. First experiments, however, show that
mixed layers of Mo and ZrN give good protection.

        The corrosion rate of ferritic as well as of ferritic-marten-
sitic steels is lower than that of austenitic qualities. However,
the considerable grain growth of the ferritic qualities at higher
temperatures prevents their technical application. Fig. 11 shows
an example.

        Fig. 11 gives another example of the excellent protection given
to a steel coated by a plasma-sprayed molybdenum layer, only 150 μm
thick.

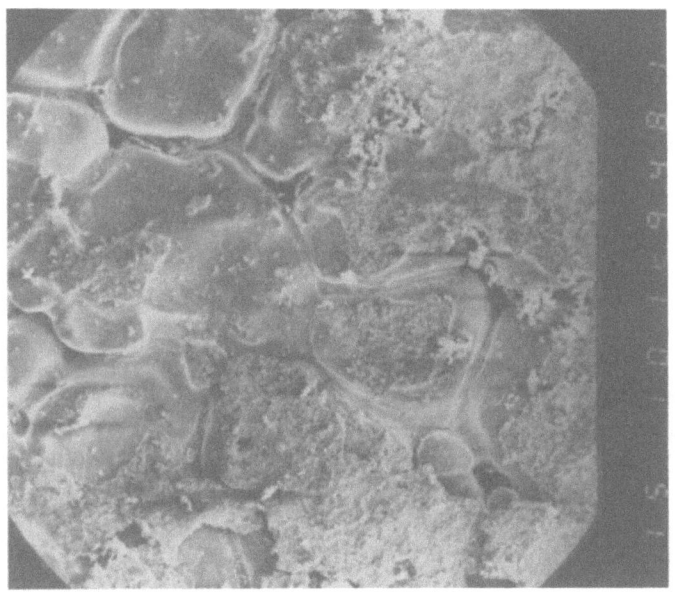

Fig. 10: Lead penetration on the grain boundaries of an uncoated
         specimen of the special steel NiCr 15 Fe (2.4640) after
         100 h in liquid lead at 1350 K; 100 : 1; SEM

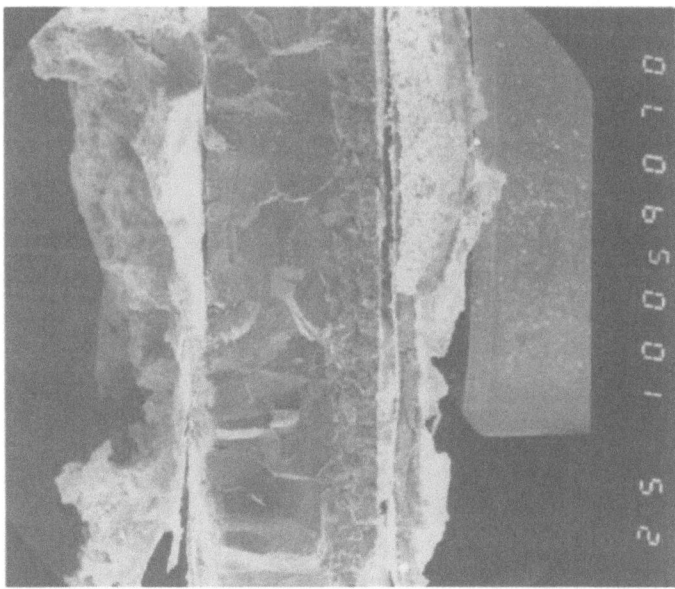

Fig. 11: Transverse section through the steel X 10 CrAl 18 (1.4742);
         considerable grain growth on the uncoated side of the
         sheet

CONCLUSIONS

The production of high-quality composites resistant to liquid lead corrosion is made possible by thermal-spraying protective layers on high temperature steels.

Plasma-spraying is the most suitable method for coating with high melting substances. Adhesion and corrosion protection are improved significantly by using mixed layers. A further increase in adhesion is achieved by high temperature annealing of the composites.

For all tested steels the corrosion resistance is enhanced by coatings with suitable layers. The best results are achieved with the austenitic-ferritic steel X 20 CrNiSi 25 4 (1.4821) coated with a plasma-sprayed molybdenum protective layer. Even at temperatures higher than 1350 K and after a testing time of 100 hrs no corrosion attack could be observed.

# INFLUENCE OF CARBON SPECIES IN SODIUM ON THE METHANE EQUILIBRIUM

Jürgen Jung, Ulrich Buckmann, Rainer Pütz

INTERATOM GmbH
Berg. Gladbach
Fed. Rep. of Germany

The carbon present in sodium partially reacts with hydrogen to form methane. By use of a $H_2$/Ar mixture flowing through the sodium, it was possible to substantially reduce the carbon content which reacts to form methane. The equilibration of hydrogen and methane partial pressures over sodium allowed to calculate the carbon activity which amounted to about $10^{-2}$ in the case of non-doped sodium. High methane partial pressures have been measured after $Na_2C_2$ addition to sodium at experimental temperature of 450 °C, for carbonate addition at 525 °C and for cyanide addition at 590 °C. The calculational basis assumes that monoatomic dissolved carbon determines the methane reaction. In some cases this leads to carbon activities of > 1. This indicates that the methane is partially or completely formed by the metastable $C_2^{2-}$ in sodium.

## INTRODUCTION

Carbon is present in sodium in different species (e.g. acetylide, carbonate, cyanide). Better knowledge of these carbon compounds is of importance to understand the carburization and decarburization processes of structural materials in sodium systems.

The available reactive part of carbon in sodium may be determined by means of the reaction of the carbon with hydrogen to form methane.

In the cover gas of sodium systems methane contents of between 0.1 and 20 vpm are normally measured; high concentrations indicate that lubricants during fault conditions have entered the sodium (1).

In sodium systems the formation of methane is possible by

- reaction of carbon present in sodium with hydrogen
- reaction of hydrogen with the carbon in the structural
  material
- interaction of high molecular hydrocarbons (oil, lubricants)
  with sodium; in addition of methane hydrogen and other hydrocarbons have been observed.

The following investigations described are restricted to the
determination of the methane formation by carbon species in sodium.

FUNDAMENTALS

Carbon dissolved in sodium may react with hydrogen to form
methane according to

$$C_{Na} + 2 \ H_2 = CH_4. \tag{1}$$

For the equilibrium constant K of this reaction

$$K = p_{CH_4} / p_{H_2}^2 \cdot a_c \tag{2}$$

$(p_{CH_4}, \ p_{H_2}$ = methane and hydrogen partial pressure respectively;

$a_c$         = carbon activity in sodium)

is valid in the temperature range 288 - 1200 $^{\circ}$K (2):

$$K = \exp \ (7.862 - 6.165 \ \lg \ T + 8314/T). \tag{3}$$

Methane and hydrogen partial pressure measurements after recirculating a carrier gas through sodium and the results of carbon
activity calculations, performed by Johnson (3), led to the conclusion
that additional mechanisms other than those described by eq. (1) are
involved for explaining the results obtained.

In addition to dissolved carbon the metastable sodium acetylide
$Na_2C_2$ is reported to be present in liquid sodium (4), ensuing
methane formation according to

$$1/2 \ Na_2C_2 + 2 \ H_2 = CH_4 + Na \tag{4}$$

also comes into consideration. This was already indicated by
Barton and Maffei (5), who investigated the inverse reaction by
addition of methane to liquid sodium.

Investigations on the solubility (6) and the stability of $Na_2CO_3$ in sodium led to the assumption (7) that the reactions

$$Na_2CO_3 + 4 \ Na = 3 \ Na_2O + C \tag{5}$$

and

$$Na_2CO_3 + 5 \ Na = 3 \ Na_2O + 1/2 \ Na_2C_2 \tag{6}$$

are possible.

The formation of methane in carbonate doped sodium is conceivable as consequent reaction according to eq. (1) and (4) respectively.

The formation of cyanide in sodium reported Hobart and Bjork (8), an enhancement of the carbon solubility by nitrogen partial pressures has been determined by Ainsley et.al. (6). Intermediate steps for the methane formation of cyanide dissolved in sodium are probably comparable to those found for carbonate, eq. (5) or (6).

EXPERIMENTAL METHODS AND RESULTS

The arrangement, Fig. 1, for the determination of the reaction of the carbon present in sodium with hydrogen to form methane basically consists of a reaction vessel for the liquid sodium. Influence of the vessel material on the reaction can be excluded by a copper lining. A copper tube for introduction of the reaction gas is immersed in the sodium. An aerosol trap is provided at the outlet of the vessel. The reaction gas, an 7.5 % $H_2$/argon mixture, was taken from a gas container; the total pressure was measured by using a spring manometer. The gas composition was specified by means of a gas chromatograph with argon-ionization detector (Lambert, A18; Firm: L'Air Liquide). The sensitivity for hydrogen and methane was 0.1 vpm, for oxygen and nitrogen 1 vpm. By means of a dilution method it was possible to determine high hydrogen concentrations beside low methane contents. A high vacuum tight diaphragm pump was used to recirculate the gas.

The total volume including gas piping, fittings and pump is ascertained to be 707 ml.

Sodium, glove-box quality (total carbon $\sim$ 10 wppm, carbonate carbon $\sim$ 3 ppm), was used for the experiments. An analytical grade of the doped carbon compounds NaCN, $Na_2CO_3$ (Merck, FRG) and a commercial grade compound of $Na_2C_2$ (Pierce Inorganics, The Netherlands) were employed. Before the arrangement was put into operation the high vacuum tight system was evacuated, then the reaction gas was supplied.

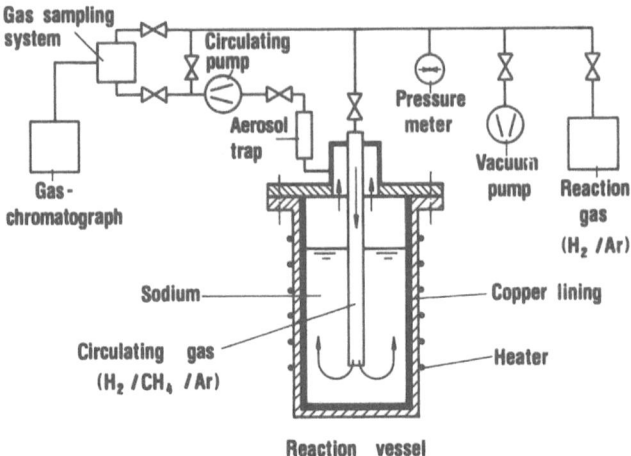

Fig. 1:   Experimental arrangement

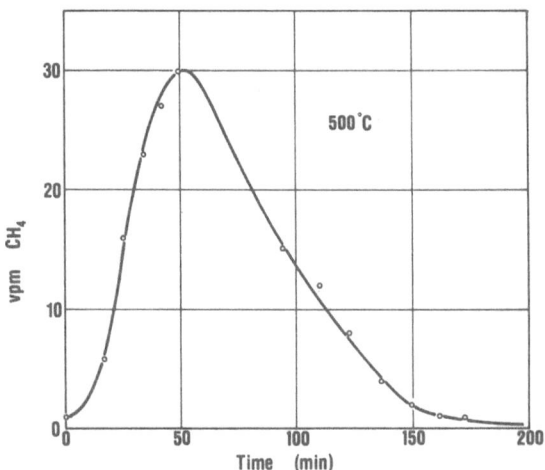

Fig.   : Methane Formation by flowing the $H_2$/Ar-reaction gas
        through sodium

In a first experiment a 7.5 % $H_2$/argon mixture was flushed through sodium (glove-box quality, mass of sodium: 46 g) at 500 °C with a gas flow of 50 ml/min and the quantity of methane formed was measured as a function of time, Fig. 2. The methane content determined reaches a maximum and finally sinks to a methane concentration of less than 1 vpm after approx. 200 min. The carbon portion reacting to form methane may thus be removed from the sodium.

In the further tests the methane-hydrogen equilibrium was adjusted by recirculating a given $H_2$/Ar-gas mixture through sodium ( 170 g). In the case of sodium in glove-box quality an equilibrium value resulted in the closed system after about 200 min at a gas circulation velocity of approx. 100 Nml/min. After the sodium had been cooled the gas phase was pumped off and a further 7.5 % $H_2$/argon gas mixture was added. It was then brought to reaction at 525 °C. By re-addition of the gas phase with hydrogen, a hydrogen level higher by a factor of 1.7 resulted, the equilibrium methane pressure increased by a factor of 3, Fig. 3.

The addition of acetylide to sodium (0.69 weight % $Na_2C_2$) resulted in a fast methane increase at 450 °C, which was higher by 2 orders of magnitude when compared to non-doped sodium, Fig. 4. A temperature increase of the sodium to 525 and 590 °C only resulted in a slight increase in the methane equilibrium value at approx. the same hydrogen partial pressure.

A carbonate doping (2.39 weight % $Na_2CO_3$) of the sodium had a different influence on the methane reaction than the acetylide.

While an equilibrium methane value was reached at 450 °C, which was in the same order of magnitude of the non-doped sodium, a large methane increase resulted at 525 °C, Fig. 5. This reaction occurred more slowly than the methane formation at 450 °C in the case of acetylide addition.

In the case of cyanide doping (2.17 weight % NaCN) a strong methane reaction only occurred at a sodium temperature of 595 °C. At the lower test temperatures of 450 and 525 °C the methane equilibrium partial pressure did not differ notably from the non-doped sodium value. The higher hydrogen pressure of the experiment at 525 °C resulted from the substitution of the gas phase by a new $H_2$/Ar-reaction gas, Fig. 6.

DISCUSSION

In the case of an $H_2$/argon mixture flowing through the sodium, Fig. 2, the methane quantity produced corresponds to a carbon content of approx. 1 ppm in sodium. When compared to the total carbon content of 10 wppm carbon (3 ppm in carbonate form) in the initial sodium it follows that only a slight portion has reacted to form methane.

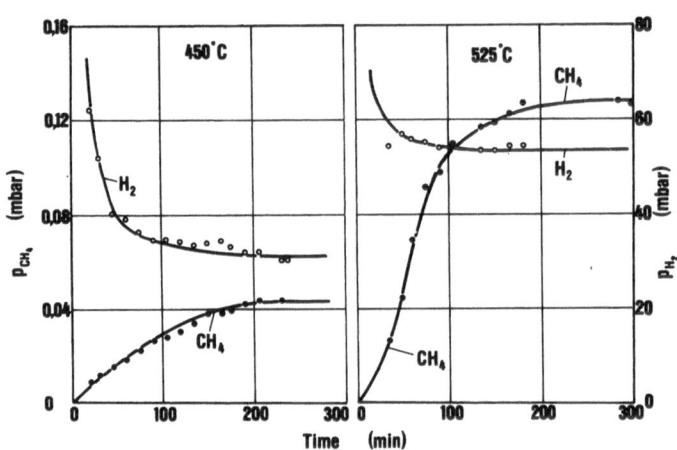

Fig. 3: Methane equilibration over sodium (glove-box quality)

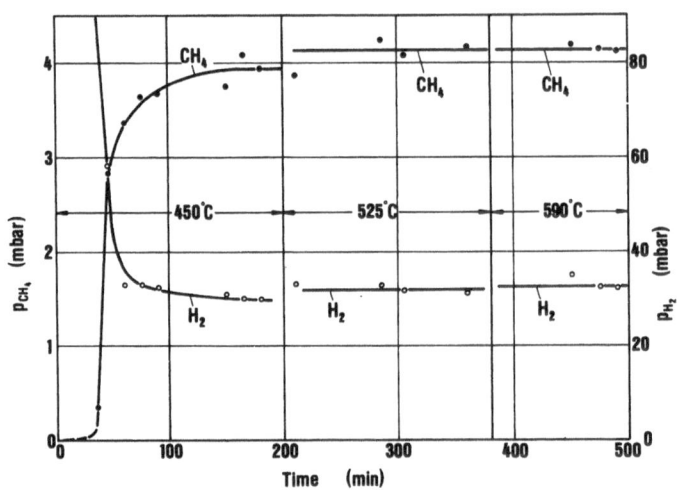

Fig. 4: Methane equilibration after $Na_2C_2$-addition (2300 wppm C) to sodium

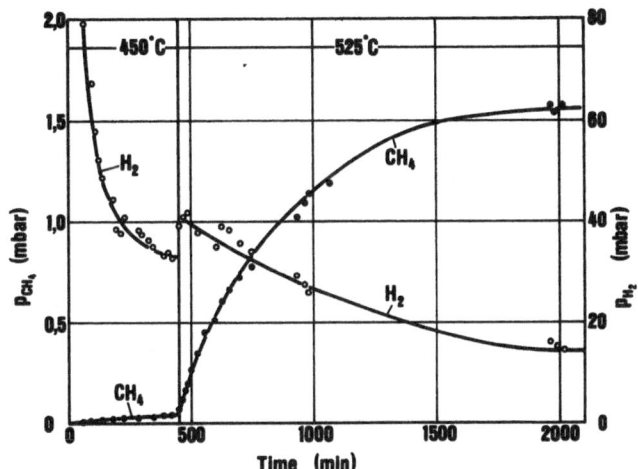

Fig. 5: Methane equilibration after $Na_2CO_3$-addition (2700 wppm C)
to sodium

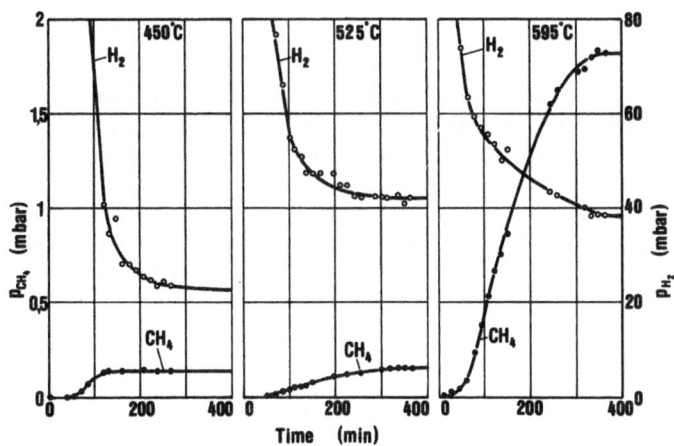

Fig. 6: Methane equilibration after NaCN-addition (5300 wppm C)
to sodium

Table 1. Carbon Activity in Sodium at Different Temperatures and Carbon Additives

| No. | Carbon additive in sodium | | $\vartheta_{Na}$ (°C) | $p_{CH_4}$ (mbar) | $p_{H_2}$ (mbar) | $a_C$ |
|---|---|---|---|---|---|---|
| 1 | (sodium: glove box quality) | total carbon: 10 wppm; carbonate carbon 3 wppm | 450 | 0.044 | 31 | $8 . 10^{-3}$ |
| | | | 525 | 0.13 | 53 | $3 . 10^{-2}$ |
| 2 | $Na_2C_2$ | 2300 wppm C | 450 | 3.9 | 30 | 0.76 |
| | | | 525 | 4.1 | 32 | (>1) |
| | | | 590 | 4.1 | 32.5 | (>1) |
| 3 | $Na_2CO_3$ | 2700 wppm C | 450 | 0.04 | 33 | $7 . 10^{-3}$ |
| | | | 525 | 1.55 | 14.5 | (>1) |
| 4 | NaCN | 5300 wppm C | 450 | 0.14 | 23 | $4,5 . 10^{-2}$ |
| | | | 525 | 0.16 | 42 | $6 . 10^{-2}$ |
| | | | 595 | 1.8 | 38.5 | (>1) |

Table 1 presents the equilibrium hydrogen and methane partial pressures over liquid sodium for the different experiments, Figs. 3 - 6. Carbon activities may be calculated according to eq. (2). In the case of non-doped sodium the values are in the order of $a_c \sim 10^{-2}$, a typical magnitude of value obtained by other methods with $Na_2CO_3$ ($\vartheta$ = 450 °C) and NaCN-addition ($\vartheta$ = 450, 525 °C) at the lower experimental temperatures. In the case of $Na_2C_2$ a carbon activity of $a_c$ = 0.76 results of the experiment at $\vartheta$= 450 °C.

On this basis of calculation carbon activities $a_c$ > 1 are in some cases obtained for the $Na_2C_2$-, $Na_2CO_3$- and NaCN addition to sodium, Table 1. This indicates that the methane over sodium is partially or completely formed by the metastable acetylide, eq. 4.

It is interesting to note that different carbon species resulted in high carbon activities at different temperatures. While in the case of the acetylide doping high methane partial pressures were measured at 450 °C, this applies to the carbonate doping at the temperature step of 525 °C and to cyanide at the experimental temperature of 590 °C.

REFERENCES

1.  J. T. Holmes, C. R. F. Smith, W. H. Olson: Proc. Int. Conf.
    on Liquid Metal Technology in Energy Production
    Champion, CONF-760503-P1 (May 3 - 6, 1976).
2.  O. Kubaschewski, E. L. Evans, C. B. Alcock
    Metallurgical Thermochemistry, Pergamon Press 1967.
3.  D. L. Johnson, NAA-SR-8448 (Sept. 1964).
4.  G. K. Johnson, E. H. Deventer, J. D. Ackermann,
    W. N. Hubbard, J. Chem. Thermodynamics 5 (1973) 57.
5.  G. B. Barton, H. P. Maffei, BNWL-SA-Z181 (1969).
6.  R. Ainsley, L. P. Hartlib, P. M. Holroyd, G. Long
    J. Nucl. Mat. 52 (1974) 255.
7.  J. P. Maupré, CEA-R-4905 (1977).
8.  E. W. Hobart, R. G. Bjork, Nucl. Application 1 (1965) 490.

THERMODYNAMIC AND EXPERIMENTAL STUDY OF SODIUM HYDROXIDE

DECOMPOSITION IN SODIUM BETWEEN 430 and 550 $^\circ$C

Claude Oberlin, Pierre Saint-Paul

Electricité de France
Ecuelles
France

INTRODUCTION

Earlier tests (1) carried out in sodium polluted (up to 10 %) by sodium hydroxide between 400 and 550 $^\circ$C have shown that the austenitic materials used (especially Alloy 800) for the steam generator tubes of fast breeder reactors cooled with sodium could induce a risk of stress corrosion cracking in this medium. However, in sodium polluted by sodium oxide, the same materials are not affected by this phenomenon in the same conditions of stress, but are only subject to slight intergranular corrosion.

Sodium hydroxide is formed immediately when a water leak occurs in a steam generator, but its stability in sodium is governed by the pressure of hydrogen in the gaseous phase. When this pressure is lower than a given value (equilibrium pressure), sodium hydroxide decomposes to yield sodium oxide. Because the low hydrogen pressure in the cover gas of a steam generator operating normally or when a safety system trips after a water leak, the sodium hydroxide formed tends to decompose into sodium oxide. However, in order to determine how long steam generator materials can stay in contact with the Na-NaOH medium following a sodium water reaction, a study of the decomposition of sodium hydroxide in sodium between 430 $^\circ$C and 550 $^\circ$C was undertaken. It consisted of finding, for a given temperature (between 430 and 550 $^\circ$C) and employing Na-NaOH mixtures of known composition (ratio of mass of NaOH to mass of Na between 5 and 50 %):

- partial hydrogen pressure in equilibrium over a quaternary mixture Na-NaOH-Na$_2$O-NaH obtained from the reactions:

- 2Na + NaOH $\rightleftarrows$ Na$_2$O + NaH
  NaH $\rightleftarrows$ Na + 1/2 H$_2$

- the kinetic properties of the decomposition of sodium hydroxide in sodium.

EXPERIMENTAL CONDITIONS

The testing device used, shown schematically in Fig. 1, is made up of a glove box allowing Na-NaOH mixtures to be inserted in the reaction chamber under a controlled atmosphere, and a stainless steel reaction chamber containing the Na-NaOH mixture, which can be heated using a tubular furnace. A first thermocouple enables the temperature of the mixture to be recorded and a second thermocouple measures the temperature of the gaseous mixture. A pumping device (subjected to reduced pressure to evacuate hydrogen formed during the decomposition of NaOH), connects the reaction chamber to the partial hydrogen pressure measurement system or to a gas analysis loop (chromatography in gaseous phase) via an appropriate set of valves. A stainless steel filter (of thickness 2 mm and porosity 8 µm) is inserted in the upper part of the reaction chamber to prevent sodium losses, but allows nevertheless free diffusion of hydrogen in the gaseous phase. The pumping unit consists of a rotary vane pump. A simplified device placed at the rotary vane pump outlet allows the volume of the gas extracted to be measured. The reference chamber of the differential pressure detector is subjected to reduced pressure during the measurements.

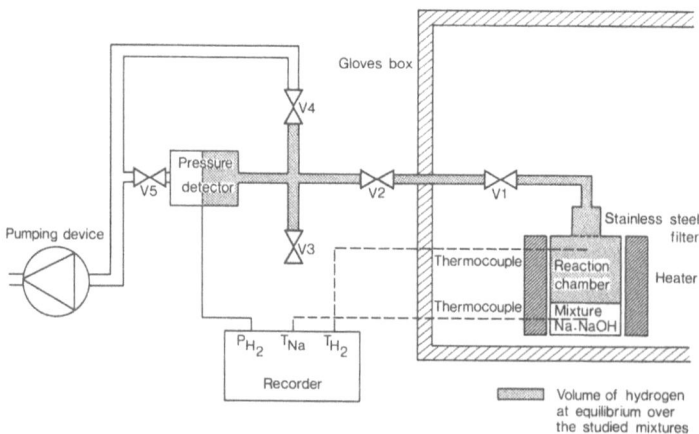

Figure 1: Schematic view of the testing equipment

Operating mode

Performing a test involves the following operations: Filling up the reaction chamber with the mixture to be studied (approximately 20 g of Na and the corresponding amount of NaOH to obtain the desired molar fraction), in controlled atmosphere. The chamber is then closed and connected to the pumping and pressure measurement devices. The unit thus formed is put under reduced vacuum (dynamic vacuum) until the pressure detector reads zero (valves $V_1$, $V_2$ and $V_4$ open, $V_3$ closed); when pressure is measured, one of the two chambers of the detecting element is connected to the pumping unit ($V_5$ open). After closing valve $V_4$, the Na–NaOH solution is heated to the desired temperature (between 430 and 500 $^\circ$C). The hydrogen formed fills up the space between the reaction chamber, valves $V_4$ and $V_3$ and the pressure detecting element (shaded part of the diagram in Fig. 1). Gaseous hydrogen is extracted when the pressure measured by the detecting element is stable (equilibrium reached), by opening valve $V_4$. The volume of hydrogen thus estracted is measured at rotary vane pump outlet. After closing valve $V_4$, the system tends to a new equilibrium. This sequence of operations is followed up until the NaOH content of the solution is close to zero (molar fraction of NaOH < 0.01).

EXPERIMENTAL RESULTS – DISCUSSION

Hydrogen pressure at equilibrium $P_e$ as a function of the molar fraction $X_{NaOH}$ of the Na–NaOH mixture

The various test carried out have allowed the respective roles of temperature and initial NaOH content to be studied. The curves shown in Fig. 2 show the pressure of hydrogen at equilibrium $P_e$ over the Na–NaOH–$Na_2O$–NaH mixtures formed from the initial Na–NaOH mixture (initial NaOH mass content is equal to 50 % of Na) at different temperatures. The following observations can be made:

The results of testing at 550 $^\circ$C are scattered, which is probably due to the fact that during the measurements of hydrogen pressure, sodium losses through vaporisation at this high temperature have altered the composition of the chemical medium under study; this phenomenon has already been observed in earlier reports (3). The general shape of the various curves is similar for the different temperatures selected. They have three distinct sections: a section in which $P_e$ increases with the NaOH content before levelling off. The corresponding level which takes the following values of $P_e$ at the different temperatures studies:

| Temperature | ($^\circ$C) $\pm$ 5 | 430 | 450 | 480 | 525 | 550 |
|---|---|---|---|---|---|---|
| equilibrium pressure | (Pa) $\pm$ 800 | 24800 | 24200 | 20500 | 13300 | 1070 |

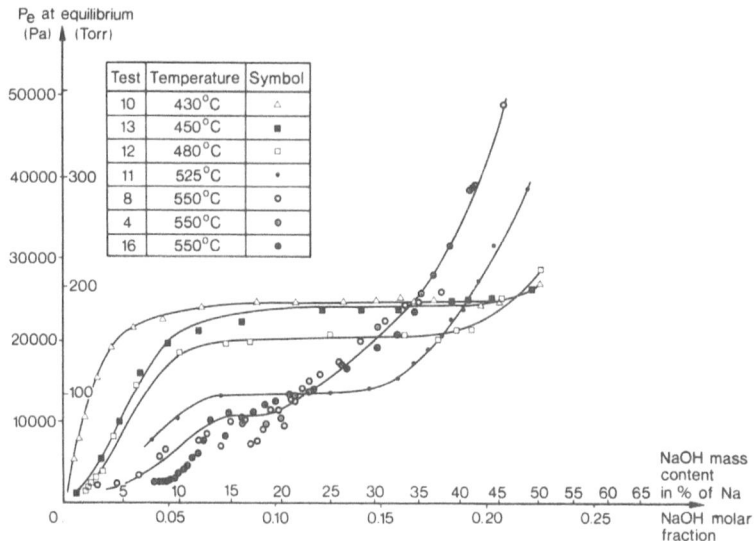

Figure 2 : Evolution of the pressure of hydrogen at equilibrium $P_e$ over the initial Na-NaOH mixture
(initial NaOH mass content equal to 50 % of Na) at different temperatures.

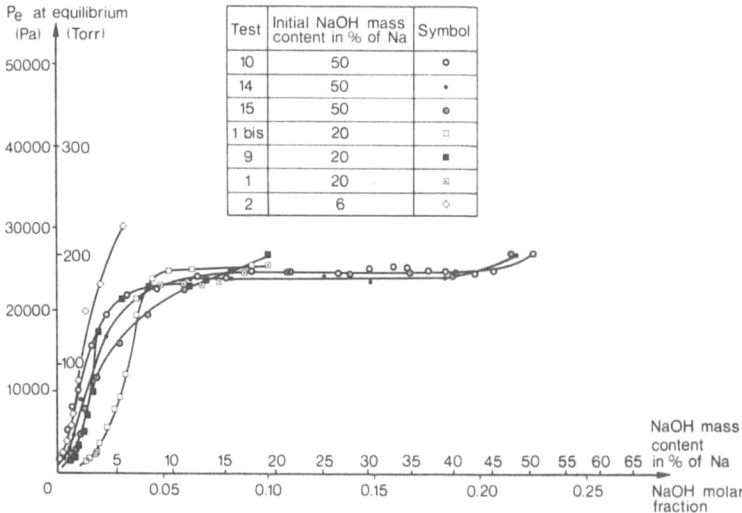

Figure 3 : Evolution of the pressure of hydrogen at equilibrium $P_e$ over Na-NaOH mixtures
(different NaOH mass content) at 430°C.

The length of the steady level decreases markedly when the temperature rises. A third section in which $P_e$ increases rapidly with the NaOH content. The NaOH content from which $P_e$ increases following the level section of the curve, declines markedly when the temperature increases, especially above 500 °C.

The curves in Figs. 3 and 4 depict the change in hydrogen equilibrium pressure $P_e$ as a function of NaOH content at temperatures in the range 430° and 550 °C for variable initial NaOH contents (5 to 50 % mass). The curves yield the following observations:

The shape of the curves is identical for initial NaOH contents of 20 and 50 % mass (existence of a pressure level), however, for an initial NaOH content of 6 %, there is no steady pressure level. The length of the steady pressure level decreases with the initial NaOH content.

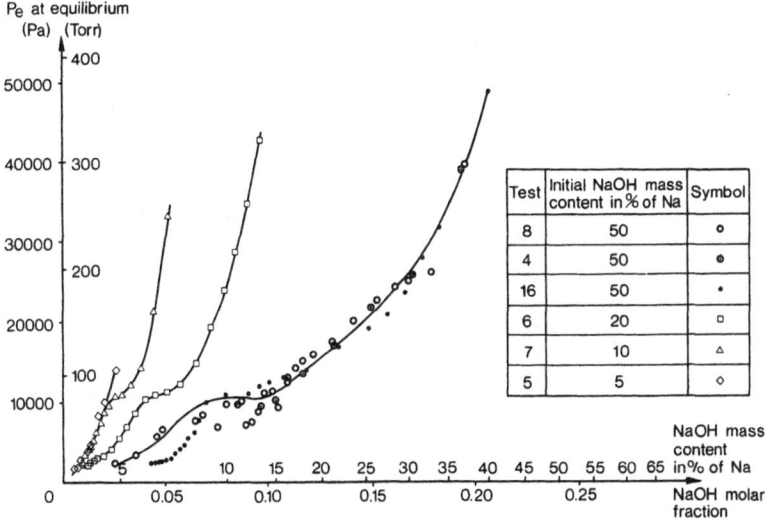

Figure 4: Evolution of the pressure of hydrogen at equilibrium $P_e$ over Na-NaOH mixtures (different NaOH mass content) at 550°C.

The curves obtained are devided into the three sections described earlier for all initial NaOH contents, except 5 % mass, for which the pressure level is almost absent. The length of the level decreases as the initial NaOH content of the Na-NaOH solution under study decreases. The maximum pressure of $P_e$ obtained for the highest NaOH contents diminishes when the initial NaOH content decreases.

A priori, the general form of the curves $P_e$ = f ($X_{NaOH}$) should be a function of the nature and quantity of the various components present in the system.

The diagram of phases in the NaOH system recently described by
Maupré (4) and which is given in schematic form in Fig. 5, was used
to depict the results obtained.

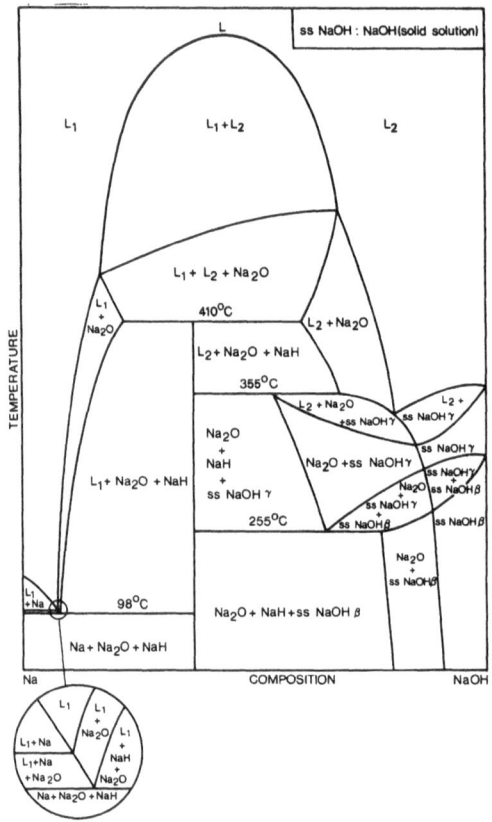

Figure 5: Na-NaOh phases diagram [3]

The diagram shows that in the testing conditions selected
(temperature ranging from 430 to 550 °C, NaOH content less than
50 % mass), three domains can be distinguished: $L_1 + L_2$, in which
$L_1$ and $L_2$ respectively represent the liquid metallic phase (NaOH
dissolved in liquid sodium) and the liquid caustic phase (Na dissolved in liquid NaOH) obtained by the demixing reaction (dissolved
$Na_2O$ can exist in this domain):

$$Na + NaOH \rightleftarrows L_1 + L_2$$

$$L_1 + L_2 + Na_2O \text{ (appearance of the } Na_2O \text{ phase)}$$

$$L_1 + Na_2O \text{ (disappearance of demixing phase } L_2)$$

The shape of the curves $P_e = f(X_{NaOH})$ obtained at different temperatures and for different NaOH contents can be interpreted as follows on the basis of Maupré (4) results: At the beginning of a test, if the NaOH content is high enough, the phase $L_1$ and $L_2$ are present. The pressure of hydrogen at equilibrium point is high and declines when NaOH content or temperature fall. As the extraction of hydrogen proceeds during the tests, a displacement to the right of the reaction:

$$L_1 + L_2 \rightleftarrows NaH + Na_2O$$

can be observed followed by the decomposition reaction of NaH:

$$NaH \rightleftarrows Na + 1/2 \ H_2$$

In these conditions, the content in phase $L_2$ decreases and the dissolved $Na_2O$ content rises. Correlatively, pressure $P_e$ falls, not only because the NaOH content decreases, but perhaps also because of the rise in dissolved $Na_2O$ contained in $L_1 + L_2$. To support this hypothesis, it can be seen that the slope of the curves $P_e = f(X_{NaOH})$ increases when the initial NaOH content decreases in the tests performed at 550 $^\circ$C (Fig. 4), which means that the lower the $NaOH/Na_2O$ ratio, the greater the interaction of the dissolved $Na_2O$ with equilibrium $NaOH-H_2$.

When precipitation of $Na_2O$ occurs at the given temperature, the three phases $L_1$, $L_2$ and $Na_2O$ (s) are in presence. Pressure $P_e$ steadies and the level shown on the curves of Figs. 2, 3 and 4 is obtained.

When $L_2$ disappears following the diminution in NaOH content, pressure $P_e$ falls again as extraction proceeds.

<u>Kinetic properties of the decomposition of sodium hydroxide in sodium</u>

Taking it that sodium hydroxide decomposes according to the general reaction: $Na + NaOH \rightarrow Na_2O + 1/2 \ H_2$, its rate of decomposition:

$$v = -\frac{d \ (N_{NaOH})}{dt}$$

can be measured by the rate of hydrogen formation $v = K \dfrac{dP}{dt}$

with $K = \dfrac{V}{2RT}$ (V: volume of the gaseous phase above the mixture,
  R: the gas constant and
  T: temperature at which volume V is determined)

In addition, after each hydrogen extraction during the tests, the change of gas pressure P over time was measured. The results obtained should thus allow study of the rate of decomposition of

NaOH in varied conditions. However, the rate of decomposition is dependent on both P and $P_e$, which, in the light of the operating method used (successive hydrogen extraction from the gazeous phase), hinders use of the resulting curves.

In a first stage, in an attempt to reach the rate of decomposition by positing that the pressure of hydrogen over sodium cannot rise (sweeping by a neutral gas in a steam generator), reference was made to the slope of the curve P = f (t) at time t = 0, after each hydrogen reaction. The results obtained were very scattering and the method was discarded.

Due to this result, the diagram showing NaOH content as a function of time was drawn for each test, taking into account the different times required for hydrogen pressure to evolve to the equilibrium pressure $P_e$ after each extraction, the duration of hydrogen extraction being considered as nil (< 40 s). This provides an approximation by excess of the time actually taken to obtain complete decomposition of NaOH on the basis of solutions of known NaOH content.

The curves thus obtained for different Na-NaOH solutions at different temperatures (Figs. 6, 7 and 8) yield the following indications:

The rate of decomposition of NaOH declines when temperature increases; tests undertaken using solutions with an initial content of 50 % mass of NaOH to Na at various temperatures (430, 450, 480, 525 and 550 $^\circ$C) illustrate this phenomenon well (Fig. 6). The scattering of the results recorded for the different tests at 550 $^\circ$C is probably due to the fact that at this temperature, as was noted earlier, evaporation of sodium in the gaseous phase is no longer negligible.

The rate of the NaOH decomposition reaction at 430 $^\circ$C (Fig. 7) is greater, the higher the initial NaOH content of the solution; this is less clear at 550 $^\circ$C (Fig. 8). The difference between the two decomposition curves starting from an initial 20 % NaOH solution at 430 $^\circ$C (tests 1 bis 9) can be explained by the higher initial $Na_2O$ content in test 1 bis.

It can be observed that the slope of the curves under study changes, and that these changes can be explained in the light of the conclusions drawn earlier from the curves $P_e$ = f ($X_{NaOH}$). The rate of decomposition of NaOH in sodium is characterized by: A relatively rapid phase, corresponding to the solutions formed by the two demixing solutions $L_1$ + $L_2$ (the test at 525 $^\circ$C is significant as regards observation of this phenomenon). A very rapid intermediary phase, corresponding to the solutions contained $L_1$, $L_2$ and precipitated $Na_2O$; this phase is almost invisible when testing at

550 $^{\circ}$C (Fig. 8). A third phase, much slower than both previous ones, is characterized by the presence of $L_1$ and $Na_2O$ in the solution studied.

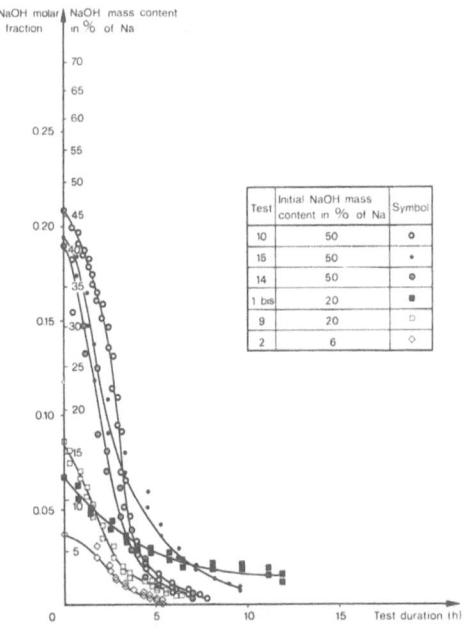

| Test | Intial NaOH mass content in % of Na | Symbol |
|------|------|------|
| 10 | 50 | o |
| 15 | 50 | • |
| 14 | 50 | ⊙ |
| 1 bis | 20 | ■ |
| 9 | 20 | □ |
| 2 | 6 | ◇ |

Fig. 6. Kinetic curves of the decomposition of NaOH in Na obtained for a Na-NaOH mixture (initial NaOH mass content equal to 50% of Na) at different temperatures.

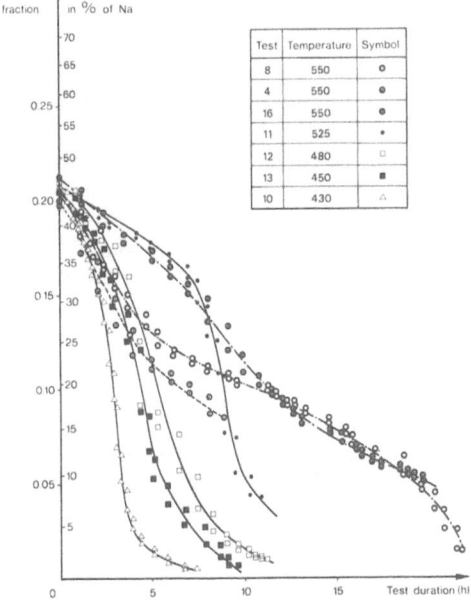

| Test | Temperature | Symbol |
|------|------|------|
| 8 | 550 | o |
| 4 | 550 | ⊙ |
| 16 | 550 | ● |
| 11 | 525 | • |
| 12 | 480 | □ |
| 13 | 450 | ■ |
| 10 | 430 | △ |

Fig. 7. Kinetic curves of the decomposition of NaOH obtained for different NaOH mixtures (different NaOH mass content) at 430°C.

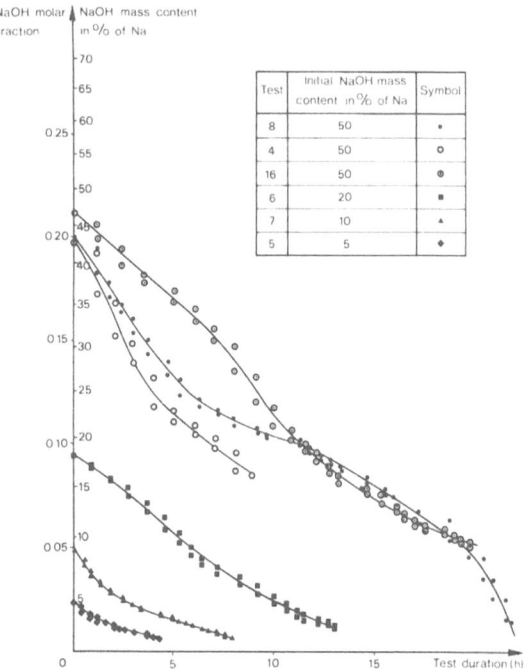

<u>Figure 8</u>: Kinetic curves of the decomposition of NaOH in Na obtained
         for different Na-NaOH mixtures (different NaOH mass content)
         at 550°C


     Austenitic materials (especially alloy 800 used for SUPER
PHENIX) are subject to risk of corrosion inducing cracking under
stress above 430 °C in sodium polluted by sodium hydroxide, but
only when the content exceeds 5 % of the solution (1); by contrast,
the same materials in contact with sodium oxide are only subject
to intergranular corrosion in the same conditions of stress (2). It
is thus of interest to assess the time needed to reach this 5 %
concentration starting with Na-NaOH mixtures with varying NaOH
contents (up to 50 %) for temperature ranging from 430 to 550 °C.
The results deduced from Figs. 6, 7 and 8 prove that when leakage
occurs in a steam generator, the sodium hydroxide formed during the
Na-H$_2$O reaction decomposes rapidly. In most cases, it takes much
less than a day to reach a solution whose NaOH content does not
entail the risk of stress corrosion cracking of the steam generator
tubes; the longest duration is obtained at 550 °C for a 50 % initial
NaOH content (about 21 hours); however, this is probably higher
than the time needed in operating conditions, since in this case,
hydrogen pressure in the gaseous phase has a steady low value or is
nil (when the safety is triggered) which was not the case during
testing. The attempt to interpret results by using the slope of

curves $P = f(t)$ for $P = 0$ always yielded much lower durations
(approximately 0.5 h), although the scattering of the results was
important.

CONCLUSION

The experimental results obtained made it possible to draw the
curves showing hydrogen pressure in equilibrium with an Na-NaOH
solution for various initial NaOH contents and at different temperatures, during the successive extractions of the hydrogen formed by
a NaOH decomposition. The shape of the curves could be explained
by the following system of reactions, displaced to the right during
hydrogen extractions:

$$Na + NaOH \rightarrow L_1 + L_2 \rightarrow L_1 + L_2 + Na_2O \rightarrow L_1 + Na_2O$$

in which $L_1$ is the phase NaOH dissolved in liquid sodium and $L_2$ is
the phase Na dissolved in NaOH in liquid form.

In addition, the study of the kinetic properties of the decomposition of NaOH in Na has enabled us to show that water leakage
from a steam generator engendering automatic switching on of the
safety system should not entail a risk of stress corrosion cracking
of Alloy 800 tubes, given the rate at which the sodium hydroxide
decomposes. In localised areas (e.g. below pins), the time it takes
to reach an NaOH content of sodium that is low enough to avoid the
occurence of stress corrosion cracking can be longer, and it would
be necessary to execute representative tests to assess the effective
risk in actual operation.

REFERENCES

1.  P. Berge, C. Oberlin, P. Saint-Paul, M. Zbinden, J. of
       Nuclear Materials, 66, 65 (1977).
2.  EDF, DER, Internal report (1980).
3.  J. L. Henry, S. A. O'Hare, M. P. Krug, USBM, 1532, (1971).
4.  J. P. Maupré, Report CEA, R 4905 (1978).

# THE EQUILIBRIUM HYDROGEN PRESSURE-TEMPERATURE DIAGRAM FOR THE

# LIQUID SODIUM-HYDROGEN-OXYGEN SYSTEM

Clive F. Knights        Andrew C. Whittingham

UKAEA                   CEGB
Harwell                 Berkeley Nuclear Laboratories
United Kingdom          United Kingdom

## INTRODUCTION

The composition-temperature phase diagram for the sodium-hydrogen-oxygen system is complex and not amenable to simple interpretation. However, the underlying equilibria in this system can be more simply represented in the form of a complementary hydrogen equilibrium pressure-temperature diagram, which has been constructed using published data supplemented by experimental measurements of hydrogen equilibrium pressures over condensed phases in this system. Possible applications of the diagram and limitations regarding its use are outlined.

## PHASE EQUILIBRIA

Previous studies e.g. (1-3) have clearly shown that the following phases can exist in the Na-H-O system:

(i)   Liquid Sodium, $L_1$, containing dissolved sodium hydride, sodium monoxide and sodium hydroxide, whose composition is close to 100 mol% sodium, $Na_{(\ell)}$.

(ii)   Liquid $L_2$, based on liquid sodium hydroxide, $NaOH_{(\ell)}$, containing dissolved sodium hydride and sodium monoxide, whose composition varies with temperature and hydrogen pressure.

(iii) Solid solutions, $\alpha L_2$ and $\beta L_2$, based on the $\alpha$ and $\beta$ phase modification of sodium hydroxide, $NaOH$ (solid sol).

(iv)   Solid sodium hydride, $NaH_{(s)}$.

287

(v)    Solid sodium monoxide, $Na_2O_{(s)}$.

(vi) Hydrogen gas, $H_{2(g)}$.

In addition, the following reversible reactions can occur between the components:

$$2Na_{(\ell)} + NaOH_{(s,\ell)} \; \underset{\leftarrow}{\rightarrow} \; Na_2O_{(s)} + NaH_{(s)} \tag{1}$$

$$Na_{(\ell)} + NaOH_{(s,\ell)} \; \underset{\leftarrow}{\rightarrow} \; Na_2O_{(s)} + 1/2H_{2(g)} \tag{2}$$

$$Na_{(\ell)} + 1/2H_{2(g)} \; \underset{\leftarrow}{\rightarrow} \; NaH_{(s)} \tag{3}$$

$$Na_2O_{(s)} + H_{2(g)} \; \underset{\leftarrow}{\rightarrow} \; NaH_{(s)} + NaOH_{(s,\ell)} \tag{4}$$

(s, ℓ, g - solid, liquid, gas)

of which any two are sufficient to define the system.

SCOPE OF PRESENT WORK

In the present work, hydrogen equilibrium pressures were measured over the following reactant mixtures, whose initial compositions are given below:

a)   Na-NaOH > 95 Mol% Na
b)   $Na_2O$-NaH 50:50
c)   $Na_{(Excess)}$-$Na_2O$-NaH (oxygen:hydrogen mole ratio unity)

at temperatures up to 470 °C. A further equilibrium reaction was studied in which hydrogen gas was reacted with pure, solid sodium monoxide to produce sodium hydride and sodium hydroxide in accordance with equation (4). Equilibrium hydrogen pressures over equimolar mixtures of $Na_2O$, NaH and NaOH were measured in the temperature range 205 - 475 °C.

Confirmation that the equilibrium given by equation (4) pertained was provided by X-ray diffraction analysis of the products at room temperature (4). Approximately equimolar proportions of sodium hydride and sodium hydroxide were the sole reaction products identified, in addition to unreacted sodium monoxide, at all temperatures below 412 °C; a separate sodium phase was only identified in experiments carried out after prolonged heating (above 412 °C).

EXPERIMENTAL TECHNIQUE AND PROCEDURE

In experiments at temperatures below 300 °C, reactants were contained in a stainless steel lined pyrex glass reaction vessel

immersed in a stirred silicone oil bath. The absorption or release
of hydrogen to equilibrium was monitored by means of a capacitance
manometer. At temperatures greater than 300 $^{\circ}$C, the reactants were
contained in an isothermal, stainless steel reaction vessel,
furnished with a nickel liner, and an iron membrane (0.25 mm thick)
which allowed hydrogen to diffuse into an evacuated dead space of
low volume, where pressure measurement was made by means of a
capacitance manometer. Hydrogen could be admitted or removed from
the vessel via a high temperature valve connected to a vacuum frame.
The use of an isothermal apparatus prevented co-condensation of
hydrogen with liquid sodium in cooler parts of the apparatus which
would have interferred with the intended equilibria.

The sodium hydroxide was Analar grade (99 %, < 1 % $Na_2CO_3$),
and the sodium hydride ($\sim$ 97 %) and sodium monoxide ($\sim$98 %) were
both supplied by Alpha Inorganics. The sodium monoxide was purified
prior to use by distillation at 480 $^{\circ}$C under high vacuum in nickel
crucibles.

RESULTS

Hydrogen equilibrium pressures for the $Na-Na_2O-NaH$ and Na-NaOH
systems were identical within experimental error at equivalent
temperatures; these are plotted in Fig. 1 as a function of reciprocal temperature. Also included are hydrogen equilibrium pressures
over the binary sodium-sodium hydride system calculated from the
recommended expression of Whittingham (5):

$$\log_{10} P(kPa) = 10.82 - \frac{6122}{T} \qquad (5)$$

At temperatures below $\sim$ 350 $^{\circ}$C, the hydrogen equilibrium pressures for the ternary Na-H-O mixtures are identical to those in the
binary Na/NaH system indicating that the presence of sodium monoxide
does not affect the binary Na/NaH equilibrium, but in the temperature
range    350 - 412 $^{\circ}$C, the hydrogen equilibrium pressure in the
ternary system is increasingly reduced relative to line h$\ell$f (Fig. 1)
and reaches an abrupt maximum of 24.1 ($\pm$ 1.6) kPa at 412 $\pm$ 2 $^{\circ}$C
($\ell$). At higher temperatures, the equilibrium pressure remains effectively constant (e-g); the formation of a plateau equilibrium
hydrogen pressure at 412 $^{\circ}$C and $\sim$ 24 kPa confirms the previous observations of Williams (6), Mitkevitch and Shikhov (3) and Myles
and Cafasso (7) and is discussed further.

Hydrogen equilibrium pressures measured over the $Na_2O_{(s)}-NaH_{(s)}$
-$NaOH_{(s, \ell)}$ system are presented in Table 1; no previous data have
been published for this equilibrium as a function of temperature.
Similar equilibrium pressures were also measured over $Na_2O-NaH$
mixtures heated below 412 $^{\circ}$C. It was found that at temperatures
below $\sim$ 360 $^{\circ}$C, hydrogen equilibrium pressures were greater than
those in the binary Na-NaH system. At temperatures > 360 $^{\circ}$C,

equilibrium pressures were progressively lower than those in the
Na-NaH system, but similar to or slightly greater than those measured
in the Na-Na$_2$O-NaH and Na-NaOH systems within the limits of experimental error. At 412 $^\circ$C, the hydrogen pressure again attained a
maximum of 25.2 ($\pm$ 0.6) kPa and remained constant at higher temperatures along e-g.

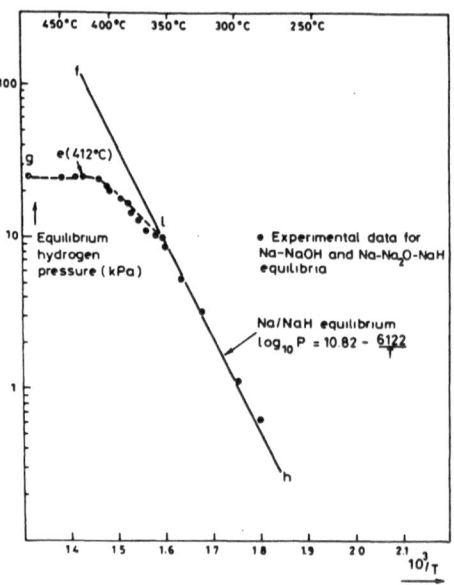

Fig. 1    Equilibrium Hydrogen Pressures for Na-Na$_2$O-NaH, Na-NaOH
          and Na-NaH Systems

Table 1    Equilibrium Hydrogen Pressures as a Function of Temperature
           over the Na$_2$O$_{(s)}$-NaH$_{(s)}$-NaOH$_{(s,\ell)}$ System

| Temperature ($^\circ$C) | $P_{H_2}$ (kPa) | Temperature ($^\circ$C) | $P_{H_2}$ (kPa) |
|---|---|---|---|
| 206 | 0.51 | 365 | 12.9 |
| 236 | 1.67 | 373 | 15.6 |
| 258 | 2.47 | 382 | 16.9 |
| 268 | 2.47 | 388 | 19.5 |
| 283 | 2.87 | 398 | 22.4 |
| 297 | 4.07 | 403 | 23.7 |
| 307 | 4.53 | 412 | 25.2 |
| 317.5 | 5.64 | 425 | 25.0 |
| 324 | 6.40 | 440 | 24.4 |
| 334 | 7.33 | 450 | 24.5 |
| 340 | 7.80 | 460 | 24.6 |
| 348 | 9.47 | 470 | 23.7 |
| 355 | 11.2 | | |

The hydrogen equilibrium pressure is plotted as a function of reciprocal temperature in Fig. 2, and the following relationship is closely obeyed in the temperature range 206 - 412 °C:

$$\log_{10} P \text{ (kPa)} = 5.386 - \frac{2723}{T} \tag{6}$$

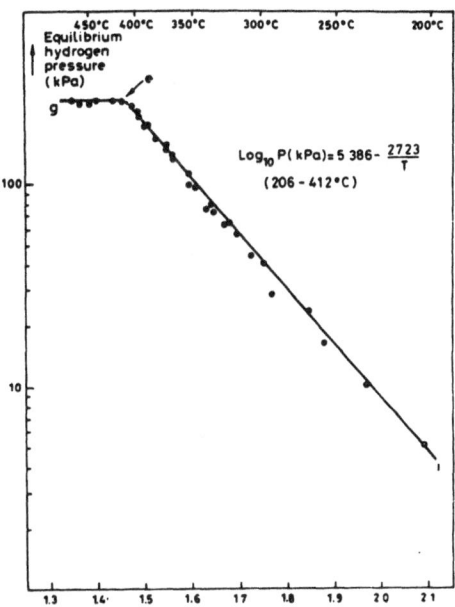

Fig. 2  Equilibrium Hydrogen Pressures over $Na_2O$-NaH-NaOH Mixtures

The equilibrium constant for reaction (4) is defined by:

$$K_4 = \frac{a_{NaOH}\, a_{NaH}}{a_{Na_2O}\, P_{H_2}}$$

Defining the ratio $(a_{NaOH} a_{NaH}/a_{Na_2O})$ by Q, then since

$$\Delta G_T^o = - RT \log K_4$$

we obtain the following expressions, calculated from free energy data (8) for reaction (4):

$$\log_{10} P/Q \text{ (kPa)} = 5.716 - \frac{2628}{T} \quad (>318\ ^oC \text{ (m.pt NaOH)}) \tag{9}$$

$$\log_{10} P/Q \text{ (kPa)} = 6.943 - \frac{3361}{T} \quad (>318\ ^oC \text{ (m.pt NaOH)}) \tag{10}$$

Since sodium monoxide does not act as a solvent for solid solution formation in this temperature range, we may put $a_{(o)} = 1$, and differences between the calculated and measured values of equilibrium hydrogen pressure suggest that the product of the sodium hydroxide and sodium hydride activities, $(a_{NaOH}, a_{NaH})$ is less than unity, varying between 0.36 at 300 $^\circ$C and 0.34 at 412 $^\circ$C; the significance of these values is discussed more fully in the following section.

EQUILIBRIUM HYDROGEN PRESSURE—TEMPERATURE DIAGRAM

Two versions of the hydrogen equilibrium pressure-temperature diagram for the Na-H-O system are presented in Fig. 3 and 4 respectively. The first (Fig. 3) is calculated from available thermochemical data (8) assuming all the condensed phases – $Na_{(\ell)}$, $NaOH_{(\ell)}$, $Na_2O_{(s)}$ and $NaH_{(s)}$ are at unit activity, and the second (Fig. 4) is the modified version based on a combination of previously published equilibrium pressure measurements (6, 7, 9, 11) and the present data.

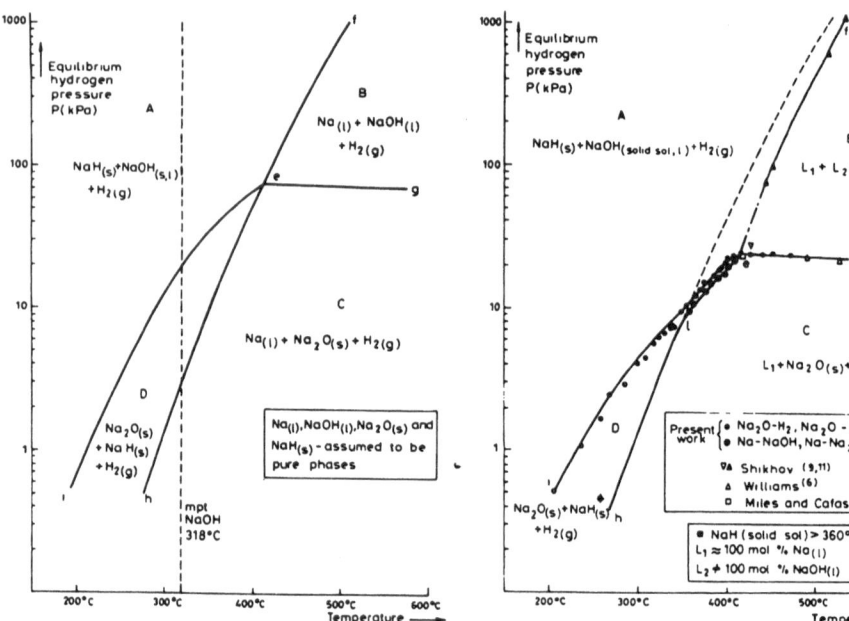

Fig. 3 Equilibrium Hydrogen
Pressure–Temperature
Diagram for the Na-H-O
System Calculated from
Thermochemical Data (8)

Fig. 4 Equilibrium Hydrogen
Pressure–Temperature
Diagram for the Na-H-O
System based on Experimental
Measurements

In each version, there are four regions A, B, C and D, each comprising two condensed phases and hydrogen gas, separated by lines e-f, e-g, e-h and e-i respectively. The phase rule states that,

$$P + F = C + 2 \tag{11}$$

where P is the number of phases, F the degrees of freedom, and C the number of components. Since $C = 3$ and hydrogen gas is always present, $P' + F = 4$, where $P'$ is the number of condensed phases in the system. Each region A, B, C, D, therefore contains two condensed phases, whereas three condensed phases are in equilibrium along the boundary lines e-f, e-g, e-h and e-i respectively. In Fig. 3 the line h-e-f is the binary Na/NaH boundary, and i-e corresponds to the equilibrium equations (9) and (10). Line e-g gives hydrogen equilibrium pressures calculated from free energy data for equation (2). For condensed phases, $Na_{(\ell)}$, $NaOH_{(\ell)}$, $Na_2O_{(s)}$ and $NaH_{(s)}$ are in equilibrium at point e, which corresponds to a class II, four phase, invariant equilibrium reaction (7):

$$Na_{(\ell)} + NaOH_{(\ell)} \quad \underset{>412\ ^\circ C}{\overset{<412\ ^\circ C}{\rightleftarrows}} \quad Na_2O_{(s)} + NaH_{(s)} \tag{12}$$

which is calculated to occur at 412 $^\circ$C and 75 kPa hydrogen pressure.

The version presented in Fig. 4 differs in that point e is displaced to a lower hydrogen pressure (24.1 $\pm$ 1.6 kPa) away from the binary Na/NaH line (dotted), albeit at the same temperature 412 $^\circ$C. This temperature and hydrogen pressure is in excellent agreement with the recent data of Myles and Cafasso (7) (412 $^\circ$C, 24.2 kPa) and Maupre (2) (410 $^\circ$C), where the equilibrium reaction is given by:

$$L_1 + L_2 \quad \underset{>412\ ^\circ C,\ >\ 24\ kPa}{\overset{<412\ ^\circ C,\ <\ 24\ kPa}{\rightleftarrows}} \quad Na_2O_{(s)} + NaH_{(s)}$$

The activity of the liquid caustic phase, $L_2$, at point e can be calculated from free energy data for equation (2) assuming sodium monoxide and liquid sodium are at unit activity.

Since,

$$\Delta G_T^o = - Rt \log\left(\frac{a_{Na_2O}\ P_{H_2}^{1/2}}{a_{NaOH}\ a_{Na}}\right) = - RT \log\left(\frac{P_{H_2}^{1/2}}{a_{NaOH}}\right) \tag{13}$$

then substituting $\Delta G_T^o = + 2152\ Jmol^{-1}$ (8) and $P_{H_2} = 0.237$ bar, we obtain $a_{NaOH} \sim 0.71$. It is of interest to note that the measured composition of $L_2$ at point e is $\sim 0.66$ NaOH : 0.28 NaH : 0.04 $Na_2O$ mol% (2, 3), suggesting that the activity coefficient for sodium hydroxide, $\gamma_{NaOH}$, defined by $(a_{NaOH}/x_{NaOH})$ is close to unity at 412 $^\circ$C for $x = 0.66$. Furthermore, since the product $(a_{NaOH} : a_{NaH})$ is calculated to be 0.34, a sodium hydride activity of 0.5 is indicated at point e.

The deviation from the binary Na/NaH line which occurs at temperatures > 350 $^\circ$C along $\ell$-e (Fig. 4) indicates a progressive reduction in sodium hydride activity from 1 to ~ 0.5; such a reduction could be due to the formation of a solid solution of sodium hydroxide in sodium hydride, $NaH_{(solid\ sol)}$. Such a solution was first proposed by Shikhov (9), and since $\beta$NaOH and NaH both have a face centred cubic structure with similar anionic radii (OH$^-$ = 0.140 mm, H$^-$ = 0.157 mm), the conditions for mutual solubility are fulfilled. However, no experimental evidence for sodium hydroxide solubility in sodium hydride has been advanced.

Although the present measurements indicate that equilibrium hydrogen pressures along i-e and h-e are very similar at corresponding temperatures between ~ 350 and 412 $^\circ$C, confirmation that these two boundary lines only converge at e is provided from the X-ray analysis of the products of reaction (4). This showed that a sodium phase was absent at temperatures below 412 $^\circ$C, whereas its presence would be required along $\ell$-e if i-e and h-e coalesced at $\ell$. Furthermore, four condensed phases would have to be in equilibrium along $\ell$-e in contradiction with the phase rule. As drawn in Fig. 4, the line $\ell$-e denotes an $L_1$-$Na_2O_{(s)}$-$NaH_{(solid\ sol)}$ co-existence but the precise nature of this equilibrium is uncertain at the present time.

The boundary e-g between areas B and C clearly denotes an $L_1$-$L_2$-$Na_2O_{(s)}$ equilibrium as recognised by Jansson (10). At temperatures > 412 $^\circ$C, hydrogen pressures greater than e-g cannot be achieved unless excess $Na_2O_{(s)}$ is either converted by hydrogenation to NaOH or dissolves in $L_2$.

Evidence for the position of the boundary e-f between areas A and B comes from the later measurements of Shikhov (11), who determined hydrogen equilibrium pressures and saturation solubilities of sodium hydride in sodium hydroxide in the presence of liquid sodium in the range 443 - 553 $^\circ$C. The equilibrium pressure obeyed the relationship:

$$\log_{10} P \text{ (kPa)} = 11.925 - \frac{7188}{T} \tag{14}$$

Extrapolation of the data to 412 $^\circ$C in Fig. 4 (dashed line) gives P ~ 27 kPa in good agreement with the equilibrium pressure at point e. Since the melting point of $L_2$ varies with dissolved sodium hydride content, the sodium hydroxide phase in area A is designated $NaOH_{(solid\ sol,\ \ell)}$ in Fig. 4.

APPLICATIONS OF EQUILIBRIUM PRESSURE-TEMPERATURE DIAGRAM

Although the diagram in Fig. 4 simplifies presentation of the underlying equilibria in the Na-H-O system, certain precautions must be observed in its application. The inter-solubilities between the condensed phases and the transfer of hydrogen to the gas phase to exert its equilibrium pressure can produce appreciable composition changes even in a closed system so that, in certain circumstances, attempts to predict hydrogen equilibrium pressures from given reactant mixtures may lead to erroneous results.

It is strongly recommended that Fig. 4 be used to indicate which condensed phases can co-exist from measured hydrogen equilibrium pressures in sodium systems, since hydrogen equilibrium pressures along the boundary lines in Fig. 4 only apply if the phases indicated to exist are in fact present.

The diagram has already been used in conjunction with Na-O-Cr and Na-O-Fe equilibrium data to predict equilibrium phases during the corrosion of steels by the products of the sodium-water reaction in LMFBR steam generators (12), and a further possible application is described below.

Periodic regeneration of fully laden secondary circuit cold ·traps in LMFBRs may be necessary to remove accumulated impurities such as NaH, $Na_2O$ and possibly NaOH. One technique currently being assessed is hydrogen removal by thermal decomposition of sodium hydride by vacuum decomposition and argon  purging at temperatures > 400 $^o$C (13). It would be useful to measure hydrogen  equilibrium pressures at the proposed regeneration temperature before commencing the process as this could give some information on possible sodium monoxide impurity levels. Under normal operating conditions, sodium hydride will be the major impurity precipitated in secondary cold traps and equilibrium hydrogen pressures lying along h-f (Fig. 3) would be expected.  However, measurement of equilibrium pressures along or below i-e-g (Fig. 4) would indicate the presence of excess sodium monoxide. Under these circumstances the re-generative technique proposed by Volchkov (14) could be applicable in which excess sodium monoxide is hydrogenated to sodium hydroxide at 430 $^o$C (cf. equation (4)), followed by draining of the molten caustic phase. Hydrogen equilibrium pressure measurements used in conjunction with Fig. 4 could enable the progress of this reaction to be monitored.

ACKNOWLEDGEMENT

This paper is published by permission of the United Kingdom Atomic Energy Authority and the Central Electricity Generating Board.

REFERENCES

1.  Y. V. Privalov, Sov.At.Energy, 48(2), 127, 1980.
2.  J. P. Maupre, Report CEA-R4905, 1978.
3.  E. M. Mitkevitch and B. A. Shikov, Russ. J. Inorg. Chem., 11(3), 343, 1966.

4.   A. J. Hooper (private communication).
5.   A. C. Whittingham, J. Nuc. Mats., 60, 119, 1976.
6.   D. D. Williams, Memo, NRL-33, 1952.
7.   K. M. Myles, F. A. Cafasso, J. Nucl. Mats., 67, 249, 1977.
8.   JANAF Thermochemical Tables, NRRDS, NB537, 1970.
9.   B. A. Shikhov, Russ. J. Inorg. Chem., 12(4), 547, 1967.
10.  S. A. Jansson, "Corrosion by Liquid Metals", Plenum Press,
        NY, 523, 1970.
11.  B. A. Shikhov, J. Appl. Chem. (USSR), 47(3), 517, 1974.
12.  C. F. Knights, R. Perkins, (private communication).
13.  J. R. Gwyther, A. C. Whittingham, this Seminar.
14.  L. G. Volchkov, F. A. Kozlov, V. A. Likharev, Y. P. Nalimov
        and B. I. Tonov, Sov. At. Energy, 45(2), 771, 1978.

# INPURITY MONITORING IN LIQUID SODIUM SYSTEMS

# BY ELECTROCHEMICAL OXYGEN AND HYDROGEN MONITORS

George J. Licina, Prodyot Roy[x] and Colin A. Smith [xx]

[x] General Electric Company, San Jose, USA
[xx] Central Electricity Generating Board,
Berkeley, United Kingdom

## INTRODUCTION

Liquid sodium is used as the coolant in both primary and secondary circuits of the liquid metal fast breeder reactor. The oxygen content of the sodium has a profound effect upon the corrosion of structural materials and upon the tribology of materials pairs. A change in the hydrogen concentration in the sodium can signal excessive water-side corrosion in the steam generators or an oil leak from a pump. Changes in oxygen or hydrogen content in the secondary circuit may also indicate a leak of water into sodium. Thus, oxygen and hydrogen detectors can be used to detect steam generator leaks and are utilized as a part of the steam generator protection system. The performance of electrochemical cells for hydrogen and oxygen under both steady-state and simulated steam generator leak conditions is described.

## SENSORS

### General Description

Both sensors utilize a solid electrolyte and liquid metal reference electrode with a stainless steel housing that is connected to the sodium loop via standard fittings. The oxygen meter's reference electrode, a mixture of indium and indium oxide, is contained in a thoria-yttria crucible brazed to a low expansion alloy holder. This holder is welded to the meter's stainless steel housing. Molybdenum wire is used for the inert lead. The meter is shown in Fig. 1.

Equation (1) as written requires that the ionic transference number ($t_{ion}$) is essentially 1.

The electrochemical oxygen meter developed by General Electric Company (1, 2) is an oxygen concentration cell of the form,

$$In, In_2O_3 \parallel ThO_2 + 7.5 \text{ mole. } \% \ Y_2O_3 \parallel O_{in \ Na}.$$

Oxygen meter characterizations in all the previous studies at GE have compared meter output to cold trap and sensor temperatures. The correlations obtained from these empirical studies have shown that cell reproducibility is excellent, drift is virtually nil, and meter-to-meter differences are generally less than 10 mV. The correlation,

$$E = 0.2607 + 5.20 \times 10^{-4} \ T - 6.62 \times 10^{-5} \ T \log [O],$$

was obtained from the empirical studies and using Smith's (4) expression for the solubility of oxygen in sodium. Cell performance agrees very well with predictions of the theoretical voltage using expressions for the free energy of formation of indium and sodium oxides from the Bureau of Mines (3) and Smith's (4) oxygen solubility determination.

The electrochemical hydrogen meter developed at Berkeley Nuclear Laboratories of the Central Electricity Generating Board (5) is a concentration cell of the form:

$$Li, \ LiH \parallel Fe \mid 10 \text{ mole } \% \ CaH_2 \text{ in Ca } Cl_2 \mid Fe \parallel H_{in \ Na}.$$

The hydrogen activities, $a_1$ and $a_2$, may be directly related to the hydrogen dissociation pressure of the Li=LiH reference electrode ($P_1$) and the equilibrium pressure of hydrogen dissolved in the sodium ($P_2$).

The expression

$$\log P_1 \ (\text{Torr}) = \frac{-9600}{T} + 11.227 \tag{2}$$

has been selected for the Li-LiH dissociation pressure, from the work of Hurd and Moore (6) as this study covered temperatures approaching the operating range of the electrochemical meter.

The equilibrium pressure of hydrogen dissolved in the sodium is derived from the Sievert's constant, $K_2$,

$$[H] = K_2 P_2^{1/2}, \tag{3}$$

where $[H]$ = hydrogen concentration in sodium (ppm). From a review of solution studies of hydrogen in sodium at low oxygen concentrations and from experimental measurements (7), a value of Sievert's constant of

$$K_2 = 5.5 \text{ ppm H torr}^{-1/2} \tag{4}$$

was chosen.

Combining equations (1-4), it may be shown that the predicted calibration equation is

$$E = -0.9524 + 1.2415 \times 10^{-3}T - 1.9841 \times 10^{-4}T \log[H] \tag{5a}$$

$$\text{or:} \quad \log[H] = 6.2570 - \frac{4800}{T} - \frac{5040}{T}E. \tag{5b}$$

Experimental calibrations covering the temperature range 300-450 °C and the hydrogen concentration range 0.05 - 1.0 ppm have been made (8). Tests at GE at temperatures from 390 °C to 470 °C and hydrogen concentrations 0.035 - 1.0 ppm (as determined from cold trap temperature) have also been performed. Both sets of data give good agreement with the theoretical prediction. Further, no deterioration in performance was observed in 12.000 hours of operation.

EXPERIMENTAL ARRANGEMENT

The test loop shown in Fig. 3 was used for meter characterization and water injection runs. This facility is a standard $\Delta T$ sodium loop with several important additions. A test vessel, rated at 150 psi (1.03 MPa), has been added upstream of the main loop heaters. The vessel head is equipped with three ports through which water/steam injectors may be loaded. Three separate overpressure protection systems are incorporated to protect the vessel and the loop. Automatic venting of the test vessel (to atmosphere) is actuated by a low-level signal in the hot leg surge tank. This venting feature is used primarily for maintaining a low pressure, without loop trip, during inert gas purging of the leak injector. High pressure in the test vessel causes automatic actuation of three sodium valves to isolate the test vessel from the loop, again without loop trip. Finally, a blowdown line with a 90 psi (620 KPa) rupture disc is included on the hot expansion tank so that reaction products may be directed to a tank outside the test building in the event that a sodium-water reaction becomes too large for the venting system. A very large cold trap (320 liter capacity) accepts full loop flow. Reaction products from more than twenty water injections have been removed from the loop since installation of this cold trap with no degradation of the trap's performance.

In the electrochemical hydrogen meter (Fig. 2), the reference electrode consists of a lithium-lithium hydride mixture, sealed in a thin-walled iron capsule, which is resistant to attack by lithium. The electrode is set in a solid electrolyte composed of a mixture of calcium hydride in calcium chloride. In order to protect the electrolyte from chemical attack, it is contained in a thin-walled iron thimble which acts as the sensing electrode of the cell when immersed in liquid sodium.

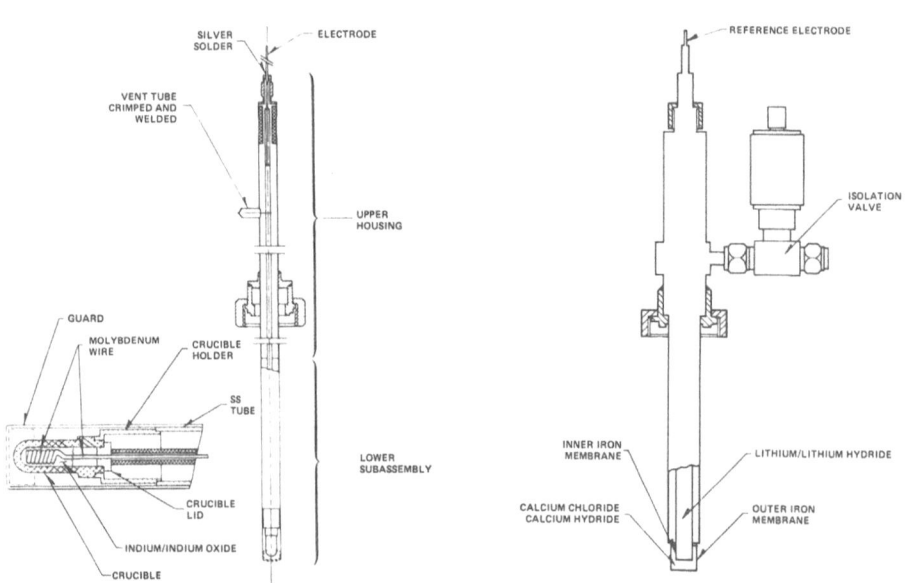

Fig. 1. Electrochemical Oxygen     Fig. 2. Electrochemical Hydrogen
        Meter                                    Meter

## Theoretical Considerations

The electromotive force E of both cells is described by the Nernst equation,

$$E = \frac{RT}{nF} \ \ln \frac{a_1}{a_2} \ (\text{Volts}), \tag{1}$$

where

R  =  gas constant

T  =  sensor temperature, $^\circ K$

n  =  number of electrons involved in the transfer reaction between the two electrodes

F  =  Faraday constant

$a_1$  =  hydrogen/oxygen activity of the reference electrode

$a_2$  =  hydrogen/oxygen activity of the test (sodium) electrode

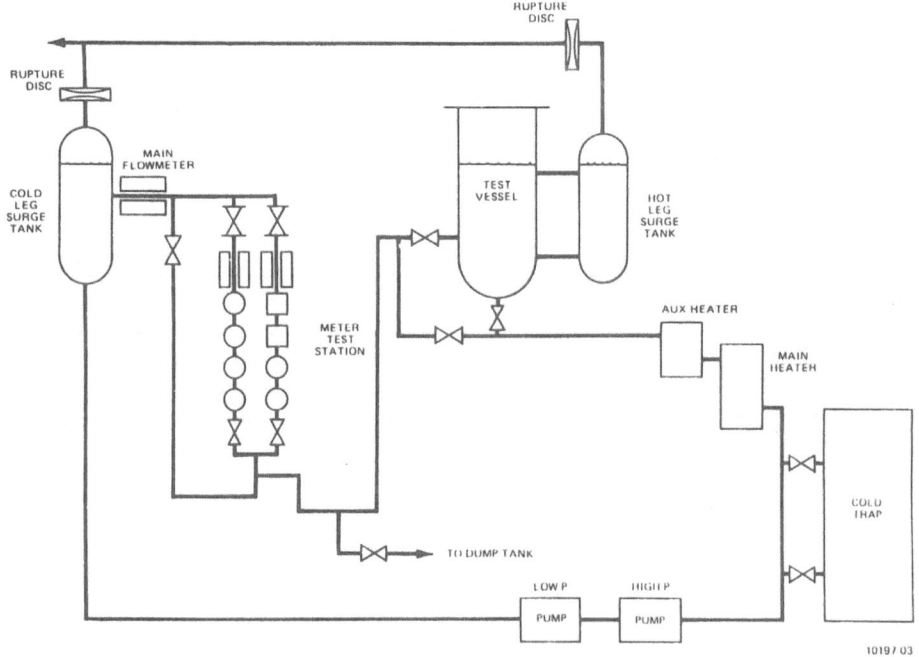

Fig. 3. Sodium Test Loop - Meter Studies

The meter test station permits simultaneous exposure of up to eight impurity monitors. The present configuration has two test legs of four stations each. Two of the stations (squares in Fig. 3) are used for the CEGB hydrogen meter. Each test leg is equipped with an electromagnetic flow meter and manually operated isolation valves for independent flow control. A bypass loop permits throttling of the flow.

The water injection system is of the same design as that used in hundreds of static pot injections. The signal from a differential pressure transducer provides a precise determination of the quantity of water transferred through the injection nozzle. The tip of the injector, fabricated from nominal 3.18 mm ODX 1.24 mm wall tubing swaged over a 0.33 mm mandrel, is located approximately 0.6 m below the sodium level in the test vessel. Oxygen and hydrogen meter outputs are monitored by a high impedance voltmeter and recorded along with other key loop parameters by an automatic data logger.

Three water injections at two different impurity background concentrations were conducted. Extensive meter characterization testing was performed in the course of the test runs to assure that the sensors were operating properly. Upon completion of the characterization phase of each run the "base" conditions were maintained for a minimum of 24 hours. The cold trap was then isolated from the system but maintained at the "base" condition.

The loop was run in this condition for another 24 hours to provide
a well-defined starting condition. The desired quantity of water
was then injected (either 413 KPa or 207 KPa pressure) and meter
responses were monitored for two to four hours. The cold trap was
then valved back into the system and meters monitored during the
reaction product clean-up.

RESULTS

The meter characterizations run prior to the water injections
verified that meters were operating in accordance with previous
results (1, 2, 8). Fig. 4 gives one example of oxygen meter EMF
plotted against reciprocal cold trap temperature. As may be seen
from Fig. 4, the oxygen dependence of meter output was in close
agreement with predictions over all meter temperatures investigated.
Hydrogen meter performance was also very good.

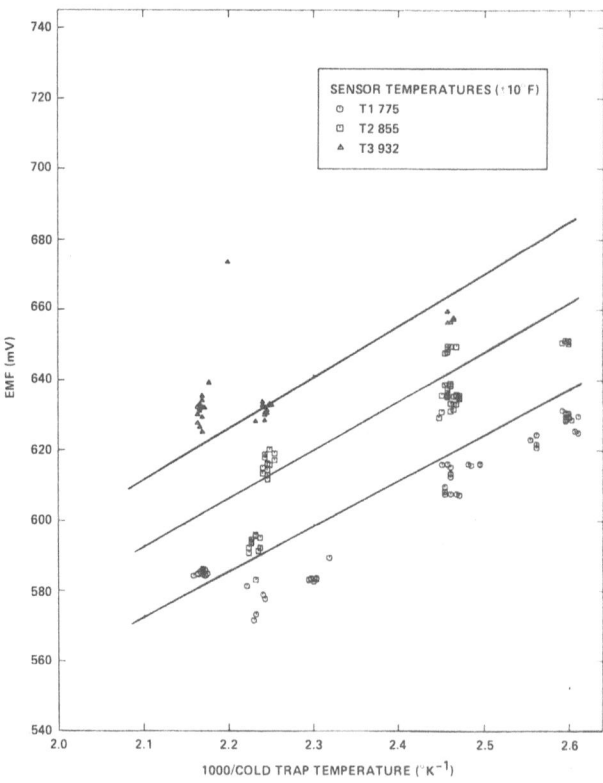

Fig. 4. Oxygen Meter    345 - Steady State Performance

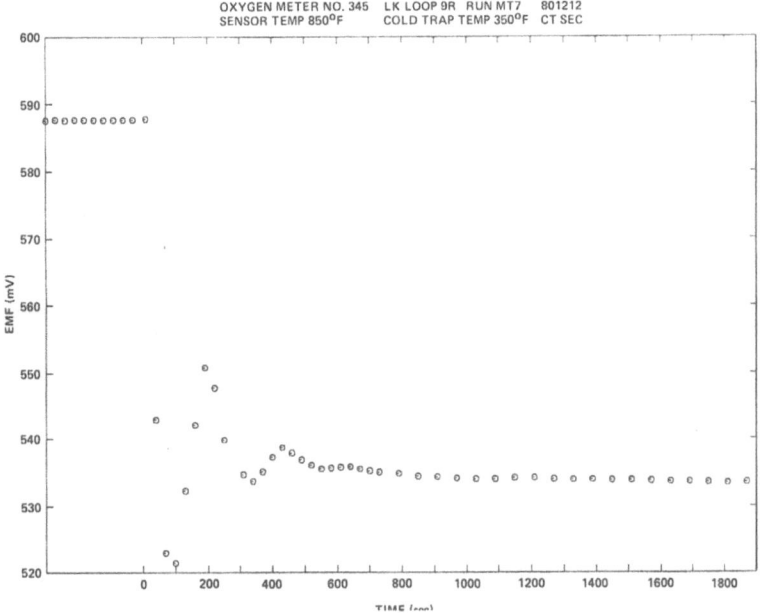

Fig. 5a. Oxygen Meter Response to Water Injection

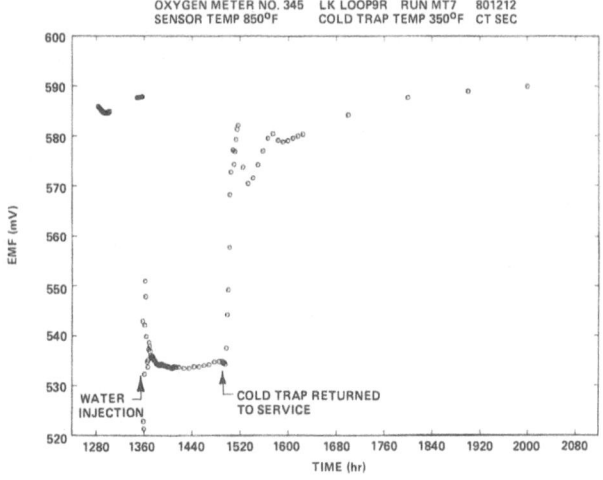

Fig. 5b. Oxygen Meter Response to Water Injection and Clean-up

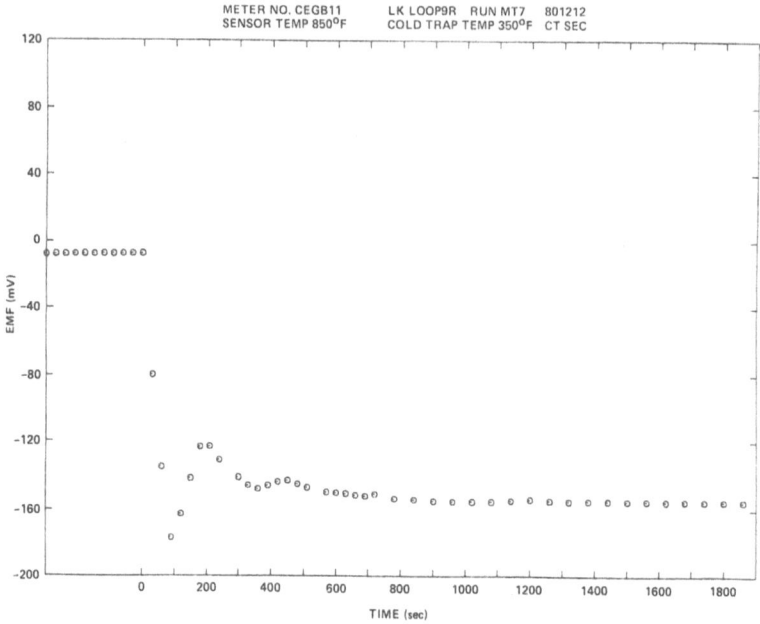

Fig. 6a. Hydrogen Meter Response to Water Injection

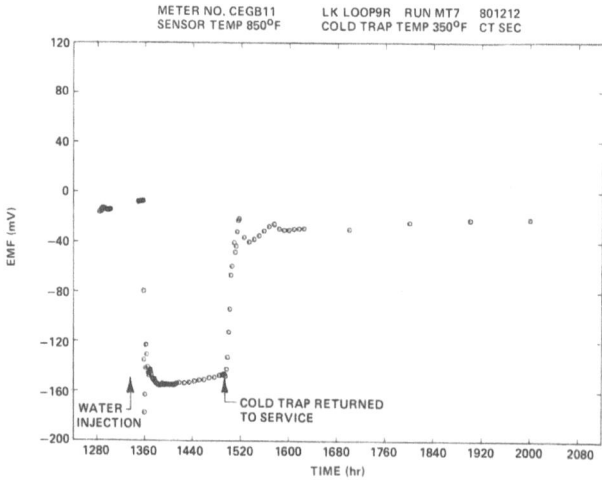

Fig. 6b   Hydrogen Meter Response to Water Injection and Clean-up

Fig. 5 and 6 illustrate the response of an oxygen meter and a hydrogen meter to injection of approximately 5 ml water at a 5 ppm background oxygen level. Other pertinent test conditions are given in Table 1. Both sensors respond virtually instantaneously to the water injection. Meter outputs decrease (reflecting the increased impurity level) rapidly, then "damp out" within about eight minutes for the oxygen meters, about fifteen minutes for the hydrogen meters, as the sodium-water reaction products become completely mixed. Upon valve-in of the cold trap, a rapid decrease in impurity level is observed followed by a very abrupt dip in the meter voltages (Fig. 5b and 6b). This dip also seems to be a mixing effect. Meter voltages then continue to rise. Within three hours, the oxygen and hydrogen sensors' outputs have returned to their initial level.

Table 1.   Meter Test Parameters - Transients

| Vessel Temperature | 875 $^{\circ}$F | (469 $^{\circ}$C) |
|---|---|---|
| Main Loop Flow | 2.3 gpm | (0.61 lmp) |
| Sensor Temperature | 850 $^{\circ}$F | (454 $^{\circ}$C) |
| Test Leg Flow (each leg) | 0.5 gpm | (0.13 1pm) |
| Cold Trap Flow | Zero | |

DISCUSSION

The steady-state performance of both oxygen and hydrogen sensors characterizes the sodium chemistry prior to the intentional introduction of impurities. Both sensors begin to respond to the water injection within thirty seconds (within the estimated trans- port time from the test vessel). The shapes of the response curves were as expected; this is, a slight overprediction of the final impurity level with a gradual leveling off to a steady valve. Mixing effects are the postulated cause of this phenomenon. Similar but more dramatic effects have been observed by others (9). The abrupt dip in the meter response curves (Fig. 5b and 6b) after the cold trap is valved back into the system is also attributed to mixing. In the latter case, the cold trap's sodium inventory, more than four times that of the loop itself, is suddenly coupled to the main loop flow. Since the large sodium volume has been main- tained at the initial background concentration, a charge of clean sodium is swept through the loop and past the meters. The meter voltage vs. time trace after this dip illustrates impurity removal by the cold trap.

For all three injections, the measured increases in oxygen level agree well with the quantity of water injection. In all three tests the measured ratio of oxygen addition to hydrogen addition is very nearly eight indicating that all hydrogen was retained in solution and not liberated as gas, in agreement with the predictions of Cafasso (10) and others (11) for this test temperature. Formation of sodium oxide and hydride apparently occurs extremely rapidly for the sodium temperature investigated here and is not dependent upon background oxygen concentration over the range of 1 to 5 ppm (0.3 to 0.5 ppm hydrogen). McKee's (9) conclusions regarding the use of electrochemical oxygen sensors for chemical leak detection have been further substantiated here.

The only discernible difference in meter performance between the 5 ppm and 1 ppm background oxygen runs was that meter-to-meter variations were less in the higher impurity case. Sensitivity and accuracy were not affected by the initial impurity concentration level.

Electrochemical oxygen and hydrogen meters have been tested under steady state conditions over the temperature range of 410 $^{\circ}$C to 454 $^{\circ}$C (oxygen meters to 516 $^{\circ}$C) with oxygen (hydrogen) levels from 0.4 (0.035) to 5 (0.46) ppm. Response to water injections into sodium at 454 $^{\circ}$C to 469 $^{\circ}$C was rapid and accurate. Both meter types also tracked impurity removal from the sodium stream on cold trapping following the water injections. Meter responses in these tests substantiate previous work toward selection of electro-chemical oxygen meters for chemical leak detection in LMFBR plants. The advantages of the BNL electrochemical hydrogen meter for sodium purity monitoring have also been demonstrated here. Sodium oxide and hydride formed rapidly and hydrogen remained in solution at the test temperatures and background impurity levels tested.

## ACKNOWLEDGEMENTS

Tests were conducted under a program sponsored by the General Electric Company. Hydrogen meters were supplied by the Central Electricity Generating Board. The dedicated contributions of R. E. Leonard and D. R. Springer to the performance of these experiments are gratefully acknowledged.

## REFERENCES

1. P. Roy and B. E. Bugbee, Nucl.Tech., 39 (7), p.216,1978.
2. G. J. Licina, P. Roy, H. Nei, A. Kakuta, Second Internatio-nal Conference on Liquid Metal Technology in Energy Production, 1980.
3. Bureau of Mines Bulletin 542, 1954.
4. D. L. Smith, Proc. International Conference on Liquid Metal Technology in Energy Production, CONF 760503-2, p. 631, 1976.
5. C. A. Smith, BNES Conference on Liquid Alkali Metals, Nottingham, p. 101, 1973.
6. D. B. Hurd and G. A. Moore, J.Am.Chem.Soc., 57, p.332, 1955.
7. A. C. Whittingham, J.Nucl.Mat., 60, p. 119, 1976.
8. P. A. Simm and C. A. Smith; unpublished dta, 1980.

9. J. M. McKee, F. A. Smith, E. R. Koehl, Trans. Am. Nuc. Soc., 33, p. 264, 1979.
10. F. A. Cafasso, K. M. Myles, A. K. Fischer, International Conference on Liquid Metal Technology in Energy Production CONF 760503-1, p. 191, 1976.
11. J. M. McKee, CONF 760503-2, p. 494, 1976.

# ANALYSIS OF NON-METALS IN SODIUM: EXPERIENCE AT THE REACTOR RESEARCH CENTRE, KALPAKKAM

N.P. Bhat, V. Ganesan, T.R. Mahalingam
and S. Rajendran Pillai

Reactor Research Centre
Kalpakkam, India

## INTRODUCTION

Measurement and control of non-metallic impurities like oxygen, hydrogen and carbon in sodium is of considerable importance in the successful long-term operation of the high temperature sodium systems of the fast reactors. In order to minimise corrosion of the structural materials, oxygen is to be maintained below 10 ppm level. Hydrogen estimation is important especially for the secondary sodium system of the reactor in order to monitor water to sodium leak in the steam generator. Control of carbon level is necessary to control carburisation/decarburisation of the structural materials. Considerable development has taken place during the last three decades in order to evolve suitable procedures for the estimation of these impurities in sodium. As a result, the laboratory methods developed in various laboratories have undergone various modifications to extend the sensitivities down to 1 ppm level. Foil equilibration methods such as vanadium wire equilibration, steel foil equilibration, and scandium foil equilibration have been developed in various laboratories to estimate oxygen, carbon and hydrogen respectively to ppb level. More recent development has been the electrochemical and diffusion type meters for on-line monitoring of these impurities in sodium. The analytical development work was started in our laboratory few years back. Several methods have been tried in keeping with the developments in this field. The evolution of the present methods have gone through several stages of modifications in order to evolve accurate, simple and fast procedures for the estimation of these impurities in sodium. This paper describes our experience in this field leading to the presently standardised procedures which are used routinely in our laboratory for the analysis of commercial sodium as well as reactor grade sodium.

ESTIMATION OF OXYGEN

The amalgamation method was standardised in the earlier stages
of development at the Bhabha Atomic Research Centre for the estimation of oxygen in sodium (1). Amalgamation was carried out under
vacuum in an apparatus similar to the one used by Champeix et al.
(2). Sodium was taken in glase ampules which were vacuum sealed
from the glass sampler (3). The method was standardised for estimation of oxygen to less than 10 ppm level. Sodium with low oxygen
content was obtained by filtration of molten sodium at 120 $^o$C
through glass frit of 5 - 15 μm pore size.

A modified procedure was utilised for estimation of oxygen in
sodium from the purification loop at RRC. Amalgamation was carried
out inside the inert atmosphere glove box following APDA procedure
(4) with potentiometric titration of sodium in the residue (5).
The piggy-back blank procedure (4) was adopted to take care of the
contamination from the glass amalgamation vessels. This method was
utilised throughout the sodium purification programme with the
experimental loop at RRC (6).

The distillation method for estimation of oxygen in sodium was
first tried by making use of the stainless steel distillation
vessel with a coldfinger, as used by Walker and France (7), with
resistance heating and vacuum distillation. Though six samples
could be distilled at a time, the distillation time was more than
six hours. It was difficult to monitor the end of complete
distillation for sodium in each crucible. The distillation vessel
was too heavy requiring difficult manipulations for handling inside
the glove box. An induction heater was since procured and a glass
distillation vessel was made (8). A schematic of the vessel is
shown in Fig. 1.

An initial coating was made on the inside surface of the
distillation vessel by distilling small amount of sodium prior to
taking actual samples. This is to avoid contamination from the
surface of the glass vessel. Sodium was taken in a nickel crucible
of 10 ml capacity. Distillation takes place at 320 - 340 $^o$C under
vacuum of $10^{-5}$ torr. Completeness of distillation was sensed by
the thermocouple. Distillation time was about 45 minutes for 8 to
10 g sample. Sodium in the residue was analysed by potentiometric
titration or by flame emission.

The method was standarised by estimation of oxygen solubility
in sodium at three different temperatures corresponding to 1, 3
and 10 ppm range respectively. Samples were taken at 125 $^o$C,
150 $^o$C and 200 $^o$C from the clear sodium in the metal pot kept
inside the inert atmosphere glove box. Oxygen values obtained at
these three temperatures are shown in Table 1 and compared with
solubility values reported in literature (9, 10). Analysis results

Table 1.  Oxygen estimation in pot sodium at 125 °C, 150 °C
          and 200 °C

| Tempe-rature °C | Oxygen in sodium (ppm) | | | | | Literature value Eichelberger | Noden |
|---|---|---|---|---|---|---|---|
| 125 | 0.94 | 1.13 | 0.96 | 0.84 | x̄ = 1.02 | | |
|     | 0.98 | 0.94 | 1.15 | 0.99 | σ = 0.10 | 1.24 | 1.46 |
|     | 0.96 | 1.13 | 1.05 | 1.16 | | | |
| 150 | 3.28 | 3.45 | 3.24 | | | | |
|     | 2.79 | 3.22 | 2.99 | | x̄ = 3.16 | 2.86 | 3.30 |
|     | 3.02 | 2.99 | 3.32 | | σ = 0.20 | | |
|     | 3.30 | | | | | | |
| 200 | 11.16 | 12.67 | 11.32 | | | | |
|     | 11.20 | 11.27 | 10.02 | | x̄ = 11.56 | 11.67 | 13.0 |
|     | 11.92 | 10.94 | 12.15 | | σ = 0.86 | | |
|     | 12.91 | | | | | | |

Table 2.  Oxygen estimation in commercial sodium and in sodium
          from experimental loops

| Commercial sodium | | | Loop sodium | | | |
|---|---|---|---|---|---|---|
| S.No. | Oxygen (ppm) | | S.No. | Oxygen (ppm) | | |
| B 145 | 4.4 | 4.5 | DFPT 3 | 3.1 | 5.0 | 5.2 |
|       |     |     |        | 3.7 | 5.7 | 6.1 |
| B 148 | 3.7 | 3.8 | DFPT 25 | 62 | 54 | 60 |
|       |     |     |         | 48 | 73 | |
| B 146 | 7.1 | 7.8 | DFPT 145 | 4.9 | 4.4 | |
| B 170 | 5.7 | 5.1 | DFPT 13 | 29.3 | 30.4 | |
| B 171 | 6.7 | 6.9 | FMCL I | 1.1 | 1.6 | |
| B 172 | 13.4 | 13.6 | FMCL II | 2.5 | 2.6 | |

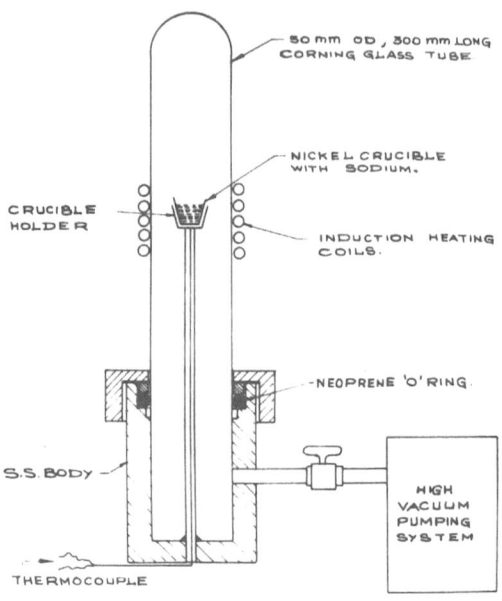

Fig. 1. The Distillation vessel for Oxygen Estimation in Sodium

of commercial sodium and samples from some of the experimental
loops are shown in Table 2. In the former case, samples from the
inner core of sodium bricks were analysed and in the latter case,
a piece of the nickel tube with sodium (flow through by-pass
sampling) was taken into the nickel crucible for distillation.

ESTIMATION OF HYDROGEN

The amalgam reflux method (11) was utilised for the estimation
of hydrogen in sodium. A schematic of the reflux vessel is shown
in Fig. 2. Refluxing was done by induction heating, the reflux
vessel being connected to high vacuum ($10^{-5}$ torr) and pure argon
system. The hydride and hydroxide hydrogen were estimated separately
by refluxing at 200 $^{\circ}$C ($10^{-5}$ torr) and 370 $^{\circ}$C (600 mm argon pressure)
respectively. The liberated hydrogen was estimated by the gas
chromatograph with thermal conductivity detector (0.2 µl $H_2$
= 60 mm peak height). Blank was reduced by successive refluxing
of the amalgam at 370 $^{\circ}$C. The vessel after reduction of blank was
taken inside the glove box and 2 to 3 gm of sodium sample added to
the amalgam. The vessel was taken out and the hydrogen liberated
during the amalgam estimated. The vessel was connected to the vacuum
system and refluxed for 15 minutes at 200 $^{\circ}$C under vacuum. The vessel

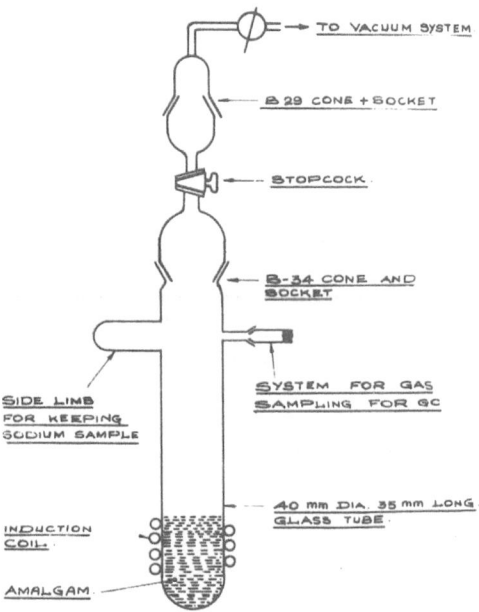

Fig. 2. The Amalgam Reflux Vessel for Hydrogen Estimation in Sodium.

was filled with argon and 5 ml of the argon sampled with a syringe
and injected to the gas chromatograph for hydrogen estimation. The
refluxing and estimation was continued till no more hydrogen was
liberated at 200 °C. Five to six refluxing and estimations were
required to liberate all the hydride hydrogen at 200 °C. Total
hydride hydrogen was calculated from all the above values including
that liberated during the amalgamation. Now the refluxing was
carried out at 370 °C with 600 mm argon pressure in the vessel.
Refluxing and estimations were repeated till all the hydroxide
hydrogen was evolved. Contamination from the glass vessel could
be avoided in our estimation by directly dropping the sample into
the amalgam instead of keeping the sample in the side limb of the
vessel, a procedure which used to give higher hydrogen values in
our case.

For standardisation of the method the pot sodium in the glove
box was sampled at 200 °C and analysed for hydrogen. The results
are shown in Table 3 and compared with the solubility of hydrogen
at 200 °C reported by Meacham et al (12). The commercial sodium
sample was also analysed by taking the inner core of the brick.
These results are also shown in Table 3.

Table 3.   Hydrogen estimation of pot sodium and commercial sodium

| Nature of sodium | Hydrogen | | | Solubility of hydrogen in sodium at 200 $^{\circ}$C (Meacham) |
|---|---|---|---|---|
| Pot sodium at 200 $^{\circ}$C | 0.69 0.66 0.59 | 0.63 0.63 | Mean = 0.64 $\sigma$ = 0.04 | 0.63 ppm |
| Commercial sodium | 0.50 0.47 0.45 | 0.43 0.49 | Mean = 0.47 $\sigma$ = 0.03 | - |

Table 4.   Carbon estimation in pot sodium (150 $^{\circ}$C)

| No. | Weight of sodium gm | Total carbon $\mu$gm | Blank $\mu$gm | Carbon in sodium (ppm) | | |
|---|---|---|---|---|---|---|
| 1 | 3.5 | 74.9 | 26.3 | 14.0 | Mean = | 13.57 |
| 2 | 4.0 | 83.5 | 26.3 | 14.3 | $\sigma$ = | 0.54 |
| 3 | 3.4 | 73.6 | 26.3 | 13.9 | | |
| 4 | 3.2 | 71.3 | 26.3 | 14.0 | | |
| 5 | 4.1 | 81.5 | 26.4 | 13.4 | | |
| 6 | 3.5 | 72.1 | 26.4 | 13.0 | | |
| 7 | 4.4 | 83.0 | 26.4 | 12.8 | | |
| 8 | 4.1 | 79.1 | 26.4 | 12.8 | | |
| 9 | 3.3 | 71.3 | 26.4 | 13.6 | | |
| 10 | 3.3 | 68.2 | 26.4 | 13.9 | | |

We could not detect any hydroxyde hydrogen in above two samples. Reason for this is not clear at present.

Subsequent experiments for hydrogen estimation after a lapse of one month gave high hydrogen values compared to the earlier results. On investigation, it was found that the hydrogen content of the argon in the glove box had increased to 1000 vpm level. This is attributed either to impurity in the argon which is used for filling the side port of the box or due to permeation of moisture into the box which reacts with pot sodium liberating hydrogen (200 °C). Incorporating a hydrogen getter in the recirculating line of the box is being planned before attempting further estimations of hydrogen.

ESTIMATION OF CARBON

In the initial stages the total combustion method with $Al_2O_3$ and $SiO_2$ flux (5) was selected for estimation of total carbon in sodium. In order to minimise the blank several hours of conditioning at 1100 °C was to be done for the boat, quartz sleeve, alumina, silica etc. The amount of sodium that could be taken for analysis was also limited to max. 0.5 gms. In spite of the acidic flux the quartz sleeve could be used only once, and also the quartz combustion tube was found to be attacked with sodium oxide requiring frequent replacement with new tube. In spite of all the precautions the blank values were in the range of 70 μgm. On the whole the method was found to be too tedious, time consuming and costly.

The distillation combustion method was tried subsequestly. Initially the sodium was distilled in the stainless steel vessel and the residue combusted at 1100 °C in oxygen and the $CO_2$ collected in a molecular sieve trap and estimated by gas chromatograph (13). As explained earlier, the manipulation of the heavy stainless steel vessel inside the glove box was difficult. Distillation took a long time and often resulted in high blanks.

The glass distillation vessel was then utilised with induction heating as in the case of oxygen for distillation of sodium. A quartz crucible (20 mm dia 30 mm height) was used for taking sodium samples for distillation. After the distillation the residue in the crucible was covered with quartz wool and combusted at 1100 °C in oxygen and the $CO_2$ formed was estimated by the low pressure method (14) modified in our case with a molecular sieve trap.

Samples from the sodium pot inside the glove box were analysed for carbon content. The results are shown in Table 4 which show a good precision. The blank remained constant at 26.3 ± 0.1 μgm C. Standard addition experiments were carried out with graphits and calcium carbonate standards. The recovery (90 - 98 %) was within the experimental error. These results are shown in Table 5.

Table 5.  Carbon estimation in sodium with standard addition

| Standard | Weight of sodium gm | Standard added μgm | Total carbon estimated μgm | Carbon recovered μgm | Percentage recovery |
|---|---|---|---|---|---|
| Graphite | 2.4 | 102.9 | 159.4 | 100.5 | 97.7 |
| " | 2.1 | 105.8 | 152.3 | 97.4 | 92.1 |
| " | 1.9 | 102.9 | 153.2 | 101.1 | 98.2 |
| " | 3.6 | 91.4 | 161.6 | 86.5 | 94.6 |
| Calcium | 2.0 | 91.0 | 140.2 | 86.8 | 95.4 |
| carbonate | 2.5 | 65.4 | 120.5 | 60.3 | 92.0 |
| " | 2.9 | 101.9 | 163.6 | 98.0 | 96.0 |

Table 6.  Carbon estimation in commercial sodium and sodium from experimental loops

| Nature of sodium | Carbon content (ppm) |
|---|---|
| Commercial sodium samples from single brick analysed | 18.8   19.6   20.7   $\bar{x}$ = 19.8<br>19.8   19.0   18.0   $\sigma$ = 1.4<br>22.6 |
| Commercial sodium-5 samples from different bricks of single consignment with duplicate results | a) 19.0   20.0 b) 20.0   22.1<br>c) 23.3   22.0 d) 19.4   18.4<br>e) 21.1   23.4 |
| Experimental loop sodium:<br>a) Mobile purification loop | a)  4.9   4.6   2.2 $\bar{x}$ = 3.9<br>$\sigma$ = 1.5 |
| b) Dummy fuel pin test rig | b)  7.4   9.2   9.2 $\bar{x}$ = 8.6<br>$\sigma$ = 0.8 |
| c) Flow meter calibration loop | c) 17.8   18.9   16.3 $\bar{x}$ = 17.8<br>18.4   $\sigma$ = 1.1 |
| d) 500 KW loop | d) 14.5   12.7   11.8 $\bar{x}$ = 12.0<br>10.7   10.3   $\sigma$ = 1.7 |
| e) 500 KW loop (final stages of cold trapping) | e) Six determinations giving only blank values |

Table 6 shows some of the carbon values estimated in commercial sodium and samples from experimental sodium loops. Samples from 500 KW loop taken after several hours of cold trapping gave only blank values indicating carbon level below detection limits of this method ($\sim$ 1 ppm).

CONCLUSION

The results obtained for oxygen, hydrogen and carbon in sodium show that the methods selected and standardised in our laboratory are sensitive enough for estimation of these impurities in reactor grade sodium. Estimations are possible down to 1 ppm level for all these impurities and the methods are simple and fast enough.

Addition of absolute standards in the case of hydrogen and oxygen was difficult at these low levels. Hence our results of solubility determination are compared with the literature values of solubility of hydrogen and oxygen. The oxygen solubility data presented in this paper agree well with the literature values within the experimental error with better precision in our values. The solubility of hydrogen determined at 200 $^{\circ}$C also agrees well with that of Meacham et al (12).

Our experience shows that the hydrogen content of the glove box is to be monitored while sampling sodium for hydrogen estimations in sodium.

The standard addition and recovery experiments with carbon show that there is no loss of carbonate or graphite carbon distillation.

ACKNOWLEDGEMENT

We wish to thank Dr. C. K. Mathews for his keen interest and encouragement for this work. We are grateful to Dr. H.U. Borgstedt, Kernforschungszentrum, for his help and discussion under the Indo-German collaboration programme. We are thankful to our colleagues Messrs. Dr. Krishnamurthy, R. Ranganathan and K. Swaminathan who had helped in the experimental work.

REFERENCES

1.  N.P. Bhat et al, The Determination of Oxygen in sodium; BARC-353, Bhabha Atomic Research Centre (1968).
2.  L. Champeix et al, Determination of Oxygen in sodium; J. Nucl. Mater.2 (1959) 113.
3.  J. Sannier and R. Vingot, Sodium sampling device on a circuit for impurity analysis, CEA-2369 (1963).
4.  S.A. Meacham and E.F. Hill, Evaluation of the proposed ASTM Method for measuring oxygen in sodium utilising a Round Robin Programme, APDA-238 (1970).

5.  R.F. Keough, Pacific North-west Laboratory Manual of Proce-
    dures for the analysis of metallic sodium, BNWL-MA-76
    (1970).
6.  R.D. Kale et al, Experience in the Operation of sodium
    purification loop in the Reactor Engineering Laboratory,
    RRC-11, Reactor Research Centre (1976).
7.  J.A.J. Walker et al, Determination of Oxygen in sodium by
    distillation; TRG Report 952 (C) UKAEA (1963).
8.  H.U. Borgstedt et al, Sodium distillation set up "NADESTAN-4"
    for analytical estimation,  KfK Report 1941 (1973)
    Nuclear Research Centre, Karlsruhe.
9.  R.L. Eichelberger, The solubility of Oxygen in liquid sodium:
    A recommended expression, Atomics International Report,
    AI-AEC-12685 (1968).
10. J.D. Noden, A General equation for the solubility of Oxygen
    in liquid sodium; J. Brit. Nucl. Energy Soc. 12 (1973) 57.
11. S.A. Meacham and E.F. Hill, The Determination of Hydrogen
    in sodium, APDA-183, Atomic Power Development Associates
    (1966).
12. S.A. Meacham et al, The Solubility of Hydrogen in sodium,
    APDA-241, Atomic Power Development Associates (1970).
13. J.F.M. Rohde, Determination of total carbon content of
    sodium, SR-TN-7009-32, TNO, Holland (1970).
14. V.T. Athavale et al, A modified apparatus for determination
    of carbon in metals by the low pressure method; Anal. Chem.
    Acta 35 (1966) 247.

# CARBON IN SODIUM MEASUREMENTS WITH VARIOUS CARBON METERS

T. Gnanasekaran

Kernforschungszentrum Karlsruhe
Fed.Rep.of Germany
(Delegated from R.R.C., Kalpakkam, India)

## INTRODUCTION

Carbon transport in sodium systems of fast breeder reactors
is of significance as it affects the mechanical properties of the
structural materials severely. This may determine the useful life
of certain components of the system. A thorough understanding of
this process in an operating sodium system requires the thermodynamic driving force(the carbon activity gradient) to be known.
Conventional methods of analysis of carbon in sodium and then
relating to the carbon activity gradient is complicated because
of the lack of clear understandig of Na-C system. Moreover the
analytical methods suffer from sensitivity requirements at the
expected levels of carbon, apart from contamination during sampling.
On-line carbon meters are of help at this juncture. They determine
the carbon activity directly and do not face the problems of
sampling. In this paper experimental results with two diffusion
based carbon meters, namely Harwell Carbon Meter (HCM) (1) and an
improved version of the carbon meter originally designed by United
Nuclear Corporation (UNC) (2) and also with an electrochemical
carbon meter (3) are presented.

## PRINCIPLES OF OPERATION AND PREVIOUS EXPERIENCE

In diffusion based carbon meters, carbon is allowed to diffuse
through a thin iron membrane, the one side of which comes into
contact with sodium under analysis. At the other side of the membrane
a decarburising reaction is made to take place. This reaction keeps
the carbon activity at low levels. Under these conditions the carbon
flux from sodium to gas is proportional to carbon activity in sodium.

The product of the decarburising reaction is analysed continuously.
In the meter of the UNC design Ar- 5 % $H_2$ - 0.5 % $H_2O$ gas mixture
is passed over the membrane. Water vapour reacts with the carbon
to produce carbon monoxide. The same gas mixture sweeps away the
CO produced. In HCM a thin previously formed FeO layer reacts with
the diffused carbon to produce CO. An argon gas flow carries the
product away from the membrane. In case of both the meters the
CO is converted to $CH_4$ and detected by flame ionisation detectors
(FID). In the UNC design the membrane is in the form of a small
cup of approx. 10 $cm^2$ area and 0.25 mm wall thickness. The operating
temperature is 750 °C. In the HCM the membrane is in the form of
coils made from thin walled iron tubes and the recommended operating
temperature is above 500 °C.

The electrochemical carbon meter tested consists of a liquid
$Na_2CO_3$-$Li_2CO_3$ eutectic (m.p. approx. 500 °C) contained in a thin
iron membrane cup of 0.25 mm wall thickness. A reference electrode
is kept immersed into the electrolyte. The membrane cup assembly
is then dipped into the sodium for analysis. Carbon in sodium
equilibrates with the iron membrane and establishes the carbon
activity in it. An emf develops across the electrolyte due to the
presence of a carbon activity difference and this is measured.

The UNC designed meter has been under testing since many
years. Experience has shown that the sensitivity of the meter at
lower carbon levels is poor and the hydrogen injection from the
decarburising gas into sodium becomes a problem in small sodium
systems, though in bigger systems this may be tolerated. Similar
problems are not to be encountered in HCM. An increase in the
sensing probe area and decrease in operating temperature of the
UNC-meter are carried out in this work.

Various reported works have shown that the emf of the electrochemical carbon meter drops with time of operation which is
accelerated by an increase of temperature. Carbon activity
instability in reference electrode due to loss of carbon and
corrosion processes in the cell are attributed as the reasons.
Evaluation of different reference electrodes have been carried
out in this work.

EXPERIMENTAL

The sensing probe for UNC-type meter was made in the form of
a coil from 2 x 0.2 x 1200 mm long tube of iron. (A longer coil
of 400 cm tube length was also fabricated and was under testing.
However after approx. 350 hours of operation this coil failed and
hence the experiment was proceeded with the smaller coil.) Both
the UNC-type meter and HCM were tested in static pot arrangements
as shown in Fig. 1. Both the pots had provisions for stainless
steel foil equilibrations. The foils used were of 25 μm thickness.

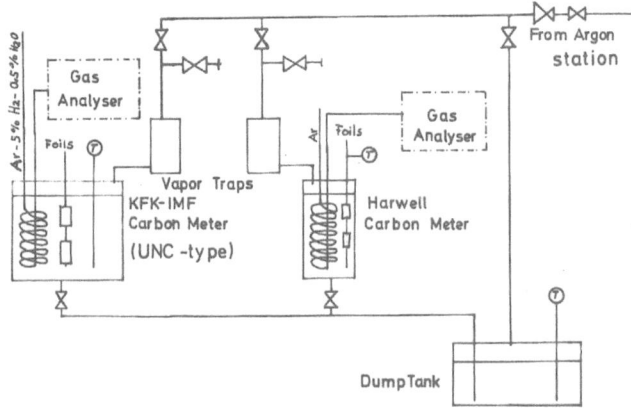

Fig. 1.   Static pot arrangement for diffusion based carbon meters.

The meters were operated at a given temperature till the required equilibration times for foils are completed. The argon cover gas of the UNC-type meter vessel was analysed at the end of each experiment using Lambert AL-7 Gas Chromatograph for $H_2$, $N_2$ and $CH_4$. This choice of time of analysis ensures the equilibrium in the system not to be disturbed during equilibration periods. The response to carbon source introduction were carried out by immersing carburised ferritic steel into sodium at 120 $^\circ$C and by subsequent heating.

The static pot arrangement for testing electrochemical carbon meter is shown in Fig. 2. The electrolyte needed for this meter was made by mixing analar grade salts and drying them under a flow of $CO_2$ at 400 $^\circ$C. The assembling of the electrode and the electrolyte were carried out in an argon atmosphere glove box. The reference electrode system of the meter is of sliding type as recommended by Hobdell et.al. (4). Changing of the reference electrodes under argon flow was also possible with this design. In order to keep the argon gas phase above the electrolyte pure, it was intermittently kept open to sodium pot by means of the connecting gas valves. Reference electrodes of pack-carburised iron rod, graphite filled carburised iron tube of 0.5 mm wall thickness and high purity graphite electrodes were tested.

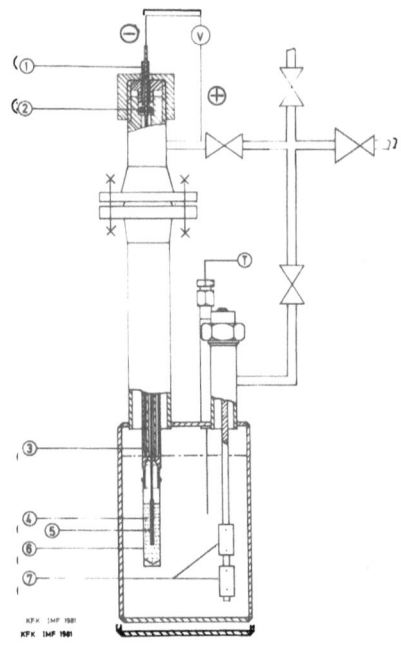

1, 3: Alumina insulators
2    : Viton O-ring seal
4    : $Li_2CO_3-Na_2CO_3$ eutectic
5    : Reference electrode
6    : Armco iron cup
7    : S. S. Foils
(T)  : Thermocouple

Fig. 2.   Electrochemical carbon meter test assembly

RESULTS AND DISCUSSION

      Fig. 3 shows the output of the UNC-type meter with respect
to time. The data represent the output during the second experiment
(550 $^{o}$C). The first experiment lasted approx. 350 hours at 600 $^{o}$C.
Fig. 4 shows the carbon activity $a_c$ (HCM) calculated from HCM
output during the start up of the experiment. The decreasing trend
of the output remained throughout all the experiments though slower
during later experiments. The removal of the carbon by stainless
steel pots and also to an extent by the meters could have lead to
this observation. Because of the changing carbon potential in the
static pot systems it is difficult to correlate the carbon activities
determined by foil equilibrations with the meter outputs. The
analysis of the cover gas showed that the hydrogen build up in the
system is severe in the case of the UNC-type meter. This is easily
understandable in case of static pot system where no trap for
hydrogen is present. The concentrations of $H_2$ and $CH_4$ at the end of
each of the experiments (Fig. 3) were 4100 and 50 ppm respectively.
Calculation of carbon activity from these values yield very high
values. Similar behaviour has been reported earlier by Schreinlechner et.al. (5).

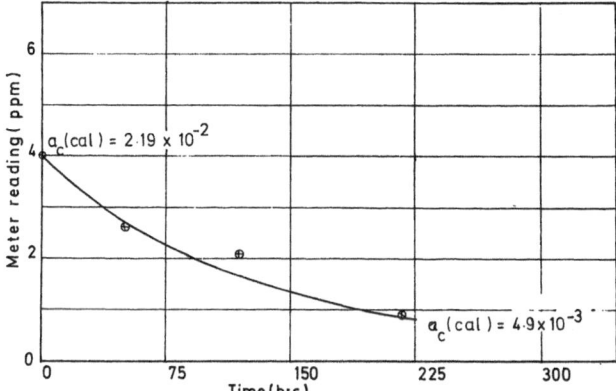

Fig. 3   Variation of output of the UNC-type meter with time
         (temp. 550 $^{\circ}$C)

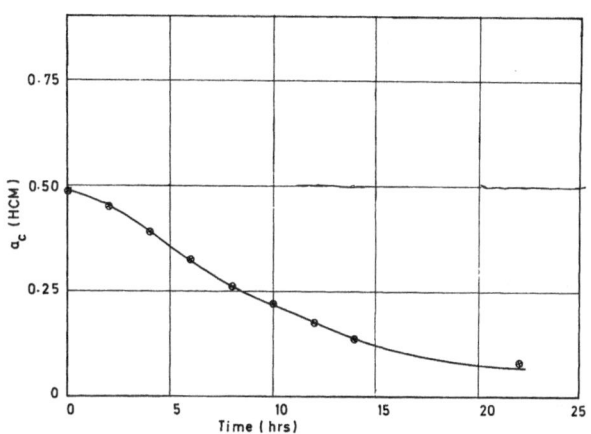

Fig. 4   Variation of HCM output with time (temp. 600 $^{\circ}$C)


        Fig. 5 and Fig 6 show the response of the UNC-type meter and
HCM for introduction of the carbon source. Their response time
could not be easily calculated since this time falls between the
time needed for heating up of the system. Fig. 7 shows the temperature dependence of the output of the UNC-type meter. The experiment
was carried out after a lapse of approx. 100 hours with the carbon
source in sodium at 550 $^{\circ}$C. It is to be noted that there is a change
in slope at approx. 570 $^{\circ}$C. Though the reason for this change is not
yet certain, compound formation of the hydrogen at lower temperatures,
for example NaOH, NaH etc., and hence having lesser influence on
carbon activity may be one among them. Further experiments to

calibrate the meters need operation in a flowing sodium system
where carbon activities are relatively constant.

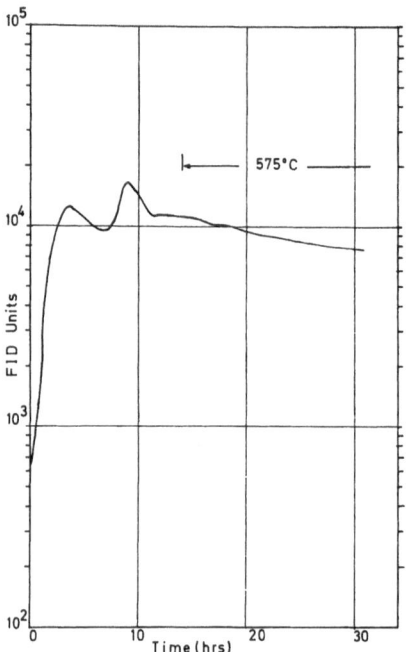

Fig. 5   Response of the HCM for introduction of carbon source

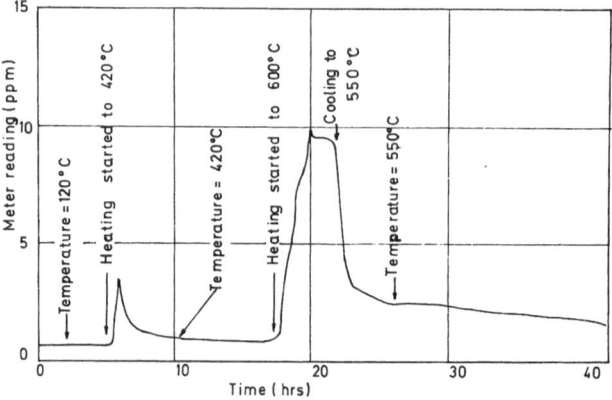

Fig. 6   Response of UNC-type meter for carbon source introduction

   Fig. 8 shows the emf of the electrochemical carbon meter with
respect to time in case of carburised iron rod and graphite filled
iron tube reference electrodes. The time plotted is the integrated
period of time during which the electrode remained dipped into the
electrolyte. As can be seen from the figure the output of the meter
drops steeply during the initial period and then slowly. The emf
measured during the first time with graphite filled carburised
iron tube reference electrode gave near theoretical emf (expected
from the result of the foil equilibration test done before changing
the electrode). Fig. 9 shows the temperature dependence of emf for
graphite filled carburised iron tube and graphite ref. electrodes.
The measurements with graphite ref. electrode were done by keeping
it always dipped into electrolyte as the attainment of equilibrium
temperature required longer times than the other two electrodes.
The difference in slope of the lines is due to different characteristics of the ref. system.

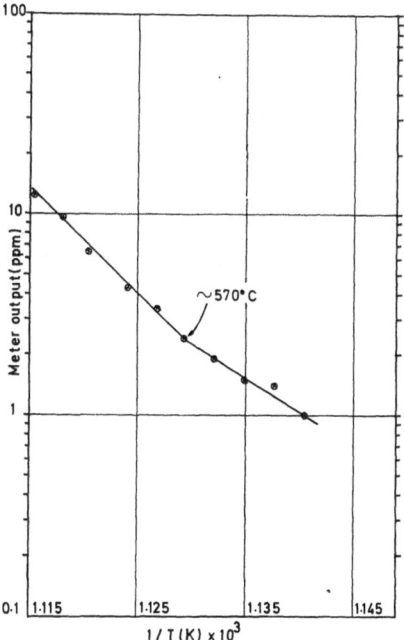

Fig. 7   Temperature dependence of output of UNC-type meter

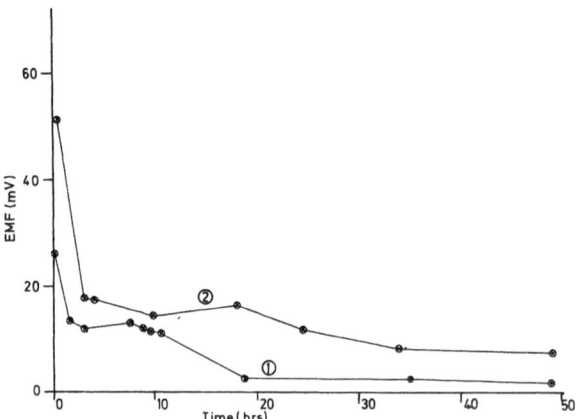

Fig. 8  EMF change of electrochemical carbon meter with time

　　1.  Carburised iron on ref. electrode:
　　　　Temperature = 650 $^{\circ}$C
　　2.  Graphite filled carburised iron tube ref. electrode:
　　　　Temperature = 550 $^{\circ}$C

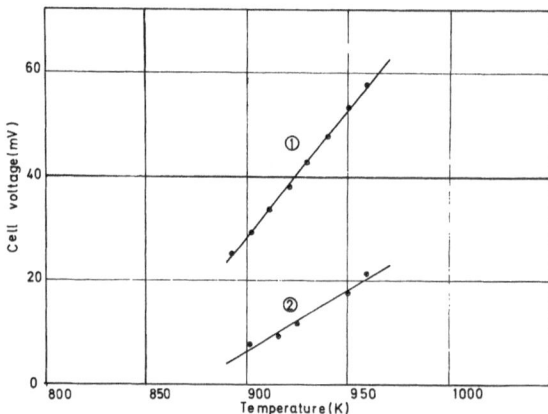

Fig. 9  Temperature dependence of EMF for two different reference
　　　　electrodes

　　1.  Graphite ref. electrode
　　2.  Graphite filled carburised iron tube  ref. electrode

CONCLUSIONS

The new design for the probe of the UNC-meter could overcome the sensitivity requirements, though operation in small system would lead to hydrogen contamination. The HCM and UNC-type meter would have to be further tested and calibrated in flowing sodium systems in order to maintain a relatively constant carbon activity. Further characterisation of the reference electrodes are needed for the reliable operation of the electrochemical carbon meter.

ACKNOWLEDGEMENTS

This work was carried out under Indo-German collaboration in chemistry of liquid metals. My sincere thanks are due to Dr. H. U. Borgstedt for his discussions and constant encouragement during this work. Thanks are due to Mr. Frees and Mr. Drechsler who helped in design and construction of the experimental set up. I thank Mr. Unk for his help during various periods of the experiment.

REFERENCES

1.  R. C. Asher et.al., in Liquid Alkali Metals 1976, Inst. Phys. Conf. Ser. No. 30, p. 561 (1977).
2.  W. Caplinger, USAEC Report UNC-5226 (1969).
3.  F. J. Salzano et.al., Nuclear Technology, Vol. 10 (1971) p. 335.
4.  M. R. Hobdell et.al., International Conference on Liquid Metal Technology in Energy Production, Champion, Pennsylvania, 1976, p. 533.
5.  I. Schreinlechner et.al., Proc. BNES Conf. on Liquid Alkali Metals, Nottingham, p. 137 (1973).

# ANALYSIS OF TRACE METALS IN SODIUM BY FLAMELESS ATOMIC ABSORPTION SPECTROPHOTOMETRY

T. R. Mahalingam, R. Geetha, A. Thiruvengadasamy
and C. K. Mathews

Reactor Research Centre
Kalpakkam    India

## INTRODUCTION

The determination of trace metals in sodium is usually carried out by distilling off the sodium in vacuum and analysing the residue by flame Atomic Absorption Spectrophotometry (1-3). The distillation is done to avoid aspirating a highly concentrated solution which would lead to problems such as clogging of the nebuliser and severe spectral interference. Moreover distillation also serves as a concentration step. A sample weight of 50 gms has been used for the analysis of nuclear grade sodium (2). In our laboratory we have followed the method of distillation and analysis by flame AAS for the commercial sodium samples. But there are certain situations where it is desirable to avoid the distillation step as in the case of the analysis of volatile impurities such as Zn, Cd and Pb which are lost partially or wholly during the distillation process. Moreover use of small sample size is desirable especially while analysing the primary sodium. With these objectives a direct method to analyse sodium by AAS using the graphite furnace atomiser has been developed in our laboratory. With the superior sensitivities ($10^{-12}$ g) attainable with very small sample volumes (20 µl) we have found that it is possible to determine various impurities such as Cu, Cr, Co, Mn and Pb directly by placing as low as 2 mg of the sample in the graphite cuvette. By employing a suitable temperature programme it has been possible to remove the bulk of the matrix before atomisation, thereby minimising the matrix interference. This paper describes the direct method developed in our laboratory for the analysis of sodium and compares it with the distillation cum flame AAS method.

EXPERIMENTAL

Apparatus

Atomic Absorption Spectrophotometer (Model IL-751) made by
M/s. Instrumentation Laboratory Inc., USA fitted with both flame
atomiser and the controlled temperature furnace atomiser (CTF 555)
has been used. It is a double-beam instrument having the $D_2$ arc
background correction facility. The graphite cuvette temperature
is measured by a temperature sensor (tungsten alloy) which is in
contact with it and acts as a resistance thermometer. The sensor
forms part of an electrical bridge circuit, which in turn switches
off the operating current when the proper temperature is reached.
The temperature is indicated on a panel meter and the correctness
of the indication was tested by checking the melting points of pure
copper and tantalum in the cuvette.

Reagents

Demineralised water, double distilled in quartz distillation
assembly has been used for all experimental work. Electronic grade
nitric acid (BDH) was double distilled in quartz and used.

Stock standard solutions containing 1 mg/ml of the elements of
interest were prepared using highly pure metals or salts (99.9 % or
better). Dilutions were done with quartz distilled water and 0.1 N
acidity was maintained. Solutions were stored only in clean
polythene bottles.

Procedure
1. Sample Preparation:

All the laboratory-ware used were thoroughly cleaned
successively in 1:1 nitric acid and 10 % EDTA solution (left
overnight). Subsequently they were thoroughly washed with quartz
distilled water and the washings checked by flameless AAS before
they were used for either standards or samples. Washing with EDTA
helped especially in the reduction of the manganese blank which was
not getting removed by just acid cleaning.

"Nuclear grade" sodium supplied by M/s. Metaux Speciaux, West
Germany and the "commercial grade" sodium supplied by M/s. Alkali
Metals, Hyderabad were studied. These sodium samples were out and
weighed in an inert atmosphere glove box. For sampling the sodium
for the analysis of Cr, Co, Mn and Pb a clean knife made of high
puritiy copper was used. For the analysis of copper, a clean
stainless knife was used.

The distillation procedure is described elsewhere (2). Sodium
was sampled in tantalum crucibles and placed in a pyrex distillation

assembly which was then closed and taken out of the box. The sodium
was then distilled off in vacuum, making use of induction heating.
After the distillation was completed, the crucible was taken out
and the residue carefully dissolved in acid and made to volume
(25 ml).

For the direct determination, sodium was sampled in a clean
quartz flask. The samples were dissolved by the drop-wise addition
of quartz distilled water under argon atmosphere. The flask was
externally cooled by ice-cold water. It took about 45 minutes to
dissolve 5 g sample. The samples were then neutralised by adding
the required amount of nitric acid, made to volume (50 ml) and
stored in polythene bottles. The nitric acid needed for the
neutralisation of roughly 5 g sample was found to be between 9 and
9.5 ml. Hence 10 ml of the acid was taken in a clean platinum dish,
evaporated to near dryness in a special hood under argon flow, made
to volume and tested for the blank by flameless AAS. For all the
elements studied, the blank has been found to be very low, viz.,
of the order of only a few picograms.

## 2.  Analysis by flame AAS:

Air acetylene flame was used and the instrumental conditions
were chosen to give the best sensitivities for the elements studied.
Effect of sodium on the atomic absorption of the elements viz., Ca,
Mg, Cr, Co, Mn etc., were checked and found to be nil upto a
concentration of 500 ppm of sodium which is the maximum amount
likely to be present in the distillation residue. Hence samples were
analysed by comparing with the calibrations obtained for pure aqueous
standards. Only calcium was analysed in the presence of 1 mg/ml of
lanthanum chloride to take care of the observed chemical interference.
The results are presented in Table 1.

## 3.  Direct determination by flameless AAS:

Sample volumes used were 20 µl for Cr, Mn, Co and Cu and
10 µl for lead. Finn pipettes having disposable polythene tips were
used for injecting the solutions. Cylindrical pyrolytically coated
graphite cuvettes were used. The purge gas used was argon. The peak
absorbance was read off from the digital display of the instrument
console. The absorbance profiles were also monitored on a strip chart
recorder.  The dry, char and atomisation programmes (Table 2) were
optimised to get the best sensitivites and the least non-specific
absorption. Sodium nitrate decomposes at 230 $^{o}$C and sodium oxide,
the decomposition product, sublimes between 900 $^{o}$C and 1400 $^{o}$C.
For all the elements studied except lead, the bulk of the matrix
could be removed prior to atomisation by the use of proper charing
temperatures. The residual matrix was found to cause a non-specific
absorption of only around 0.1 to 0.2 absorbance units which was
taken care of by the use of simultaneous $D_2$ arc background
correction.

Table 1.   Analysis of Sodium by Flame AAS

| Element | "Nuclear grade" sodium concentration (ppm) | "Commercial grade" sodium concentration (ppm) | Precision of replicate measurements (RSD) | Precision of the method | Detection limit (5g sample (ppm) |
|---------|------|------|------|------|------|
| Ca | < 0.5  | 63      | 0.9 | 9 | 0.5  |
| Mg | < 0.05 | 0.2     | 0.9 | 7 | 0.05 |
| Cr | < 0.3  | < 0.3   | –   | – | 0.3  |
| Mn | < 0.1  | < 0.1   | –   | – | 0.1  |
| Co | < 0.3  | < 0.3   | –   | – | 0.3  |
| Cu | < 0.2  | < 0.2   | –   | – | 0.2  |

Table 2.   Instrumental Conditions for Flameless AAS

| Element | Wave length (nm) | Band pass (nm) | Furnace Time – Temperature Programme | | DRY | | CHAR | | ATOMISE | |
|---------|------|------|------|------|------|------|------|------|------|------|
| Cr | 357.9 | 1,0 | Temp °C | | 80 | 125 | 850  | 1350 | 2200 | 2200 |
|    |       |     | Ramp Time Sec. | | 20 | 45 | 25 | 25 | 0 | 5 |
| Mn | 279.5 | 0.5 | Temp °C | | 80 | 125 | 800  | 1350 | 2200 | 2200 |
|    |       |     | Ramp Time Sec. | | 20 | 45 | 20 | 20 | 0 | 5 |
| Co | 240.8 | 0.3 | Temp °C | | 80 | 125 | 900  | 1250 | 2200 | 2200 |
|    |       |     | Ramp Time Sec. | | 20 | 45 | 30 | 25 | 0 | 5 |
| Cu | 324.7 | 1.0 | Temp °C | | 80 | 125 | 1100 | 1100 | 1900 | 1900 |
|    |       |     | Ramp Time Sec. | | 20 | 45 | 30 | 25 | 0 | 5 |
| Pb | 283.3 | 1.0 | Temp °C | | 80 | 125 | 500  | 550  | 1800 | 1800 |
|    |       |     | Ramp Time Sec. | | 20 | 45 | 30 | 25 | 0 | 5 |

In the case of lead, charing temperature higher than 600 °C could not be used as it resulted in serious loss of signal. Hence to bring down the matrix concentration, the sample was diluted ten times and only 10 µl volume was used. The background absorption was found to be low of the order of 0.1 absorbance unit. In all cases multiple standard addition were carried out and graphs were plotted to get the concentration.

## Matrix Effect

To study the effect of matrix on the absorption of these elements, one mixed standard was prepared with the concentration of sodium nitrate maintained at the same level as samples. Spec-pure sodium nitrate was used for this purpose. There was reduction in

Table 3.  Matrix Effect

| Element | Volatility loss % | Matrix effekt % |
|---------|-------------------|-----------------|
| Cu | 80 | 10 |
| Mn | 20 | 46 |
| Co | 75 | 10 |
| Cr | 14 | 43 |
| Pb | Nil | 25 |

Table 4.  Analysis of "Nuclear Grade" Sodium by Flameless AAS Direct Method

| S. No. | Element | Conc. ppm | Precision of replicate measurements (RSD) | Precision of the method (RSD) | Detection limits (5g sample) ppm | Analysis after distillation Conc.in ppm |
|--------|---------|-----------|---------------------------------------------|-------------------------------|----------------------------------|------------------------------------------|
| 1. | Cu | 0.15 | 6 | 10 | 0.08 | 0.12 |
| 2. | Mn | 0.04 | 6 | 10 | 0.02 | 0.03 |
| 3. | Co | <0.08 | - | - | 0.08 | <0.01 |
| 4. | Cr | 0.17 | 5 | 15 | 0.06 | 0.16 |
| 5. | Pb | 0.16 | 8 | 12 | 0.07 | - |

the absorbance signal due to two reasons: (1) consequent to the use
of higher charing temperature to remove the matrix, the impurity
elements themselves are volatalised off which has been termed as
volatility loss (this was studied using pure standard) and (2) the
additional reduction in signal strength due to the presence of
sodium matrix which has been referred to as "matrix effect"
Table 3). The total loss in signal was about 60 % to 90 % for the
various elements studied, but as the inherent sensitivity of the
furnace AAS is quite high, these elements could still be estimated
successfully. The precision of the replicate measurements have been
found to be between 5 % and 8 % RSD.

Precision and Accouracy

     To assess the accuracy of the method, the results obtained
have been compared with the results got by analysing the distillation
residue (Table 4). The detection limits, the precision of the
replicate measurements and the precision of the method (average of
three samples) are also given in the same table (Table 4).

     The results obtained by the direct method compare well with
the analysis of the distillation residue. The detection limits
attainable (0.02 to 0.08 ppm) are quite good and better than
achievable with flame. Even volatile elements such as lead which are
lost in vacuum distillation could be done by this method. Further
work is being done to check the accuracy of this method for the
analysis of lead and also to extend the method to the analysis of
other volatile elements such as zinc and cadmium.

REFERENCES

     1.  The determination of impurities in nuclear grade sodium
         metal, Louis Silverman, Pergamon Press, 1971.
     2.  Determination of metallic impurities at the ppb-ppm level
         in sodium, T.P. Rachandran et al, ANL-7668, March 1970.
     3.  Interim methods for the analysis of sodium and cover gas,
         ANL-ST-6, January 1971.

# THE KINETICS OF HYDROGEN REMOVAL FROM LIQUID SODIUM

John R. Gwyther and Andrew C. Whittingham

C.E.G.B., Berkeley Nuclear Laboratories

United Kingdom

## INTRODUCTION

Thermal decomposition of sodium hydride and removal of hydrogen under vacuum or flowing argon appears a possible technique for the regeneration of hydride laden LMFBR cold traps. However, little is known about the kinetics of these processes, and this paper describes some measurements of the rate of hydrogen removal from sodium/sodium hydride mixtures by continuous evacuation in the temperature range 260 – 420 $^{\circ}$C and by argon purging in the range 293 – 360 $^{\circ}$C. Some observations on the kinetics of hydrogen desorption from unsaturated solutions of sodium hydride in liquid sodium at temperatures up to 400 $^{\circ}$C are also presented.

## THEORY

Hydrogen dissolves in liquid sodium monatomically in accordance with the equilibrium:

$$1/2 \ H_2 \ (g) \ \rightleftarrows \ H_{(solution)} \tag{1}$$

In a closed system of gas volume, V, and sodium surface area, S, the rate of approach to equilibrium is given by the general expression:

$$\frac{dV}{dt} = k_a \ S \ (P)^x - k_d \ S \ (n_H)^y \tag{2}$$

where dV/dt is the rate of hydrogen absorption, P the hydrogen pressure, $n_H$ the dissolved hydrogen concentration in sodium (ppm by weight), and $k_a$ and $k_d$ are the respective rate constants for the absorption and desorption processes of order x and y. At equilibrium $dV/dt = 0$ so that:

$$k_a (P_e)^x = k_d (n^e_H)^y \tag{3}$$

It has been shown that the absorption process is first order with respect to hydrogen pressure i.e. $x = 1$, and that $k_a$ obeys the following expression in the temperature range $337 - 404 °C$ (1):

$$\log_{10} K_a (mm^3 \ mm^{-2} \ . \ s^{-1} . Pa^{-1}) = - 0.28 - \frac{3640}{T} \tag{4}$$

Since the order of the desorption process is unknown $K_d$ will have the units $mm^3 \ mm^{-2} \ . \ s^{-1} \ ppm^{-y}$. For saturated solutions of hydrogen in sodium, $n_H$ is constant at a given temperature ($=n_{H(sat)}$) and $k_d n_H^y$ may be replaced by a constant $k_d'$. Under continuous evacuation at low hydrogen pressure, where $P \sim 0$, the volume of hydrogen gas removed per unit time, $V_d$, will be given by:

$$- \frac{dV}{dt} = V_d = k_d (n_H^{sat})^y S = k_d' S \tag{5}$$

However, under conditions of flowing argon where $P \gtrless 0$, the net hydrogen removal rate, given by the product of the hydrogen partial pressure in argon, P, and the argon flow rate, F, will be the difference between the desorption and absorption reactions:

$$V_d = FP = k_d' S - k_a PS \tag{6}$$

EXPERIMENTAL

The hydrogen pressure over liquid sodium/sodium hydride mixture. (typically 0.2 kg Na, 20 g NaH) was maintained at a negligible fraction of the sodium hydride dissociation pressure by continuous evacuation of the hydrogen from the reaction vessel containing the mixture. The reaction vessel (65 mm i.d. x 200 mm high) was located in an electromagnetic stator (2) which provided a rotating magnetic field of adjustable strength to stir the sodium.

The steady state hydrogen flow rate was measured as a function of temperature at atmospheric pressure and temperature at the exit of the vacuum pump by means of calibrated capillary flow meters. Under these conditions the reverse reaction of hydrogen absorption by liquid sodium was effectively eliminated. The amount of hydrogen removed from the sodium was a very small fraction of the sodium hydride present in excess of its terminal solubility in liquid sodiu at the experimental temperature. Sodium vapour was condensed in two stainless steel mesh packed vapour traps to prevent carry over into the vacuum pump. Experiments were performed with static and stirred liquid sodium in the range $260 - 420 °C$.

The second technique consisted of passing known flow rates of pre-heated argon (up to $10^4$ $mm^3 s^{-1}$) at ambient pressure over liquid sodium/sodium hydride mixtures in the reaction vessel described above at different temperatures. The exit gas was analysed for hydrogen partial pressure in argon using a gas chromatograph (Pye Series 104) with a katharometer detector. Hydrogen evolution rates as a function of argon flow rate over the sodium were thus determined for static and stirred liquid sodium at temperatures in the range 293 - 360 °C. It should be stressed that in all these experiments, the argon flowed directly over the sodium surface and was not bubbled through the liquid sodium.

RESULTS

The measured hydrogen flow rates under dynamic vacuum are presented for static and stirred sodium as a function of temperature in Table 1.

Table 1.  Hydrogen Evolution Rate under Dynamic Vacuum as a Function of Temperature

| Unstirred Sodium[x)] | | Stirred Sodium | |
|---|---|---|---|
| Temperature °C | Hydrogen Evolution Rate $(mm^3 s^{-1})$ | Temperature °C | Hydrogen Evolution Rate $(mm^3 s^{-1})$ |
| 293 | 8.2 | 261 | 6.5 |
| 316 | 16 | 300 | 38.3 |
| 330 | 28 | | |
| 341 | 35 | 340 | 217 |
| 355 | 116 | 380 | 1183 |
| 380 | 670 | 400 | 2100 |
| | | 410 | 3420 |
| | | 417 | 5950 |
| | | 420 | 7260 |
| | | 421 | 7780 |

[x)] Geometric Sodium Surface Area = 3300 $mm^2$.

The salient features of the results are given below:

(i)    The rate of hydrogen evolution at a given temperature was increased by stirring, but was independent of the amount of solid sodium hydride present i.e. solid dissolution was not rate controlling.

(ii)   The rate of hydrogen evolution increased markedly with increasing temperature.

(iii) Independent experiments showed that extensive "blubbing" and turbulence occurred at the surface of hydrogen-saturated liquid sodium under dynamic vacuum, especially at temperatures > 400 °C, where hydrogen equilibrium pressures approach 1 bar.

It was shown in separate experiments that the measured differences in hydrogen evolution rate between stirred and static sodium could be accounted for, within experimental error, by differences in the nominal liquid sodium surface area.

The following least squares analysis of data below 400 °C yields the expression:

$$\log_{10} V_d \ (mm^3 \ s^{-1}) = 13.13 - \frac{6935}{T} \tag{7}$$

where $V_d = k_d'S$ for nominal liquid sodium surface area of 3300 $mm^2$.

The hydrogen evolution rates, calculated from the product of the hydrogen partial pressure in argon and the argon flow rate, increased with argon flow rate at each temperature studied as shown in Fig. 1. Hydrogen evolution rates also increased with sodium stirring rate at a given temperature and argon flow rate.

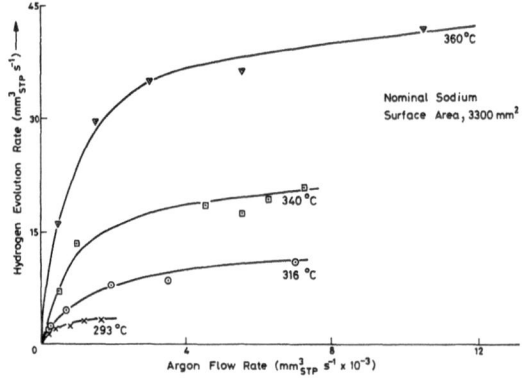

Fig. 1   Hydrogen Evolution Rate as a Function of Argon Flow Rate
         at Different Temperatures

Re-arrangement of equation (6) and substitution for $k_a$ (= $k_d'/P_{e,sat}$) yields the expression:

$$\frac{1}{P} = \frac{F}{Sk_d'} + \frac{1}{P_{e,sat}} \tag{8}$$

where $P_{e,sat}$ is the hydrogen equilibrium pressure over saturated solutions of sodium hydride in liquid sodium. Plots of 1/P against F are linear as shown in Fig. 2, and the reciprocal slopes give the maximum value of the hydrogen removal rate i.e. $\lim(FF) = k_d'S$ at each temperature. Least

$$P \to 0$$
$$F \to \infty$$

squares analysis of the data thus derived gives the expression:

$$\log_{10} V_d \ (mm^3 \ s^{-1}) = 10.34 - \frac{5485}{T} \tag{9}$$

for a nominal sodium surface area of 3300 $mm^2$.

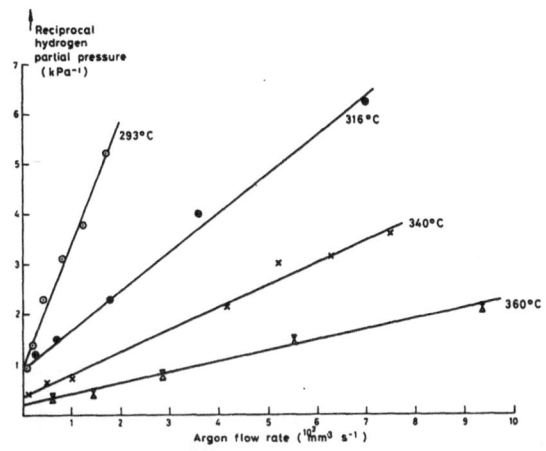

Fig. 2    Reciprocal Hydrogen Partial Pressure as a Function of Argon Flow Rate

Since $k_d' = k_a P_{e,sat}$, a 'theoretical' hydrogen evolution rate in the absence of hydrogen bubble formation can be calculated from the known temperature dependence of $k_a$ and $P_{e,sat}$; the latter is given by (1):

$$\log P_{e,sat} \ (Pa) = 13.62 - \frac{6122}{T} \tag{10}$$

Teh calculated evolution rate is then:

$$\log_{10} V_d \ (mm^3 \ s^{-1}) = 17.06 - \frac{9762}{T} \tag{11}$$

for a sodium surface area of 3300 $mm^2$. Values of $V_d$ calculated from equations (7), (9) and (11) are compared in Fig. 3.

Agreement between the data obtained from the argon purge experiments and those derived from hydrogen absorption data is reasonable above 350 °C, but both are appreciably lower than those obtained under dynamic vacuum at corresponding temperatures. Since hydrogen

bubble formation is eliminated in the presence of argon at atmospheric
pressure, the above comparison suggests that hydrogen bubble formation below the sodium surface contributes appreciably to the
measured hydrogen evolution rate at all temperatures under dynamic
vacuum.

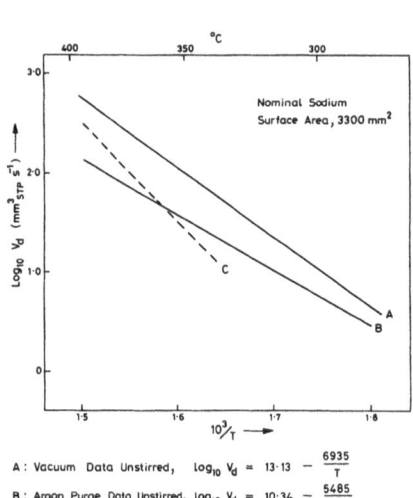

A : Vacuum Data Unstirred,  $\log_{10} V_d = 13\cdot13 - \dfrac{6935}{T}$

B : Argon Purge Data Unstirred,  $\log_{10} V_d = 10\cdot34 - \dfrac{5485}{T}$

C : Derived from Hydrogen,  $\log_{10} V_d = 17\cdot06 - \dfrac{9762}{T}$
    Absorption Data.

Fig. 3 Comparison of Hydrogen
       Removal Rates as a
       Function of Reciprocal
       Temperature from Vacuum,
       Argon Purge and Hydrogen
       Absorption Data

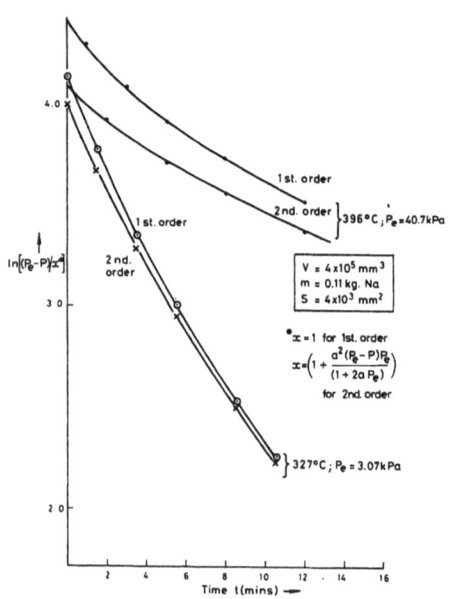

Fig. 4 Plots of ln $\left[ (P_e - P)/x \right]$
       as a Function of Time for
       Desorption of Hydrogen from
       Unsaturated Solution in
       Liquid Sodium

APPLICATION TO IN-SITU COLD TRAP REGENERATION

     The above data may be used to give an estimate of the timescales
necessary to regenerate a hydride-laden cold trap by making certain
simplifying assumptions:

     (i)    The cold trap is not drained of residual sodium and has a
free sodium-cover gas interface from which hydrogen gas could be
removed.

     (ii)   The hydrogen removal rate is directly proportional to the
sodium surface area (this could involve an extrapolation over three
orders of magnitude from the present experiments to a full size
LMFBR cold trap).

(iii) The sodium hydride burden is 100 kg and the sodium surface area is normalised to 1.0 m$^2$ (equivalent to a cold trap vessel
of 1.1 m internal diameter).

Regeneration timescales for removal under vacuum and argon
calculated on the basis of equations (7) and (9) respectively are
compared in Table 2 at various temperatures. It should be noted that
because of the contribution from hydrogen bubble formation in the
vacuum case, which cannot be quantified, the data in Table 2 will
give minimum evolution rates i.e. maximum regeneration times.
Although hydrogen removal by argon purging appears slower, its ef-
feciency could be improved by sparging with small argon bubbles at
the base of the trap improve the contact time and surface area.

Table 2.    Calculated Regeneration Times for Hydrogen Removal (100 kg
            NaH) as a Function of Temperature under Dynamic Vacuum and
            Argon Purging for Sodium Surface Area 1.0 m$^2$

| Temperature | Vacuum | | Argon Purging | |
|---|---|---|---|---|
| | Regeneration Rate $kgH_2\ m^{-2}\ day^{-1}$ | Time days | Regeneration Rate $kgH_2\ m^{-2}\ day^{-1}$ | Time days |
| 350 $^{\circ}$C | 0.23 | 18 | 0.082 | 51 |
| 375 $^{\circ}$C | 0.62 | 6.7 | 0.18 | 23 |
| 400 $^{\circ}$C | 1.56 | 2.7 | 0.37 | 11 |
| 425 $^{\circ}$C | 3.64 | 1.2 | 0.73 | 5.7 |
| 450 $^{\circ}$C | 8.03 | 0.5 | 1.35 | 3.1 |

Inspection of Table 2 shows that hydrogen removal by either
technique could give acceptable timescales for regeneration at
temperatures > 350 $^{\circ}$C, although further work on full size cold traps
is clearly necessary to validate the technique for in-situ regeneration.

HYDROGEN DESORPTION FROM UNSATURATED SOLUTIONS OF HYDROGEN IN SODIUM

In an attempt to determine the order, y, of the desorption
process, experiments were carried out with unsaturated solutions of
hydrogen in sodium. The technique consisted of expanding hydrogen
gas, initially in equilibrium with an unsaturated solution of hydrogen
in sodium, into an evacuated dead space of known volume and measuring
the rate of approach to a new, lower equilibrium pressure, $P_e$. An
isothermal apparatus fitted with a capacitance manometer for pressure measurements was used to avoid hydrogen loss due to co-condensation with sodium in cold spots.

The desorption process is likely to be first or second order in dissolved hydrogen since either of the following steps could be rate determining:

1st order   $H_{(solution)} \rightarrow H_{(surface)}$                    (12)

2nd order   $H_{surface} + H_{surface} \rightarrow H_2 \text{ (g)}$              (13)

Integration of equation (2), and substituting for $k_a$ in equation (3), yields the following expressions for $k_d$ for first and second order kinetics respectively:

$$- \ell n \ (P_e - P) = (1 + aP_e) \frac{K_s RTS.k_d t}{2VP_e^{1/2}} + \text{constant} \qquad (14)$$

$$- \ell n \left[ (P_e - P) / \left( 1 + \frac{a^2 (P_e - P)P_e}{(1 + 2aP_e)} \right) \right] = \frac{(1 + 2aP_e)K_s^2 RTS.k_d t}{2V}$$

$$+ \text{ constant} \qquad (15)$$

where P is the hydrogen pressure at time t, $P_e$ is the hydrogen equilibrium pressure and a is a constant equal to $2V/RTmK_s P_e^{1/2}$ where R is the gas constant, m the mass of sodium, and $K_s$ is Sievert's constant for hydrogen in sodium at temperature T. Plots of these two functions against time for hydrogen desorption runs at 327 $^{\circ}$C and 396 $^{\circ}$C respectively are shown in Fig. 4. In each case there is some deviation from linearity which is more marked at 396 $^{\circ}$C. This deviation is probably a consequence of the hydrogen "bubbling" which occurs at the liquid metal/gas interface; this effect is enhanced at higher temperatures where values of $P_e$ are greater and although bubbling will diminish as the equilibrium pressure is approached, changes in liquid metal surface area may be occurring during a desorption run. An additional problem encountered in these experiments was that since a << 1, there was little difference between the expressions on the left hand side of equations (14) and (15) even for $P_e$ >> P, so that distinction between first and second order kinetics would have been difficult even if linear plots had been obtained.

Further experiments will therefore be necessary to elucidate the order of the desorption reaction under conditions where the formation of hydrogen bubbles are eliminated e.g. by studying hydrogen removal rates from unsaturated solutions of hydrogen in sodium under flowing argon.

ACKNOWLEDGEMENT

This paper is published by permission of the Central Electricity Generating Board.

REFERENCES

1.  A. C. Whittingham, J. Nuc. Mats., $\underline{60}$(2), 119, 1976.
2.  D. J. Hayes, M. R. Baum and M. R. Hobdell, J. Brit. Nuc.
    Energy Soc., April 1971.

A THERMODYNAMIC ASSESSMENT OF THE BEHAVIOR OF CESIUM AND

RUBIDIUM IN REACTOR FUEL ELEMENTS

Rajiv Kohli

Österreichisches Forschungszentrum Seibersdorf

Wien, Austria

INTRODUCTION

The nature of the chemical environment in the vicinity of the PCI-failure region in LWR fuel rods is crucial to understanding the mechanisms involved in such failure. This, in turn, requires detailed knowledge of the transport and reaction behavior of the fission products with each other, the fuel, and the zircaloy cladding. The purpose of this paper is to present just such a detailed thermodynamic model of fission product reaction behavior.

During irradiation large amounts of oxygen and fission products are generated. Since the affinity of the fission elements for oxygen is less than uranium (the rare earths form more stable oxides; however, they are soluble in $UO_2$ and will not exist as separate phases), all the oxygen released is not combined with the fission products. Thus, the chemical states of the fission products are effectively set by the oxygen potential ($\Delta \bar{G} = RT \ln p\text{-}O_2$) of the system. Also, for a reaction to occur at the cladding, migration of volatile fission product species must take place. Thus, the oxygen potential and the temperature gradient in the oxide fuel (2500 K at the centerline to ca. 900 K at the fuel periphery) are the two key parameters in this model.

Using thermochemical data for the species involved and the in-reactor oxygen potential range of - 100 to - 130 kcal/mol (1, 2), equilibrium thermodynamic calculations were performed for the most reactions of the fission products. The emphasis here is on the behavior of cesium and rubidium, since cesium

(and, to a lesser extent, rubidium) has always been observed
in the fuel-cladding gap in LWR fuel rods. In addition, the alkali
metals form some of the most stable compounds known. This suggests
that the chemical activity of other embrittling or potentially
embrittling fission products such as iodine and bromine (3),
tellurium and selenium (4), and molybdenum (5) might well be
controlled by cesium and rubidium. Except to note the fact, the
thermodynamics of another well-established aggressive species,
cadmium (6), and a potentially embrittling species, ruthenium (7),
will not be considered. Detailed thermodynamic analyses of the
behavior of cadmium and ruthenium will be published separately
(8, 9).

COMPOSITION OF THE EQUILIBRIUM ALKALI-METAL VAPOR

        Although Cs (g) and Rb (g) form a number of oxides, the
calculations show that the partial pressures of the various
gaseous oxide species are less than $10^{-27}$ atm. The equilibrium
form of the vapor should, therefore, be predominantly elemental
cesium and rubidium.

FUEL REACTIONS

        In the vapor state Cs (g) and Rb (g) can react with the fuel
forming a series of uranates (10). But at the oxygen potentials
of interest to LWRs only $Cs_2UO_4$ (s) and $Rb_2UO_4$ (s) will be stable.
The activities of all other uranates are calculated to be neglegible.
The oxygen potentials associated with the formation of $Cs_2UO_4$ (s)
and $Rb_2UO_4$ (s) for $p_{Cs}$ and $p_{Rb}$ of $10^{-2}$ and $10^{-3}$ atm, respectively,
and for $UO_{2+x}$ (s) are shown in Fig. 1 and 2. For oxygen potentials
above the respective lines in Fig. 1 and 2, the uranates can exist
as separate phases.

FORMATION OF MOLYBDATES

        Cs (g) and Rb (g) react with Mo to form several molybdates
(10), some of which are very stable. The oxygen potentials
associated with the formation of $Cs_2MoO_4$ (s) and $Rb_2MoO_4$ (s) are
also shown in Fig. 1 and 2 (other molybdates are estimated to
have very low activities). Inspection of the values indicates
that at higher temperatures (> 900 K) $Cs_2MoO_4$ is more stable than
$Cs_2UO_4$ (s). This suggests that co-migration of cesium and molybdenum
as $Cs_2UO_4$ (g) should occur and it should condense in the fuel-
cladding gap ($Cs_2MoO_4$ (s) melts at 1213 K). However, $Cs_2MoO_4$ (s)
is less stable than $Cs_2UO_4$ (s), thereby implying that $Cs_2UO_4$ (s)
will be the more prevalent condensed phase in the fuel-cladding
gap. On the other hand, for cesium pressures below $10^{-10}$ atm
$Cs_2MoO_4$ (s) is thermodynamically more stable than $Cs_2UO_4$ (s) over
the entire temperature range. The thermodynamics of $Rb_2MoO_4$ (s)
will be similar.

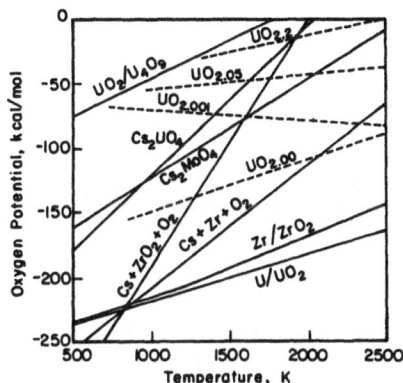

Fig. 1   Oxygen Potentials Associated with the Formation of Various
Ternary Cesium Compounds for a Cesium Partial Pressure of
$10^{-2}$ atm

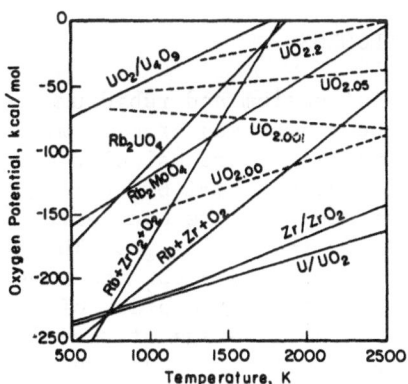

Fig. 2   Oxygen Potentials Associated with the Formation of Various
Ternary Rubidium Compounds for a Rubidium Partial Pressure
of $2 \times 10^{-3}$ atm

FORMATION OF ZIRCONATES

Although Zr forms a very stable oxide which is in solid
solution with $UO_{2+x}$ (s), its high yield could also lead to the
formation of the alkali-metal zirconates. In fact, both zirconate-
forming reactions, $Cs(Rb)+Zr+O_2$ and $Cs(Rb)+ZrO_2+O_2$, are thermodynamically much more favorable than either the uranates, or the
molybdates at temperatures below about 1500 K, as Fig. 1 and 2
show. This, of course, also means that at the fuel periphery and
in the fuel-cladding gap, cesium and rubidium zirconates will be
the major condensed phases present.

REACTIONS WITH IMPURITIES IN THE FUEL

Two manufacturing impurities in the fuel, $Al_2O_3$ and $SiO_2$,
could affect the chemistry of cesium and rubidium. A comparison
of the oxygen potentials associated with the formation of $CsAlO_2$
($RbAlO_2$) and $Cs_2SiO_3$ ($Rb_2SiO_3$) suggest, that only the silicates
are thermodynamically more stable than the uranates or the
molybdates, but less stable than the zirconates (the aluminates
are thermally unstable above about 1373 K (11)). This is shown
clearly in Table 1 which lists the equilibrium cesium partial
pressures associated with the various ternary cesium compounds.
Thus, deliberate additions of Si or $SiO_2$ to the fuel could inhibit
the migration of cesium and rubidium. On the other hand, the
simultaneous addition of Al or $Al_2O_3$ to the fuel does not enhance
the effectiveness of Si or $SiO_2$ additions to the fuel.

REACTIONS WITH THE HALOGENS

The large excess of cesium and rubidium over the halogens
ensures that all of the iodine and bromine will be tied up as the
respective halides. The latter are calculated to be the only
significant species in the $Cs(Rb)-I(Br)-Zr-UO_{2+x}$ system. Hence,
the halides will evaporate from the hot center of the fuel and
condense in the cooler fuel-cladding gap. The pressures of free
iodine calculated for the $CsI(s) + UO_2$ reaction for conditions
in the fuel-cladding gap are of the order of $10^{-21}$ atm, large
enough for monolayer iodine coverage of the zircaloy cladding (12).
This is probably not sufficient to cause the observed PCI failures
in zircaloy-clad fuel rods and the role of halogens seems to be
less significant than is generally accepted.

REACTIONS WITH THE CHALCOGENS

Fission-product tellurium and selenium should be tied up by
cesium and rubidium. However, the alkali-metal chalcogenides might
be thermally unstable above 1000 K (13), and other fission product
chalcogenides are thermodynamically less stable. Tellurium and
selenium should, therefore, migrate in the elemental state to

the cladding. Lack of thermodynamic information on complex chalcogenides, however, makes detailed predictions difficult.

Based ond the above analysis, cesium and rubidium are expected to transport the halogens as well as some of the molybdenum to the fuel-cladding gap, but not tellurium and selenium which will migrate elementally. A number of solid phases will also be present simultaneously. The relevance of such a chemical environment to PCI-failure needs to be explored further. The role of the halogens seems to be a relatively minor one.

Though the analysis as proposed here is comprehensive, it is based on equilibrium calculations using mainly room temperature thermochemical data. For more accurate calculations, data on most of the complex fission-product compounds is needed in the temperature range 300 - 1000 K. A research programm has been initiated to obtain high temperature thermodynamic data on cesium and rubidium chromates, molybdates, zirconates, and the chalcogenides. First results (14) appear to be in good agreement with the sparse data available in the literature.

Table 1.  Partial Pressures of Cs (g) Associated with Various Termary Compounds

| Compound | Temp. | $P_{Cs}$(atm) at $\overline{\Delta G}(O_2)$ kcal/mol | |
|---|---|---|---|
| | K | - 110 | - 130 |
| $Cs_2UO_4$ | 600 | $3.7 \times 10^{-13}$ | $1.6 \times 10^{-9}$ |
| | 1000 | $7.2 \times 10^{-4}$ | 0.11 |
| $Cs_2MoO_4$ | 600 | $3.8 \times 10^{-19}$ | $7.3 \times 10^{-12}$ |
| | 1000 | $3.9 \times 10^{-6}$ | $9.2 \times 10^{-2}$ |
| $Cs_2ZrO_3$ | 600 | $5.1 \times 10^{-40}$ | $1.5 \times 10^{-34}$ |
| (from Zr) | 1000 | $6.9 \times 10^{-19}$ | $1.3 \times 10^{-15}$ |
| $CsAlO_2$ | 600 | $3.2 \times 10^{-5}$ | $2.1 \times 10^{-3}$ |
| | 1000 | 1.2 | 13.8 |
| $Cs_2SiO_3$ | 600 | $1.7 \times 10^{-14}$ | $1.2 \times 10^{-12}$ |
| | 1000 | $3.8 \times 10^{-9}$ | $1.1 \times 10^{-4}$ |

ACKNOWLEDGEMENTS

The author would like to thank Professor Peter Ettmayer and
Dr. Pat O'Hare for useful discussions, and Carolyn Hayes for
assistance with manuscript preparation.

REFERENCES

1.  H. Kleykamp, J. Nucl. Mater. 84 (1979) 109.
2.  M. G. Adamson, E. S. Darlin, and J. H. Davies, "Thermo-
    migration and Reactions of Mobile/Active Fission
    Product Species in LWR Fuel Systems", in Specialists'
    Meeting on Internal Fuel Rod Chemistry, Erlangen, West
    Germany, Jan. 1979, IAEA, Vienna.
3.  B. Cox, Rev. Coatings Corros. 1 (1975) 367.
4.  R. Kohli, "Reaction Behavior of Zircaloy with Simulated
    Fission Products", Report LBL-12069, Lawrence Berkeley
    Laboratory, University of California, Jan. 1981.
5.  R. Kohli and F. Holub, Nucl. Technol. 48 (1980) 70.
6.  W. T. Grubb and M. H. Morgan III, Proc. ANS Topl.Mtg.
    Water Reactor Fuel Performance, St. Charles,
    Illinois, May 1977.
7.  R. Kohli and F. Holub, "Embrittlement of Zircaloy-2 by
    Ruthenium", submitted to Journal of Less-Common Metals.
8.  R. Kohli, "Thermodynamic and Transport Behavior of Cadmium
    in LWR Fuel Elements", Paper in Preparation.
9.  R. Kohli, "The Thermodynamics of Metallic Fission Products
    in LWR Fuel Rods", Paper in Preparation.
10. H. R. Hoekstra, Inorg. Nucl. Chem. Letters, 9 (1973) 1291.
11. G. Langlet, "Synthèse et étude cristallographique des
    composés du système $Cs_2O-Al_2O_3-SiO_2$", Report CEA-R-3853,
    Saclay, France, May 1969.
12. D. Cubicciotti, R. L. Jones, and B. C. Syrett, "Chemical
    Aspects of Iodine-Induced Stress Corrosion Cracking
    of Zircaloys", Proc. 5th ASTM Conf. on Zirconium in
    the Nuclear Industry, Boston, Aug. 1980.
13. Private Communication, H. Kleykamp, Kernforschungszentrum
    Karlsruhe, West Germany, Jan.1981.
14. R. Kohli and W. Lacom, "High-Temperature Enthalpies of
    $Cs_2MoO_4$ and $Rb_2MoO_4$ in the Temperature Range 373
    - 850 K", Unpublished Report, Austrian Research
    Center Seibersdorf, Dec. 1980.

THERMOCHEMISTRY IN THE SYSTEMS Na-O-C, Na-Al-O-C AND Na-C-N WITH

RESPECT TO INTERACTIONS OF OXYGEN, CARBON AND NITROGEN IN SODIUM

Hans Migge

Hahn-Meitner-Institut für Kernforschung Berlin GmbH
Berlin
Fed.Rep. of Germany

INTRODUCTION

Basic knowledge of the impurities in liquid sodium as well as
in lithium, especially of oxygen, carbon, nitrogen and hydrogen,
is a prerequisite for understanding the chemical behaviour of these
liquids in their environments. A lot of thermodynamic data of these
nonmetallic impurities have therefore been measured and collected
(1, 2). The solubilities of the nonmetals in sodium (and lithium)
are the main properties of interest, and there is also early knowledge about interaction between different impurities in the molten
metals. However, the results on this subject are inconsistent. By
thermochemical analysis oxygen and carbon were shown to be able to
form $Na_2CO_3$ in liquid sodium at T > 960 K (3-5). Experimental results
on the solubility of $Na_2CO_3$ in Na in the temperature range 423 K to
673 K (6) are in contradiction to this, where the carbonate should
not dissolve, but decompose (3,4). The interaction of nitrogen with
carbon in sodium is known in principle from the measurements of the
solubility of NaCN (7) and from the detection of this compound in
cold traps of sodium loops. The interaction was also demonstrated
by comparing the solubilities of carbon, influenced by different
concentrations of nitrogen in sodium (4). On the other hand nitrogen
was shown to have an extremely low solubility in Na (7).

Thermochemistry makes it possible to find out something about
the stabilities of such compounds in liquid sodium and lithium and
also about the interactions of their impurities (8,9). Systematic
thermochemical calculations have been performed in the systems Na-O-C,
Na-Al-O-C, Na-N-C and Na-O-N. The Results of the first three will
be dealt with here. The results of Na-O-N have been reported
previously (8), showing that the interaction of oxygen with nitrogen

in sodium will be low. The results are the basis for further calculations, for example in the system Na-O-N-C.

DATA AND METHOD OF CALCULATION

Calculations of the three component systems were made according
to Hirschwald et al.(10) and Jansson (11). Table 1 shows the chemical
species and the number of equilibria which have to be considered.
The gas pairs used are also cited. The corresponding oxygen and
carbon pressures $p_{O_2}$ and $p_C$ and the carbon activity $a_c$ were also
calculated.

Table 1.  Substances and origin of their free energy data

| System | condensed phases considered | gaseous Na-species above Na considered | number of possible heterogeneous equilibria | gas pair used for calcu-lations |
|---|---|---|---|---|
| Na-O-C | Na, $NaO_2$, $Na_2O_2$ $Na_2C_2$, C, $Na_2CO_3$ | - | $\binom{5}{2}= 10$ and Boudouard | $CO/CO_2$ |
| Al-O-C | Al, $Al_2O_3$, $Al_4C_3$, C | - | $\binom{3}{2}= 3$ and Boudouard | $CO/CO_2$ |
| Na-Al-O-C | Na, $Na_2O$, $Na_2C_2$, C, $Na_2CO_3$, Al, $Al_2O_3$, $Al_4C_3$, $NaAlO_2$ | - | 10 + 3 + 11 and Boudouard | $CO/CO_2$ |
| Na-N-C | Na, $Na_3N$[x), $Na_2C_2$, C NaCN, $Na_2NCN$[x) | Na, NaCN, $Na_2(CN)_2$ | $\binom{5}{2}= 10$ and Boudouard | $N_2/C_{gas}$ |

[x) means estimated data

All pressures in this paper are expressed in the units of atm.
Where it is not otherwise indicated the free energies of the pure
substances were taken from Janaf (12) or Barin and Knacke tables
(13), the data of $Na_2C_2$ were taken from Johnson et al. (14).

In contrast to $Li_3N$ (15) the only known thermodynamic quantity
of $Na_3N$ is the formation enthalpie, which was calculated from a
Born-Haber-cycle to be - 36 Kcal/mol (16). This value was used to
estimate free energies according to $G(T) = \Delta H^f - TS^{298}$. The formation

entropy was taken equal to that of $Li_3N$ (15). Corresponding estimations for $Li_3N$ using its $\Delta H^f$ and S-298 values are about 3 % too
low at 500 K, 11 % too low at 700 K and 23 % too low at 1000 K as
can be seen by comparing them with the experimentally determined
values (15). Corrections were therefore made at these temperatures.
The estimated G values of $Na_3N$ are listed in Table 2. They may be
too negative. This is concluded from the formation enthalpy of
$Li_3N$ calculated the same way as $Na_3N$ (16), resulting in - 76 Kcal/mol,
while the experimental value is about - 50 Kcal/mol (15).

Table 2.   Estimated G values for $Na_3N$ and $Na_2NCN$

| Temperature K | G value of $Na_3N$ Kcal/mol | G value of $Na_2NCN$ Kcal/mol | measured equilibrium pressures (17) | calculated pressures of nitrogen |
|---|---|---|---|---|
| 500 | - 42.7 | | | |
| 700 | - 48.7 | | | |
| 800 | - 52.8 | - 83.2 | $5.92 \times 10^{-2}$,(773 K) | $5.13 \times 10^{-2}$ |
| 900 | - 56.1 | - 93.0 | $5.92 \times 10^{-2}$,(923 K) | $7.82 \times 10^{-2}$ |
| 1000 | - 60.2 | -102.6 | $1.18 \times 10^{-1}$,(973 K) | $1.09 \times 10^{-1}$ |

Free energies of $Na_2NCN$ were estimated using the lowest equilibrium pressures reported by Sakurazawa et al. (17) from the synthesis
of $Na_2NCN$ by the reaction $2NaCN + 2Na + N_2 = 2 Na_2 NCN$ in the temperature range 723 K - 973 K. The authors always observed reaction
mixtures of $Na_2NCN$, NaCN and Na. Although they did not analyse the
gasphase, it must have consisted mainly of nitrogen, since other
gaseous species like NaCN, $Na_2(CN)_2$ and sodium vapor result in total
pressures below or nearly equal to those reported by the authors.
These nitrogen pressures were found to increase remarkably at
$Na_2NCN$ contents above about 10 mol% obiously indicating solution
effects. Such large $Na_2NCN$ contents do not appear to be relevant
in liquid sodium technology and can therefore be neglected here.
The equilibrium pressures taken from Sakurazawa (17) and the estimated values of the free energies of $Na_2NCN$ are listed in Table 2
together with the calculated pressures of nitrogen from equilibria
$2NaCN + 2Na + N_2 = 2Na_2NCN$.

An important advantage of the metallurgical phase diagrams for
pure substances is the possibility of making them gradually more
complex (18). So the quaternary system Na-Al-O-C was derived by
superimposing the two systems Na-O-C and Al-O-C. Near the area of
existence of Na the sodium aluminate $NaAlO_2$ then has to be considered.
This results in 11 additional equilibria (see Table 1). They form
the quadrupelpoints of the system, which can be used for selecting
the new relevant equilibria. An easier case is a quaternary system

with three nonmetals including graphite. This has been demonstrated for Li-O-H-C with respect to the stability of $Li_2O$ in helium, which is contaminated by CO, $CO_2$, $H_2$, $H_2O$ and $CH_4$ (19). It must be pointed out again that all these diagrams are calculated with the data of the pure substances, which implies neglecting the mutual solubilities of coexisting phases. However, often these effects are less important especially if no other information exists about a complex system than the free energies of the pure substances.

RESULTS AND DISCUSSION

Altogether five equilibria were graphically selected as being relevant up to a temperature of 1100 K in the Na-O-C system. They limit the fields of existence of the condensed phases in Figs. 1 to 3. $Na_2O_2$ does not appear, since it is stable at very high oxygen potentials (8). The solid line $C_{sol}$ marks the systems' maximum activity of carbon $a_c$ = 1, which can be obtained from the equilibrium $C_{sol}$ + $CO_2$ = 2CO. The area at the left of this graphite line has no physical sense. In accordance with the experimental results (14), $Na_2C_2$ cannot be in equilibrium with Na and consequently does not appear, but graphite ($C_{sol}$) does. As is obvious the only phases which can be in equilibrium with sodium at lower temperatures are graphite and $Na_2O$. However, the fields of existence of Na and $Na_2CO_3$ touch each other at a temperature of 932 K. At this quadruple point Na, $Na_2O$, $Na_2CO_3$ and graphite are in equilibrium simultanously. The pressure of carbon monoxide here is $3.83 \times 10^{-6}$ atm. So at temperatures T > 932 K $Na_2CO_3$ can be in equilibrium with Na. Hofer (3), using older data for $Na_2C_2$, found this quadruple point at T = 960 K consisting of the phases Na, $Na_2C_2$, $Na_2CO_3$ and $Na_2O$. The appearance of equilibria between Na and $Na_2CO_3$ is in contrast to the nonappearance of $Li/Li_2CO_3$ equilibria in the Li-O-C-system (9).

It is reasonable to conclude that the solubilities of oxygen and of carbon in liquid sodium will influence each other to a small extent at temperatures where $Na_2CO_3$ cannot be in equilibrium with Na. This conclusion is in agreement with the experimental results (4). Unfortunately the authors do not report experiments below the temperature of the quadruple point (932 K). From their theoretical considerations concerning the dilute solutions of carbon and oxygen in sodium they expect $Na_2CO_3$ not to form a separate phase below about 873 K, even if graphite is present. So there could be a decrease in the temperature of the quadruple point of about $60^{o}$, if solution effects are taken into account. There are no direct measurements of $Na/Na_2CO_3$ equilibria including the pressures of CO and $CO_2$, which would allow to calculate more precisely the temperature of the quadruple point. However, equilibria between Na and $Na_2CO_3$ below the temperature of about 873 K are not possible (4), but decomposition of $Na_2CO_3$ occurs. This was even reported for 823 K (20). Therefore, the determination of the solubility of $Na_2CO_3$ in Na using an excess of $Na_2CO_3$ is not a reasonable experiment at

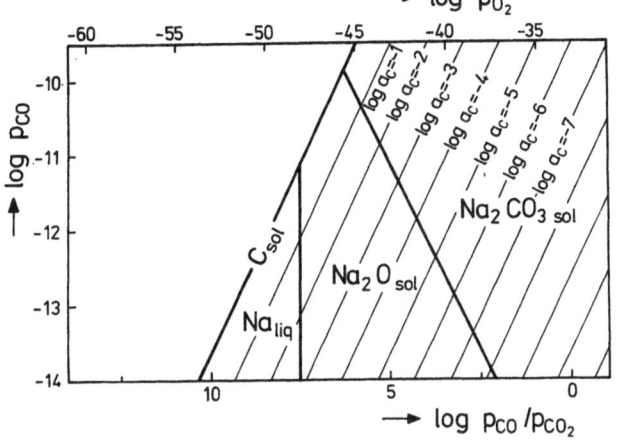

Fig. 1: Na–O–C system
at 700 K

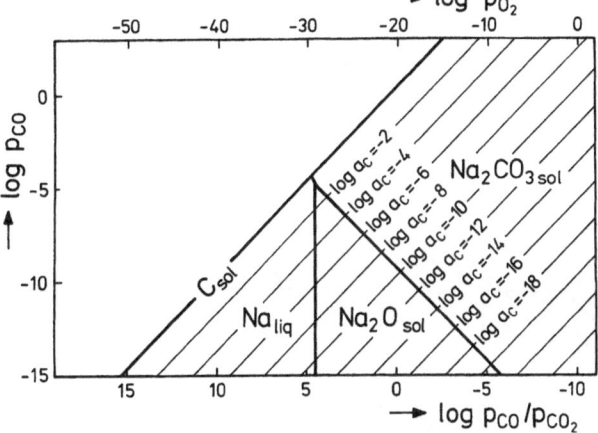

Fig. 2: Na–O–C system
at 1000 K

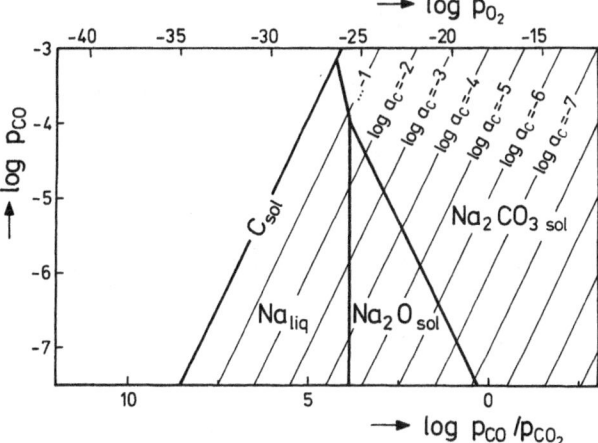

Fig. 3: Na–O–C system
at 1100 K

temperatures between 423 K and 673 K (6). The situation becomes more complex if nitrogen is also present. In the system Na-O-C-N a lot of additional compounds appear, which could influence the area of stability of sodium and so also the solubilities of the nonmetallic impurities O, C, N. Indeed, there are early experimental indications, which confirm this suggestion (21, 22).

In contrast to the above mentioned experiments of Ainsley (4) Longson and Thorley (5) observed no influence of oxygen on the solubility of carbon in sodium. Also in contrast to Ainsley these authors kept the sodium in aluminia crucibles. Although they suggested that sodium aluminate could form thereby acting as a 'getter' for oxygen they didn't treat this problem consistently. The possibility of forming aluminium carbide $Al_4C_3$ was even excluded. Therefore, the system Al-O-C was calculated, which is a simple one with only one triple point. It was then superimposed with the Na-O-C system.

For a better understanding of the Na-Al-O-C system, which results from the superimposing of Na-O-C and Al-O-C, Fig. 4 shows the system Al-O-C at 1000 K. As is obvious alumina can coexist with Al, $Al_4C_3$ or with graphite depending on the conditions.

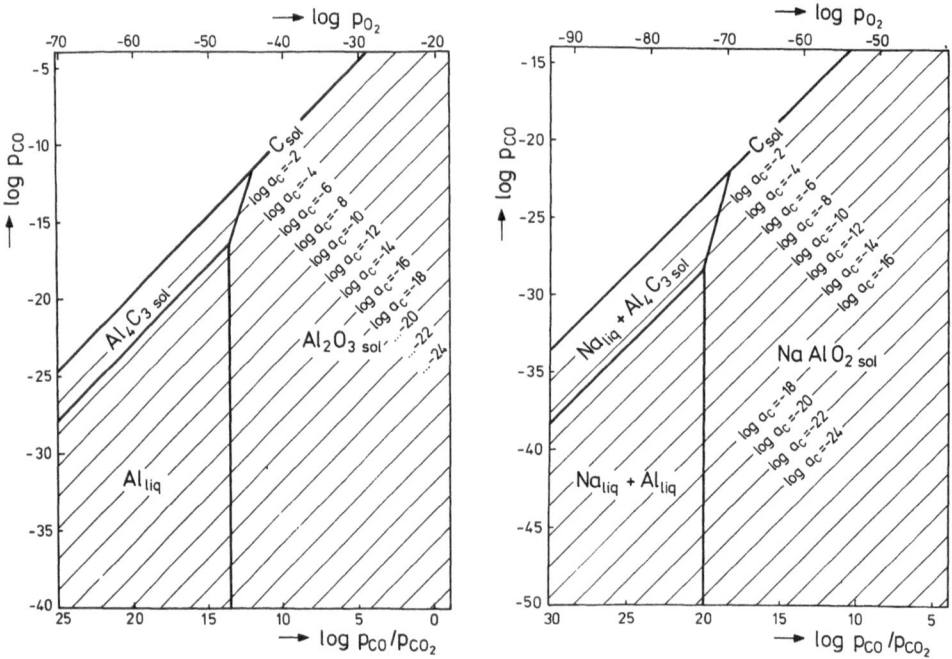

Fig. 4  Al-O-C system at 1000 K    Fig. 5  Na-Al-O-C system at 700 K

The diagrams at 700 K and 1100 K are in principle similar to Fig. 4. If this is superimposed with Fig. 3, bearing in mind the equal axis and neglecting $NaAlO_2$ then $Al_2O_3$ appears to be stable in liquid sodium down to very low oxygen potentials. However, taking into account the aluminate $NaAlO_2$, then practically the whole field of $Al_2O_3$ is replaced by $NaAlO_2$, since the triplepoint $Al/Al_4C_3/Al_2O_3$ is very close to the quadruple point $Na/Al/Al_4C_3/NaAlO_2$. This is obvious from the Figs. 5 and 6. The only phases which can be in equilibrium with Na now are Al, $NaAlO_2$, $Al_4C_3$ and graphite, while $Na_2O$ and $Na_2CO_3$ cannot, not even at 1100 K. The equilibria, which limit the field of Na are

1) $Na + Al + 2CO_2 = NaAlO_2 + 2CO$
2) $Al_4C_3 + 3CO_2 \rightleftharpoons 4Al + 6CO$
3) $4Na + Al_4C_3 + 11CO_2 = 4NaAlO_2 + 14\ CO$
4) $C + CO_2 \rightleftharpoons 2CO$

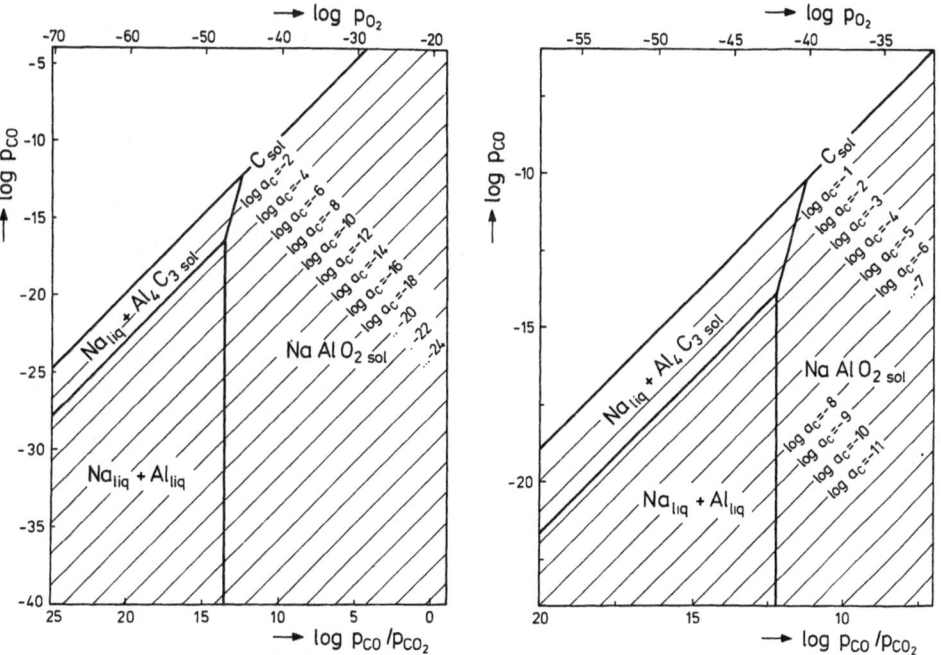

Fig. 6 Na-Al-O-C system at 1000 K    Fig. 7 Na-Al-O-C system at 1100 K

By a rough comparison of Figs. 1 to 3 with Figs. 5 to 7 it appears that the oxygen pressures of the equilibria $Na/Na_2O$ are about 48, 18 or 16 orders of magnitude larger than the oxygen pressures of the equilibria $Na + Al/NaAlO_2$. In other words: The field of existence of Na is extremely reduced to lower oxygen pressures, if aluminium or alumina is present, since the formation of $NaAlO_2$ at the surface cannot be excluded. The effect is similar to that which occurs if chromium is present forming $NaCrO_2$ (23), but the effect of Al is much more drastic from the thermodynamic point of view. Thus in turn it is impossible to realize high oxygen concentrations in sodium if $NaAlO_2$ is present.

These observations of Ainsley et al. lead to a second important result of the Figs. 5 to 7, namely that the aluminium carbide $Al_4C_3$ is stable in sodium even in the presence of $NaAlO_2$ and of carbon. Taking the information of Ainsley and of Longson on the carbon activities they used, it may be assumed that both the authors probably worked under the conditions of the equilibrium $Na + Al_4C_3/NaAlO_2$, when applying $Al_2O_3$ crucibles. Therefore it cannot be excluded that to a certain extent $Al_4C_3$ was formed. Probably the first reaction was $3Na + 2Al_2O_3 \rightarrow Al + 3NaAlO_2$, forming a thin film of $NaAlO_2$ on the surface of the crucible. The liberated Al then reacted with the dissolved carbon to form $Al_4C_3$ also inside the sodium. Therefore the "carbon containing species" described by Ainsley et al. probably consisted of $Al_4C_3$. The influence of the $Al_4C_3$ on the results of Longson and Thorley (5) on the solubility of carbon in sodium is probably as follows:

The $Al_4C_3$ particles contribute to the total content of carbon in the sodium samples. Since this was determined by combustion of the sodium in oxygen the $Al_4C_3$ is transformed to $Al_2O_3$ and $CO_2$, and this $CO_2$ adds to that formed from the dissolved carbon. So the solubility is measured too high. This suggestion agrees with the results of Longson and Thorley reported in their Tab. IV, which shows a decreased solubility, if no $Al_2O_3$ crucibles are used.

However, if one compares the results of Longson and Thorley using $Al_2O_3$ crucibles and of Ainsley et al. using no $Al_2O_3$ crucibles, it is surprising to find practically no difference in the solubilities. So from the analysis given here it cannot be excluded that the solubilities given by Ainsley et al. are also somewhat too high. The most probable explanation for this are the values of the carbon solubility in nickel (carbon source in the experiments) chosen by Ainsley et al. for the evaluation of their experimental data. In fact they use the largest values out of seven determinations by different authors. This is obvious from a comparison given by Natesan and Kassner (24). If Ainsley et al. had used values lower by a factor of two their results would agree with the results of Gehri (25) and the calculated solubilities of Johnson et al. (14).

This Na-C-N system has not yet been treated thermodynamically. Three compounds $Na_3N$, $NaCN$ and $Na_2NCN$ are known, which could in principle limit the field of existence of sodium. Sodium carbide $Na_2C_2$ cannot limit it, since it is less stable in sodium than carbon (14). Sodium nitride is also thought to be unstable in sodium. In contrast to $Li_3N$ (15) there are no thermodynamic quantities of $Na_3N$ except a calculated value of the standard formation enthalpy.

Sodium cyanide is known to appear in sodium loops, particularly in cold traps and is soluble in sodium (7). In contrast to this the disodium-cyanamide $Na_2NCN$ has never been considered in liquid sodium technology, and there are no thermodynamic data of this well-known compound. However, from the equilibrium observed in a synthesis of $Na_2NCN$ by Sakurazawa et al. (17) the cyanamide can be expected to be stable in sodium.

Seven equilibria were graphically selected as being relevant in the temperature range 800 K - 1000 K. They limit the fields of existence of the condensed phases in Figs. 8 to 10. The area above the carbon line has no physical meaning. As is obvious, all the phases considered can coexist with liquid sodium under appropriate conditions, even the disodiumcyanamide at sufficiently low carbon activities $a_c$ : The upper limit of $a_c$ is defined by the triple point $Na/NaCN/Na_2NCN$. The relations for this point are

$$\log a_c = 0.979 - 3.59 \cdot 10^3 \frac{1}{T} \qquad \qquad \text{a)}$$
$$\qquad \qquad \qquad \qquad \qquad (800 - 1000 \text{ K})$$
$$\log p_{N_2} = 0.357 - 1.32 \cdot 10^3 \frac{1}{T} \qquad \qquad \text{b)}$$

These relations should also represent the conditions of the synthesis of $Na_2NCN$ performed by Sakurazawa et al. (17), since they always observed reaction mixtures consisting of Na, NaCN, and $Na_2NCN$. Therefore, the nitrogen pressure of relation b) should agree with the equilibrium pressures reported by Sakurazawa et al. (17). As can be seen from Table 2, there is fairly good agreement, and the G values of $Na_2NCN$ are therefore thought to be reasonable. So it seems feasible to draw further conclusions from the Figs. 8 to 10.

If the carbon activity is higher than given by equ. a) and the nitrogen pressure is lower than given by equ. b), then NaCN can be in equilibrium with Na. This holds up to $a_c = 1$, where Na is in equilibrium with carbon and NaCN simultanously. If, however, the carbon activity is lower than given by equ. a) and the nitrogen pressure is higher than given by equ. b), then $Na_2NCN$ could be in equilibrium with sodium. This holds up to the triple point $Na/Na_3N/Na_2NCN$ (see below). Inside these limits the equilibrium Na/NaCN is given by

$$\log a_c = 1.157 - 4.245 \times 10^3 \frac{1}{T} - \frac{1}{2} \log p_{N_2} \quad (800 - 1000 \text{ K})\text{c)}$$

and the equilibrium $Na/Na_2NCN$ is given by

$$\log a_c = 1.333 - 4.92 \times 10^3 \frac{1}{T} - \log p_{N_2} \quad (800 - 1000 \text{ K}). \quad \text{d)}$$

Equations for the concentrations of solutions of NaCN or $Na_2NCN$ in sodium should correspond to the relations c) or d). This implies that the solubilities not only depend on temperatures but possibly on further parameter $a_c$ and $p_{N2}$. There is one determination of the terminal solubility of NaCN in Na in the range 675 K to 923 K (7). The experiments were carried out in Ta crucibles taking an excess of NaCN and a helium atmosphere. So one can assume $p_{N_2} < 10^{-4}$ atm and $a_c \ll 1$. Comparing these conditions with the Figs. 8 to 1C it is, however, safe to assume that $Na_2NCN$ was not formed. Ainsley et al. (4) measured the influence of nitrogen on the solubility of carbon at about 980 K using initial carbon activities between 0.4 and 1.0 and nitrogen pressures between 1 atm and 18 atm. If these pressures equilibrate with the sodium then the formation of $Na_2NCN$ is more probable than the formation of NaCN, which was expected to form. Since - as reported - their carbon source (carbon dissolved in nickel) was depleted in carbon during the experiment, then the condition of equ. a) for the formation of $Na_2NCN$ at 980 K could also have been fulfilled. In fact Ainsley et al. report $CN^{--}$ values which are far below those expected from the above cited data of Veleckis (7). Because of slow kinetics the authors suggest it to be unlikely that equilibrium conditions with respect to NaCN formation were achieved in their ampoule experiments during 30 h. However, an undetected formation of $Na_2NCN$ could have also lowered the yield of NaCN.

Measurements on the solubility of nitrogen in sodium were carried out in the temperature range 723 - 873 K by using a constant pressure of 10 atm of nitrogen (7). It was enriched in $N_2$-15 in order to provide the requisite sensitivity for the measurements and to distinguish molecular from atomar dissolution. The dissolved nitrogen was stripped with helium and measured, assuming that stable nitrides of impurities which had formed, namely $Ca_3N_2$, do not decompose during sparging with helium. The carbon activity of the system was not considered and the possible formation of NaCN or $Na_2NCN$ not discussed. From the Figs. 8 and 9, however it can be concluded that a pressure of 10 atm nitrogen results in the formation of $Na_2NCN$, if the carbon activity is in accord with equ. d) or higher If the carbon activity is lower, then the sodium is unsaturated in $Na_2NCN$. It is possible to decide which case applied, but the measurements appear to be near the equilibrium $Na/Na_2NCN$. This is also supported by the results: The solubility is found to be extremely low in contrast to the solubility of nitrogen in lithium (15). Furthermore the calculated solubility according to the hard-sphere model was found to be about $10^5$ times greater than that measured. Both these discrepancies could not be explained satisfactorily. Possibly there are different equilibria on which the solubilities of nitrogen in lithium and in sodium are based. The dilithiumcyanamide $Li_2NCN$ can be regarded as stable in Li (26, 8), but its thermodynamic data are unknown and those of LiCN are only estimated (8). So it is not possible to compare the systems Na-C-N and Li-C-N in detail.

If the results of Veleckis et al. on the solubility of nitrogen are based on the equilibrium $Na/Na_2NCN$ then the solubility of $Na_2NCN$

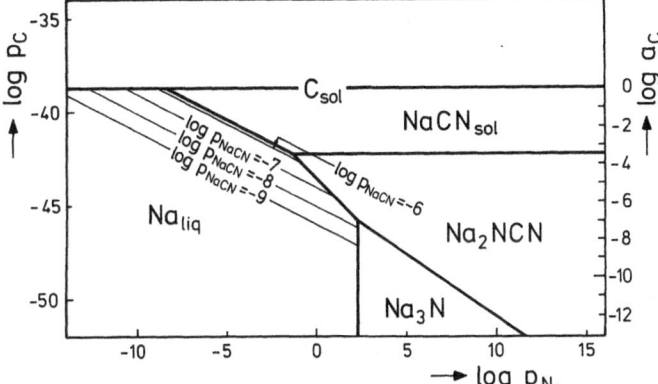

Fig. 8:
Na-C-N system
at 800 K

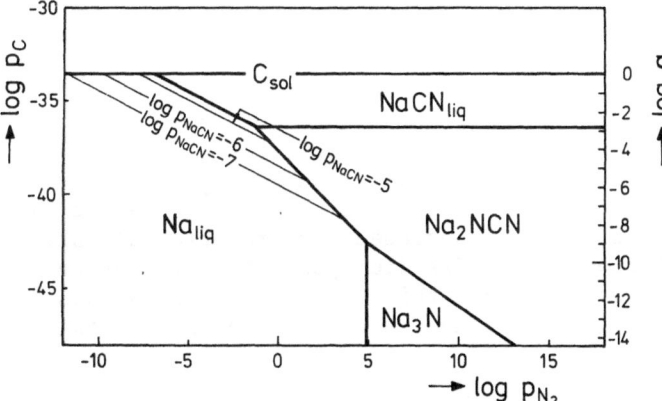

Fig. 9:
Na-C-N system
at 900 K

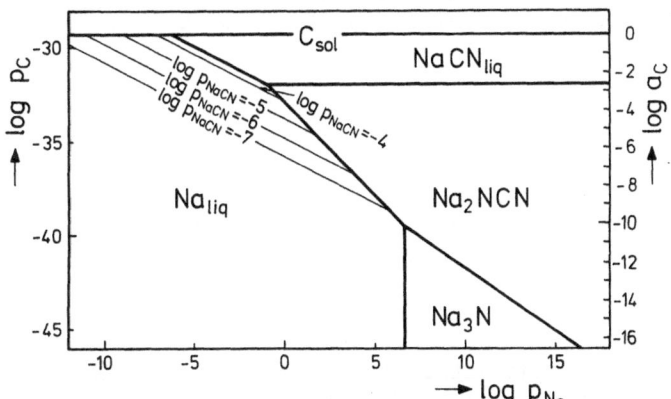

Fig. 10:
Na-C-N system
at 1000 K

in Na can be concluded as extremely low and the solubility of
nitrogen in sodium based on an equilibrium $Na/Na_3N$ can be regarded
as unknown. This equilibrium is characterized by the decomposition
pressure of $Na_3N$, which is estimated from the data of Table 2 to be

$$\log p_{N_2} = 23 - .16000 \frac{1}{T} .\qquad\qquad\text{e)}$$

This pressure also limits the field of existence of $Na_3N$ with
respect to Na (Figs. 8 to 10). If relation e) is reasonable, then
the decomposition pressure of $Na_3N$ equals the 10 atm (7) at a temperature of 740 K. But it is unlikely that $Na_3N$ could have been
formed below this temperature, since the carbon activities necessary are very low. Furthermore it should be born in mind that the
G values of $Na_3N$ might be too negative, and therefore the decomposition pressures derived from relation e) several orders of magnitude
too low. In this case the field of existence of $Na_2NCN$ would be
much larger, since the triple point $Na/Na_3N/Na_2NCN$ would go down
to lower values of the carbon activity.

Since the relations a) - d) have been obtained from the values
of Table 2, they are relevant in the temperature range 800 - 1000 K.
But some more or less qualitative extrapolations to other temperatures seem possible: Since carbon activities below $10^{-4}$ are not expected in sodium loops (24) it is concluded from relation a) that
below a temperature of about 700 K disodiumcyanamide will not be
formed in cold traps. However, above this temperature the $Na_2NCN$
may be potentially both a carburizing and a nitriding agent like
NaCN.

This should also be taken into account with respect to nitrogen
as a covergas. Although it is known to have an extremely low solubility in sodium (see above) it could react to form $Na_2NCN$ in the
case of low carbon activities in sodium. There are no investigations
on the transportation of carbon and nitrogen in sodium by this process.

To the authors knowledge the gasphase above liquid sodium has
never been analysed with respect to carbon and nitrogen containing
species. Therefore the Figs. 8 to 10 include the pressures of NaCN
over liquid sodium, which result from equilibria $2Na_{liq} + 2C + N_2 = 2NaCN_{gas}$ and depend on the nitrogen and carbon pressures. The
pressures of $Na_2(CN)_2$, which is also known as a gaseous species
(13), are nearly the same at the triple point $Na/NaCN/Na_2NCN$ but
differ in their dependence on the nitrogen and carbon pressures.
As is obious from the Figs. 8 to 10, the pressure of NaCN reaches
about $10^{-4}$ atm at 1000 K near the saturation of NaCN in Na. So
$NaCN_{gas}$ cannot be ignored as a carbon and nitrogen containing species,
which transports these impurities across the gasphase.

The vapor pressures of sodium are $5.42 \times 10^{-3}$ atm (773 K); $7.21 \times 10^{-2}$ atm (923 K); $1.42 \times 10^{-1}$ atm (973 K). So they are below or nearly equal to the lowest pressures (Tab. 2) reported by Sakurazawa et al. (17), which can therefore be regarded as the nitrogen pressures of the equilibrium $2NaCN + 2Na + N_2 = 2Na_2NCN$.

REFERENCES

1. E. Fromm, and E. Gebhardt (Editors), "Gase und Kohlenstoff in Metallen", Springer Verlag 1976.
2. E. Fromm, H. Jehn, G. Hörz, "Gases and Carbon in Metals", Physics Data 5-1, 1976.
3. G. Hofer, Z. Metallkd. 60 (1969) 457.
4. R. Ainsley, Linda P. Hartlib, P. M. Holroyd, G. Long, J.Nucl.Mat. 52 (1974) 255.
5. B. Longson and A. W. Thorley, J. appl. Chem. 20 (1970) 372.
6. H. T. Carmichael, S. A. Meacheam, APDA-184 (1968).
7. E. Veleckis, K. E. Anderson, F. A. Cafasso, H. M. Feder, Proc. Int. Conf. Sodium Techn., ANL 7520 (1968).
8. H. Migge, Reaktortagung 1980, 25.-27.3.1980, Berlin, ISSN 0173-0924, Tagungsbericht S. 918.
9. H. Migge, 2. Int. Conf. Liquid Metal Techn. in Energy Production, April 20-24, 1980, Richland, WA, USA.
10. W. Hirschwald, O. Knacke, P. Reinitzer, Erzmetall X, (1957) 123.
11. S. A. Jansson, "Corrosion by Liquid Metals", edited by I.E. Draley and I. R. Weeks, Plenum Press, N.Y., London 1970, p. 523.
12. JANAF Thermochemical Tables in J. Phys. and Chem. Reference Data 7 (1978) 793.
13. J. Barin, O. Knacke, Thermochemical Properties of Inorganic Substances, Springer Verlag 1973, Supp. 1976.
14. G. K. Johnson, E. H. van Deventer, J. P. Ackerman jr., W. N. Hubbard, J. Chem Thermodyn. 5 (1973) 57.
15. E. Veleckis, R. M. Yonco, V. A. Maroni, Thermodynamics of Nucl. Mat. 1979, Proc. Int. Symp., Vol. II, Jülich 29. Jan. - 2. Febr. 1979, IAEA Wien 1980.
16. T. Moody, J. Chem. Education 43 (1966) 205.
17. K. Sakurazawa, H. Handa, R. Hara, J. Soc. Chem. Ind. Japan 37 (1934) 701.
18. S. A. Jansson, J. Vacuum Sci. Techn. 7 (1970) 55.
19. H. Migge, 11. Symp. Fusion Techn., Oxford 15.-19.Sept.1980.
20. ANL-7325 (1966).
21. ANL 7125 (1965).
22. E. W. Hobart, R. D. Bjork, Nucl.Applications, 1, (5) (1965) 490.
23. S. A. Jansson, E. Berkey, Corrosion by Liquid Metals, edited by J. E. Draley and I. R. Weeks, Plenum Press, N.Y., London 1970, p. 479.

24. K. Natesan, T. F. Kassner, Nucl. Techn. 19 (1973) 46,
    Fig. 12.
25. AI-AEC-12859 (1969)
26. M. G. Down, M. J. Haley, P. Hubberstey, R. J. Pulham, A. E.
    Thunder, J.C.S. Dalton (1978) 1407.

ADDENDUM

   After the seminar the author got information on the work of
J. P. Maupré and J. Trouvé "Contribution à l'étude du système Na-
$Na_2O-Na_2CO_3-C$ diagramme de phases", IAEA Specialists' Meeting on
Carbon in Sodium, November 27-30, 1979, Harwell, U.K.. Several
experimental results of this paper agree well with the results of
calculations of the ternary Na-O-C system presented here:

| | temp. T' of stable pair reversal | equilibria at T < T' | equilibria at T > T' | equilibrium at all temp. considered |
|---|---|---|---|---|
| Maupré exper. | 963 K | $C-Na-Na_2O$; $C-Na_2O-Na_2CO_3$; $Na_2O-C$ Figs. 3&5 | $Na-Na_2O-Na_2CO_3$; $C-Na-Na_2CO_3$; $Na-Na_2CO_3$; Figs. 3&5 | $Na_2O-Na_2CO_3$ Fig. 4 |
| this work calcul. | 932 K | same results Fig. 1 | same results Figs. 2&3 | same results Figs. 1-3 |

# THERMODYNAMIC AND KINETIC ASPECTS OF OXYGEN - HYDROGEN

# INTERACTIONS IN LIQUID SODIUM

Colin A. Smith and Andrew C. Whittingham

Central Electricity Generating Board
Berkeley
United Kingdom

## INTRODUCTION

Sodium hydroxide is an important intermediary in the sodium-water reaction (1):

$$Na + H_2O \rightarrow NaOH + 1/2\ H_2 \tag{1}$$

The kinetics of its decomposition in sodium and the magnitude of the interaction between hydrogen and oxygen in solution in sodium to give an association product which is normally assumed to be sodium hydroxide are of interest in LMFBR technology.

$$2Na + NaOH\ (solid,\ liquid) \rightarrow Na_2O\ (soln.) + NaH\ (soln.) \tag{2}$$

$$H\ (soln.) + O\ (soln.) = OH\ (soln.) \tag{3}$$

In a previous study (2) the reaction sequence between sodium and sodium-hydroxide was deduced to be

$$2Na + NaOH\ (solid, liquid) \xrightarrow{fast} Na_2O\ (solid) + NaH\ (solid) \tag{4}$$

$$Na_2O\ (solid) \xrightarrow{slow} O\ (soln.) \tag{5}$$

$$NaH\ (solid) \xrightarrow{slow} H\ (soln.) \rightleftharpoons 1/2\ H_{2(gas)} \tag{6}$$

with a contribution from the reaction

$$Na + NaOH\ (solid, liquid) \xrightarrow{rapid} Na_2O\ (solid) + 1/2\ H_{2(gas)} \tag{7}$$

365

Estimates of the equilibrium constant, K, for reaction (3), which is given by:

$$K = \frac{(OH)}{(O) \cdot (H)} \tag{8}$$

were found to be independent of the sodium temperature, in the range 370 - 500 °C and to have a value of 7 x $10^3$ mole fraction$^{-1}$ (0.17 ppm$^{-1}$). This indicates that, at a hydrogen background of 1 ppm, approximately 17 % of the total oxygen present is in the form of hydroxide.

The present work continues our study of the Na-H-O system and describes measurements of the kinetics of hydrogen gas release from the sodium-sodium hydroxide reaction and the interaction between hydrogen and oxygen in solution in sodium.

The interaction has generally been considered previously by the above approach and this has been adopted here to interpret the experimental results since measurements made have been of hydrogen and oxygen concentrations in sodium. A more fundamental approach would consider the chemical activities of the species involved in the interaction but has not been used in this instance.

THE SODIUM-SODIUM HYDROXIDE REACTION

Experiments were carried out to examine the constribution of reaction (7) in the production of solutions of sodium-monoxide and hydride from the reaction between sodium and sodium-hydroxide.

Experimental

Pellets of sodium hydroxide, typically 100 mg, were dropped onto the surface of liquid sodium (0.1 - 0.2 kg) contained in an evacuated stainless steel vessel of known volume via a 9 mm i.d. ball valve (Hoke 7223G8Y) (2). Subsequent hydrogen pressure changes were monitored by a pressure transducer (0 - 0.75 bar, Bell and Howell Type 4-326L100) attached to the reaction vessel. Pellet additions were carried out in the temperature range 270 - 500 °C, each addition being made to cold trapped sodium (<10 ppm oxygen, <1 ppm hydrogen).

Discussion and Results

The variation of hydrogen pressure with time for a typical experiment is shown in Fig. 1 for the sodium-sodium hydroxide reaction at 335 °C. There was an increase of hydrogen pressure for the first 150 s of reaction, followed by a slower absorption of hydrogen to equilibrium by the liquid sodium. The dashed lines OAC and OC represent the predicated hydrogen pressure variation

with time if reaction proceeded in accordance with equation (7)
followed by hydrogen absorption in sodium and equations (4) - (6)
respectively, calculated for the particular volume and sodium surface area of the reaction vessel used. At 335 $^{o}$C, the maximum
hydrogen pressure recorded is  11 % of the total hydrogen pressure
which would have been measured if all the hydrogen (added as sodium
hydroxide) had appeared in the gas phase. The variation of this
maximum pressure with temperature is shown in Fig. 2, from which it
can be seen that the contribution from reaction (7) increases from
 3 % at 300 $^{o}$C to  16 % at 500 $^{o}$C. The time to reach the maximum
pressure was also observed to decrease with increasing temperature
as expected (  50 s at 500 $^{o}$C).

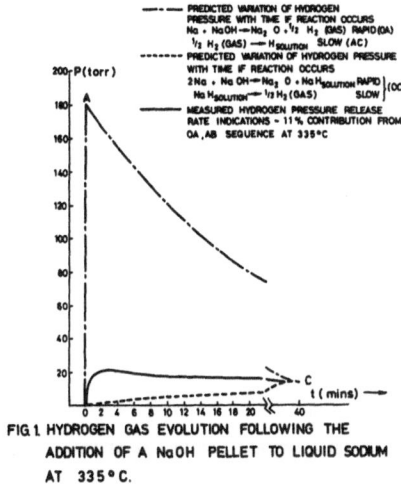

FIG 1. HYDROGEN GAS EVOLUTION FOLLOWING THE
ADDITION OF A NaOH PELLET TO LIQUID SODIUM
AT 335°C.

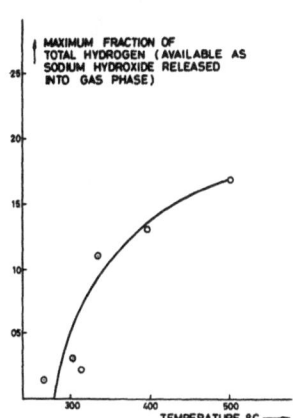

FIG 2. FRACTION OF TOTAL HYDROGEN
RELEASED AS GAS FROM NaOH

This study suggests, therefore, that even at 500 $^{o}$C, sodium
hydroxide decomposes in excess liquid sodium predominantly in
accordance with equations (4) - (6) as suggested by our previous
study (2).

THE O+H=OH REACTION

Experimental

The apparatus used in the study of the interaction between
dissolved hydride and oxide is shown schematically in Fig. 3. It
consisted of a stainless steel reaction vessel which contained
galvanic cell hydrogen and oxygen meters and two traps, one
containing uranium foil and the other yttrium foil. The traps
could be isolated from the reaction vessel by valves. The vessel
held ~ 1 kg of sodium which could be pumped electromagnetically
through either trap and the sodium was stirred by pumping through
a recirculating side arm. Hydrogen or oxygen gas could be added
to the vessel from a reservoir connected to the vessel through a
valve in the lid.

The hydrogen and oxygen meters were based on cells 9 and 10

$$Li, \; LiH|Fe|CaH_2 - CaCl_2|Fe|Na(H) \tag{9}$$

$$Air, \; Pt|ThO_2 - 7.5w/oY_2O_3 \; Na(O) \tag{10}$$

and have been described previously (2).

The vessel was maintained with the temperature of the valve in the gas space and the vessel walls above the sodium > 15 $^{\circ}$C higher than that of the sodium to prevent sodium condensation in the valve or on these surfaces. Thus any hydrogen or oxygen gas added to the vessel would react only with the liquid sodium surface and consequently, quantitative additions could be made.

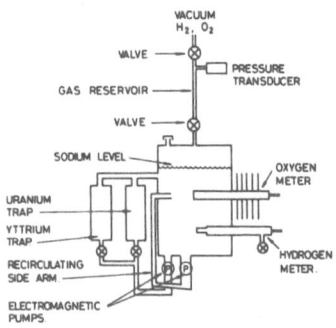

FIG. 3. SCHEMATIC DIAGRAM OF APPARATUS.

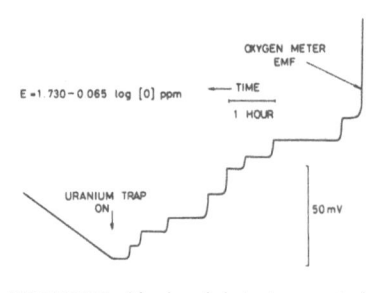

FIG. 4. OXYGEN METER CALIBRATION AT 435°C.

## Procedure

The interaction was monitored by noting the meter responses following the addition of aliquots of either oxygen to solutions of hydrogen in sodium or hydrogen to solutions of oxygen in sodium. Initially it was necessary to calibrate the meters at each temperature in order to enable quantitative interpretation of the experimental data. Calibrations were carried out by adding measured amounts of hydrogen or oxygen gas to the sodium, which initially contained low concentrations of these elements following yttrium or uranium trapping. Hydrogen was always removed by the yttrium at low oxygen backgrounds to prevent the formation of surface oxide films which might have kinetically impaired the ability of yttrium to remove hydrogen. Uranium was not effective in removing hydrogen and oxygen could be trapped readily at high hydrogen backgrounds.

The apparent volume of the gas reservoir to the reaction vessel was measured at each temperature with argon using a mercury filled gas burette. The quantities of hydrogen or oxygen added to the

sodium in the calibration or interaction experiments were measured
by filling the reservoir to a known pressure before rapidly opening
and closing the valve to the sodium vessel which was maintained
under vacuum throughout the experiment. The pressure in the reservoir
was noted immediately after the gas addition and the pressure change
in the known volume of reservoir could be related to the mass of
hydrogen or oxygen added.

The reaction vessel had been filled with a known weight of
sodium and a correction to this weight was made to allow for the
sodium in the hot traps. The volume of the traps and associated
pipework was less than $\sim$ 6 % of the total volume of sodium in the
vessel. The pipework to the traps was narrow bore and errors in
the results through diffusion of hydrogen or oxygen into the hot
traps are considered to be negligible.

Results

The oxygen meter emf change following an oxygen addition was
consistent with the formation on the sodium surface of an oxide
film which slowly dissolved since a period of time was required
before the emf stabilised. On opening the valve to the reservoir
the oxygen pressure immediately fell to zero. Periods of $\sim$ 1 hour
were required at 340 $^\circ$C and $\sim$ 20 minutes at 470 $^\circ$C for the emf to
reach its equilibrium value.

The results of an oxygen meter calibration are shown in Fig.4.
Emf variations with concentration were within 92 - 105 % of those
predicted theoretically and calibrations were in general agreement
with those obtained previously by sodium-monoxide pellet addition
(2). This gives supporting evidence that the above method of
adding oxygen to sodium is quantitative.

Following the addition of oxygen to sodium in a calibration
the meter emfs were stable for periods of hours, which suggests
that the rate of sodium chromite formation is sufficiently slow
on the timescale of the interaction (see below) and calibration
experiments to be negligible. Following uranium trapping to low
oxygen concentrations (<< 1 ppm) there was a slow decrease in
oxygen meter voltage with time when the uranium trap was isolated
indicating an increase in oxide concentration. This may have arisen
from the decomposition of sodium chromite or small oxygen sources
in the reaction vessel. Typical concentration changes were from
< 0.1 ppm hr$^{-1}$ at $\sim$0.5 ppm and 470 $^\circ$C to 5 x 10$^{-3}$ ppm hr$^{-1}$ at
$\sim$ 0.05 ppm and 340 $^\circ$C. These oxygen sources were considered to be
negligible over the time taken to examine the oxygen-hydrogen
interaction.

Hydrogen meter calibrations took a longer period of time than
for the oxygen meter as a result of the slower kinetics of the

sodium-hydrogen reaction (7). One day was required for the determination of each point at 300 $^{\circ}$C. Thus a complete calibration took
approximately one week at this temperature. At 435 $^{\circ}$C the reaction
was sufficiently rapid to obtain a calibration point within 2 hours
and a complete calibration in  2 days. The rate constants for the
first order hydrogen absorption reaction were found to be in good
agreement with those derived from an expression given by Whittingham
(3).

## Oxygen-Hydrogen Interaction

Reaction (3) was axamined by adding aliquots of hydrogen to
solutions of oxygen in sodium (initially with low hydride concentrations) or oxygen to solutions of hydrogen (initially with low
oxide concentrations) and noting the meter responses. The latter
technique was the more rapid and gave some qualitative indication
of the rate of attaining equilibrium between hydride, oxide and
hydroxide in solution. When hydrogen was added to solutions of
oxygen the initial oxygen concentrations were between 14 and 55 ppm
and hydrogen additions of $\sim$ 0.5 to 1 ppm were made to a total added
hydrogen concentration of up to $\sim$ 6 ppm. When oxygen was added to
solutions of hydrogen, initial hydrogen concentrations were between
$\sim$ 1 and 5 ppm with oxygen aliquot additions of $\sim$ 7 - 20 ppm being
made to a total added oxygen concentration of up to $\sim$ 100 ppm. All
experiments were performed with concentrations of hydrogen and
oxygen below the saturation value.

Fig. 5 illustrates the hydrogen and oxygen meter response to
the addition of oxygen to a solution of $\sim$ 1 ppm hydrogen in sodium
at 300 $^{\circ}$C. Approximately 1 1/2 hours was required for the oxygen
meter emf to sabilise, indicating complete dissolution of the
added oxygen. The hydrogen meter emf change mirrored that of the
oxygen meter and the initial response of the meters following an
oxygen addition was virtually simultaneous. Similar meter emf
behaviour was observed for oxygen additions to hydrogen solutions
at higher temperatures although equilibrium was achieved more
rapidly. Thus the rate of hydrogen-oxygen interaction in sodium is
rapid in comparison with the rate of oxide dissolution.

In any experiment the amount of gas added was known, as were
the initial and final concentrations of hydrogen and oxygen. Therefore, it was possible to deduce the equilibrium hydroxide concentration from the differences between the initial and predicted concentrations and the measured final concentrations of hydrogen and
oxygen. Equilibrium constants could thus be evaluated together with
the atom ratio of the interaction.

Results of experiments are given in Table 1 which indicates
the temperature of measurement, the equilibrium constant, the atom
ratio and the experimental method employed.

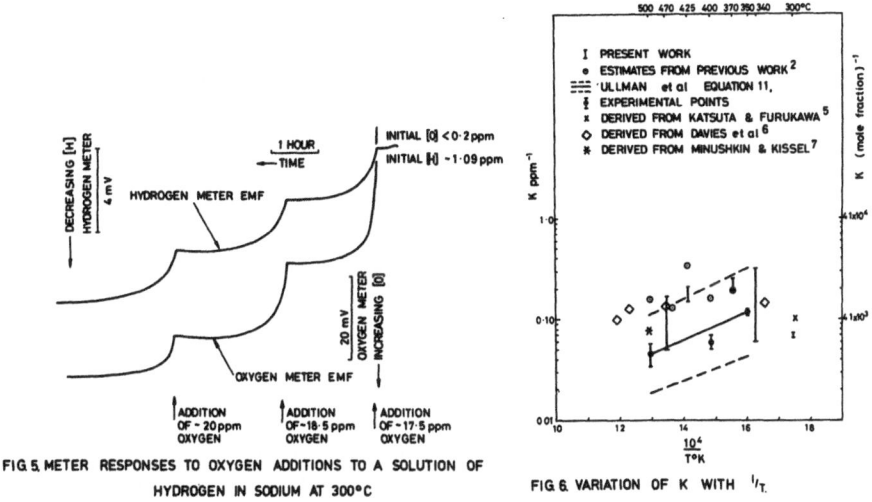

FIG. 5. METER RESPONSES TO OXYGEN ADDITIONS TO A SOLUTION OF
HYDROGEN IN SODIUM AT 300°C

FIG. 6. VARIATION OF K WITH $1/_T$.

Table 1.  Results of Measurements on the O+H=OH Interaction

| Temperature °C | Equilibrium Constant K | | Atom Ratio (O) : (H) |
|---|---|---|---|
| | Mole fraction$^{-1}$ | ppm$^{-1}$ | |
| 300[xx) | 2.80+0.1 x 10$^3$ [xxx) | 0.069+0.003[xxx) | – |
| 340[xx) | 7.77+5.3 x 10$^3$ | 0.19 +0.13 | 3.3 +0.2 : 1 |
| 370[x) | 9.00+1.2 x 10$^3$ | 0.22 +0.03 | 1.20+0.04: 1 |
| 435[x) | 7.37+1.2 x 10$^3$ | 0.18 +0.03 | 1.50+1.0 : 1 |
| 470[xx) | 4.50+2.5 x 10$^3$ | 0.11 +0.06 | 1.65+1.0 : 1 |

x)    hydrogen addition to solution of oxygen in sodium.
xx)   oxygen addition to solution of hydrogen in sodium.
xxx)  equilibrium constant derived from hydrogen meter emf changes.

DISCUSSION

The product of the association reaction has generally been
assumed to be hydroxide since hydroxide can ixist as a separate
phase in equilibrium with sodium under suitable conditions of
hydrogen pressure and temperature. Although measurements on the
Na-K-O-H system were consistent with an interaction in solution
involving approximately one atom of hydrogen and one atom of oxygen
at 380 °C, this relationship has not been previously established
for the Na-H-O system. The present results, apart from the points
at 340 °C, suggest that the atom ratio in sodium is close to 1:1
and the assumed equilibrium, (3), is justified. The reason for the

high value of the 340 °C point is not clear at the present time
but it may be related to drifts in the oxygen meter calibration at
this low temperature.

The equilibrium constants in Table 1 are compared graphically
in Fig. 6 with our previous estimates (2), with values derived
from other work on the Na-H-0 system (5, 6, 7) and with the recent
measurements of Ullmann, Kozlov et al, (8, 9, 10) who derived the
equation:

$$\log K \text{ (atom fraction}^{-1}) = 1.53(\pm 0.21) + \frac{1340(\pm 140)}{T} \qquad (11)$$

at temperatures between 350 °C and 500 °C. There is broad agreement between all values of the equilibrium constant which has little
apparent variation with temperature. Its magnitude is such that
between ∼ 5 and 15 % of the total oxygen in solution in sodium is
present as hydroxide at a hydrogen background of 1 ppm and 0.5
- 1.4 % at a hydrogen background of 0.1 ppm hydrogen in the range
300 - 500 °C. Thus hydroxide formation is unlikely to be of any
practical importance in a sodium system unless there are high
hydride and oxide impurity backgrounds.

The variation with temperature of the free energy of formation
of hydroxide in sodium, calculated from the equilibrium constants
in Table 1 is compared in Fig. 7 with the equation given by Ullmann,
Kozlov et al (9)

$$\Delta G^O_{OH} \text{ (J mole}^{-1}) = -(25967 \pm 2681) - (28.762 \pm 4.136)T \qquad (12)$$

The two sets of data are in excellent agreement.

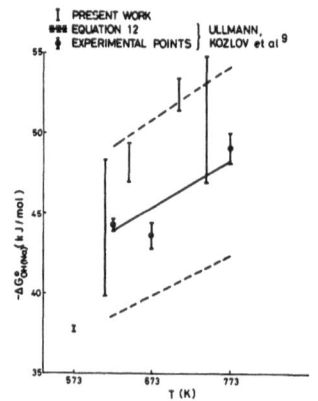

FIG. 7 VARIATION OF $\Delta G^O_{OH\,(Na)}$ WITH TEMPERATURE

ACKNOWLEDGEMENT

Experimental assistance by R. J. Smith and R. Wilkins in
these experiments is gratefully acknowledged. This paper is
published by permission of the Central Electricity Generating
Board.

REFERENCES

1.  R. J. Pulham, P. A. Simm, 1973, "Liquid Alkali Metals",
    BNES Conference, Nottingham, p 1.
2.  A. C. Whittingham, C. A. Smith, P. A. Simm and R. J. Smith,
    1980, ANS Conference Richland, Session XVI.
3.  A. C. Whittingham, 1976, J. Nucl. Mats., $\underline{60}$, 119.
4.  M. N. Arnol'dov, M. N. Ivanovskii, V. A. Morozov and
    S. S. Pletenets, 1970, Teplofizika Vysokikh Temperatur
    $\underline{8}$, 88.
5.  H. Katsuta and K. Furukawa, 1976, Nucl. Technol. $\underline{31}$, 219.
6.  R. A. Davies, J. L. Drummond and D. W. Adaway, 1973,
    "Liquid Alkali Metals", BNES Conference, Nottingham,
    p. 93.
7.  B. Minushkin and G. Kissel, 1971, IMS Symposium, Detroit,
    p. 333.
8.  H. Ullmann, F. A. Kozlov et al, 1980, ANS Conference,
    Richland, Session XVI.
9.  H. Ullmann, F. A. Kozlov et al, 1979, IAEA Symposium on
    Thermodynamics of Nuclear Materials, Paper SM-236/80.
10. K. Teske, H. Ullmann, F. A. Kozlov and E. V. Kuznecov,
    1979, Kernenergie 22, p. 25.

THE REACTIONS OF OXYGEN AND HYDROGEN WITH LIQUID SODIUM -

A CRITICAL SURVEY

Helmut Ullmann

Central Institute for Nuclear Research
Rossendorf
German Democratic Republic

INTRODUCTION

The importance of the sodium-water reaction for the second circuit of a fast reactor induced many authors to investigate the interactions of oxygen, hydrogen and their compounds such as water vapour or hydroxide with liquid sodium. The results are used for the development of effective cleaning processes of sodium, for measures to prevent corrosion and mass transfer, for the correction of the reading of measuring devices for hydrogen and oxygen concentrations in sodium, and, finally, for the production of sodium and its compounds.

We can suppose the existence of negatively charged nonmetal ions solvated by partially positively charged metal ions in the alkali metal solution (1). This was proved particularly by resistivity measurements (2).

Whittingham (17) reviewed the work on the system Na-O-H in 1973. Since then many new results have been published. The results on the saturation solubilities were critically evaluated for the system Na-O (3) and for the system Na-H (4). Only partial results are available as to the kinetics of phase transitions between gaseous or condensed phases on the one hand and the liquid sodium on the other hand [NaOH (5); $H_2O$ (6); $H_2$ (7); $Na_2O$ (8)].

Earlier investigations of the phase diagram Na-O-H (9) have been supplemented by further experiments (11) and thermodynamic calculations (10, 12).

There is a lack of knowledge on the reactions between the dissolved components within the range of homogen solution in an excess
of liquid sodium, because these investigations are very complicated.
So far the measurement of the hydrogen equilibrium pressure has been
the only possibility to determine the activity changes of hydrogen
in a dilute sodium solution. In this way Pulham (6), Kozlov (13)
and Davies (14) investigated the influence of oxygen on the activity
of hydrogen in sodium. Now that sufficiently reliable electrochemical
oxygen meters have been developed, the measurement of oxygen activity
changes affected by hydrogen dissolved in sodium becomes practicable
(15, 16, 8).

## FUNDAMENTALS OF THE SOLVATION CHEMISTRY

In a solvent consisting of an excess of quasi-free electrons
and positively charged alkali ions the existence of negative simple
or complex covalent bonded nonmetal anions solvated by partially
charged sodium ions is supposed (1, 2). Solid alkali metals form
cubic body-centered lattices with a coordination number of 8. On
melting the grade of order is nearly maintained, which is noticed
by a small increase in volume (Na: 2.5 %). These are the starting
conditions for theoretical solution models (24-29). The model calculations yield energy values, which are relevant in relation to
the values of other ions or compounds, if the starting conditions
of the model have been chosen correctly. The energy value calculated
for the solvation of oxygen in sodium, e.g. is obtained correctly
if an $O^{2-}$ ion is assumed to exist in the solution (24, 28, 29).

In the presence of further condensed and/or gaseous phases the
following components will take part in the chemical reactions:

$$1/2 \ H_2 + 1 \ e = H_{Na}^- \qquad (1a) \ H_{Na}^- + Na^+ = NaH_s \qquad (1b)$$

$$1/2 \ O_2 + 2 \ e = O_{Na}^{2-} \qquad (2a) \ O_{Na}^{2-} + 2Na^+ = Na_2O_s \qquad (2b)$$

$$H_2O = 2 \ H_{Na}^- + O_{Na}^{2-} \qquad (3a) \ OH_{Na}^- + Na^+ = NaOH_{s,1} \qquad (3c)$$

$$H_{Na}^- + O_{Na}^{2-} = OH_{Na}^- + 2 \ e \qquad (3b)$$

$$NaH + NaOH_{s,1} = NaH_{NaOH} \qquad (4a) \ Na_2O + NaOH_{s,1} = Na_2O_{NaOH} \qquad (4b)$$

Experimentally it was found that $p_{H_2}^{1/2}$ and $c_H$ gives a linear
dependence of the type

$$K_{1a} = c_H / p_{H_2}^{1/2} \qquad (5)$$

which proves the existence of monatomic hydrogen in the sodium solution. The equilibrium constant of this reaction is usually expressed in the literature as the Sievert's constant. An analogous relation is assumed to be valid for the dissolution of oxygen in sodium:

$$K_{2a} = c_0/p_{O_2}^{1/2} \qquad (6)$$

Oxygen and hydrogen capture electrons from the conduction band of the liquid metal to form negatively charged ions. This was shown by measurements of the electrical resistivity, which increases with an increasing concentration of the nonmetal (2, 6). The dissolution of small amounts of water causes an increase in resistivity according to the sum of the effects of $O^{2-}$ and $H^-$ (reaction 3a). The dissolution of greater amounts of water causes an increase in resistivity smaller than the sum of the $O^{2-}$ and $H^-$ effects, because of the incomplete dissociation of the $OH^-$ formed in the reaction (3b).

To balance the enthalpies of the reactions (1) to (3) we use Born Haber cycles. For the reaction (1) we obtain

$$\Delta H_r^o = \Delta H_f^o + \Delta H_s^o \qquad (7)$$

assuming the validity of Henry's law up to the saturation concentration ($\Delta H_r^o$ enthalpy of reaction, $\Delta H_f^o$ enthalpy of formation, $\Delta H_s^o$ enthalpy of solution). The tendency of release the gaseous component from the solution with increasing temperature increases with increasing exothermicity of the formation of the corresponding sodium-nonmetal compound. The dissolution of the compound in the liquid metal (reactions 1b and 2b) is an endothermic process, i.e., the dissolution takes place preferably at increasing temperature. By application of experimental enthalpy values (Table 1), we obtain from equation (7) for the reaction of hydrogen (1a) $\Delta H_r^o = -4.67$ kJ/mole, for the reaction of oxygen (2a) $\Delta H_r^o = -375$ kJ/mole. These values are compared to other experimentally determined or calculated values in Table 1. Deviations of < 40 kJ/mole must be accepted because of the errors in the experiments and simplifications in model calculations.

The balance of the enthalpy values for the reaction (3) of oxygen with hydrogen in sodium solution is given by the Born-Haber cycle (16)

$$\Delta H_{s(O)}^o + \Delta H_{s(H)}^o + \Delta H_{r(OH)}^o + \Delta H_{f(Na_2O)}^o + \Delta H_{f(NaH)}^o$$
$$= \Delta H_{f(NaOH)}^o + \Delta H_{s(NaOH)}^o + 2\,\Delta H_{m(Na)}^o \qquad (8)$$

When using the experimental value of Table 1 we obtain an enthalpy value of $\Delta H_{r(OH)}^o = -9.68$ kJ/mole for the reaction (3b).

Table 1.  Enthalpies of formation $\Delta H_f^o$, of reaction $\Delta H^o$, of solution $\Delta H_s^o$, and of melting $\Delta H_m^o$ in the system Na-O-H ($\Delta H$ values in kJ/mole

| Enthalpy of | selected experimental values | calculated values |
|---|---|---|
| form. of Na$_2$O | $-420.6 + 5.02$ (21) | |
| form. of NaH | $- 59.9$ (22) | |
| form. of NaOH | $-428.3 + 5.02$ (20) | |
| sol. of Na$_2$O | $+ 45.72 + 0.68$ (3) | |
| sol. of NaH | $+ 55.27$ (4) | |
| sol. of NaOH | $+ 22.8$ (30; 33.9 + 2.3(18) | |
| melt. of Na | $+ 2.64 + 0.04$ (20);371K | |
| melt. of NaOH | $+ 6.37 + 0.33$ (20);593K | |
| react. of O$_2$/Na | | $-332$ (28);$-375$ $-375$ (this work) |
| react. of H$_2$/Na | $- 2.34 \pm 0.17$ (4) | $+ 43$ (28); $+ 55$ (25) $- 4.67$ (this work) |
| react. of O+H/Na | $- 25.7$ (16) | $- 9.86$ (this work) |

The oxygen or hydrogen activity in sodium related to the state of saturation is given by

$$a_O = \frac{x_O^o}{x_{O,sat}} \cdot f_O^H \quad (9a) \; ; \quad a_H = \frac{x_H^o}{x_{H,sat}} \cdot f_H^O \quad (9b)$$

The Raoult's activity coefficient $\gamma = 1/x_{sat}$ is the factor for the different chemical efficiencies of the dissolved component in the binary solution Na-O or Na-H, compared with the state of saturation. The validity of the Raoult's law for binary solutions of non-metals in liquid alkali metals can be supposed within the whole region of concentration up to the saturation due to the low saturation concentrations. In multinary solutions reactions between the dissolved components can occur. As a result the initial concentrations of the dissolved components can be reduced. We express the reduction of the free concentration of a component as the interaction coefficient f.

The interaction coefficient of a component depends on the concentration of the second dissolved component and vice versa. This dependence is expressed by the interaction parameter $\varepsilon$ (23):

$$\varepsilon_O^H = \frac{\partial \ln f_O^H}{\partial x_H^o} \quad (10a) \qquad \varepsilon_H^O = \frac{\partial \ln f_H^O}{\partial x_O^o} \quad (10b)$$

In relation to mole fraction concentrations $\varepsilon_O^H = \varepsilon_H^O = \varepsilon$ is valid. From this we obtain for the reaction of two components in diluted solution (16)

$$K_{diss} \sim - 1/\varepsilon. \tag{11}$$

METHODS OF INVESTIGATION AND KINETICS OF THE REACTIONS

For the investigation of reactions and equilibria between various condensed phases mixtures of sodium, hydroxid or oxide were heated at a constant temperature and then analyzed chemically (9) or by X-ray analysis (11), or thermal effects were observed (18). The total concentration is estimated by sampling and chemical analysis of oxygen and hydrogen. The chemical activity of the dissolved components is determined by equilibrium pressure or electrochemical measurements (15, 16, 17).

In order to determine the equilibrium of oxygen- and/or hydrogen-bearing compounds in an excess of sodium, gaseous oxygen and hydrogen, water or solid samples of sodium oxide or hydroxide were introduced into the liquid sodium of a reaction vessel or circuit. The dissolution of impurities from cold traps is another way of altering the concentration of nonmetals in liquid sodium.

For the estimation of equilibria between components which emerge from various phases it is necessary to observe the kinetics of phase transitions. The reaction of oxygen with liquid sodium (2a) runs down quickly. The absorption of hydrogen in liquid sodium is more slowly (1a). Since the reaction of water with sodium to form oxide and hydride (3a) proceeds quickly, excess hydrogen will be released first, followed by a slow absorption on the sodium surface to form hydride and hydroxide (6, 7).

The dissolution of sodium hydroxide samples at temperatures of 200 ... 280 $^{\circ}$C proceeds relatively slowly (5, 17). The rate law constant is given by

$$lg \; k_{sol} = (4.1 \pm 0.3) - (4060 \pm 160)/T$$

($k_{sol}$ in g/cm$^2$ . s) with an activation energy of the dissolution process of 78 kJ/mole (5). The dissolution of sodium monoxide pellets was investigated by electrochemical measurements (17):

$$lg \; k_{sol} = (0.6 \pm 0.52) - (3674 \pm 346)/T$$

($k_{sol}$ in g/cm$^2$ . s). The activation energy of the process is 70.4 $\pm$ 6.6 kJ/mole. Consequently, Na$_2$O dissolves somewhat more slowly than NaOH.

THE HYDROGEN EQUILIBRIUM PRESSURE

The hydrogen equilibrium pressures over unsaturated and saturated solution in sodium are shown in Fig. 1. The few experimental
values for unsaturated solution (4, 13) confirm a low temperature
dependence. Vissers obtained his results from equilibrium pressure
measurements during loop operation at a constant temperature of the
cold trap, where oxygen and hydrogen are present in nearly equimolar
parts ($10^{-5}$ mole fraction) in the sodium solution. Kozlov obtained
his results after injection of hydrogen into a loop, the sodium of
which had previously been cleaned by cold trapping at 130 $^\circ$C (H:O
= 30 ... 200:1). If the mole ratio H:O is too low, as was the case
in the experiments of Vissers, the results may turn out incorrect
due to the influence of the oxygen-hydrogen reaction. This must be
taken into account when considering the temperature dependence of
the Sievert's constant given by Vissers

$$lg\ K_{1a} = -\ 2.72 - 122.0/T$$

(K in mole fr. $Pa^{-1/2}$; T 640 ... 773 K). From this a reaction
enthalpy of $-$ 234 $\pm$ 0.17 kJ/mole was derived. From the equilibrium
pressure values of Kozlov we calculated a concentration dependent
value of lg K (K in mole fr. $Pa^{-1/2}$; T = 573 ... 723 K)

$$lg\ K_{1a} = -\ 4.40 - 0.645\ lg\ x_H + (639\ lg\ x_H - 572)/T.$$

The reaction enthalpy rises from $-$ 1.6 to + 7.2 kJ/mole in the
concentration range of 2.5 to 15 ppm H in sodium. The temperature
dependence of the Sievert's constant can be neglected.

The equilibrium pressure of hydrogen over saturated solution

$$lg\ p_{sat}^{1/2} = -\ 5.50 + 3080/T$$

(p in Pa) has an enthalpy of solution of 59.06 kJ/mole. By means
of simultaneously dissolved oxygen the solubility of hydrogen will
be increased or the hydrogen equilibrium pressure will be lowered,
respectively (6, 13-16, 31). According to the results of several
authors (Fig. 2) it can be observed that the Sievert's constant at
an oxygen concentration of zero is difficult to decide correctly.
But an increase with oxygen concentration is clearly indicated.

SATURATION SOLUBILITIES

Solubility data by several authors are shown in Fig. 3. The
following typical curve sets are obtained: The curve for hydrogen in
the form of sodium hydride has a steeper slope than the curve for
oxygen in the form of monoxide. After dissolution of impurities from
the cold trap the electrochemical analysis of the free oxygen content
in the sodium yields either the slope typical of oxygen in the form

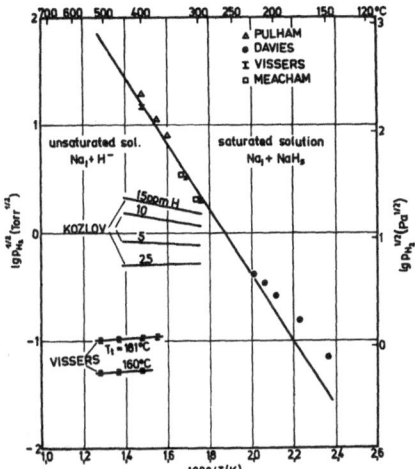

Fig. 1: Equilibrium pressures versus reciprocal temperature of
unsaturated and saturated solutions of hydrogen in liquid
sodium

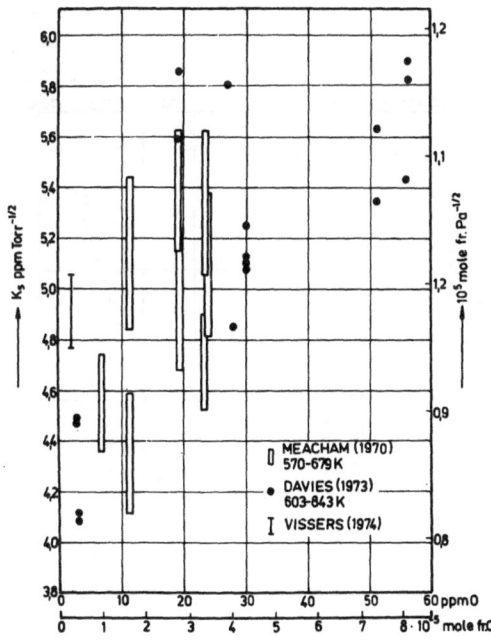

Fig. 2: Influence of oxygen dissolved in sodium on the Sievert's
constant

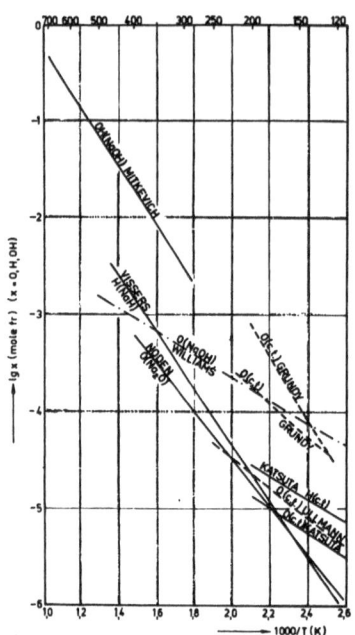

Fig. 3: Solubility concentrations in liquid sodium
versus reciprocal
temperature

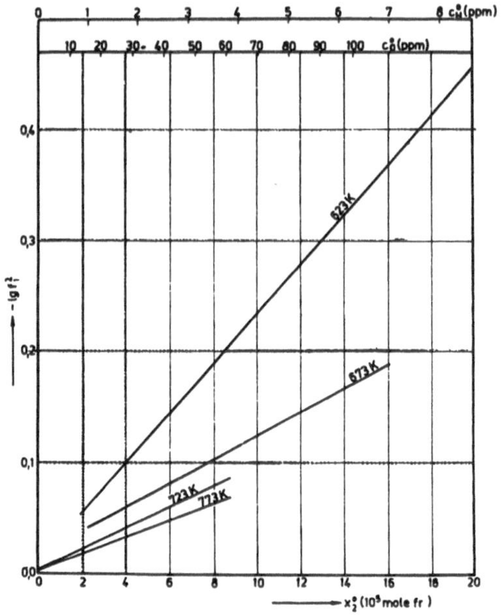

Fig. 4: Activity coefficients of O in
the presence of H, $f_O^H$, and of
H in the presence of O, $f_H^O$,
in sodium as functions of
concentration

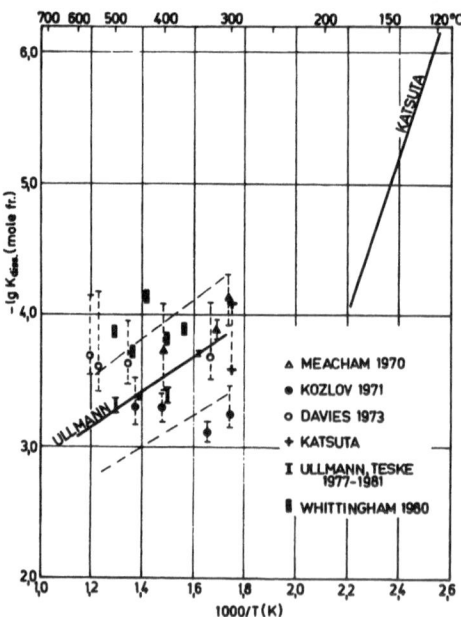

Fig. 5: Equilibrium constant of the reaction $OH^- = O^{2-} + H^+$ in
liquid sodium as a function of temperature

of monoxide or - especially at low temperatures - a curve of a
smaller slope (32, 19, 15). Even the hydrogen concentration analysis
yields analogous results (15), i.e. more oxygen and hydrogen are
dissolved than would be expected from the solubilities of monoxide
or hydride at the cold trap temperature. The slopes of these curves
agree with the one found for the solubility of sodium hydroxide by
Williams (30), but the values of the solubilities are lower. With
the exception of the earlier results of Mitkevich (9) who mixed
sodium hydroxide in great amounts with sodium at high temperature,
all the results fit into this picture. Table 2 gives an overview
of the solubility data, determined graphically or taken from the
authors. We obtain three typical values of the solvation enthalpies:
   50 kJ/mole for oxygen as $Na_2O$, $\sim$55 kJ/mole for hydrogen as NaH
and   23 kJ/mole for OH as NaOH. Solubilities obtained after release
of impurities from cold traps fit often in the range of the low
slope (23 ... 33 kJ/mole), depending on the cold trap content (predominantly oxygen or oxygen with hydrogen). The concentration of the
free (active) oxygen must be determined by two different solubility
equilibria in the two cases: on the one hand by the equilibrium O
in $Na_1$ - $Na_2O_s$, on the other hand probably by the equilibrium O in
$Na_1$ - $Na_2O$ in $NaOH_{l,s}$. Recently Privalov (12) recalculated the phase
diagram Na-O-H. Especially the sodium-rich region was supplemented
by him. The phase diagram is characterized by two liquid phases $L_1$
and $L_2$ on the basis of sodium and sodium hydroxide, respectively,
and by five solid phases: sodium, monoxide, hydride and solid
solutions $\alpha L_2$ and $\beta L_2$ on the basis of sodium hydroxide. The formation of solutions on the basis of sodium hydroxide had been proved
previously by Mitkevich and Schichlov (9). They found that these
solutions begin to solidify below 623 K. Above 681 K they observed
the formation of monoxide and hydride in the sodium hydroxide. The
amounts of oxide and hydride in the sodium hydroxide decrease with
increasing temperature. All investigations on the formation of new
phases (6,9,10,11,12) led to the statement of a "transformation
temperature" of 643 ... 707 K for the reaction of oxide with hydride
to form hydroxide.

   In opposition to the statement that $Na_1$ and $NaOH_{s,1}$ cannot coexist at low temperatures we observe below 500 K the precipitation
or dissolution of a phase with an enthalpy of solution similar to
that of sodium hydroxide. These contradictions might be explained
by the coexistence of the solid solutions  $L_2$ and $\beta L_2$ on the basis
of NaOH with liquid sodium. Thermodynamic or X-ray data of these
phases are not yet available, but they will be different from the
data of the pure phases.

THE CHEMICAL EQUILIBRIUM BETWEEN O AND H IN SOLUTION

   Several authors have observed a reduction of the hydrogen
equilibrium pressure in the presence of oxygen in sodium (1,6,13,14).

Table 2.  Saturation solutions of oxygen, hydrogen and hydroxide in
liquid sodium (X in mole fractions)

| author | solution function | temperature region (Kelvin) | heat of solution (kJ/mole) |
|---|---|---|---|
| Noden (3) | $\lg x_{O(Na_2O)} = 0.415-2445/T$ | 387...828 | 46.88 |
| Vissers (4) | $\lg x_{H(NaH)} = 1.428-2880/T$ | 400...673 | 55.22 |
| Mitkevich (9) | $\lg x_{OH(NaOH)} = 2.73-3000/T$ [1] | 550...978 | 57.53 |
| Kozlov (13) | $\lg x_{O(H_2O)} = 1.72-3000/T$ [1] | 473...673 | 57.53 |
| Williams (30) | $\lg x_{O(NaOH)} = -1.26-1190/T$ [1] | 373...623 | 22.82 |
| Grundy (32) | $\lg x_{O(c.t.)} = 3.83-3300/T$ [1] | 400...450 | 63.28 |
| " | $\lg x_{O(c.t.)} = -0.08-1750/T$ [1] | 420...673 | 33.56 |
| Katsuta (15) | $\lg x_{O(c.t.)} = -2.20-1270/T$ [1] | 393...473 | 24.35 |
| " | $\lg x_{H(c.t.)} = -2.05-1190/T$ [1] | 393...473 | 22.82 |
| Ullmann (19) | $\lg x_{O(c.t.)} = -1.28-1600/T$ [1] | 383...523 | 30.68 |

[1] our calculations from experimental results of the authors
conversion factors: 1 ppm H = 2.30 . $10^{-5}$ mole fraction
1 ppm O = 0.144 . $10^{-5}$  "    "
1 ppm OH = 0.135 . $10^{-5}$  "    "

On the other hand, a decrease in oxygen activity in the presence of
dissolved hydrogen has been detected by electrochemical measurements (15,16,17). The determination of the activity coefficients and
equilibrium constants becomes possible by combined activity- and
concentration measurements.

The interaction coefficients of oxygen in the presence of
hydrogen and vice versa in sodium estimated by us (16) at temperatures of 623 to 773 K are shown in Fig. 4. We find a linear increase
of lg f with increasing concentration of the second component.

By using the equations (10) and (11) on the basis of our results
we obtain the temperature function of the dissociation constant

$$- \lg K_{diss} = (1.53 \pm 0.21) + (1340 \pm 140)/T.$$

From this we obtain the temperature function of the free energy of formation

$$\Delta G^o_{OH} = (25660 \pm 2680) + (29.30 \pm 4.02)T.$$

The value of the enthalpy of formation of $- 25.66$ kJ/mole agrees well with the value calculated by a Born-Haber cycle (see Table 1). The absolute values of $K_{diss}$ derived from the results of various authors are in sufficiently good agreement (Fig. 5). The temperature dependence is low, on the basis of our results the dissociation increases with temperature.

REFERENCES

1.  C. C. Addison, Liquid Metals 1976, Inst.Phys.Conf.Ser.No.30
    (1977) p. 357; Chemistry in Britain 10 (1974) 331.
2.  P. Hubberstey, Liquid Metals 1976, Ins.Phys.Conf.Ser.No.30
    (1977) p. 539.
3.  J. D. Noden, J. Brit. Nucl. Energy Soc. 12 (1973) 329.
4.  D. R. Vissers, J. T. Holmes, L. G. Bartholme, P. A. Nelson,
    Nucl.Technology 21 (1974) 235.
5.  F. A. Kozlov, G. P. Sergeev, A. R. Sednev, W. M. Makarov,
    Atomn. Energija 44 (1978) 88.
6.  R. J. Pulham, P. A. Simm, Liquid Alkali Metals, Proc.Conf.,
    Nottingham 1973, p. 1.
7.  H. Isaacs, W. Becker, Report BNL-50205 (1969) p. 112.
8.  A. C. Wittingham, A review of the previous work on the
    liquid sodium-hydrogen-oxygen systems, Report CEGB/RD/B/N
    -2546 (Febr. 1973).
9.  E. M. Mitkevich, B. A. Schichov, Zh. neorg. khim. 11 (1966)
    633, 639.
10. S. A. Jansson in: J.E. Dralay, J.R. Weeks (ed.): Corrosion
    by Liquid Metals, AIME, Plenum Press, New York-London
    (1970), p.523.
11. K. M. Myles, F. A. Cafasso, J.Nucl.Mater. 67 (1977) 249.
12. Ju.V. Privalov, Atomn. Energija 48 (1980) 108.
13. F. A. Kozlov, E. K. Kuznecov, G. P. Sergeev, L. G. Wolckov,
    W. W. Sotov, IAEA specialists meeting on the Na/H$_2$O
    interaction, Melekess, May 18-21, 1971 (russ.)
14. R. A. Davies, J. L. Drummond, D. W. Adaway in: Liquid
    Alkali Metals, Proc.BNES Conf., Nottingham (1973), p.93.
15. H. Katsuta, K. Furukawa, Nucl.Technology 31 (1976) 218.
16. H. Ullmann, K. Teske, F. A. Kozlov, E. K. Kiznecov, Kern-
    energie 20 (1977) 80; 22 (1979) 25; IAEA-SM-236/80
    (1979); Report ZfK-422 (1980) p. 123.
17. A. C. Whittingham, C. A. Smith, P. A. Simm, R. J. Smith,
    Sec.Int.Conf. on Liquid Metal Technology, Richland, WA.,
    April 1980.
18. Ju.V. Privalov, The chemical equilibrium in the system
    Na-O-H (russ.), Report ZfK-337 (1977) p. 9.

19. H. Ullmann, Dissertation B, Adademie der Wissenschaften
    der DDR, 1977.
20. O. Kubaschevski, E. L. Evans, C. B. Alcock, Metallurgical
    Thermochemistry, 4th Edition, Pergamon Press 1967.
21. C. B. Alcock, G. P. Stavropoulos, Canad.Metallurg.Quarterly
    10 (1971) 257, (see Brewer and Kelley).
22. M. R. Hobdell, A. C. Whittingham, Report CEGB/RD/B/N-2545
    (1973).
23. C. Wagner, Thermodynamics of Alloys, Cambridge/Mass. 1952,
    p. 51.
24. R. Thompson, J.Inorg. Nucl. Chem. 34 (1972) 2513; Liquid
    Alkali Metals, Proc. Int. Conf. BNES, Nottingham 1973,
    p. 47.
25. P. J. Gellings, G. B. Huiscamp, E. G. Van den Broek,
    J. Chem. Soc. Dalton (1972) 151.
26. P. J. Gellings, A. Van der Scheer, W. J. Caspers, J. Chem.
    Soc. Faraday Trans. II, 70 (1974) 531.
27. A. Mainwood, M. Stoneham, J. Less.-Common Metals 49 (1976)
    271; Philisophical Magazine 37 (1978) 255, 263.
28. M. Osterbroek, H. P. Van de Braak, P. J. Gellings in: Liquid
    Metals 1976, Inst. Phys. Conf. Ser. No. 30 (1977) p.547.
29. D. A. Greenwood, V. K. Ratti, Liquid Alkali Metals, Proc.
    Int. Conf. BNES, Nottingham 1973, p. 43.
30. D. D. Williams, J. A. Grand, R. R. Miller, J. Phys. Chem.
    63 (1959) 68.
31. S. A. Meacham, E. F. Hill, A. A. Gordus, Report APDA-241
    (1970).
32. B. R. Grundy, Transact. ANS 18 (1974) 101.

INVESTIGATION METHODS FOR THE DETERMINATION OF THERMODYNAMIC

PROPERTIES OF LITHIUM ALLOYS

Ferdinand Sommer

Max-Planck-Institut für Metallforschung,
Institut für Werkstoffwissenschaften
Stuttgart, Fed. Rep. of Germany

INTRODUCTION

The objective of thermodynamic investigations of alloys is
the determination of the partial and integral Gibbs free energies
of mixing ΔG, of the mixing enthalpies ΔH as well as of the mixing
entropies ΔS, with their respective dependences on the concentration
x, temperature T and pressure p, in correlation with the phase
diagram. The three quantities of state quoted here are inter-
connected with each other by the Gibbs-Helmholtz relationship

$$\Delta G = \Delta H - T \Delta S \qquad (1)$$

and are defining the equilibrium state of an alloy at a given outer
pressure. The enthalpies as well as the specific molar heats of a
system can be directly obtained using calorimetric measuring methods
whilst the partial Gibbs free energies as well as the integral Gibbs
free energies, obtained herefrom by the Gibbs-Duhem integration,
are directly accessible by EMF as well as by vapor pressure
measurements.

The mixing enthalpies can also be determined from the temperature dependence of the free enthalpies. The enthalpy values
obtained in this manner, however, are, as a rule, affected by
larger measurement uncertainties than the according values gained
calorimetrically. A general review of thermodynamic measurement
methods is given in the papers of K. L. Komarek (1).

In the course of the last years, a series of measuring methods, especially for the determination of the thermodynamic properties of the highly reactive lithium alloys have been developed further. They shall be presented subsequently. Special interest in lithium alloys has been called forth by the new possibilities for their technological application, for instance in high temperature electrical batteries or as light metal alloys, as well as because of their possible usefulness in thermonuclear reactors. Furthermore, in the liquid lithium alloys distinct short range ordering effects are occurring.

CALORIMETRIC MEASURING METHODS

For the determination of mixing enthalpies of liquid alloys, an isothermally operating calorimeter is peculiarly suitable. The experimental arrangement described subsequently, allows for a direct measurement of ΔH on the mixing of the liquid components at temperatures up to 1300 K. For this, two crucibles are necessary which are initially containing the two liquid components at the measuring temperature and are arranged one above the other in a manner which enables the mixing process for the measurement. Both crucibles are inserted in a massive thermax (stainless steel) block consisting of several segments in order to equalize any possible temperature gradients occuring in the central part of a long furnace. The temperature gradient in the interior of the furnace can already be minimized by three separately, the one above the other, arranged heating coils which can also be regulated separately. A stirrer guarantees for the thorough and rapid mixing of the components. Temperature equality between the upper and the lower crucible can be controlled by thermocouples. The temperature change occuring during the mixing process is determined by a thermocouple within the alloy melt and a reference thermocouple inserted into the metal block or a thermopile mounted beneath the bottom of the reaction (i.e. lower) crucible. The area below the ΔT-time curve is a measure for the heat production or consumption of the reaction. Calibration can be effected by the addition of cylindrical metal samples which don't react with the alloy melt and with known heat contents (f.i. Mo, Ta). Calibration samples are dropped through a calibration tube into the lower (reaction) crucible. The calorimeter block is suspended in a reaction tube made of thermax steel which is closed below, over which the furnace is drawn. With this reaction tube material possible reactions of alkali or earth alkali metal vapors with the tube material can be avoided. Furthermore, this measurement arrangement can be peculiarly well kept free of oxygen by evacuation, heating out and flushing with argon. For the investigation of lithium alloys, this last point has a special importance due to the high formation enthalpy of $Li_2O$ ($ΔH = -595$ kJ/mol).

For the measurement of ΔH of the liquid Li-Mg alloys (2), as the ΔH values and, herewith, the measuring effects which are to be expected are very small, a special arrangement of the crucibles has been chosen, which has stood the test also during other corresponding investigations of liquid earth alkali metal systems(3). Bottom and covering of the upper crucible are consisting of tantalum foils. By piercing the tantalum foils with the stirrer, the mixing process of the liquid metals is started. Using this arrangement of the iron-crucibles, oxidation as well as vaporization of the components during the sample preparation period and the heating process can be largely prevented.

For lithium alloys which are exhibiting bigger $\Delta H$ values (f.i. Li with Ag, Pb, Tl, In and Bi) an experimental procedure has been chosen which is less time consuming (4, 5). In order to achieve this, the calibration tube is replaced by an iron tube closed at the lower end which serves as the reaction crucible. The real reaction crucible is filled with liquid lead in order to get a good heat transfer. The solid alloy components are added successively from the outside and a series of measurements can thus be effected at different concentrations in turn. Calibration is carried out before the beginning of the addition of the second component by repeated addition of samples of the initial (bath) component. If, for the addition of the second component, small concentration steps are chosen ( 1 at. %), it is possible to determine, additionally, the derivative of $\Delta H$, $d\Delta H/dx$ as a function of the concentration. The latter quantity is particularly useful for the investigation of chemical short range ordering phenomena in liquid alloys (6).

The determination of formation enthalpies $\Delta H^f$ of solid lithium alloys can be carried out, by solution calorimetry, in an analogous manner like the one used for the determination of $\Delta H$. The solid alloy samples as well as those of the components, are dropped through an open tube into a given bath (f.i. consistent of liquid tin). From the measured solution enthalpy of the alloy at infinitely low concentration $\Delta H^s$, as well as of the components $\Delta H^s_A$ and $\Delta H^s_B$ for $\Delta H^f$, at the measurement temperature results

$$\Delta H^f_{A-B} (T) = x_A \, \Delta H^s_A (T) + x_B \, \Delta H^s_B (T) - \Delta H^s_{A-B} (T) \qquad (1)$$

The solution calorimeter used for the determination of $\Delta H^s$ of lithium alloys (7) is shown in Fig. 1. It is representing a further development of an experimental arrangement which has been already described (8). In addition to the calorimeter arrangement depicted in Fig. 1, a closed protective gas circuit has a particular importance. The argon protective gas is continuously purified from oxygen by conducting it through a molecular sieve and over Zr at 1000 K. In addition, the whole apparatus is operated at a slight surplus pressure of the protective gas in order to prevent the introduction of oxygen into the system, especially on the dropping of the samples.

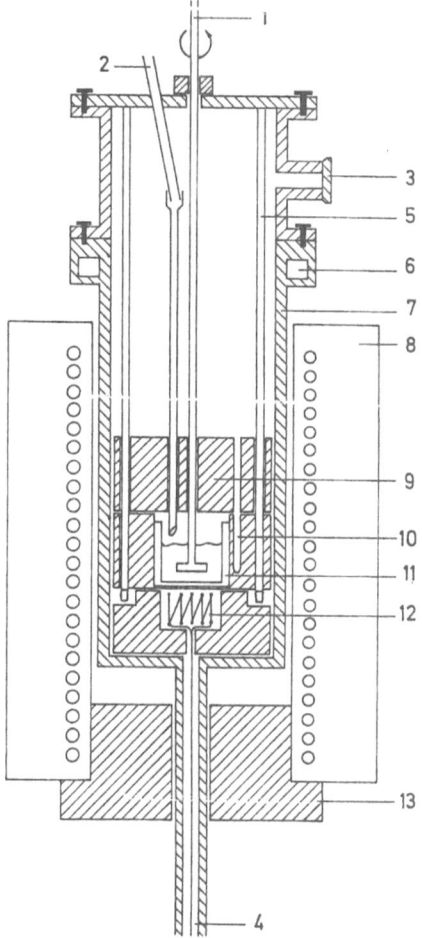

Fig. 1.   Schematic representation of a solution calorimeter

    1.  Motor-driven stirrer
    2.  Sample addition tube topped by a valve chamber
    3.  Protective gas inlet (closed circuit)
    4.  Thermopile output leads and protective gas outlet
    5.  Support rods
    6.  Water cooling
    7.  Thermax reaction tube
    8.  Furnace
    9.  Thermax blocks
   10.  Thermocouple (for the measurement of the temperature)
        boring
   11.  Reaction crucible
   12.  Thermopile
   13.  Alumina plug

The specific heats of liquid and solid lithium alloys as well
as their heats of fusion can be obtained by heat contents measurements using a drop calorimeter (9). For this, a sample which is
gas-tightly enclosed in a molybdenum container, is heated up to a
given temperature in the region of the liquid or solid state and
afterwards is thrown into a calorimeter being at room temperature.
In order to obtain valid measurement data, however, it is necessary
that the liquid and solid alloys are attaining the same final state
after the cooling process.

ELECTROCHEMICAL MEASURING METHODS

For the determination of the partial free enthalpies of
lithium $\overline{\Delta G}_{Li}$, and, herewith, of the lithium activities $a_{Li}$, EMF
measurements are feasible, as vapor pressure measurements of
lithium alloys for the determination of the thermodynamic activities
are, as a rule, leading to less exact results due to the high
chemical reactivity of lithium. The EMF cell for these measurements
can be arranged in the following manner:

$$-\left| \begin{array}{l} \text{Li metal (1)} \\\\ \text{Bi (1) saturated with Li}_3\text{Bi (s)} \\ \qquad \text{or} \\\\ \text{Al (s); LiAl (s)} \end{array} \right| \begin{array}{l} \text{LiCl-KCl} \\\\ \\ \text{LiCl-LiF} \end{array} \left| \text{Li-alloy (1)} \right| +$$

where the equilibrium voltage of the cell E obeys the following
relationship with the thermodynamic quantities

$$- E \cdot F = \overline{\Delta G}_{Li} = RT \ln a_{Li} \tag{2}$$

where R is the General Gas Constant and F the Faraday Constant. If,
as the reference electrode, a two phase alloy is used, to the cell
voltage obtained the cell voltage between lithium and the respective
reference electrode with its dependence on the temperature has to
be added. A two phase reference electrode should be given preference
to a pure lithium electrode, as the first one exhibits a much lower
solubility of lithium in the electrolyte. For low temperatures, as
liquid electrolyte an eutectic mixture of LiCl and KCl offers good
opportunities, whilst for temperatures above 774 K an eutectic
mixture of LiCl and LiF is to be preferred in order to prevent the
reduction of KCl by Li, the consequence of which would be the
escape of K vapor and the shifting of the alloy concentration. The
liquid alloy, which f.i. is contained within a tantalum capsule,
should be stirred, if possible, during the investigation in order
to provide for a zero temperature gradient and to guarantee for a
homogeneous alloy. The electrolyte is contained in a porous BeO
crucible which is submerged in the alloy bath (10). Immersed within
the electrolyte, the two phase electrode sheathed in a tantalum tube

is resting. In the case when pure lithium serves as the active substance of the negative electrode, porous stainless steel is soaked with the lithium (10). The application of clean water-free electrolytes as well as the performance of all necessary manipulations in a glove box which is as clean as possible, are the absolute requirements in order to get reliable measurement results.

Finally, the fact should be mentioned, that, with the coulometric titration technique (11) with continuous change of the alloy composition, using the electrochemical cell described above, the different ranges of existence of closely neighbouring intermetallic phases can be investigated especially well.

Using the experimental arrangements described above, all thermodynamic quantities of state of lithium alloys are accessible, whereat, naturally, for higher temperatures (T > 1200 K) and, in the case when very exact results are necessary (that means, with an accuracy better than $\pm$ 8 %) the experimental problems are increasing considerably.

The available thermodynamic data for binary systems of lithium until 1975 have been reviewed and summarized (12).

ACKNOWLEDGEMENT

I would like to thank Dr. I. Arpshofen for the translation of the text.

REFERENCES

1. K. L. Komarek: Z. Metallkde., 1973, Vol. 64, pp.325 – 341, pp. 406 -- 418; Ber. Bunsenges. Phys. Chem., 1977, Vol. 81, pp. 936 – 950.
2. F. Sommer: Z. Metallkde., 1979, Vol. 70, pp. 359 – 361.
3. F. Sommer and B. Predel: Phys. cond. Matter, 1974, Vol. 17, pp. 249 – 265.
4. B. Predel, G. Oehme and W. Vogelbein: Z. Metallkde., 1978, Vol. 69, pp. 163 – 166.
5. B. Predel and G. Oehme: Z. Metallkde., 1979, Vol. 70, pp. 450 – 453; pp. 618 – 623.
6. F. Sommer: to be published.
7. F. Sommer, B. Fischer and B. Predel: In these proceedings.
8. W. Vogelbein, M. Ellner and B. Predel: Thermochem. Acta, 1981 (in print).
9. F. Sommer, D. Eschenweck, B. Predel and R. W. Schmutzler: Ber. Bunsenges. Phys. Chem., 1980, Vol. 84, pp. 1236 – 1239.
10. M. L. Saboungi and M. Blander: J. Electrochem. Soc., 1975, Vol. 122, pp. 1631 – 1634.

11. W. Weppner and R. A. Huggins: J. Electrochem. Soc., 1978, Vol. 125, pp. 7 - 14.

12. J. F. Smith and Z. Moser: J. Nucl. Mat., 1976, Vol. 59, pp. 158 - 174.

# DETERMINATION OF THE FORMATION ENTHALPIES OF SODIUM ALLOYS OF LITHIUM WITH In, Tl, Sn, Pb and Bi

Ferdinand Sommer, Bernd Fischer, Bruno Predel

Max-Planck Institut für Metallforschung, Institut für
Werkstoffwissenschaften, Stuttgart, and
Institut für Metallkunde der Universität Stuttgart
Fed. Rep. of Germany

## INTRODUCTION

The formation enthalpies in systems with lithium as the one
alloy component are, at present, only known in a few cases, and
in these cases they are affected with very large uncertainties (1).
Doubtlessly this is due to experimental difficulties. However, the
exact knowledge of the thermodynamic properties of such phases is
important for the investigation of the formation conditions and
the stability of these alloys. Therefore, a solution calorimeter
has been constructed which is peculiarly suitable for the investigation of highly reactive alloys.

## EXPERIMENTAL PROCEDURE AND RESULTS

The purity of the metals used amounted to at least 99.9 %.
The samples were fused in iron crucibles under argon and tempered
closely beneath the decomposition temperature for several hours.
Preparation of the samples was carried out in an oxygen-free glove
box and afterwards the samples were introduced into the solution
calorimeter by transporting them, without interruption, under
protective gas. As all of the alloys are very brittle, the production
of cylindrical or spherical samples has involved great difficulties.
The solution enthalpies were determined at 800 K. For this, the
solid samples of the alloys as well as of the pure components are
dissolved in a liquid tin bath with a mass of about 20 g which is
present in the calorimeter. Before each of the measurements,
calibration was done by dropping of 10 cylindrical tin samples from
room temperature into the calorimeter. The heat contents of tin are
sufficiently well known (2). For the alloys as well as for the

according components, 5 - 10 measurements were carried out, respectively. The arrangement of the solution calorimeter has already been
presented in the same proceedings (3). For the evaluation of the
observed data the concentration dependence of the solution enthalpy
has been represented as an exponential series. By the extrapolation
of the best fit curves obtained, the solution enthalpies at
infinitely low concentrations $\Delta H^S$ are determined. The $\Delta H^S$ values
obtained with their respective mean quadratic margins of error are
given in Table 1. The solution enthalpies of Tl, Pb, Bi and In in
tin are already known with sufficient precision (1, 2). In order
to check the accuracy of the newly constructed calorimeter in
comparison with literature data, the solution enthalpy of In in
tin has been measured in addition. From the $\Delta H^S$ values of the
components, by subtraction of the respective heat contents, the
partial mixing enthalpies $\overline{\Delta H}^0$ of the components in tin can be
determined. For liquid In in tin at 800 K, results:

$$\overline{\Delta H}^0_{In} = 16.87 - 17.82 \text{ kJ/mol} = -0.954 \text{ kJ/mol}.$$

M.J.Pool (1) has obtained for $\overline{\Delta H}^0_{In}$, at 750 K: $-0.878 \pm 0.125$
kJ/mol, whereas Orr and Hultgren (1), at 700 K, got a value of
$-0.945 \pm 0.210$ kJ/mol. Within the respective margins of error,
the values which we have determined are in agreement with the
literature data.

From the solution enthalpies of the alloys at 800 K, the
according formation enthalpies $\Delta H^f_{A-B}$ at the same temperature can
be obtained applying the following relationship:

$$\Delta H^f_{A-B} = x_A \Delta H^S_A + x_B \Delta H^S_B - \Delta H^S_{A-B}. \tag{1}$$

The values gained in this manner are given in Table 1. For
the Li-Sn system, the concentration dependence of the formation
enthalpies is shown in Fig. 1.

DISCUSSION

The formation enthalpies obtained by solution calorimetry are
considerably differing from those given by Kubaschewski et al.(4)
(see Table 1). They have determined the formation enthalpies by
reaction calorimetry. Applying the latter method, it cannot be
guaranteed at all, in addition to other possible sources of error,
that the state of equilibrium has been attained, so that these
results are affected with great errors.

With the aid of known mixing enthalpies $\Delta H$ of the liquid alloys,
based on the condition that, at the melting point, the free
enthalpies of the solid and the liquid phase have to be equal, the
melting enthalpies $\Delta H^m_{A-B}$ and the melting entropies $\Delta S^m_{A-B}$ can be
calculated using the following relationship:

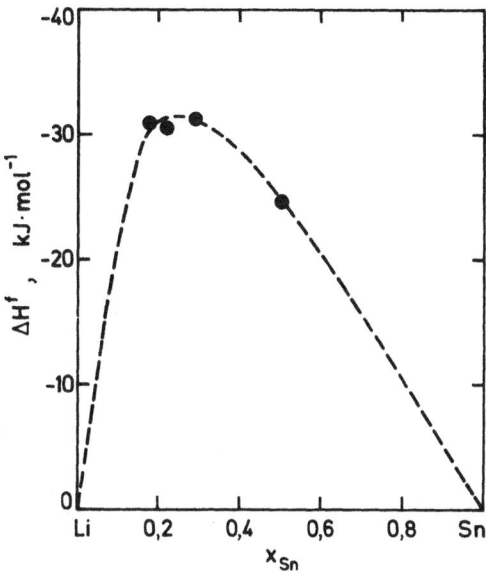

Fig. 1.  Concentration dependence of the formation enthalpies
of lithium-tin-alloys.

$$\Delta H^m_{A-B} = x_A \Delta H^m_A + x_B \Delta H^m_B + \Delta H - \Delta H^f_{A-B} \, 1 \qquad (2)$$

$$S^m_{A-B} = \frac{H^m_{A-B}}{T}. \qquad (3)$$

The melting enthalpies of the pure components $\Delta H^m_A$ and $\Delta H^m_B$
are known (2); the values obtained for $\Delta H^m_{A-B}$ are quoted in Table 2.
The excess entropies $\Delta S^{ex}_{A-B}$ of the intermetallic phases can be
determined, with the already known excess entropies $\Delta S^{ex}$ of the
liquid alloys, from

$$T(\Delta S^{ex} - \Delta S^{ex}_{A-B}) = x_A \Delta S^m_A (T^m_A - T) + x_B \Delta S^m_B (T^m_B - T) +$$

$$+ \Delta H - \Delta H^f_{A-B} \qquad (4)$$

$T^m$ is the melting temperature of the compound, $\Delta S^m_A$, $\Delta S^m_B$ are
the entropies of fusion of the pure components, $T^m_A$ and $T^m_B$ the
respective melting temperatures. The results obtained are cited
in Table 2.

Table 1.  Experimentally obtained solution enthalpies $\Delta H^s_{A-B}$ at 800 K and formation enthalpies $\Delta H^f_{A-B}$ (measured as well as calculated according to eq. (5)) and mixing enthalpies $\Delta H$ of lithium alloys.

| | $\Delta H^s_{A-B}$(800 K) | $\Delta H^f_{A-B}$(800 K) | | $\Delta H^f_{A-B}$(Eq.5) | $\Delta H$ | |
|---|---|---|---|---|---|---|
| | kJ/mol | kJ/mol | | kJ/mol | | |
| In | 16.87+ 4  % | | | | | |
| Li | −31.8 ∓ 2.2% | | | | | |
| Li Tl | 15.6 ∓ 5.6% | −19.9 | −26.8 (4) | −27.5 | −18.8 | (8) |
| Li$_5$Tl$_2$ | 9.2 ∓ 3.8% | −25.3 | | −19.2 | −20.3 | (8) |
| Li$_3$Tl | 5.86∓ 1.4% | −23.9 | | −17.6 | −19.6 | (8) |
| Li$_3$Pb | 20.9 ∓ 2.4% | −24.3 | −30.5 (4) | −38.6 | −21.9 | (8) |
| Li$_7$Pb$_2$ | 12.2 ∓ 3.3% | −31.4 | −35.1 (4) | −21.1 | −28.7 | (8) |
| Li Sn$_2$ | 19.6 ∓ 2.3% | −24.8 | −25.1 (4) | −31.6 | | |
| Li$_5$Sn$_2$ | 14.8 ∓ 6.1% | −31.4 | −38.9 (4) | −22 | | |
| Li$_7$Sn$_2$ | 10.7 ∓ 4  % | −30.7 | −40.2 (4) | −17.3 | | |
| Li$_{22}$Sn$_5$ | 9.4 ∓ 4  % | −31.3 | −39.2 (4) | −14.7 | | |
| Li In | 17    ∓15  % | −24.5 | | −22.8 | −19.2 | (8) |
| Li$_3$Bi | 30.5 ∓ 2  % | −47.9 | −57.7 (4) | −26 | | |
| Li$_3$Bi | 32.5 ∓ 1.3% | −35.5 | −37.6 (4) | −41 | −35 | (8) |

Table 2.  Melting temperatures T, melting enthalpies $\Delta H^m_{A-B}$, melting entropies $\Delta S^m_{A-B}$ and excess entropies $\Delta S^{Ex}_{A-B}$ of solid lithium alloys as well as excess entropies $\Delta S^{Ex}$ of liquid lithium alloys.

| | $\Delta H^m_{A-B}$ | $\Delta S^m_{A-B}$ | $\Delta S^{Ex}_{A-B}$ | $\Delta S^{Ex}$ | |
|---|---|---|---|---|---|
| T K (1) | kJ/mol | J/mol K | J/mol K | J/mol K | |
| Li Tl | 780 | 4.7 | 6 | − 5.1 | −6 | (11) |
| Li$_5$Tl$_2$ | 721 | 8.3 | 11.5 | −11.4 | −8 | (11) |
| Li$_3$Tl | 720 | 7.6 | 10.5 | −11.3 | −7.5 | (11) |
| Li$_3$Pb | 755 | 6.3 | 8.3 | − 5.8 | −4.7 | (12) |
| Li$_7$Pb$_2$ | 999 | 6.1 | 6.1 | − 7.6 | −8.4 | (12) |
| Li In$_2$ | 908 | 8.4 | 9.3 | − 8.8 | −7 | |

Most of the lithium phases investigated here can be classified amongst the group of the Zintl phases. However, this relation is not so clear in the case of $Li_7Pb_2$ and the compounds containing Sn, as these phases are crystallizing in structures quite untypical for Zintl phases. The heteropolar and, herewith, salt-like character of these phases is expressing itself in their high brittleness.

Regarding the $\Delta H^f$ values given in Table 1, one sees that the negative maxima in the Li-Tl, Li-Pb and Li-Sn systems are positioned respectively, at compositions belonging to the lithium rich compounds.

Qualitatively, the trend of the change of the bonding conditions on alloy formation can be understood, when the contributions originating in the difference of the atomic radii of the components $\Delta r$, the electronegativity difference $\Delta\Phi$ and the structural difference, are regarded, simplifyingly, as to be additive and independent from each other (5). The first constituent results in a positive contribution being proportional to $\Delta r$, whilst the second one is yielding a negative contribution being proportional to $(\Delta\Phi)^2$.

Miedema (6) has formulated a quantitative relationship, based on similar conceptions, as follows:

$$\Delta H^f_{A-B} = f\ (c^s_A,\ c^s_B)\ g\ \cdot\ p\ -\ e(\Delta\Phi^*)^2 + \frac{Q_o}{P}\ (\Delta n^{1/3})^2 \qquad (5)$$

where $\Phi^*$ and $n^{1/3}_{ws}$ are, respectively, the electronegativity and the electron density parameter which characterize a metallic element, $\Delta$ denoting the respective differences between the two elements in regard, e is the elementary charge and P, $Q_o$ and g are empirical parameters, which have been determined for different groups of alloys based on a great number of experimental results. Already regarding the 1:1 compounds, large deviations between $\Delta H^f$ values calculated according to eq. (5) and those ones experimentally obtained are to be observed, while the concentration dependences are not at all correctly reproduced (see Table 2). In a lot of other systems, however, in particular, when substitutional solid solutions are regarded, much better accordances between the experimental and the calculated (based on equation (5)) values are resulting (7). Lithium-rich alloys are, apparently, energetically preferred due to strong chemical short range ordering effects. This is demonstrated by the high formation enthalpies and the large negative excess entropies in the according concentration regions. There, the peculiar interatomic interactions are so strong that also the liquid alloys are exhibiting distinct chemical short range ordering effects, a fact, which can be shown regarding to concentration dependence of $\Delta H$ (8) and $\Delta S$ (11, 12) as well as the results of investigations on the structure (9) and on structure sensitive physical quantities (10). A better agreement between the calculated and the experimentally obtained $\Delta H^f$ values can be reached, if the symmetrical concentration dependence given by Miedema in eq. (5) is replaced by the concentration dependences of the mixing enthalpy experimentally found in liquid alloys.

ACKNOWLEDGEMENT

We are grateful to the "Deutsche Forschungsgemeinschaft" for the financial support of this work. Also, we would like to express our thanks to Dr. I. Arpshofen for the translation of the text.

REFERENCES

1.  R. Hultgren, R. D. Desai, D. P. Hawkins, M. Gleiser and K. K. Kelley: Selected Values of the Thermodynamic Properties of the Binary Systems, ASM, eds., Metal Park, Ohio, 1973.

2.  R. Hultgren, R. D. Desai, D. P. Hawkins, M. Gleiser, K. K. Kelley and D. D. Wagman: Selected Values of the Thermodynamic Properties of the Elements, ASM, eds., Metals Park, Ohio, 1973.

3.  F. Sommer: In these proceedings.

4.  O. Kubaschewski and W. Seith: Z. Electrochem., 1937, Vol. 43, pp. 743 - 749, 1941, Vol. 47, pp. 623 - 666; Z. Metallkde., 1938, Vol. 30, pp. 7 - 9.

5.  O. Alpaut and Th. Heumann: Acta Met., 1965, Vol. 13, pp. 543 - 548.

6.  A. R. Miedema and P. F. de Chatel: Theory of Alloy Formation, 1979, Proceeding of a Symposium of the AIME Annual Meeting, ed. L.H. Bennet, New York, 1980, pp. 344 - 387.

7.  W. Vogelbein, M. Ellner and B. Predel: Thermochim. Acta, 1981 (in print).

8.  B. Predel and G. Oehme: Z. Metallkde., 1979, Vol. 70, pp. 450 - 453; pp. 618 - 623.

9.  P. Chieux and H. Ruppersberg: J. de Physique Colloque C-8, 1980, pp. 145 - 152.

10. C. van der Marel, B. P. Albas, J. Hennephof, G.J.B. Vinke and W. van der Lugt: In these proceedings.

11. S. P. Yatsenko and E. A. Saltykova: Zhur. Fiz. Khim., 1975, Vol. 49, pp. 507 - 8; Elektrokhim., 1975, Vol. 11, pp. 580 - 581.

12. M. L. Saboungi, J. Marr and M. Blander: J. Chem. Phys., 1978, Vol. 68, pp. 1375 - 1384.

# PHYSICAL MEASUREMENTS ON LIQUID LITHIUM ALLOYS

Cees van der Marel, Jean Hennephof, Gerrit J. B. Vinke,
Ben P. Alblas and Willem van der Lugt

University of Groningen
The Netherlands

## INTRODUCTION

During the last seven years we have carried out a number of
physical measurements on the following series of liquid alloys
systems: Li-Na, Li-Mg, Li-Cd, Li-In, Li-Pb and Li-Sn. The thermodynamic properties of these alloys systems change systematically
and drastically in the order given above: Li-Na has a large
miscibility gap accompanied by strong critical fluctuations at the
consolute point, whereas in Li-Sn there is a strong tendency to
compound formation. Electron transfer from the Li atoms to the Sn
atoms takes place and gives rise to strong Coulomb attractions
between the two kinds of ions. As a consequence, at compositions
close to $Li_4Sn$, diffusive motion of the conduction electrons occurs,
which can be considered as the first stage of a Mott-Anderson
transition.

The physical properties investigated experimentally were: the
electrical resistivity $\rho$, the thermopower S, the Knight shift K
and the density d, while neutron diffraction measurements are in
progress. Not all of these experimental techniques were applied to
all of the alloys systems mentioned above: a choice was made
depending on the knowledge already available in the open literature
and on the expected contribution of an experiment to our understanding of the physical processes in these alloys. The experiments
were all carried out on a laboratory scale at temperatures up to
850 $^{\circ}$C. Considerable materials compatibility problems were
encountered. The solution of these problems was not the aim of our
experiments but an essential condition to success. Consequently,
some materials compatibility research was carried out, albeit in
a rather superficial way, e.g. by determining weight losses or by
looking at discolouration.

     From a point of view of materials compatibility, the experi-
ments can be divided into two groups. For the first group, which
includes density, resistivity (metallic conduction range) and
diffraction measurements, the refractory materials are metallic,
in the second group, including Knight shifts and thermopowers,
additionally electrically insulating refractories had to be used.
The second group provided most of the difficulties. Looking down
the column of the alkali metals one expects that, as a consequence
of decreasing ionization energy, the overall reactivity increases
from Li to Cs but, somehow, Li is an exception. It is as if lithium
ions, due to their smallness, easily penetrate ceramic materials.

     Both NMR (Knight-shift) and thermopower measurements are
rather critical with respect to material problems. For NMR one has
to make a fine dispersion of metal particles in an isolating matrix.
In our case, particles of approximately 20 μm were mixed with an
isolating powder. Due to the large contact area between the metal
and the powder, contamination takes place easily. Thermopowers are,
as is well-known, very sensitive to the presence of impurities.
Below follows a survey of our experiences.

ISOLATING REFRACTORIES

  1.  BeO (commercial, immersed in liquid Li at 350 $^\circ$C during
      48 hours
      96 % BeO: dissolved completely
      98.25 % BeO: serious discolouration
      99.8 % BeO (HP grade) still some discolouration, although
      distinctly less than for 98.25 % BeO

      Kendall (1) has successfully used BeO for thermopower
      measurements in Li up to 600 $^\circ$C and his results agree
      fairly well with ours (which were performed in LiF, see
      below). The poisonousness of BeO kept us from using it
      as long as other candidates were available.

  2.  $Al_2O_3$ (99.5 %, Degussit). Tentative thermopower measure-
      ments on liquid lithium gave irreproducible results above
      300 $^\circ$C.

  3.  $MgO.Al_2O_3$ (HP grade, Battles, Argonne National Lab.) was
      found to fall apart rapidly into small particles when
      immersed in liquid lithium at 400 $^\circ$C.

  4.  BN (HP grade, Carborundum). The weight loss of a cylindri-
      cally shaped piece (h = 10 mm, ∅ = 7 mm), immersed in
      liquid lithium at 320 $^\circ$C during 48 hours, was found to
      be only 0.6 %. But NMR measurements at 320 $^\circ$C on a
      dispersion of Li-metal (particles size ᴠ 20 μm) in BN
      powder showed that rapidly a substantial amount of $Li_3N$
      was formed.

5.  Silicon Nitride ($Si_3N_4$) showed strong discolouration and
    decomposes after a while in liquid Li at 250 °C.

6.  Finally, following an advice of den Hartog, LiF was
    chosen as an isolating refractory. Below 600 °C it is
    virtually insoluble in liquid Li and liquid Na (2,3,4).
    Still, hot-pressed and/or polycrystalline LiF was unusable,
    because it falls apart into small particles when immersed
    in liquid Li and because it suffers from bad machinability.
    So we made single crystals of LiF by pulling them from
    the melt and, next, zonemelting them. Crucibles and other
    parts were made by lathing the crystals with a kind of
    dental drill, a technique requiring some special skill,
    particularly because the crystals break easily.


METALLIC REFRACTORIES

    Electrical resistivity measurements up to 700 °C on pure
liquid lithium (5) and on the alloys systems Li-Na (5), Li-Mg (6),
Li-Cd (7) and Li-In (8) have been carried out in a metallic tube,
consisting of stainless steel AISI 321, shunted to the liquid
metal wire. This material could not be used for measurements on
liquid Li-Sn and Na-Sn alloys. Sn very easily forms alloys with
many other metals; the combination lithium and tin forms a most
agressive liquid alloys system. Finally, a tungsten tube was used
for these alloys.

    For X-ray diffraction measurements, Be windows are usually
most suitable due to their small scattering power. But thin Be
windows were perforated by liquid lithium if the latter is
contaminated with small amounts of oxygen or LiOH. For neutron
diffraction measurements in Li-Sn vanadium is not usable. At 810 °C
TZM molybdenum in $Li_{78}Sn_{22}$ gave a weight increase of only 0.1 at%
in 24 hours, and may be suitable for neutron diffraction measurements.

    Below follows a survey of the experiments, together with some
recent results.

RESISTIVITY ($\rho$)

    The apparatus for the resistivity measurements in liquid
Li-Sn and Na-Sn is shown in Fig. 1a. A tube of tungsten (1) was
made by forcing a plate (thickness .3 mm) at a temperature of
700 °C. The seam along the cylinder axis was closed by means of
electron beam welding.

Six Pt 10% Rh wires ($\emptyset$ = .2 mm) are spotwelded to the tube, approximately 25 mm apart (I, III, IV, V, VII and VIII). On top of the lower and the fourth wire (I and V) wires of pure platinum ($\emptyset$ = 0.2 mm) are spotwelded; in this way two Pt/Pt 10% Rh thermocouples were obtained. The wires are electrically insulated from each other and from the tungsten tube by means of alumina capillaries. The tube sections between I and V are used for the actual resistivity measurements, whereas the contacts VII and VIII are used for fluid level control. In order to get a homogeneous temperature across the tube sections between I and V, this part of the tube is supplied with a cylinder of pure copper (6) fitting closely around the capillaries and in the furnace. The furnace consists of two separate coils. The current through the upper coil is regulated electronically in such a way that the difference in temperature, measured by the Pt/Pt 10% Rh thermocouples spotwelded on the tube (I, II and V, VI) is less than $0.3^{\circ}$C. The heater coils are wrapped in a 25 mm thick layer of Refrasil tape for thermal insulation. The furnace takes approximately 160 Watt at 800 $^{\circ}$C and has a stability better than $0.3^{\circ}$C per minute. The measurements proceed as follows. The alloy to be investigated, contained in a crucible of tungsten or molybdenum, is heated in a separate furnace. As soon as the alloy is molten and homogeneous the furnace is pressed against the bottom of furnace B. Subsequently, by lowering the pressure in the tube, the fluid level is raised up to a point somewhere between VII and VIII. A current of 2 A is passed through the tube. From the voltage across a tube section the specific resistivity $\rho$ can be calculated, provided that the resistivity of the empty tube section and its geometrical factor are known. Comparison of the results obtained from the three sections provides a test for possible inhomogeneities (gas bubbles, incomplete mixing) in the liquid.

The equipment is supplied with a HP 9825 S microprocessor, used for process control, data storage and the necessary calculations. The connections from and to the microprocessor are shown schematically in Fig. 1b. The voltage and thermocouple leads on the tungsten tube are connected to a 24 pole motor driven switch. The microprocessor can read the position of the switch and change it. The output of the switch is connected to a HP 3455 A digital voltmeter (nominal accuracy $\pm$ 1$\mu$V), which can be read out by the microprocessor. Furthermore, the microprocessor controls the setpoints of the furnace A and B. After filling the tungsten tube with liquid metal the measurements occur completely automatically. For $\rho$ = 100 $\mu\Omega$ cm the error in the measurements is approximately $\pm$ 0.2 %, but the error is, in this configuration, proportional to $\rho$. As an example, we present in Fig. 2 recent results for the Na-Sn system (Li-Sn is not yet entirely completed) revealing very large metallic resistivities (which are accompanied by strongly negative values of d$\rho$/dT) reflecting the strong chemical interactions occurring in this system.

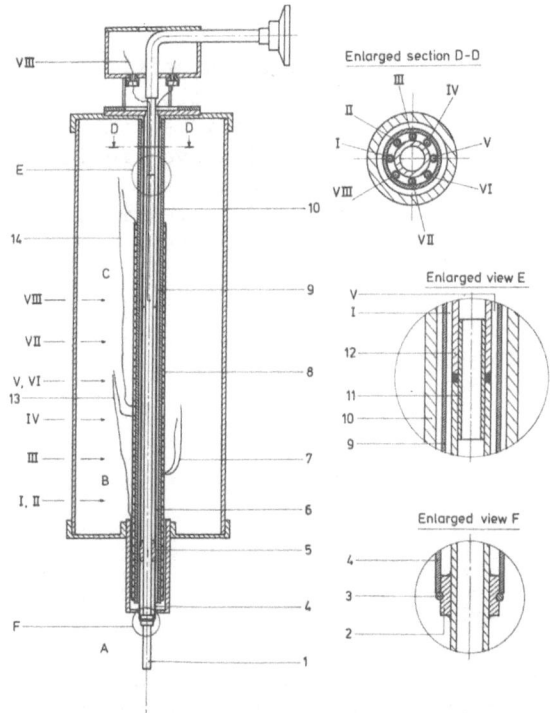

Fig. 1a.

Apparatus for electrical resistivity measurements; for the meaning of the numbers we refer to the text.

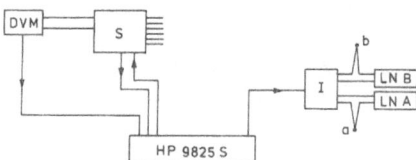

Fig. 1b.

Read and Listen connections from the HP 9825 micro-processor to digital volt-meter HP 3455, switch S and interface I.

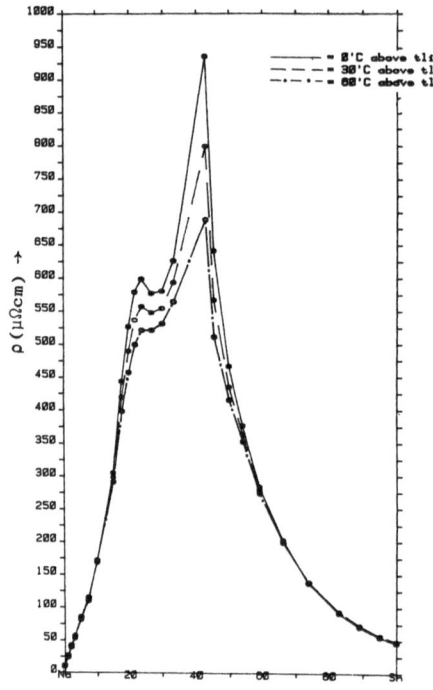

Fig. 2. Electrical resistivity ρ of
liquid Na-Sn alloys:
——— at liquidus temperature
----- 30 °C above liquidus
       temperature
-.-.- 60 °C above liquidus
       temperature

THERMOPOWER OF LIQUID Li-Na ALLOYS

       As mentioned before, monocrystalline LiF was chosen as containe
material. The cell used for the measurements is sketched in Fig. 3.
It consists of a tube of LiF 1, supplied with three calibrated
copper-constantan thermocouples 2 and 3 and a stainless steel wire
4 used as a fluid level indicator. The commercially obtained thermocouples are sheated with a wall of stainless steel AISI 316. The
LiF tube was made by drilling a hole through the entire length of
a monocrystalline rod of LiF. By means of springs the top end of
the tube is connected to a stainless steel manipulation rod 5.
By pressing the manipulation rod, the LiF cell can be lowered into
a crucible containing the metal to be investigated. The cell is
immersed just as deep as necessary to achieve electrical contact
between the indicator wire and the thermocouples. Crucible and
cell are heated by means of an electronically controlled furnace.
The furnace is supplied with auxiliary coils to establish the
desired temperature gradient across the sample. The measurements
were based upon the "small $\Delta t$" method, described more extensively
in ref. 5. The total absolute inaccuracy in S is less than 0.4
μV/°C. Fig. 4 shows a recently obtained result: the sudden increase
of the thermopower when the consolute point is approached. This
increase is caused by critical composition fluctuations. A quantitative explanation has not yet been given.

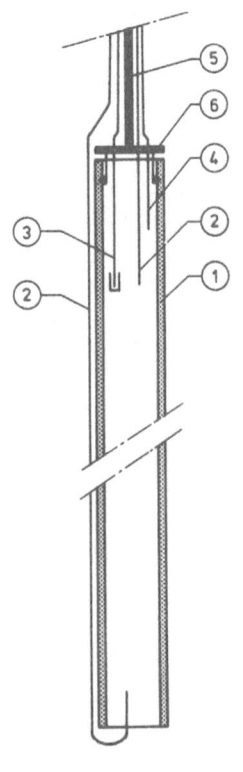

Fig. 3. Cell for the thermopower measurements.

1: LiF tube (length 140 mm, diameter 10 mm)
2: Copper-Constantan thermocouple, sheeted
with stainless steel, with grounded hot
junction
3: see 2, but insulated hot junction
4: indication wire for fluid level
5: manipulation rod
6: stainless-steel disc with alumina
feedthroughs for thermocouples

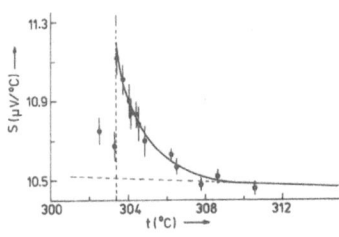

Fig. 4. Thermopower S of a Li-Na alloy
with the critical composition,
65 at% Li, as a function of
temperature, close to $T_c$.

## KNIGHT SHIFT

For the NMR measurements a Varian VF 16 wide-line spectrometer
was used. In the standard probe of this spectrometer a cylindrical
hole of only 17 mm is available for sample and furnace. The sample
has to be heated up to 800 °C, whereas the entire probe should be
kept below 60 °C. We largely adopted the design of a water-cooled
furnace from Styles and Sheffield (9), the main difference being
that we used alumina instead of quartz glass. Alumina has the advantage of being mechanically stronger; furthermore, it can be
used at higher temperatures and it is found not to react with the
platinum heater wire at high temperatures like quartz glass does (10).

The furnace assembly is sketched in Fig. 5. The inside of the
Varian probe is covered by a nylon liner 5 which is folded back
at each end over a brass sleeve inserted into the cover plate of
the probe. Water passes upwards in the space between the nylon
liner and the outer alumina tube 3. The heating element consists
of 1.20 m of 0.3 mm platinum wire, bent in a series of hairpins
and wrapped around the inner alumina tube 1, to form a non-inductive
element in which the current flow is parallel to the axis of the
tube. The element is held in place by winding it with Refrasil
cord, which also provides a 2 mm thick layer of thermal insulation
between the element and the cool outer alumina tube 3. The furnace
required approx. 180 Watt to stabilize at 800 °C. A dispersion of
the liquid with LiF powder was made and the entire sample was
placed in a single-crystal LiF sample tube with cover. Evaporation
problems were often serious and limited the time available for the
measurements. As an example, we present in Fig. 6 the Knight shift
results for Li-Sn (11). The deep dip near the stoichiometric
composition is a consequence of the tendency to localization of the
conduction electrons.

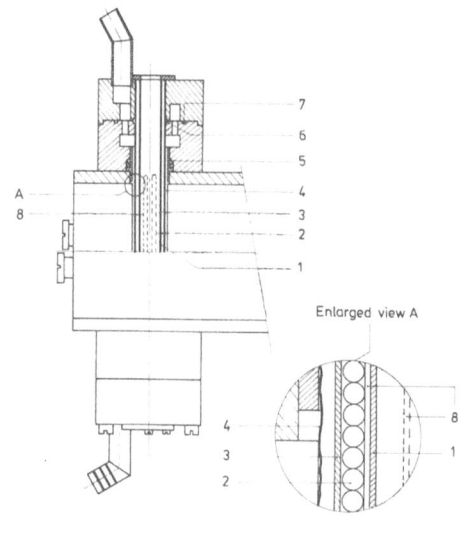

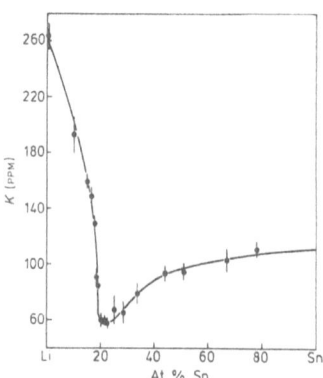

Fig. 5. NMR probe with water-
        cooled furnace

1: inner alumina tube
2: Refrasil cord for thermal
   insulation
3: outer alumina tube
4: nylon liner
5,6 and 7: O-rings
8: heating element (Pt wire,
   Ø 0.3 mm)

Fig. 6.  $^{7}$Li Knight shift in liquid Li-Sn

Density measurements were carried out in a pycnometer of
stainless steel (content 8 cm$^{3}$). As an example we give, in Fig. 7,
the results for the mean atomic volume of liquid Li-In alloys. An
appreciable volume contraction occurs as a consequence of the charge
transfer from Li to the less electropositive In.

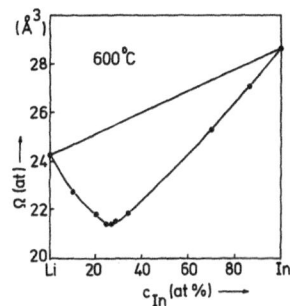

Fig. 7. Mean atomic volume $\Omega$ in liquid Li-In at 600 °C

Of the alloys systems investigated, Li-Pb has been proposed as a breeder and neutron multiplier in fusion reactors (12). In near future, similar investigations as described above will be undertaken for Li-Bi, also a candidate for this purpose.

ACKNOWLEDGEMENTS

The authors want to acknowledge the contributions of Dr. P.D. Feitsma and Dr. H.W. den Hartog in an early stage of this investigation.

This work forms part of the research program of the "Stichting voor Fundamenteel Onderzoek der Materie" (Foundation for Fundamental Research on Matter-FOM) and was made possible by financial support from the "Nederlandse Organisatie voor Zuiver Wetenschappelijk Onderzoek" (Netherlands Organization for the Advancement of Pure Research - ZWO).

REFERENCES

1. P. W. Kendall, Phys.Chem.Liq.1 (1968) 33.
2. E. G. Groff and G. M. Faeth, Ind.Eng.Chem.Fundam. 17 (1978) 328.
3. M. A. Bredig, J. W. Johnson and W. T. Smith Jr., Am.Chem. Soc. 77 (1955) 307.
4. M. A. Bredig and H. R. Bronstein, J.Phys.Chem. 64 (1960) 64.
5. P. D. Feitsma, J. J. Hallers, W. van der Lugt and T. Lee, Physica 93B (1978) 47.
6. P. D. Feitsma, T. Lee and W. van der Lugt, Physica 93B (1978) 52.
7. C. van der Marel and W. van der Lugt, J.Phys.F: Metal Phys. 10 (1980) 1177.
8. C. van der Marel, E. P. Brandenburg and W. van der Lugt, J.Phys.F: Metal Phys. 8 (1978) L273.
9. G. A. Styles and T. B. Sheffield, J.Phys. E9 (1976) 223.
10. J. McQuillon, J.Sc.Instr. 26 (1949) 329.

11. C. van der Marel, W. Geertsma and W. van der Lugt, J.Phys.
    F: Metal Phys. 10 (1980) 2305.
12. D. K. Sze, R. Clemmer, E. T. Cheng, 4th ANS Topical
    Meeting on the Technology of Controlled Nuclear Fusion,
    King of Prussia, 14-17 Oct. (1980).

# SOLUBILITY OF MANGANESE AND IRON IN SODIUM

G. Periaswani, V. Ganesan, S. Rajan Babu and C.K. Mathews

Reactor Research Centre
Kalpakkam
India

## INTRODUCTION

Neutron irradiation of the structural materials in a fast reactor core gives rise to activation products such as 54 Mn, 59 Fe, 51 Cr, 58 Co and 60 Co. These radionuclides may be released into the coolant, get transported to various parts of the primary coolant system and deposited there resulting in high radiation fields. This will render maintenance and repair very costly, necessitate longer waiting periods and lead to higher reactor down time which will affect reactor operating economios. Among the nuclides, 54 Mn is the most abundant and the most troublesome as it gets readily transported to cooler parts of the primary circuit (1). Data on the solubility of manganese in sodium will help in efforts to model the behaviour of 54 Mn in sodium systems and to devise means of minimising its transport. A program of experiments to measure the solubility of manganese in sodium is underway at Reactor Research Centre, Kalpakkam. Our initial experiments (2) had indicated strong dependence of the manganese solubility on oxygen concentration in sodium. The present experiments were carried out both in static capsules and in a small pumped isothermal loop.

The importance of accurate solubility data for iron in sodiu.. in explaining the corrosion processes in sodium-stainless steel systems is well recognised (3). Available data in literature on solubility of iron in sodium do not show enough agreement and they can be categorised into high solubility data and low solubility

data. High solubility data were obtained from experiments in which
iron vessels were used for equilibration and wet chemical techniques
were used for estimation of iron. Low solubility data experiments
were carried out in non-iron vessels and radiochemical methods
were employed in the analysis. Surface effects such as a high
iron and oxygen concentration at the sodium-cover gas interface
and error in wet chemical analysis are held responsible for the
high solubility values in the former experiments (3). In order
to verify these observations, experiments on the solubility of
iron were also carried out simultaneously with those for manganese.

MANGANESE SOLUBILITY EXPERIMENTS

Capsule Experiments

In this set of experiments, 4 to 5 g of gettered sodium was
equilibrated with manganese chips. Samples were taken through a
filter and analysed for manganese after distilling off the sodium.

Capsule Design: Figure 1 gives the design of the capsule.
This has two stainless steel tubes forming the inner and outer
compartments. Equilibrations are carried out in the inner stainless
steel crucible. The outer crucible is made of tantalum and is used
for collecting the samples. They are separated by a "porosint"
welded wire mesh filter of 5 $\mu$ pore size. A metal sheathed chromel-
alumel thermocouple is used for measuring the crucible temperature
during the equilibration. The capsule is heated using metal
sheathed heaters. A stepless proportional type controller is used
for controlling the capsule temperature. "Fiberfrax" is used as
the thermal insulation.

Experimental Procedure: About 5 grams of nuclear grade sodium
from a bath kept inside an inert atmosphere box was taken in the
stainless steel crucible. Chips of pure manganese metal were added
to it. About 30 mg of magnesium ribbon was also added as the getter
for oxygen. The capsule was assembled with the tantalum crucible
in the inverted position. All these operations were carried out
inside the inert atmosphere box. The capsule was brought outside
and pressurised with argon to about 1 $kg/cm^2$ above atmospheric
pressure. The heater and monitor thermocouple were applied and
over this thermal insulation was added. Equilibrations were carried
out for more than 24 hours. The capsule was then inverted at the
same temperature and pressurised with pure argon to 2 $kg/cm^2$. The
filtered sodium was collected in the tantalum crucible. After
cooling, the capsule was disassembled inside glove box and sample
was removed. The sample was then distilled under a vacuum of
$10^{-5}$ torr using an induction heater. The residue was dissolved out
using 6 M HCl-$HNO_3$ mixture prepared from quartz distilled acids.
This was later analysed for manganese by flameless atomic
absorption spectrophotometry.

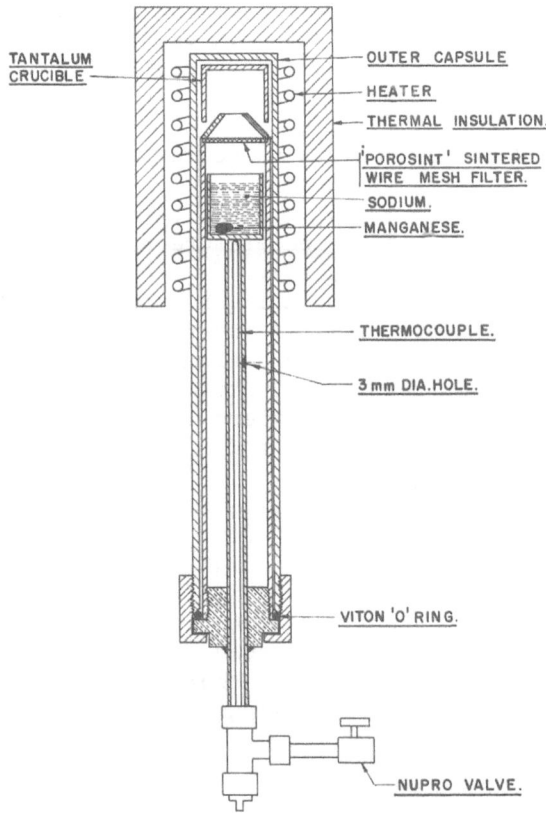

Fig. 1: Capsule for Mn solubility studies

Selection of Crucible Material: Stainless steel was selected as the crucible material as it has been found to be one of the least sorbing for 54 Mn thus indicating no marked affinity for manganese (4). So it is presumed that it will not serve as a sink for manganese and interfere in the experiments. Baus and Bogard as described in (3) did not observe isotopic exchange between 59 Fe in sodium and iron in stainless steel in these temperatures. So it can be assumed that the iron release into static sodium is not significant enough to modify the manganese surface.

Selection of Oxygen Getter: The oxygen level in sodium taken from the centre of the bath kept in the inert atmosphere box is expected to be around 1 to 2 ppm. The crucible surface in contact with sodium and the degassing from the remaining surfaces of the vessel in the course of the experiment will also add to the oxygen content of sodium. Gettering this oxygen using uranium metal at a high temperature ( 700 °C) for four hours was not adopted as it can lead to supersaturation and formation of particulates when cooled to the equilibration temperature. Distillation of sodium may also occur at this high temperature. In view of the above, an alternative getter material, viz., magnesium was chosen. The reasons for selecting magnesium are:

1. It has limited solubility in sodium at the temperatures of the experiments (5).
2. There is very low mutual solubility between iron and magnesium in the above temperature range and so it is not likely to interfere (5).
3. Solubility of manganese in liquid magnesium is very low in the temperature range of interest (6). Owing to this fact, a small amount of magnesium present in sodium is not likely to alter the solvation characteristics of sodium towards manganese. There are no intermetallics involving Mg and Fe, or Mg and Mn.

The use of magnesium getter will lead to very low oxygen levels and can minimise the free surface effects where the oxygen and manganese concentration will be higher than in the bulk sodium.

Equilibration Time: Equilibration times higher than 24 hours were employed as equilibrations in such low volumes are expected to near completion in as low as four hours (3). Temperature gradients within sodium cannot be significant owing to the high thermal conductivity of sodium.

Loop Experiments

As manganese is capable of being present in various oxygen associated forms in sodium (7), experiments to measure the solubility of manganese as a function of temperature and oxygen activity in a small isothermal loop were initiated.

Loop Design: The schematic of the loop is given in Fig. 2. This has a DC electromagnetic pump, an electrochemical oxygen meter and a sampling port. The pump was constructed using a permanent magnet of 0.6 T strength. A current of 25 amperes was passed across sodium to effect the flow. The functioning of the pump was verified in a separate experiment. Loop temperature was controlled within $\pm$ 1 $^{\circ}$ using a proportional controller. Samples were obtained by dipping a tantalum crucible into sodium and withdrawing it. These were distilled under vacuum and the residues were taken for analysis.

Oxygen Meter: The oxygen meter used in this loop was assembled using a $ThO_2$–7.5 % $Y_2O_3$ tube of 1/4" OD and 6" length (Fig. 3). The reference electrode selected was $In-In_2O_3$. The tungsten lead was taken out through a ceramic to metal seal. The electrolyte tube was secured to the upper portion using a "viton" O ring.

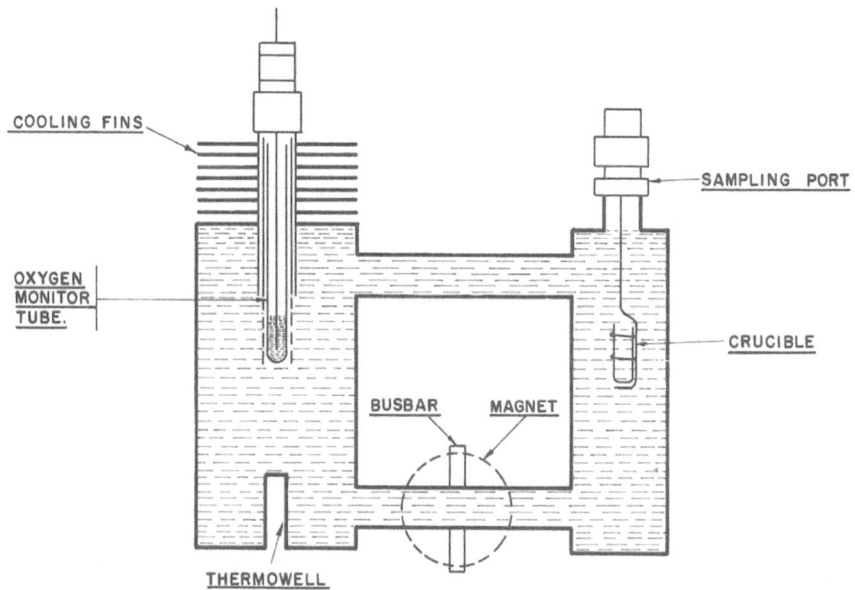

Fig. 2:   Manganese solubility loop

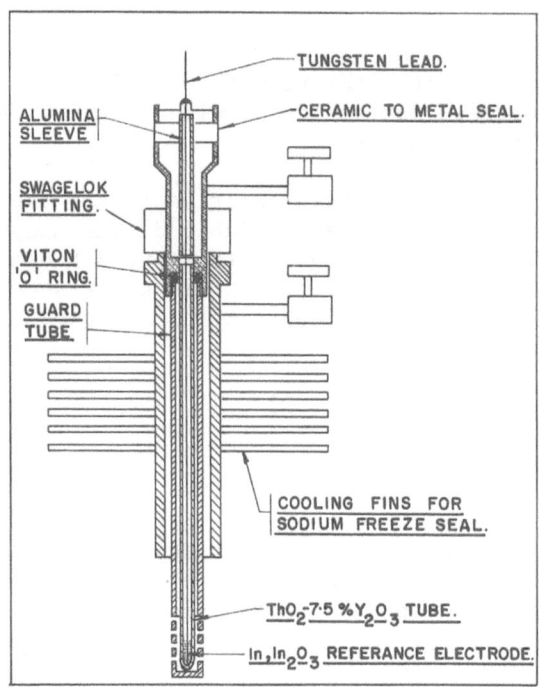

Fig. 3: Monitor for oxygen in sodium

The EMF output was measured using a Keithley Model 616 digital
electrometer. This meter was tested outside using Na-Cr-O system.
The EMF output and the oxygen levels in the system calculated
using the standard free energy of formation of NaCrO$_2$ as determined by Shaiu et al (8) are given in Table 2.

## EXPERIMENTS ON SOLUBILITY OF IRON IN SODIUM

These experiments also were carried out both in capsules and
in small sodium loop. Pieces of pure iron plates were added to
the sodium contained in stainless steel crucible and equilibrated.
Samples were obtained by filtration and were analysed for their
iron content by Atomic Absorption Spectrophotometry.

## RESULTS AND DISCUSSION

Initial capsule experiments on the solubility of manganese in
sodium were carried out in a double compartmented capsule constructed of AISI 316 stainless steel (2). It was possible to obtain
overflow samples both for oxygen and manganese in sodium by
inverting it after equilibration. Table 1 gives the data obtained
from these experiments. We concluded from these initial results
manganese solubility is a function of both temperature and oxygen
level. The scatter in the values may be explained to be due to
particulates  as there was no provision for filtering the samples.
Moreover the oxygen levels were not kept low.

The later experiments, the results of which will now be
discussed, were carried out at low levels of oxygen and with
provision for filtering the sample.

The results from the capsule experiments for the solubility
of manganese in sodium are given in Table 3. The manganese solubility ranges from 0.07 ppm at 549 K to about 3.5 ppm at 811 K. The
data are plotted in Fig. 4 where the solid curve shows the least-
square-fit. The equation for solubility of manganese in sodium
may be given as:

$$\log S \text{ (PPM)} = 3.6406 - \frac{2601.73}{T.K}$$

The correlation co-efficient is 0.825. In spite of the scatter
in the data the temperature dependence is clearly brought out. The
relative standard deviation for the entire procedure of sampling,
distillation, leaching, dilution and final analysis by flameless
atomic absorption spectrophotometry obtained by repeated analysis
of manganese in nuclear grade sodium has been estimated to be $\pm$ 12 %
in the concentration ranges encountered (9). The scatter in the data
is larger than this and could be due to the possible presence of
particulates of sizes smaller than the pore size in the filter
employed (5 $\mu$).

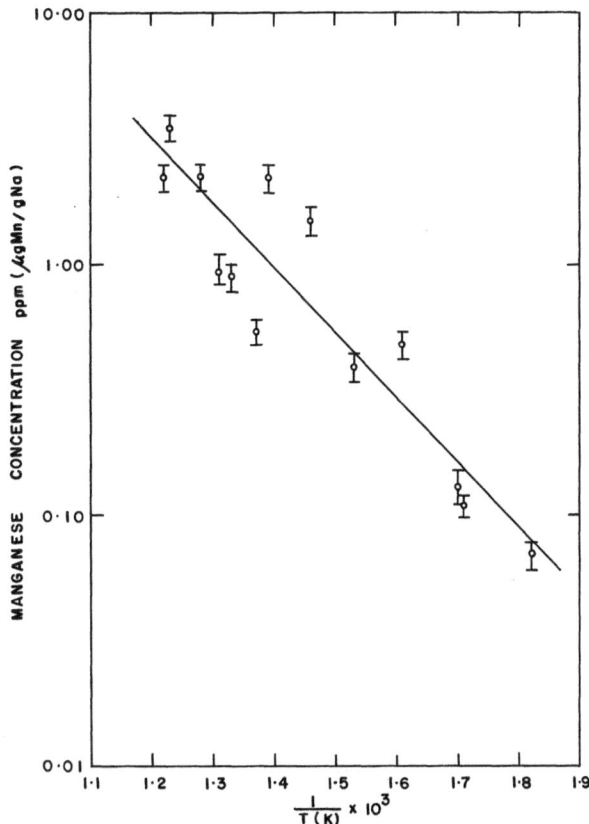

Fig. 4:   Solubility of Mn in low oxygen Na

Results from the loop experiment are rather high when compared to capsule experiments. These are given in Table 4. These samples were not filtered. Further experiments with filtration sampling are being planned. Oxygen levels also will be varied to measure the oxygen effect.

The preliminary results on the solubility of iron are given in Table 5. The data are available only in the temperature range 718 to 846 K. As can be seen from the table, the data are quite scattered and do not show any pattern. The possible sources of error are being investigated. These include the method of measurement, viz. atomic absorption spectrophotometry. The results reported above have been obtained using flame atomisation method which was not sensitive enough for measurements of solubility below 718 K. The blanks were, however, negligible compared to the measured values. For electrothermal atomisation, the blanks were found to be high and variable. However, efforts are underway to control the blank in this more sensitive measurement technique so that lower levels of iron can be measured. Further experiments are planned with the use of a finer filter for sampling.

Table 1: Solubility of Manganese in Sodium

| Temp. (K) | Manganese (ppm) | Oxygen (ppm) |
|-----------|-----------------|--------------|
| 476       | 1.0             | 13.2         |
| 585       | 2.8             | 26.6         |
| 587       | 5.4             | 39.0         |
| 613       | 2.9             | 87.5         |
| 633       | 6.0             | 43.0         |
| 673       | 4.2             | 31.0         |
| 775       | 10.0            | 53.0         |
| 819       | 10.5            | –            |

Table 2: Testing of Oxygen Meter in Na-Cr-O System

| Temp. (K) | EMF (Volts) | Oxygen (calculated) (ppm) |
|-----------|-------------|---------------------------|
| 680       | 0.596       | 0.35                      |
| 705       | 0.600       | 0.52                      |
| 731       | 0.604       | 0.76                      |
| 740       | 0.604       | 0.86                      |
| 753       | 0.606       | 1.02                      |
| 758       | 0.613       | 1.09                      |
| 769       | 0.614       | 1.24                      |
| 777       | 0.608       | 1.39                      |
| 801       | 0.619       | 1.86                      |
| 824       | 0.620       | 2.41                      |

Table 3: Solubility of Manganese
in Low Oxygen Sodium

| Temp. (K) | Manganese (ppm) |
|-----------|-----------------|
| 817 | 2.22 |
| 811 | 3.50 |
| 779 | 2.25 |
| 764 | 0.94 |
| 758 | 0.89 |
| 732 | 0.54 |
| 720 | 2.22 |
| 718 | 2.22 |
| 687 | 1.50 |
| 653 | 0.39 |
| 623 | 0.48 |
| 588 | 0.13 |
| 584 | 0.11 |
| 549 | 0.07 |

Table 4: Results from Loop Experiment

| Temp. (K) | Manganese (ppm) | Iron (ppm) |
|-----------|-----------------|------------|
| 581 | 0.46 | 10.0 |
| 641 | 0.66 | – |
| 686 | 1.41 | 104.4 |
| 744 | 0.96 | 6.1 |
| 712 | 0.74 | 22.2 |

Table 5: Results from Iron Solubility
Experiments

| Temp. (K) | Iron (ppm) |
|-----------|------------|
| 846 | 1.82 |
| 806 | 2.04 |
| 779 | 1.75 |
| 764 | 2.36 |
| 740 | 4.91 |
| 732 | 5.90 |
| 729 | 2.00 |
| 720 | 3.82 |
| 718 | 4.85 |

ACKNOWLEDGEMENT

     The authors gratefully acknowledge the help of Mr. T.R.Maha-
lingam and Miss R. Geetha for the atomic absorption analyses,
Mr. R. Saikumar for the regression analysis of the data and
Mr. D. Krishnamurthy for the distillations.

REFERENCES

   1. W. F. Brehm et al, "Radioactive material transport in
      sodium cooled nuclear reactors", Second International
      Conference on Liquid Metal Technology in Energy
      Production, Richland, USA, April 1980.
   2. C. K. Mathews et al, "Sodium chemistry research at Reactor
      Research Centre, Kalpakkam", Second International
      Conference on Liquid Metal Technology in Energy
      Production, Richland, USA, April 1980.
   3. G. Skyrme, "Solubility measurements of iron in sodium-A
      critical review", CEGB Report RD/B/M-3924, October 1977.
   4. H. H. Stamm et al, "Removal of radionuclides from liquid
      sodium by sorption on metallic surfaces", Second
      International Conference on Liquid Metal Technology
      in Energy Production, Richland, USA, April 1980.
   5. W. G. Moffat, The Handbook of binary phase diagrams,
      General Electric, 1978.
   6. M. Hansen, "Constitution of binary alloys", 2nd Ed.
      McGraw Hill, New York, 1958.
   7. M. G. Barker et al, "The solubility of ternary oxydes in
      liquid sodium and the formation of refractory metal
      nitrides and carbides in liquid lithium", Second
      International Conference on Liquid Metal Technology
      in Energy Production, Richland, USA, April 1980.
   8. B. J. Shaiu et al, "Thermodynamic properties of the double
      oxides of sodium with the oxides of chromium, nickel
      and iron", Journal of Nuclear Materials, (67) p.13,
      1977.
   9. T. R. Mahalingam et al, "Direct determination of trace
      metals in sodium by flameless atomic absorption
      spectrophotometry", This Conference.

# THE SOLUBILITY OF TRANSITION METALS, Mn AND Co IN LIQUID SODIUM

W. Peter Stanaway, Roy Thompson

AERE, Harwell

United Kingdom

## INTRODUCTION

A programme studying the solubility of metals in sodium has been in progress at Harwell for a number of years. The solubility of manganese and cobalt is being investigated at present. This paper presents the status of the investigation and describes the results so far obtained.

The importance of information on the solution chemistry of manganese in sodium arises from it being an activated corrosion product in the primary circuit of a fast reactor. It exists as 54Mn from the n, p reaction of 54Fe in natural iron. The manganese isotope is present in very low concentrations in iron containing materials but, nevertheless, is observed to transfer rapidly from the hot core regions of a sodium circuit, to low temperature areas where it is readily taken up by metal surface and may give rise to decontamination problems during maintainance of some reactor components.

In a reactor, active cobalt is formed from the 58Ni (n, p) 58Co and 59Co (nγ) 60Co reactions in materials in the core region. Mass transfer of cobalt has been observed in sodium systems with a cobalt deposit forming almost immediately after the hot region (1). Knowledge of the solubility of cobalt in sodium is required to help explain this behaviour.

PREVIOUS WORK

Mass transfer of 54Mn does occur in sodium circuits and this indicates that manganese should have a measureable solubility in liquid sodium. Eichelberger and McKisson (2), using their crucible inversion technique reported data, obtained at 700 - 900 °C, between 1 and 4 ppm; however, the results do not appear to be temperature dependant. Work on the solubility of Mn in liquid sodium was undertaken at Brookhaven (3) and Atomics International (4) about 12 years ago, but no results have been published.

Some data on the solubility of cobalt in liquid sodium is reported by Claar (5). Data obtained by Grand et al (6) and Lee and Berkey (7) are presented which indicated that the solubility is at the ppm level. However, the data are too limited and scattered to derive a solubility temperature relationship.

EXPERIMENTAL

The technique used in previous solubility studies at Harwell (8, 9) i.e. sealed nickel capsules sampled at temperature, has been developed further. The method now uses a samling technique in which only the lid of the capsule is pieced, this allows inert liners such as alumina to be used. The new sampling technique also enables samples to be filtered at temperature.

SAMPLE PREPARATION AND SAMPLING TECHNIQUES

Manganese

Nickel cans, 22.5 mm dia x 30 mm tall with a 0.25 mm wall were degreased in acetone and cleaned in a hot 5 % solution of Decon 75 prior to being hydrogen fired in moist hydrogen at 750 °C for 24 hrs; the dissolved hydrogen was pumped off under a vacuum of - $10^{-5}$ torr at 750 °C. An aluminia crucible liner, 20 mm dia x 25 mm tall, cleaned in boiling aqua regia, was fitted inside each nickel can. A small manganese ingot was used as the manganese source with $MnO_2/MnO/Na_2O_2$ added as an oxygen source as required. If the oxygen in the sodium was to be reduced by gettering, a slip of uranium foil, 2 mm x 15 mm, was inserted. These foils were prepared by electropolishing in 50 % $H_3PO_4/C_2H_5OH$. The crucible was then partly filled with 3- 3.5 gm of Reactor Grade sodium, dispensed by pressure from a large pot through a 1.5 μm mean pore diameter nickel filter frit and stainless steel tube at ∼ 150 °C. A nickel lid was then welded onto the can. The filling and welding operations were done in a high purity argon atmosphere. The oxygen level of the 'as poured' sodium is presently under investigation but it is estimated that the sodium in the cans, as sealed, was mainly between 15 - 20 μg/g but may have been higher on occasions.

The filled and welded can was heated at the temperature of
interest for 45 hours in a non-inductively wound copper furnace
block under an argon atmosphere. A sample was taken by piercing
the can lid with a molybdenum needle and withdrawing ∼ 2 g of
sample through a 1.5 µm mean pore diameter, treated nickel frit,
at 650 torr negative pressure. The sample was collected in a
weighed alumina crucible - again, previously cleaned in boiling
aqua regia. See Fig. 1 - for diagram of apparatus.

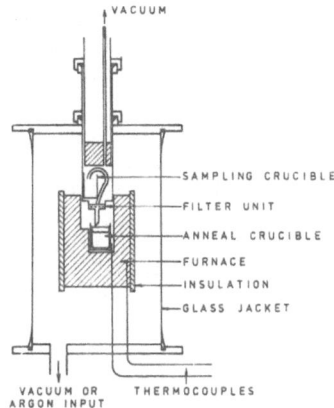

Fig. 1.    Schematic Diagram of Apparatus
for Equilibration and Sampling

The nickel filter frits used for sampling were treated by
immersion in a manganese sodium solution for 24 hours at 650 °C.
The frits, cut to size, were stood in an alumina rack in an alumina
crucible containing a manganese ingot. The crucible was put into
a cleaned and hydrogen fired nickel can, filled with sodium and
sealed as described above. After heat treatment at 650 °C, the can
lids were cut off and the sodium remelted at 120 °C. The rack and
frits were removed with the minimum amount of attached sodium and
transferred to a glass apparatus for distillation at ∼ 350 °C under
a vacuum of ∼ 10$^{-5}$ torr. The frits were exposed for the minimum time
in air whilst mounting in the sampling apparatus.

## Cobalt

A similar technique has been used for cobalt, but using an
active 60Co source with a specific activity of 1.85 x 10$^8$ Bq/gm;
the small foils (∼ 0.2 cm$^2$ and 12 mg in weight) replacing the
manganese ingot. Only 'as poured' sodium has been used so far.
The nickel filter frits were again pretreated before use but with
an active cobalt foil.

ANALYTICAL METHOD

By atomic absorption spectroscopy

The sodium sample was dissolved in methanol and the manganese
separated from the bulk sodium by a wet chemical co-precipitation
technique. The precipitate was dissolved in the minimum volume of
concentrated hydrochloric acid and made up to a standard volume
before being compared with standard solutions of manganese similarly
prepared. The analysis was made using a Perkin Elmer model 272
Atomic Absorption Spectrophotometer with an HGA 500 graphite furnace
and AS1 auto sampling unit attachments. Metal deposited on the walls
of the aluminia sample collection crucible was dissolved in aqua
regia and determined on the Atomic Absorption Spectrophotometer.

By radiochemical counting for 60Co

The total sample, i.e. sodium + crucible, was counted for a
time to give a sufficient number of counts to give at least 10 %
accuracy. This was compared to the activity of a standard cobalt
solution of the same geometry. The sample was subsequently used for
chemical analysis.

RESULTS

The results for manganese are presented graphically in Figs.
2 to 4. Fig. 5 shows the cobalt results.

Fig. 2 shows the results for the solubility of manganese in
'as poured' sodium. The circular points at each temperature are
an average; the vertical bars denote upper and lower limits in all
the determinations. The oxygen concentration was estimated at
15 - 20 µg/gm. The triangular points are results from capsules in
which the oxygen level was kept low by incorporating uranium foil
getters.

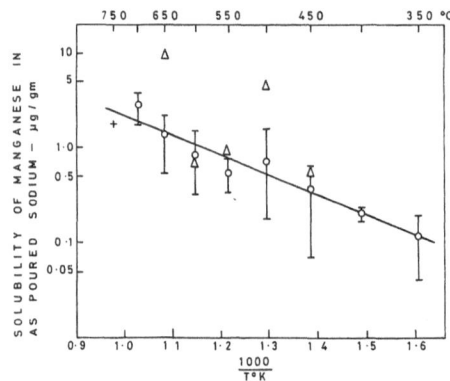

Fig. 2.   Solubility of Manganese in 'As Poured" Sodium vs.
          Temperature

Figs. 3 and 4 show the results from capsules to which oxygen had been added either as $MnO_2$, $MnO$ or $Na_2O_2$. Experiments were done at two temperatures, 650 $^{\circ}$C and 450 $^{\circ}$C as shown. Fig. 5 shows the cobalt results.

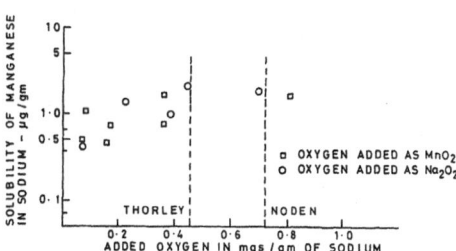

Fig. 3.   Solubility of Manganese in Oxide Rich Sodium at 450°C

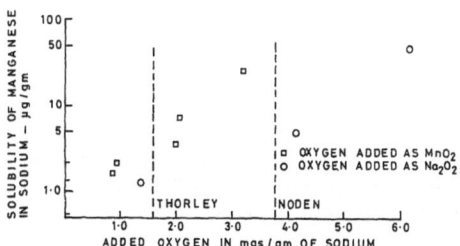

Fig. 4.   Solubility of Manganese in Oxide Rich Sodium at 650°C

DISCUSSION

The results of the solubility of Mn in sodium at various temperatures, as shown in Fig. 2, give a regression line through the average points with the relationship:

$$Log_{(Sol\ \mu g/gm)} = 2.325 \pm 0.043 - \frac{2017 \overset{+}{-} 213}{T}$$

with a correlation coefficient of 0.96. T is the absolute temperature.

Although the considerable spread in results is much reduced from earlier results obtained without filtering (10), the continuing disparity may be caused by sub-micron particles of manganese in the system breaking through the filter.

The results of the solubility of Mn in sodium gettered with
a small piece of electropolished uranium are shown on Fig. 2 as
the triangular points. The data at 450, 550 and 600 $^\circ$C fall within
the scatter band of the 'as poured' results, but the data at 500
and 650 $^\circ$C are high.

The results of adding oxygen, either as $MnO_2$, MnO or $Na_2O_2$
are shown in Fig. 3 and 4. The apparent solubility of manganese
in sodium is shown to increase with oxygen content of the sodium
at 450 $^\circ$C and 650 $^\circ$C. The effect is particularly marked at 650 $^\circ$C.
The actual form in which the oxygen was added appears to have
little effect, indicating that oxygen from either source reacts
in the system. Work by Barker et al (11) indicates that MnO may
be the stable oxide in a Mn/Na/O system up to about 25 $\mu$g/gm
oxygen content, with the formation of $NaMnO_2$ then $Na_4Mn_2O_5$ as the
oxygen level in the sodium is increased. Manganese thus follows
along similar lines to that shown by Thompson (8); for iron the
manganese work is still at a preliminary stage and a thermodynamic
study may be needed before the results can be fully understood.
The vertical dotted lines on Figs. 3 and 4 indicate the oxide
saturation level, as reported by Thorley (12) and Noden (13).

The results of the solubility of cobalt in sodium are shown
in Fig. 5. Results obtained by the radiochemical technique are a
factor of 10, or more, lower than those obtained by chemical analysis. Some temperature dependence is indicated particularly in
the radiochemical results. However, both sets of data are at the
lower limit of each of the analytical techniques used ($10^{-7}$ –
$10^{-8}$ gm) and no firm conclusions can be drawn.

CONCLUSIONS

The solubility of manganese in 'as poured' sodium measured
in aluminia crucibles contained in nickel cans over a temperature
range of 350 – 750 $^\circ$C, is given by the relationship:

$$\text{Log } S_{(\mu g/gm)} = 2.325 - \frac{2017}{T}$$

The presence of oxygen in sodium has been shown to increase
the manganese found in solution at both 450 $^\circ$C and 650 $^\circ$C.

No firm conclusions, as to the solubility of cobalt can be
drawn from the work at present.

ACKNOWLEDGEMENT

We wish to record our thanks for the valuable analytical
assistance given by Messrs. A. J. Upfold and P. E. Moss of
Newcastle Polytechnic.

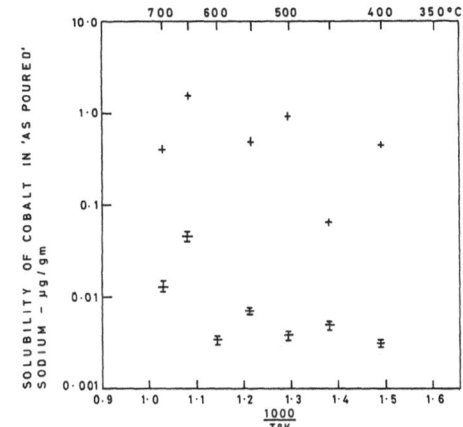

Fig. 5.   Solubility of Cobalt in "As Poured" Sodium vs Temperature

REFERENCES

1.  I. H. Newson et al - Second Int. Conf. on Liquid Metal
    Technology in Energy Production, Richland, Washington,
    April 1980.
2.  R. L. Eichelberger, R. L. McKisson - AL AEC 12955, 1970.
3.  J. R. Weeks, Brookhaven, Reported in Claar T.D. Reactor
    Tech. 13, 2, 1970.
4.  J. K. Balkwill - Liquid Metal Eng. Centre. Reported in
    Claar T. D. ibid.
5.  T. D. Claar - Reactor Technology, Vol. 13. No. 2,
    Spring 1970, p124.
6.  J. A. Grand et al - J. Phys. Chem. 63, 1192 (1959).
7.  P. K. Lee, E. Berkey - USAEC Report ANL-7520 Part 1,
    p299.
8.  R. Thompson - Solubility behaviour of Iron in
    Sodium/Oxygen solutions. AERE R 9172.
9.  R. Ainsley et al - J. Nuc. Materials 52, 255 - 276, 1974.
10. W. P. Stanaway, R. Thompson - Second Int. Conf. on Liquid
    Metal Technology in Energy Production - Richland,
    Washington, April 1980.
11. M. G. Barker et al - USA Paper 1979.
12. A. W. Thorley, A. C. Raine - Experimental techniques for
    the purification of sodium in the Alkali Metals, an
    Int. Symp. Nottingham. The Chemical Society (London)
    1967, p374.
13. Noden, Jnl. of the Int. Nucl. Energy Soc., January 1973,
    Vol. 12 No. 1 p.

# CHEMICAL REACTIONS IN LIQUID SODIUM AND LIQUID LITHIUM

Richard J. Pulham

University of Nottingham

United Kingdom

## INTRODUCTION

Lithium shows some significant differences from Na in its solvent properties and also in the chemical reactions which occur between dissolved solutes in the liquid metal. The elements N, H, C, Si and Ge are more soluble in lithium (N and H form eutectics) and the compounds which crystallize are more stable than their Na counterparts. Thus for compounds M = Li, Na, $-\Delta G_{298}$ values (kJ mol$^{-1}$) are $M_2O$ 561.4 (1), 378.7 (1) or 375.6 (2); $M_3N$ 128.6 (3), (87) (4); MH 70.1 (5), 38 (6); and $M_2C_2$ (56) (7), $-21.0$ (8). Values in parenthese are estimates and the present work gives $\Delta G_{873}$ $Li_2C_2 = -89$ kJ mol$^{-1}$. The corresponding solubilities (appm non-metal) at 500 °C are $M_2O$ 800 (9), 1696 (10), $M_3N$ 34200 (11), $< 10^{-4}$ (12); MH 34400 (11), 8273 (13); and $M_2C_2$ 210 (14), 1 (15). Consequently Li forms a more stable nitride which is soluble and reactive towards both non-metals and transition metals $Li_5TiN_3$ (16), $Li_2ZrN_2$ (17), $Li_7MN_4$ (M = V, Nb, Ta) (18), $Li_9MN_5$ (M = Cr, Mo, W) (19) and $Li_3FeN_2$ (20) are known and Li converts LiOH entirely to $Li_2O$ and LiH thereby virtually removing the equilibrium between OH$^-$, $O^{2-}$ and H$^-$ which exists in Na (21). The present work describes some reactions of C in Na and Li.

## REACTION OF CARBON WITH NITROGEN IN LIQUID SODIUM AT 700 °C

The main difference in the reaction between C and N is that they form CN$^-$ (NaCN) in Na (22) but NCN$^{2-}$ ($Li_2NCN$) in Li (23). The reaction in Na as followed by $N_2$ pressure is shown in Fig. 1. Graphite (0.19 mol) was added to stirred Na (2.13 mol) in a stainless steel vessel and the absorption of $N_2$ followed with time. There was no measurable absorption at 400 or 600, but reaction

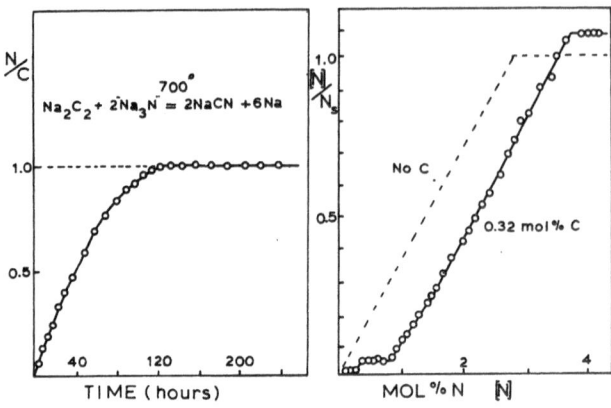

Figure 1   Na + C + N   Figure 2   Li + C + N

commenced ad 700 and was more rapid at 800 $^{\circ}$C. Reaction ceased
after 120 h, and after allowance for absorption of N by the steel,
the stoichiometry (N:C = 1.05) corresponded to the formation of
NaCN even though excess $N_2$ was available. Metallic Na was distilled
off under vacuum at 600 $^{\circ}$C leaving a solidified (mp. 564 $^{\circ}$C) globule
of NaCN which was identified by x-ray diffraction. These findings
are consistent with the observed increase in C solubility due to
the formation of CN$^-$ by an overpressure of $N_2$ (15), and the recovery
of NaCN unchanged from Na in which it dissolves (0.6 at 402 rising
to 1442 appm at 650 $^{\circ}$C) (12). The reaction is not unexpexted in
view of the probable unstable nature of $Na_3N$ and the metastable
nature of $Na_2C_2$ compared with the relatively greater thermodynamic
stability of NaCN.

$$Na_2C_2 + 2Na_3N \rightarrow 2NaCN + 6Na$$

REACTION OF CARBON WITH NITROGEN IN LIQUID LITHIUM AT 475 $^{\circ}$C

The reaction produces the Li salt of cyanamide, $Li_2NCN$,
irrespective of the order in which the reagents are added to Li
or whether N or C is in excess (24). Removal of Li at 600 $^{\circ}$C under
vacuum leaves tetragonal crystals of $Li_2NCN$ (mp > 600 $^{\circ}$C). The
corresponding solid state reaction is

$$Li_2C_2 + 4Li_3N \rightarrow Li_2NCN + 10\ Li$$

which means that $\Delta H^{\circ}Li_2NCN$ must be more negative than $-$ 359 kJ
$mol^{-1}$ when calculated from the values in Table 1.

The reaction has been followed by changes in the electrical
resistivity of the metal in which dissolved $Li_3N$ gives a large
conduction electron scattering coefficient, $dp/dc = 7.2 \times 10^{-8}$ $\Omega$m
(mol% N)$^{-1}$ and thereby provides an accurate measure of nitride in
solution. Carbide and NCN$^-$ species are too insoluble or scatter
electrons too feebly to greatly increase the resistivity so that

there is little ambiguity in interpreting the resistance changes. The absorption of $N_2$ by stirred Li containing 0.32 Mol% C is shown in Fig. 2 where the fraction of N added which will dissolve is plotted against the total amount added. Initially the nitride consumes the carbide but thereafter the excess dissolves. Assuming that the product does not influence the nitride solubility, then the lateral and vertical displacements are a measure of the N combined with C. The ratio is $CN_2$ (sometimes $CN_4$) and $Li_2NCN$ is recovered from the metal.

Table 1.   Standard enthalpies of formation, $\Delta H^o (kJ\ mol^{-1})$

| | | | | | | | |
|---|---|---|---|---|---|---|---|
| $NaCN^a$ | - 90.6 | $Li_2NCN$ ca | - 360 | $Na_3N^d$ (-150) | | $Li_2C_2^d$ (-59) | |
| $KCN^a$ | -113.4 | $CaNCN^b$ | - 350.3 | $Ca_3N_2^b$ | -430.5 | $Na_2C_2^e$ | 20.1 |
| $Ca(CN)_2^b$ | -184.3 | $BaNCN^b$ | - 256.9 | $Ba_3N_2^b$ | -362.8 | $CaC_2^b$ | -59.8 |
| $Ba(CN)_2^b$ | -222.4 | $Li_3N^c$ | - 164.6 | $Ba_2N^b$ | -217.4 | $BaC_2^b$ | -75 |

a, ref. 1; b, V.B. Parker, D.D. Wagmann and W.H. Evans, Selected Values of Chemical Thermodynamic Properties, NBS Techn. Note 270-6, 1971; c, ref. 3; d, ref. 3; e, ref. 8.


REASONS FOR THE FORMATION OF DIFFERENT PRODUCTS

It can be argued using the data in Table 1 that the stability of the MCN and $M_2CN_2$ solid compounds decrease in the order NaCN >$Na_2NCN$ and $Li_2NCN$>LiCN to rationalize the formation of the observed products. The solid state enthalpies differ from those in solution only by the small positive enthalpy of solution of the salts so that the same stability order probably holds in the metallic solutions. The cyanides decrease in stability KCN>NaCN indicating that $\Delta H$ LiCN is ca. - 70 kJ mol$^{-1}$ (a value of -126.8, which is the hydration enthalpy, has been erroneously used previously (23, 25), and this sequence is supported by Ba $(CN)_2$ and $Ca(CN)_2$. Therefore LiCN is probably far less stable than $Li_2NCN$. Consequently any LiCN which might be formed initially in Li is converted to $Li_2NCN$ by $Li_3N$.

$$LiCN + Li_3N \rightarrow Li_2NCN + 2Li$$

Similarly the stability of NaNCN may be inferred from the sequence shown by the Group II compounds. The heavier metal Ba forms the less stable MNCN and the more stable $M(CN)_2$. If this trend holds for the alkali-metal compounds also, then conversion of $CN^-$ to $NCN^{2-}$ is not expected to occur in Na.

REACTION OF CARBON WITH NITROGEN IN SODIUM-BARIUM SOLUTION

Several reactions in mixed metal solvents have been described. The reaction of $N_2$ with solutions of Li in Na produces single crystals of $Li_3N$ (26); with Si in Li to form $Li_8SiN_4$ (24); with Ba in Na to form $Ba_2N$ (27, 28). It can be shown now that the reaction between C and N in Li to form $NCN^{2-}$ can be duplicated in Na by adding Ba (or Ca or Sr). An earlier report (22) on the reaction of C with N in Na-Ba (ca. 7 mol% Ba) presented a complex picture. The terminal Ba:C:N ratios obtained in the sodium medium at 400 °C still stand (Table 2) but additional information has been gleaned which simplifies the interpretation:

(i)     Barium in Na reacts with $N_2$ which dissolves in the alloy up to a limiting ratio $Ba_4N$. Further N does not form as first supposed the $Ba_3N$ species but precipitates $Ba_2N$, and the protective Na matrix prevents the usual conversion to $Ba_3N_2$ (27, 28). A solution of Ba (or Ca) in Na, therefore, holds more N (or O) in solution than does Na alone when there is a deficit of non-metal. Excess of non-metal consumes the solvating Ba (or Ca) to precipitate $Ba_2N$ (or BaO, $Ca_2N$, CaO) thereby accomplishing the necessary purification of reactor sodium.

(ii)     $BaC_2$ persists in the products of the reaction of $N_2$ with mixtures of C with Na-Ba only when the C content rises above the ratio $CN_2$.

(iii)   The $NCN^{2-}$ ion has been identified in all Na-Ba-C-N mixtures.

Table 2  Stoichiometries in the Na-Ba-C-N reaction

| BA | : | C | : | N | Suggested composition |
|----|---|---|---|------|-----------------------|
| 2.0 | | - | | 1.00 | All $Ba_2N$ |
| 30.00 | | 1 | | 13.35 | $BaCN_2$ + $Ba_{29}N_{11.35}$ |
| 10.91 | | 1 | | 7.04 | $BaCN_2$ + $Ba_{9.91}N_{5.04}$ |
| 6.00 | | 1 | | 4.43 | $BaCN_2$ + $Ba_{5.00}N_{2.43}$ |
| 3.02 | | 1 | | 2.80 | $BaCN_2$ + $Ba_{2.00}N_{0.81}$ |
| 0.52 | | 1 | | 0.09 | All $BaC_2$ |
| 0.48 | | 1 | | 0.08 | All $BaC_2$ |

It is believed now that dilute solutions of C in Na-Ba contain $C_2^{2-}$ ($BaC_2$) which reacts with $N^{3-}$ ($Ba_2N$) to form $NCN^{2-}$ (BaNCN) and that the excess $N_2$ converts the excess Ba to $Ba_2N$. If every C atom is converted to BaNCN (Table 2), then the residual Ba can be accounted for as $Ba_2N$. It is instructive to view this from the N-rich angle

(Fig. 3). In the absence of C, all Ba forms $Ba_2N$ (N:Ba = 1/2) as
one limit. Adding C enhances the N uptake by forming BaNCN. The
other limit is when all C has been converted to BaNCN which is now
the sole product (N:Ba = 2, C:Ba = 1).

$$BaC_2 + Ba + 2N_2 \quad 2BaNCN$$

The line (Fig. 3) joining these limits depicts the intermediate
composition of this two-product mixture. The experimental points
lie very close to this line. Compositions richer in C than $CN_2$ contain
unreacted $BaC_2$ because insufficient Ba is available to produce the
$N^{3-}$ necessary for reaction. Thus at C:Ba = 2, there is no mechanism
for converting $N_2$ to $NCN^{2-}$ which is confirmed by the lower two ratios
in Table 2. Although these reactions occur in Na, the processes are
governed by the more stable compounds of Ba. Thus $BaC_2$ forms rather
than $N_2C_2$ and BaNCN rather than $Na_2NCN$. Barium (and Ca, Sr), however,
resemble Li in that they stabilize $CN_2^-$ rather than $CN^-$ (Table 2)
even in Na medium.

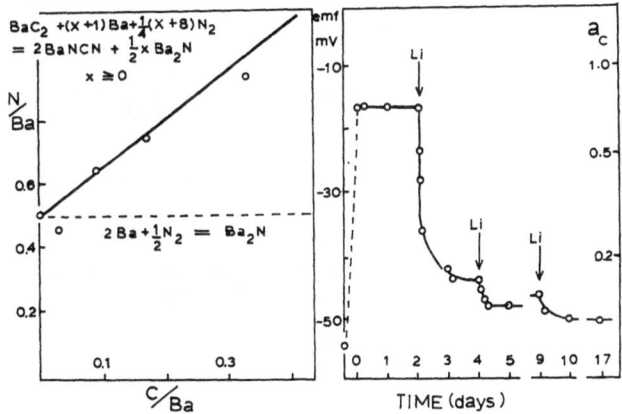

Figure 3 Na + Ba + C + N        Figure 4 Na+ C + Li

REACTION OF CARBON WITH LITHIUM IN SODIUM-LITHIUM SOLUTION

The reactions of C described above were determined indirectly
through its effect on N, itself measured by pressure or resistance
changes. Recently, however, a version of the electrochemical carbon
concentration cell (29) has been used in collaboration with Berkeley
Nuclear Labs. to follow the reactions of C directly in Na-Li and
Na-Ca solution. Filtered Na ($2 \times 10^{-4} m^3$) was contained in an Fe
beaker which was decarburized prior to use by moist $H_2$ at 530 °C
for 50 h, followed by soaking in Li + Ti at 600 °C for 300 h. The
outer stainless steel was equipped with an iron-sheathed thermocouple pocket and couplings to accommodate the cell, getter and

the addition of solutes. The cell dipped into the Na and comprised
an Fe thimble electrode (wall thickness 0.25 mm) filled with molten
$Na_2CO_3$-$Li_2CO_3$ electrolyte in which was immersed a central carburized
iron (Fe₃C) reference electrode. The carbon activity, $a_c$, in the
Na was controlled by judicious use of carburized iron rod and by
316 stainless steel getter which removed C as $(FeCr)_{23}C_6$. The emf
across the electrodes was measured by a high impedance electrometer.
After 3000 h a steady emf ($a_c$ = 0.24) equivalent to near saturation
concentration was achieved at 873 $\pm$ 1 °C (Fig. 4). Lithium (2 mol%)
was added in stages and emfs were measured continually. Lithium
progressively lowered $a_c$ due to the formation of $Li_2C_2$ but not by
the expected amount (< 2 appm C were removed by 20000 appm Li).
The precipitated $Li_2C_2$ established an equilibrium carbon activity
with the majority of Li which remained in solution

$$Li_2C_2(s) \overset{K}{\rightleftharpoons} 2Li \text{ (in Na)} + 2C \text{ (in Na)}$$

This was achieved more rapidly than the cell could respond. The
data (Table 3) can be used to evaluate $\Delta G$ for $Li_2C_2$ since $\Delta G$ =
$-nEF = -4EF = -RTlnK$, and $a_{Li} = \gamma X_{Li} = 1.0045 \ X_{Li}$ where $X_{Li}$ is
virtually the mol fraction added, in the relationship $K = a_{Li}^2 \cdot a_c^2 / a_{Li_2C_2}$/
$a_{Li}^2 \cdot a_{Li}^2 \cdot a_c^2 = 1/(a_{Li} \cdot a_c)^2$. The mean value of $\Delta G_{873}Li_2C_2$
is - 89 kJ mol⁻¹ which, though not readily converted to
$\Delta G_{298}^o$, shows that $Li_2C_2$ is considerably more stable than $Na_2C_2$.
The value may be compared with an estimated $\Delta G_{298}Li_2C_2$ =
- 60 $\pm$ 33 kJ mol⁻¹.

      The electrochemical carbon concentration cell  successfully
followed the reaction of C in this particular sodium alloy (and
in Na-Ca to be described later). The main disadvantage under the
prevailing conditions of many extraneous carbon sources was the
long times necessary to achieve the initial low carbon activities.

Table 3.   Carbon activities in sodium-lithium (2 mol%) at 600 °C

| Li (mol%) | emf | $a_c$ ($a_c$ graphite = 1) | $\Delta G_{873}$ (kJ mol⁻¹) |
|---|---|---|---|
| 1.14 | - 43.7 | 0.177 | - 89.9 |
| 1.43 | - 48.0 | 0.139 | - 90.1 |
| 2.07 | - 50.2 | 0.126 | - 86.6 |
| $(FeCr)_{23}C_6$ | - 54 | 0.1 | |

REFERENCES

1. JANAF Thermochemical Tables, Second edn. NSRDS-NBS 37 1971.
2. D. R. Frederickson and M. G. Chasonov, J. Chem. Thermodynamics 1973, 5, 485.
3. D. W. Osborne and H. E. Flotow, J. Chem. Thermodynamics,
   1978, 10, 675.
4. G. J. Moody and J. D. R. Thomas, J. Chem. Educ. 1966,
   43, 205.
5. H. R. Ihle and C. H. Wu, J. Inorg. Nuclear Chem. 1974,
   36, 2167.
6. W. M. Latimer, "The Oxidation States of the Elements and
   their Potentials in Aqueous Solution". Prentice Hall,
   2nd Ed. 1952.
7. JANAF Thermochemical Tables, 1974 Supplement.
8. G. K. Johnson, E. H. Van Deventer, J. P. Ackerman,
   W. N. Hubbard, D. Osborne and H. E. Flotow, J. Chem.
   Thermodynamics, 1973, 5, 57.
9. R. M. Yonco, V. A. Maroni, J. E. Strain and J. D. DeVan,
   J. Nucl. Mater, 1979, 79, 354.
10. J. D. Noden, J. Brit. Nucl. Soc. 1973, 12, 57.
11. P. F. Adams, M. G. Down, P. Hubberstey and R. J. Pulham,
    J. Less-Common Metals, 1975, 42, 325.
12. E. Veleckis, K. E. Anderson, F. A. Cafasso and H. M. Feder,
    Proc. Internat. Conf. Sodium Technol. and Fast Reactor
    Design, ANL-7520, Part I, 295, 1968.
13. A. C. Whittingham, J. Nucl. Mater. 1976, 60, 119.
14. R. M. Yonco and M. I. Homa, Trans. Amer. Nucl. Soc. 1979,
    32, 270.
15. R. Ainsley, A. P. Hartlib, P. M. Holroyd and G. Long,
    J. Nucl. Mater. 1974, 52, 255.
16. R. Juza, H. H. Weber and E. Meyer-Simon, Z. anorg.allgem.
    Chem. 1953, 273, 48.
17. A. P. Palisaar and R. Juza, Z. anorg. allgem. Chem. 1971,
    384, 1.
18. R. Juza, W. Gieren and J. Haug, Z. anorg. allgem. Chem.
    1959, 300, 61.
19. R. Juza, W. Gieren and J. Haug, Z. anorg. allgem. Chem.
    1961, 309, 276.
20. M. Fromont, Rev. Chim. Minerale, 1967, 4, 447.
21. R. J. Pulham and F. A. Simm, Proc. Internat. Conf. Liquid
    Alkali-Metals, April 4 - 6, Nottingham University,
    BNES London 1973, p 1.
22. C. C. Addison, B. M. Davies, R. J. Pulham and D. P. Wallace,
    The Alkali-Metals, Special Publn. No. 22, The Chemical
    Society, London 1967.
23. M. G. Down, M. J. Haley, P. Hubberstey, R. J. Pulham and
    Anne E. Thunder, J.C.S. Dalton, 1978, 1407.

24. R. J. Pulham, P. Hubberstey, Anne E. Thunder, A. Harper and A. T. Dadd, Second Internat. Conf. Liquid Metal Technol. Energy Production, April 20 - 24, Richland, Wa. ANS 1980.

25. H. Migge, Second Internat. Conf. Liquid Metal Technol. Energy Production, April 20 - 24, Richland, Wa. ANS 1980.

26. M. G. Down and R. J. Pulham, J. Crystal Growth, 1979, 47, 133.

27. C. C. Addison, R. J. Pulham and E. A. Trevillion, J.C.S. Dalton, 1975, 2082.

28. C. C. Addison, G. K. Creffield, P. Hubberstey and R.J. Pulham, J.C.S. Dalton, 1976, 1105.

29. M. R. Hobdell and J. R. Gwyther, J. Appl. Electrochem. 1975, 5, 263.

THERMODYNAMIC POTENTIAL OF NITROGEN, CARBON, OXYGEN AND

HYDROGEN IN LIQUID LITHIUM AND SODIUM

Norbert Rumbaut, Florint Casteels, Martin Brabers

Studiecentrum voor Kernenergie
Mol
Belgium

## INTRODUCTION

The study of dissolved non-metals such as carbon, nitrogen, oxygen and hydrogen in liquid alkali metals is of technological interest, as these species enhance corrosion and mass transfer in cooling circuits for future fusion reactors (1).

As chemical activities and not concentrations are the driving factors, there is need for well-established relationships between these thermodynamic quantities and the existing concentrations in dilute solution. In this paper activity-concentration relationships are deduced from "terminal" solubility equations for H, C, O and N in liquid lithium and sodium. The latter were obtained from a critical analysis of experimental data supplied by various authors. Related thermodynamic quantities such as partial molar entropies and enthalpies are deduced and indicate a similar solvation pattern for these non-metals in lithium as compared to sodium.

## TERMINAL SOLUBILITY DATA

Table 1 gives the saturation or "terminal" solubility data for H, C, O and N in liquid lithium and sodium.

Most of the equations are the result of a critical analysis on selected data obtained under reliable experimental conditions. The equation for hydrogen in sodium is based on the results of five different publications (2 to 6).

For carbon in the same metal, only the most reproducible four sets of determinations from the many published data were retained (7 to 10).

Table 1.  Terminal solubility of H, C, O and N in liquid lithium
          and sodium

$$\ln x = A + \frac{B}{T} \quad (T \text{ in } K) \quad (x = \text{mole fraction})$$

| non metal | lithium | | | sodium | | |
|---|---|---|---|---|---|---|
| | A | B | Ref. | A | B | Ref. |
| hydrogen | LiH 3.33 | −5208 | (24,21) | NaH 4.375 | − 7118 | (2-6) |
| carbon | $Li_2C_2$ 0.618 | −3855 | (24) | $Na_2C_2$ 2.99 | −12991 | (7-10) |
| nitrogen | $Li_3N$ 2.80 | −4714 | (13,14) | $Na_3N$ − | − | − |
| oxygen | $Li_2O$ 1.791 | −6310 | (15.) | $Na_2O$ 0.960 | − 5631 | (11) |

The equation for oxygen in sodium is that of Noden (11) who
based it on 268 determinations. For N in sodium reliable data for
the terminal solubility are lacking. However, the Sieverts constant
was determined as a function of temperature by Veleckis et al. (12),
from which solvation properties can be deduced. The solubility of
hydrogen in lithium was derived by Adams (13) and Maroni (21). For
carbon  in lithium, the publications show widely divergent results.

The relationship that we retained correlates best with the
other data and stems from Grishin (24). The equation for nitrogen
in lithium is from Adams et al. (13) and Yonco et al. (14) and is
based on 33 determinations between 181 and 435 $^{o}$C.

For the Li-O solubility, we used the equation of Adams et al.
(15), who combined 15 measured values of three different groups
of authors. The equation is in accordance with data of Yonco et al
(28).

The solubilities are graphically represented in Fig. 1a.
Table 2 shows the partial molar solution enthalpies and −entropies
for C, H, O and N, calculated from these solubility equations. As
can be seen on Fig. 1b, $a\Delta S_{sol} - \Delta \overline{H}_{sol}$ graph displays only a faint
correlation between the thermodynamic solution properties of the
non-metals in sodium as compared to lithium. Therefore, Gibbs free
energies of solution of the gaseous non-metals were derived from
activity-concentration relationships.

Table 2.  Enthalpies and entropies of solution of C, H, O and N-
          compounds in liquid lithium and sodium

| | NaH | $Na_2C_2$ | $Na_2O$ | $Na_3N$ | LiH | $Li_2C_2$ | $Li_2O$ | $Li_3N$ |
|---|---|---|---|---|---|---|---|---|
| $\Delta \overline{S}_{sol}$ (cal/mole K) | −8.69 | −5.94 | −1.91 | − | −6.62 | −1.23 | −3.55 | −5.56 |
| $\Delta \overline{H}_{sol}$ (cal/mole) | 14143 | 25813 | 11189 | − | 10348 | 7660 | 12538 | 9367 |

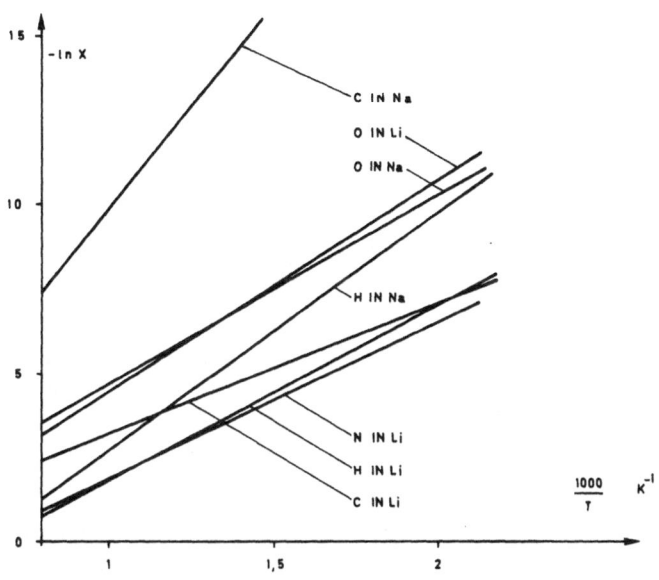

FIG.1a  SOLUBILITY OF C, H, O AND N IN LIQUID
SODIUM AND LITHIUM

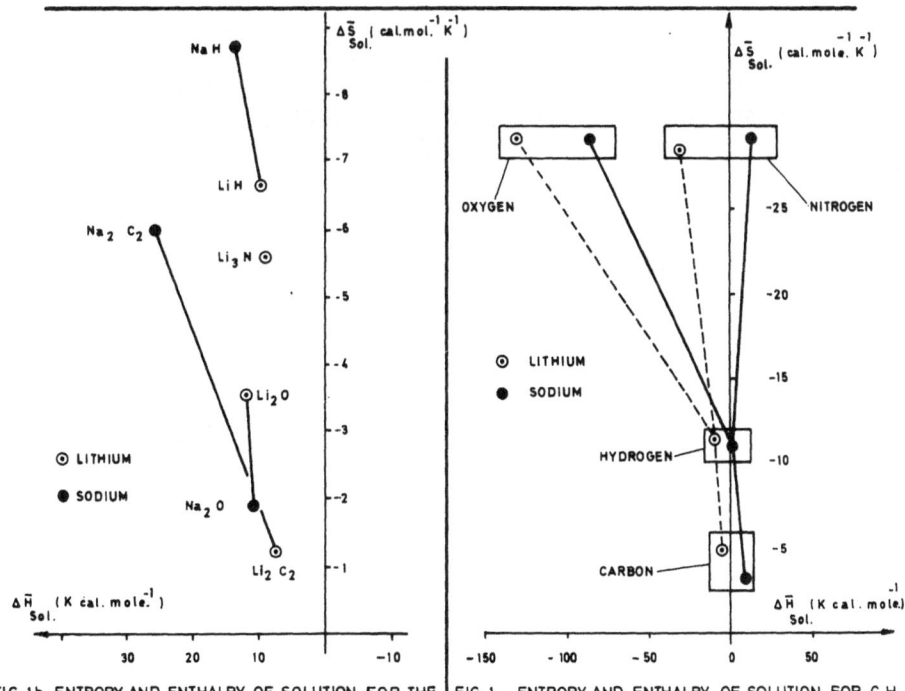

FIG. 1b ENTROPY AND ENTHALPY OF SOLUTION FOR THE
C, H, O AND N-COMPOUNDS IN LIQUID Na AND Li

FIG. 1c ENTROPY AND ENTHALPY OF SOLUTION FOR C, $H_2$,
$O_2$ AND $N_2$ IN LIQUID SODIUM AND LITHIUM

SOLUTE-SOLVENT INTERACTION OF GASEOUS NON-METALS

It can be proved that the activity $a_{(S)}^{2m}$ of a dissolved non metal (S) in liquid lithium or sodium is given by (27):

$$a_{(S)}^{2m} = (x_S/x_{S,sat}) \exp (\Delta G_f^o/RT) \tag{1}$$

Here $x_S$ is the mole fraction of the non metal, $x_{S,sat}$ the value at saturation, and $\Delta G_f^o$ is the Gibbs free energy of formation for the precipitating solid in equilibrium with the saturated solution.

m = 0.5 for H, O and N and m = 1 for C.

The activities of the non metals H, C, N and O in liquid lithium and sodium have been calculated with the aid of eq. (1).

Table 3 shows the Gibbs free energies which have been used to obtain the data of Table 4.

Table 3. Gibbs free energy of formation for the precipitating solids at 1 atm $\Delta G_f^o$ = A + B T   (T in K, $\Delta G_f^o$ in cal/mol)

|  | A | B | Ref. |  | A | B | Ref. |
|---|---|---|---|---|---|---|---|
| $Na_2C_2$ | -10581 | 12.57 | (16) | $Li_2C_2$ | -14200 | 11.2 | (19,20) |
| $NaH$ | -13630 | 19.70 | (17) | $LiH$ | -19500 | 18.05 | (21,22) |
| $Na_2O$ | -96945 | 30.8 | (18) | $Li_2O$ | -144350 | 32.8 | (23) |
| $Na_3N$ | – | – |  | $Li_3N$ | -39847 | 34.09 | (27) |

Table 4. Activities of C, H, O and N in liquid lithium and sodium
$$\ln a = \ln x + A + \frac{B}{T} \text{x)} \quad (T \text{ in K})$$
Activities based on the state at 1 atm.

|  | sodium | | lithium | |
|---|---|---|---|---|
|  | A | B | A | B |
| carbon | 1.67 | 3833 | 2.51 | - 1630 |
| hydrogen | 5.54 | 258 | 5.75 | - 4605 |
| oxygen | 14.54 | -43158 | 14.76 | - 66337 |
| nitrogen | 14.88 x x) | 6402 x x) | 14.36 | - 15390 |

x)   For carbon the equation is: $\ln a = 1/2 \ln x + A + \frac{B}{T}$

xx)   Calculated from the Sieverts constant as determined by Veleckis et al. (12). See text.

$RT \ln a_{(S)}^{2m}$ for $x_S = 1$ represent the non-mechanical work for bringing the gas or non-metal out of the standard state into the dilute solution. It can be regarded as the solvation free energy for the neutral atoms of molecules:

$$\Delta \bar{G}_{solv.} = RT \ln a_{(S)}^{2m} \tag{2}$$

From

$$\Delta \bar{G}_{solv.} = \Delta \bar{H}_{solv.} - T\Delta \bar{S}_{solv.} \tag{3}$$

the partial molar enthalpies and entropies of solvation of these neutral species are deduced and tabulated in Table 5. Although no terminal solubility for N in sodium is reported, it is possible to obtain the Sieverts constant and related quantities for this element from the gas solubility, determined as a function of temperature by Veleckis et al. (12). These authours obtained for the gas content at 10 atmospheres:

$$\log S = - 7.18 - \frac{2780}{T} \tag{4}$$

S is the solubility in gramme of total nitrogen per gramme of sodium and per atmosphere. It was also proven that most of the nitrogen dissolved in the molecular form $N_2$. In order to obtain $a_{(N)}$ from this equation, one has to consider the (minor) dissociation into free N atoms in the liquid sodium

$$1/2\ N_{2(g)} \leftrightarrows 1/2\ N_{2(sol)} \leftrightarrows (N) \tag{5}$$

from this equilibrium, it follows for dilute solutions that

$$a_{(N)} = x_{(N)} = P_{N_2}^{1/2} \tag{6}$$

Since for dilute solutions $\gamma$ is the inverse of the Sieverts constant equation (4) can be transformed and $a_{(N)}$ introduced in Table 4.

Table 5.  Partial molar enthalpies and -entropies of solvation[x)] of C, H, O and N in liquid lithium and sodium

| | | sodium | | | | lithium | | | |
|---|---|---|---|---|---|---|---|---|---|
| | | C | H | O | N | C | H | O | N |
| $\bar{S}_{solv}$ | (cal/mole.K) | -3.32 | -11.0 | -29.0 | -29.5 | -4.98 | -11.5 | -29.3 | -28.5 |
| $\bar{H}_{solv}$ | (cal/mole) | 7616 | 513 | -85755 | 12720 | -3239 | -9150 | -131800 | -30850 |

[x)] This term refers here to the interaction of C, $H_2$, $O_2$ and $N_2$ with liquid lithium and sodium.

DISCUSSION

The nearly equal solution-energy changes for the process
$Li_2O$ going into Li-O solution as compared to the process $Na_2O$
going into Na-O solution (Fig. 1b) suggest a similar rearrangement
of bonds.

Since oxygen has the same co-ordination sphere in the solids
$Na_2O$ and $Li_2O$ (fluorite structure) it is probable that this
similarity is retained in liquid solution. The differences in
entropy (Fig. 1b) could then be the result of the difference in
"free" volume disposable to the non metal atom in the solid as
compared to the liquid. This is further supported by Fig. 1c where
the entropy changes involved in the solvation of oxygen molecules
in the two metals are very similar, while on the other hand the
energy differences are much larger because of the difference in
electro negativity of the alkali metals involved. These statements
are in accordance with the calculations of Thompson (26) who
regarded an octahedrally arranged co-sphere as the most probable
for oxygen in both, lithium and sodium. The very similar energy
and entropy changes in dissolving gaseous hydrogen in Li and Na
(Fig. 1c) suggest an analogous binding pattern in solution. It is
possible that the H-coordination in the solids is retained in
solution taking into account the rather small enthalpy difference
in Fig. 1b. The large difference in thermodynamic solution values
for $Na_2C_2$ and $Li_2C_2$ (Fig. 1b) may be caused by the difference in
crystal structure of these solids at higher temperatures. Indeed,
Grishin et al. (23) and Fedorov and Mein Tsung Su (25) report
allotropic modifications for $Li_2C_2$ at 410, 440 and 560 $^\circ$C. On the
other hand, the thermodynamic quantities of solution for solid
carbon in the alkali metals indicate also a similar coordination
pattern in the two alkali metals.

The interpretation of the solute-solvent interactions for N
can only be based on the solvation properties as displayed by
Fig. 1c, since the terminal solubility equation in sodium is
unknown.

But in view of the similarity in entropy changes during solvation, it can be assumed that this non-metal is bonded in an analogous
manner in the two solvents. However, the fact that the enthalpy of
solvation has a positive value of + 12,750 cal/mole N indicates a
very loose binding of N to sodium. Indeed, as already mentioned,
a great part of the total dissolved gas has been shown to be in
the molecular form (12).

CONCLUSIONS

Although considerable differences in bond energy exist for C,
H, O and N in liquid Na as compared to Li, the thermodynamic solution
functions indicate a similar solvation pattern.

ACKNOWLEDGEMENTS

We authors thank Euratom for its financial support.

REFERENCES

1.  J. H. De Van, J. E. Selle, A. E. Morris, ORNL-TM-4927/1970.
2.  D. W. Mac Clure and G. D. Halsey, J. Phys. Chem., 69
    (1965) p. 3542-3547.
3.  S. A. Meacham, E. F. Gordus and A. A. Hill, Rep.APDA-241,
    Atomic Power Development Associates, 1973.
4.  R. J. Pulham and P. A. Simm, in: Liquid Alkali Metals,
    Proc. Int. Conf., Nottingham, England. BNES London,
    p. 1 - 4.
5.  O. R. Vissers, J. T. Holmes, C. G. Bartholme and P. A.
    Nelson, Nucl. Technol., 21 (1974) p. 235 - 244.
6.  A. C. Wittingham, British Rep. RD/B/N 2550, Central
    Electricity Generating Board, 1974.
7.  D. C. Gehri, Rep. AI/AEC-12826. Atomic International,
    Atomic Energy Commission, 1970.
8.  B. Longson and A. W. Thorley, J. Appl. Chem., 20 (1979),
    p. 372.
9.  G. K. Johnson, E. H. van Deventer, J. P. Ackermans,
    W. N. Hubbard, D. W. Osborne and H. E. Flotow, J. Chem.
    Thermodyn., 5 (1973) p. 57.
10. R. Ainsley, L. P. Hartlib, P. M. Holroyd and G. Long,
    J.Nucl.Mater., 52 (1974), p. 3542.
11. J. D. Noden and K. Q. Bagley, Rep. ANL 7399, 1967, p. 137.
12. E. Veleckis, K. E. Anderson, F. A. Cafasso and H. M. Feder,
    Rep. ANL 7520, part I, Proc.Int.Conf.on Sodium Techn.
    Nov. 1968.
13. P. F. Adams, M. G. Down, P. Hubberstey and R. J. Pulham,
    J. Less-Common Met., 42 (1975) p. 325.
14. R. M. Yonco, E. Veleckis and V. A. Maroni, J.Nucl.Mat.57
    (1975) p. 317.
15. P. F. Adams, P. Hubberstey and R. J. Pulham, J.Less-Common
    Met., 42 (1975) p.1.
16. C. E. Wicks and F. E. Block: "Thermodynamic Properties of
    65 Elements", Bureau of Mines, Bull. 605. U.S. Govern.
    Print Off., Washington, U.S. 1963.
17. W. M. Mueller, J. P. Blackledge and G. G. Libowitz,
    "Metal Hydrides", Academic Press, New-York, 1968.
18. C. B. Alcock, G. P. Stavropoulos, Can.Metall.Quart., 10
    (1971) p. 257.
19. A. N. Krestovnikov et al., "Handbook on Calculating the
    Equilibrium of Metallurgical Reactions", Moscow,
    Metallurgizdat, 1963.
20. N. M. Beskorovainyi, V. K. Ivznov and M. T. Zuev, in:
    "High Purity Metals and Alloys", Plenum Publ.Corp.,
    New-York, 1967.

21. V. A. Maroni, W. F. Calaway, E. Veleckis and R. M. Yonco,
    Int.Conf.on Liquid Metal Technol., ANL 760 503 p.437.
22. C. E. Messer, Rep. NYO 9470 (1960) p. 64.
23. D. L. Smith and K. Natesan, Nucl.Technol., 22 (1974),p.392.
24. V. K. Grishin et al. "Properties of Lithium" Moscow
    Metallurgizdat, 1963.
25. P. I Fedorov and M. T. Su, Hua Hseuh Pao (J. Chinese
    Chem. Soc.) 23 (1957) p. 30.
26. R. Thompson in "Liquid Alkali Metals", Proc.Int.Conf.
    4 - 6 april 1973, BNES, p. 47.
27. N. A. Rumbaut, "Chemisch-thermodynamisch gedrag van niet-
    metalen in dynamisch lithium", PH.D. Thesis, Univ.
    Leuven (in preparation).
28. R. M. Yonco, V. A. Maroni, J. E. Strain and J. H. Devan,
    J.Nucl.Mater., 79 (1979) p. 354.

# SOLUBILITIES OF NON-METALS IN LIQUID LITHIUM - A REVIEW

Peter Hubberstey, Andrew T. Dadd and Peter G. Roberts

Nottingham University
Nottingham
United Kingdom

## INTRODUCTION

Interest in the chemical technology of liquid lithium has been stimulated recently by its candidature for tritium breeding and primary coolant applications in D-T fuelled thermonuclear reactors (1, 2). Owing to its extreme reactivity, liquid lithium contains dissolved impurities, particularly hydrogen, tritium, carbon, silicon, nitrogen and oxygen, which, because of their pervasivity and chemical reactivity, have a profound influence on, inter alia, corrosion and embrittlement processes, tritium recovery chemistry and impurity control and purification techniques. Obviously an essential prerequisite to an understanding of the basic chemistry of these processes is a detailed knowledge of the solubility of reactive impurities in liquid lithium. In this paper we review those solubility data presently available, propose recommended solubility expressions and, where possible, attempt to rationalise solubility trends using thermodynamic concepts.

Solubility data for the non-metals in liquid lithium were previously reviewed in 1975 (3), when information was generally meagre and results often inconsistent. Solubilities of hydrogen, deuterium, carbon, silicon, germanium (and lead), nitrogen and oxygen in liquid lithium have since been redetermined, principally in continuing, complementary lithium technology programmes at Argonne National Laboratory (USA) and at Nottingham University (UK), and recommended solubility expression can now be provided for all these solutes.

SOLUBILITY DATA

The solubility of non-metal in a liquid alkali metal is given by the hypereutectic liquidus of the metal-metal salt phase diagram. For dilute solutions, the temperature (T/K) dependence of the concentration ($x_X$) of this phase boundary is given by equation (1). All recommended solubility values are

$$\ln x_X = A - B/T \tag{1}$$

summarised in Table 1.

SOLUTIONS OF HYDROGEN ISOTOPES

The solubilities of LiH ($523 \leq T/K \leq 775$) (4) and of LiD ($549 \leq T/K \leq 724$) (5) in liquid lithium have been determined using the equilibrium resistivity technique; that of LiD ($472 \leq T/K \leq 771$) (6) has also been measured by a direct sampling technique. Analysis of the samples was achieved by degassing the sample at 1123 K for 10 days in a fused silica tube connected to a Toepler pump; the envolved gas was collected in a known volume and its deuterium content assessed by mass-spectrometry.

The data are compared in Fig. 1(a) in the form of ln x vs. 1/T plots. There are minor differences in the Li-LiD solubilities, especially for very dilute solutions ($x_D = 0.004$), the resistivity method giving the lower values. It has been suggested that these differences are caused by small amounts of hydrogen in the lithium, which would affect the resistivity data more than the direct sampling data. The validity of this suggestion can be questioned since the Li-LiH data would be similarly affected; consideration of Fig. 1(a) shows that this is obviously not the case. The Li-LiH data are directly comparable with the Li-LiD solubilities obtained by the direct sampling method, LiH being marginally more soluble than LiD, thus correlating with the observed phase relationships (mono-tectic horizontal, consolute temperature) at higher temperatures (7). Consequently, the recommended solubilities for LiH and LiD (Table 1) are based on these two sets of data.

Extrapolation of the Li-LiH and Li-LiD solubilities to the Li-LiT system and extension of the observed high temperature phase relationships for all three systems indicates that solubility will decrease in the order LiH>LiD>LiT, but that there will only be marginal quantitative differences.

SOLUTIONS OF CARBON, SILICON, GERMANIUM AND LEAD

Solubility data have been obtained for the Li-Li$_2$C$_2$ system ($477 \leq T/K \leq 908$) (8) using the direct sampling method. Analysis was effected by quantitative gas chromatography of the gases ($C_2H_2$)

Table 1.  Recommended solubility expressions for non-metals in
          liquid lithium

| Solute | A | B | Temperature Range T/K | Solute | A | B | Temperature Range T/K |
|--------|---|---|-----------------------|--------|---|---|-----------------------|
| Hydrogen | 3.507 | 5314 | 523 - 775 | Germanium | 5.459 | 6630 | 530 - 715 |
| Deuterium | 3.101 | 5082 | 472 - 771 | Lead | 5.717 | 6722 | 550 - 670 |
| Carbon | -1.100 | 5750 | 477 - 908 | Nitrogen | 2.976 | 4832 | 468 - 723 |
| Silicon | 5.548 | 6775 | 500 - 700 | Oxygen | 1.428 | 6659 | 530 - 715 |

evolved on hydrolysis of the sample; the aqueous solutions were
also assayed volumetrically for lithium. Similar data for the
$Li-Li_{22}Si_5$ $(500 \leq T/K \leq 700)$(9), $Li-Li_{22}Ge_5$ $(530 \leq T/K \leq 827)$(9), and
$Li-Li_{22}Pb_5$ $(550 \leq T/K \leq 670)$ (10) systems have been determined using
the equilibrium resistivity method. The remarkable self-consistency
of the four sets of data, particularly for the heavier Group IV
elements, is evident from a consideration of the minimal scatter
of the data points on the ln x vs. 1/T plots shown in Fig. 1(b).

The data differ quite markedly (by up to a factor of $10^2$) from,
and are to be preferred to, the previously reported solubilities,
the validity of which was questioned in the earlier review (3).
The recommended solubilities are collected in Table 1.

SOLUTIONS OF NITROGEN

Solubility data for $Li-Li_3N$ solutions have been determined
using both the equilibrium resistivity techniques (473  T/K  708)
(4) and direct sampling methods (468  T/K  714) (11). The samples
were hydrolysed and analysed for nitrogen (micro-Kjeldahl), lithium
(volumetrically), iron and nickel (colorimetrically).

The correlation between the two sets of data, which are compared
in the form of a ln x vs. 1/T plot in Fig. 1(c), is amazingly good;
the recommended solubility expression (Table 1) is a least squares
analysis of all 36 points. These data also correlate satisfactorily
with the high concentration hypereutectic liquidus reported by
Fedorov et al (12) and discussed in the earlier review (3). They
differ, however, from the expression for dilute solution solubility
(shown as a dashed line in Fig. 1 (c)), which was recommended in the
earlier review (3) on the basis of a combination of two sets of
preliminary results (13, 14).

SOLUTIONS OF OXYGEN

The solubility of $Li_2O$ in liquid lithium $(468 \leq T/K \leq 1007)$ (15)
has been determined by the direct sampling techniques. The samples

were analysed for oxygen content by a direct neutron activation method. The solubility values are lower than those recommended in the earlier review (3) on the basis of data obtained previously by Hoffman (13) and Konovalov (16); the data are compared as ln x vs. 1/T plots in Fig. 1(d). It has been pointed out (15) that the observed solubilities increase with increasing filter pore size from the A.N.L. data (2 μm.) (15) through those of Hoffmann (20 μm.) (13) to those of Konovalov (30 - 40 μm.) (16). The A.N.L. data are preferred for the recommended solubility equation (Table 1) since the authors quote a large number of self-consistent points, which were obtained over an extended temperature range, using ultrafine (2 μm. pore size) filters, allied to sensitive neutron activation methods.

SOLUTIONS OF HALOGENS

The information available for these solutions has not been extended since the previous review. Thus, high temperature phase relationships in the Li-LiX (X = F, Cl, I) systems have been the subject of a preliminary study (17); the results indicate that solubility increases from LiF to LiI. The hypereutectic liquidus has only been studied in detail for the Li-LiCl system. Two independent studies gave widely differing solubilities (15, 18); in  neither case is the experimental evidence substantial enough to permit a preferred data choice.

DISCUSSION

The recommended solubility expressions are compared in Fig. 2(a); values at 473 K (probable reactor cold-trap operating temperature) and 875 K (probable reactor coolant operating temperature) are quoted in Table 2. At 473 K, solubilities decrease in the order:

$$N > H \sim D > Pb > Ge > Si > O > C$$

carbon and oxygen being far less soluble than the other solutes; obviously purification of reactor lithium by cold-trapping techniques will only be effective for these two impurities.

To analyse these data, the hypereutectic liquidus must be considered in the context of the metal-salt phase diagram. General phase relationships fall into two categories; (i) systems for which the components are mutually soluble in the liquid phase (e.g. $Li-Li_3N$, $Li-Li_{22}M_5$ (M = Si, Ge)), and (ii) systems which exhibit liquid immiscibility (e.g. Li-LiH, $Li-Li_2C_2$, $Li-Li_2O$, Li-LiX (X = Halogen)).

The magnitude of the solubility of non-metals in category (i) systems is dependent on the stoichiometry of the salt and the

Table 2.   Solubility values for non-metals in liquid lithium
           at 473 and 873 K

| Solute | appm | | wppm | |
|---|---|---|---|---|
| | 473 K | 873 K | 473 K | 873 K |
| Hydrogen | 440 | 75800 | 63 | 11800 |
| Deuterium | 480 | 65800 | 138 | 19900 |
| Nitrogen | 720 | 77400 | 1450 | 145000 |
| Oxygen | 3 | 2030 | 7 | 4670 |
| Carbon | 2 | 459 | 3 | 800 |
| Silicon | 150 | 110000 | 625 | 333000 |
| Germanium | 192 | 120000 | 2000 | 588000 |
| Lead | 205 | 140000 | 6080 | 829000 |

temperature dependence of its thermodynamic stability (free energy)
relative to that of the liquid. The effect of stoichiometry is
self-evident. The free energy effect is considered in Fig. 2 (b)
(upper diagram), in which three schematic, isothermal, free energy-
composition curves (increasing in stability from A to C) for an
$M_a X_b$ salt are compared with that for the corresponding liquid.
As the stability of the salt is increased, the composition of the
liquid in equilibrium with the salt decreases from A' to C'. The
three solubility curves obtained from this type of analysis are
plotted conventionally in Fig. 2 (b) (lower diagram). Obviously
the solubility of the salt is reduced (and its melting point
increased) with increasing thermodynamic stability of the salt with
respect to that of the liquid.

     The magnitude of the solubility of the non-metal in category
(ii) is dependent of the extent of liquid immiscibility as well as
the factors outlined above. This is considered in Fig. 2 (c)
(upper diagramm) in which schematic, isothermal, free energy-
composition curves for three liquids (increasing in immiscibility
from D to F) are compared with that for the salt at the monotectic
temperature (the relative energies of the salt and the liquid at
the salt composition are assumed constant). As the extent of
immiscibility increases, the composition of the metal-rich liquid
in equilibrium with the salt decreases from D' to F'. This is
manifest in an extension of the monotectic horizontal from D" to
F" (Fig. 2 (c) (lower diagram)), which is equivalent to a decrease
of the salt solubility in the liquid metal. Immiscibility is
related to positive deviations from ideality caused by positive
enthalpies of mixing; such positive values imply that in the liquid,

similar species attract themselves much more strongly than dissimilar species. This cohesivity (the so-called internal pressure) of the components of the mixture may be assessed from their enthalpies of vaporisation (a measure of the binding energies) and atomic volumes. Indeed, Hildebrand and Scott (19) have quantitatively related the onset of immiscibility to these parameters.

It may be concluded that solubilities in metal-salt systems are extremely sensitive to the precipitating phase and that solubility trends can only be profitably discussed for identical systems. This may be examplified by consideration of solutions of Group IV elements in the liquid alkali metals. In lithium, the precipitating phases fall into two types, $Li_2C_2$ and $Li_{22}M_5$ (M = Si, Ge and Pb) with widely differing solubilities (Fig. 2 (d) upper diagram). Similarly, in sodium the precipitating phases fall into three types, $Na_2C_2$, NaGe and $Na_{15}M_4$ (M = Sn and Pb), again with widely differing solubilities (Fig. 2 (d) lower diagram). Within the $Li_{22}M_5$ and $Na_{15}M_4$ types, however, solubility data are very similar and hence an analysis of the solubility trends can be attempted.

Following the previous discussion, the marginal decrease in solubility from $Li_{22}Pb_5$ to $Li_{22}Si_5$, which precipitate from the mutually soluble $Li-Li_{22}M_5$ systems may be attributed to a corresponding increase in the free energy difference between $Li_{22}M_5$ and the appropriate solution. Unfortunately, this assertion cannot be verified since appropriate thermodynamic data are not available for the solutions. The preliminary $\Delta G$ ($Li_{22}M_5$), 750 K) data presently available: - 495 (for $Li_{22}Si_5$) (20) - 535 (for $Li_{22}Ge_5$) (21) - 766 (for $Li_{22}Sn_5$) (22) and - 812 kJ.mol$^{-1}$ (for $Li_{22}Pb_5$) (22) decrease from $Li_{22}Si_5$ to $Li_{22}Pb_5$, thereby necessitating an even greater decrease in the free energies of the corresponding solutions. Such a trend may possible be rationalised by the stabilising influence of the metal-non metal transition which occurs in these liquids (at $x_{Li} \sim 0.8$) and which is most marked in Li-Pb solutions.

Solubility trends may also be considered for solutions of LiX (X = Halogen) in lithium since the phase relationships for these systems are identical, being dominated by an extensive miscibility gap. Although the solubility data are meagre they show a decrease from LiI to LiF. Similar trends are observed in the corresponding potassium systems. The data are so similar in the sodium systems that a trend cannot be ascertained. Following the previous discussion, the solubility trends in these systems may be rationalised by consideration of the enthalpies of vaporisation of MX (Table 3). The encrease from MI to MF corresponds to an increase in cohesivity (internal pressure), which is manifest in an increase in the extent of immiscibility and a concomitant decrease in solubility.

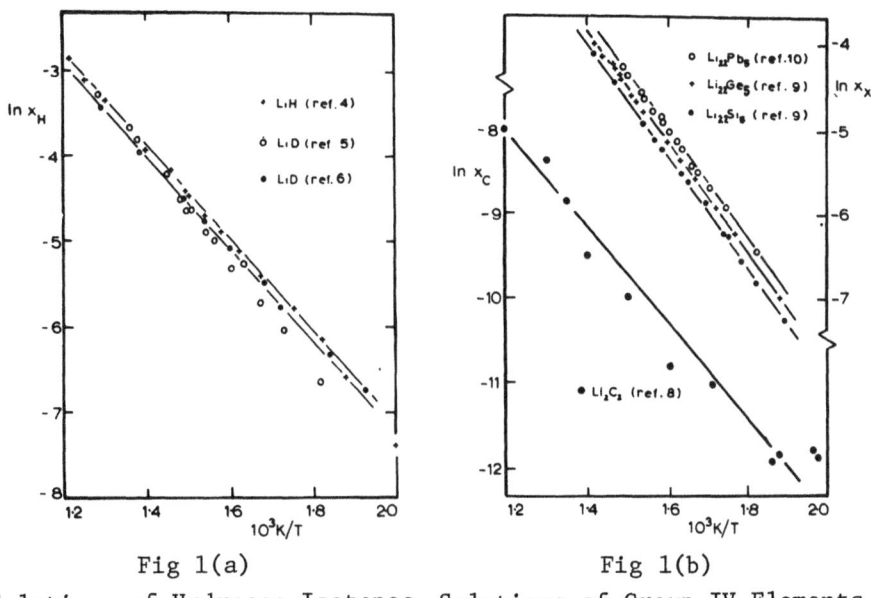

Fig 1(a)                                Fig 1(b)

Solutions of Hydrogen Isotopes    Solutions of Group IV Elements

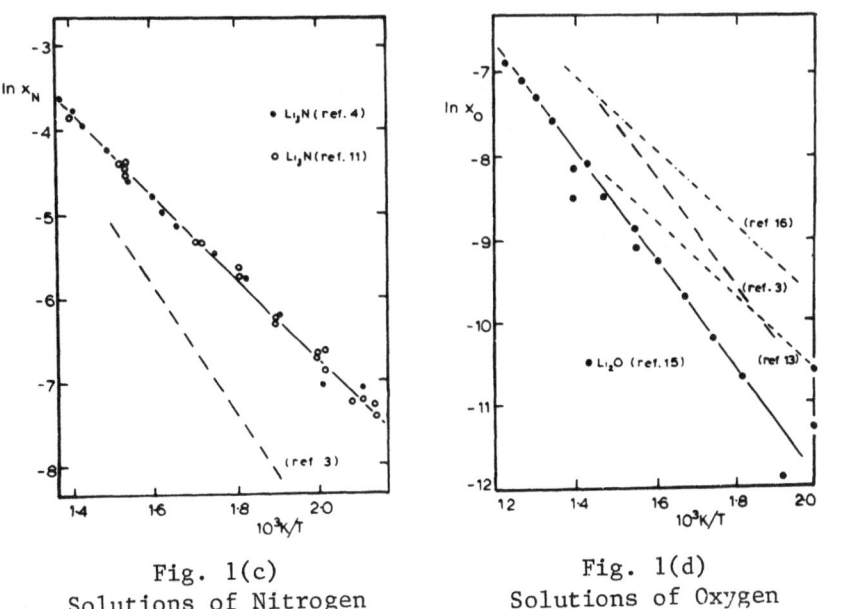

Fig. 1(c)                               Fig. 1(d)

Solutions of Nitrogen              Solutions of Oxygen

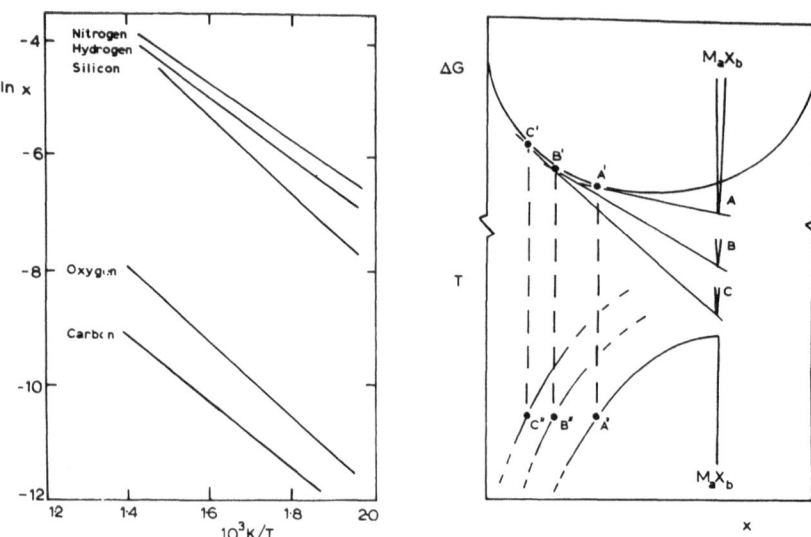

Fig 2(a)  Solubilities of non metals
          in liquid lithium

Fig 2(b)  Schematic Free
          Energy-Composition
          Curves

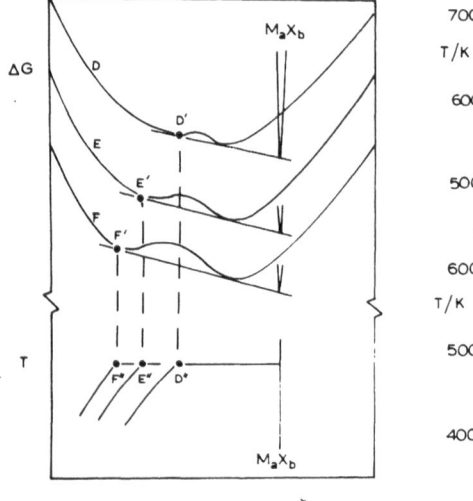

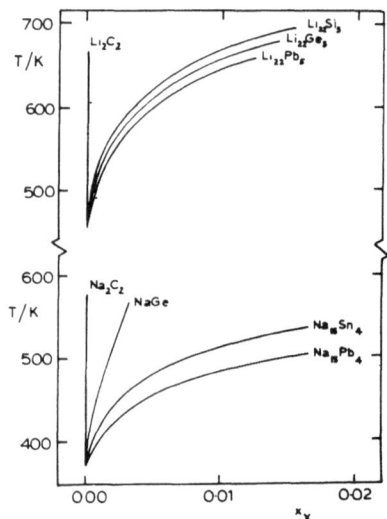

Fig 2(c)  Schematic Free
          Energy-Composition
          Curves

Fig 2(d)  Solubilities of Group
          IV Elements in Li and Na

Table 3. Enthalpies of vaporisation $\Delta H_v/kJ.mol^{-1}$ for alkali metal halides (23)

|     | LiX | NaX | KX  |
| --- | --- | --- | --- |
| MF  | 278 | 281 | 242 |
| MCl | 214 | 235 | 226 |
| MBr | 197 | 224 | 214 |
| MI  | -   | 216 | 206 |

We would like to thank the S.R.C. for the award of maintenance grants (to A.T.D. and P.G.R.).

REFERENCES

1. J. H. DeVan, J.Nucl.Mater., 85/86 (1979) 249.
2. R. E. Gold and D. L. Smith, Proc. 2nd. Int.Conf. Liquid Metals in Energy Prod. Apr. 21-25th., 1980, Richland, Wa., in press.
3. P. F. Adams, P. Hubberstey and R. J. Pulham, J.Less-Common Metals, 42 (1975) 1.
4. P. F. Adams, M. G. Down, P. Hubberstey and R. J. Pulham, J.Less-Common Metals, 42 (1975) 325.
5. P. F. Adams, P. Hubberstey, A. E. Thunder and R. J. Pulham, J.Less-Common Metals, 46 (1976) 285.
6. E. Veleckis, R. M. Yonco and V. A. Maroni, J.Less-Common Metals, 55 (1977) 85.
7. F. J. Smith, J. F. Land, G. M. Begun and A. M. Batistoni, J.Nucl.Chem.Soc., 41 (1979) 1001.
8. R. M. Yonco and M. I. Homa, Trans.Amer.Nucl.Soc., 32 (1979) 270.
9. A. T. Dadd and P. Hubberstey, J. C. S. Faraday Trans. I, in press.
10. P. Hubberstey and P. G. Roberts, unpublished data.
11. R. M. Yonco, E. Veleckis and V. A. Maroni, J.Nucl.Mater, 57 (1975) 317.
12. K. A. Bolshakov, P. I. Fedorov and L. Stepina, Izv. Vyssh. Uchebn. Zaved. Tsvetn. Metall., (1959) 52.
13. E. E. Hoffman, USAEC Rept. ORNL-2894 (1960).
14. M. N. Arnol'dov, M.N. Ivanovskii and B. A. Shmatko, Teplofiz. Vyssh. Temp., 5 (1967) 380.
15. R. M. Yonco, V. A. Maroni, J. E. Strain and J. H. DeVan, J.Nucl.Mater., 79 (1979) 354.
16. E. E. Konovalov, N. I. Seliverstov and V. P. Emel'yanov, Izv. Akad. Nauk. SSSR, Met., 3 (1968) 77.
17. A. S. Dworkin, H. R. Bronstein and M. A. Bredig, J.Phys. Chem., 66 (1962) 572.

18. T. Nakajima, R. Minami, K. Nakanishi and N. Watanabe, Bull.Chem.Soc.Japan, 47 (1974) 2071.

19. J. H. Hildebrand and R. L. Scott, "The Solubility of Non-Electrolytes", 3rd. Edn., Reinhold, New York, N.Y. (1950).

20. R. A. Sharma and R. N. Seefurth, J.Electrochem. Soc., 123 (1976) 1763.

21. Calculated by interpolation between Si, Sn and Pb data.

22. J. F. Smith and Z. Moser, J.Nucl.Mater., 59 (1976) 158.

23. G. H. Aylward and T.J.V. Findlay, S.I. Chemical Data, Wiley, London, (1971).

THE STABILITY OF METAL PARTICLES AND PARTICLE-PLATE INTERACTIONS

IN LIQUID METALS

Roy Thompson

AERE, Harwell

United Kingdom

## INTRODUCTION

   This paper is an attempt to examine the stability of small
particles in liquid metals using the classical methods of colloid
chemistry. These methods were originally developed by Derjaguin,
Landau and Verwey, Overbeek (the DLVO theory), to treat colloids
in ionic solutions. The work reported here only considers electrically conducting solids in liquid metals. Such systems may have technological significance in areas such as; metallurgical casting
techniques, where nucleation and dispersion of particulate alloy
phases are important to the properties of the final product, the
stability of magnetic ferro-fluids based on liquid metals, and in
the understanding of corrosion and mass-transfer mechanism in large
liquid metal cooled systems, particularly liquid-metal fast reactors
(LMFRs), and also some proposed designs of solar energy collectors
with high surface heat fluxes, as well as possible future thermonuclear reactor systems.

   Over many years studies have been made of the corrosion of
structural materials in liquid sodium on behalf of various LMFR
development programmes. This work has been mainly aimed at assessing
material losses but with some information on the distribution and
deposition of the resultant corrosion products. Metallic particles
of upto 2 μm diameter have been found as deposits (1) and it is
generally suspected that smaller particles may be present in the
liquid as a stable suspension. It is a well reported phenomenon
that carbon appears to exist in sodium as non-reactive particulates
(2). However, knowledge is limited as to the form, nature and behaviour of particulate material in liquid metals, such as sodium.

ATTRACTIVE POTENTIAL BETWEEN SOLID METALS IN A LIQUID METAL

If solid bodies approach close to one another in a liquid metal, as in any other medium, Van-der-Waals' forces will act between them. A comprehensive review of the literature on the theory and calculation of these forces has been made elsewhere (3). There are two approaches generally used for the calculation of these and the resultant attractive potential between macroscopic bodies.

The first method is the original Hamakar method (4) based on the summation of the attractive potentials for all atomic or molecular pairs between the two adjacent bodies, using a London dispersion force relationship. The expression for two plates, of thickness $\delta$, separated by a distance 2d, gives the attractive potential $V_A$ per unit area as,

$$_H V_A = - \frac{A_H}{48\pi} \left( \frac{1}{d^2} + \frac{1}{(d + \delta)^2} - \frac{2}{(d + \frac{\delta}{2})^2} \right) \tag{1}$$

For thick plates, ($\delta \gg d$) this reduces to the more familar form,

$$_H V_A = - \frac{A_H}{48\pi d^2} \tag{2}$$

In these equ. $A_H$ is the Hamaker constant which is calculated using fundamental properties of the atoms and the material itself. One way of achieving this calculation is to use the relationship,

$$A_H = \frac{3}{4} \pi^2 n_o h\nu a^6$$

In this $n_o$ is the number density of the atoms, a the atom radius (it is assumed here that the polarizability of an atom is proportional its volume). $\nu$ is a characteristic frequency of the atoms, for a metal this is taken as the free electron plasma frequency $\nu_p$ so that atoms deeper than just surface atoms can be considered. Above this frequency metals are generally regarded as being transparent to electromagnetic radiation. In practice this transparency extends through a relatively thin skin approximately equal to $c/2\pi\nu_p$ (5) where c is the velocity of the wave (light) whose intensity decays expotentially in this distance. So for metal plates an effective constant thickness should be used for $\delta$ in equ.1, equal to half the skin depth given above. The value of $\delta$ for most metals is between 600 and 1800 mm which will be within, or a little beyond, the range, of the separation distances to be considered. For this conditions a further simplified form of (1) may be used,

$$_H V_A = - \frac{A_H}{48\pi} \left( \frac{1}{d^2} - \frac{7}{\delta^2} \right) \tag{2a}$$

where $\delta = c/4\pi\nu_p$

The second method is the macroscopic approach pioneered by Lifshitz (6), which considers interactions between two semi-infinite media. The attraction is calculated from macroscopic properties of the two media, and is assumed to arise from fluctuating electro-magnetic fields set up in the gap separating them. These fields arise from spontaneous electric and magnetic polarizations within the media.

An expression for the Van-der Waals' attractive potential between two metal plates has been derived using the macroscopic theory (7). It is,

$$_L V_A = \frac{h\nu_p}{256\pi} \frac{1}{2} \frac{1}{d^2} \left( \frac{69}{64} - \frac{6.187\lambda}{d} + \frac{34.576\lambda^2}{d^2} - \frac{156.45\lambda^3}{d^3} \right.$$
$$\left. + \frac{407.3\lambda^4}{d^4} - \cdots \right) \tag{3}$$

where
$$\lambda = \frac{V_F}{16\sqrt{3\pi\nu_p}}$$

and $V_F$ is the Fermi velocity of the electrons in the metal. At relatively large separations (3) reduces to,

$$_L V_A = - \frac{A_L}{48\pi d^2} \tag{3a}$$

where $A_L = h\nu_p/7$. Equ. 3a is the same form as equ. 2 above. A comparison between $A_H$ and $A_L$ can be made from the values in Table 1 calculated for different metals at 500 K.

Table 1.  Hamaker Constants for Various Metals (at 500 K)

|  | Li | Na | Cr | Mn | Fe | Co | Ni | Cu | Mo |
|---|---|---|---|---|---|---|---|---|---|
| $A_H \times 10^{19}$ J | 2.436 | 1.925 | 5.750 | 4.970 | 5.147 | 5.553 | 5.700 | 5.443 | 5.071 |
| $A_L \times 10^{19}$ J | 1.780 | 1.448 | 3.436 | 3.408 | 3.493 | 3.575 | 3.616 | 3.475 | 3.067 |
| $\nu_p \times 10^{-15}$ s$^{-1}$ | 1.88 | 1.53 | 3.63 | 3.60 | 3.69 | 3.77 | 3.82 | 3.67 | 3.24 |
| $2\delta$ μm | 2.54 | 3.12 | 1.31 | 1.33 | 1.29 | 1.26 | 1.25 | 1.30 | 1.47 |

As can be seen from Table 1 the constants are of the same order
using both methods. The microscopic summation method gives values
about 40 % higher than the macroscopic method. However calculated
values of the attractive potential at separations greater than
0.1 μm may not be so different since the macroscopic method equ.3
and 3a, take into account only the surface and subsurface layers
of the metal, whereas equ. 2a should be used for metals in the
summation method calculations, as discussed above. Nonetheless the
macroscopic approach is generally considered physically more
satisfactory and so only this will be used further.

Equ. 1 and 3 apply to interactions in a vacuum, but in most
situations the bodies are separated by some other medium. The
effect the separating medium has on the attractive forces, in this
case a liquid metal, must be considered. Classically the modification
of the Hamaker constant is given by,

$$A = A_{12} + A_{33} - A_{13} - A_{23} \tag{4}$$

Where the constants for the various interacting materials - in
vacuo - are identified by the subscripts; 1 and 2 are the interacting
media and 3 is the separating medium. The interaction between the
plate and the separating medium, e.g. $A_{13}$ is assumed $(A_{11}A_{33})^{1/2}$.
The effective Hamaker constant for interactions between similar
materials separated by a second, becomes,

$$A = (A_{11}^{1/2} - A_{33}^{1/2})^2 \tag{5}$$

It has been further suggested that this should be divided by the
dielectric constant of the separating medium. For liquid metals
using frequences above $v_p$ this equals one.

The constant may need further modification since the forces
have to be transmitted through the separating medium electromagnetically and the properties of the medium may give rise to
attenuation and retardation of the forces. If, as in this case, the
medium is metallic, the fluctuating electric dipoles will be screened
by the electrons in the metal usually within distances around $1/q$.

This will give rise to a more rapid attenuation of the attractive
potential with distance than is the case in non-metallic media. However as explained above, if only frequencies above $v_p$, of the separating metal, are considered then the screening will be less effective
and the attenuation not so rapid. To take full account of the effect
of the separating medium it is suggested that a function be added
to the equ. for the attractive potential in the same way as it has
been suggested to allow for retardation effects at large separations
(8). So that equ. 3a takes the form,

$$V_A = - \frac{A}{48\pi} \frac{1}{d^2} f(d) \tag{6}$$

For liquid metal systems the appropriate function is assumed to be an exponential of distance and skin depth $2\delta$, so that (6) becomes,

$$V_A = - \frac{A}{48\pi d^2} \exp - \frac{bd}{2\delta} \tag{7}$$

where b is a constant assumed to be close to unity, and the $2\delta$ values are of the order given in Table 1. A similar modification has been shown to apply for attraction between plates separated by strong electrolyte solutions (9).

## REPULSIVE POTENTIALS

It is well known that when two metals touch a potential difference is set up between them. This is the historic Volta effect and arises from electrons transfering between the metals to equalize their electrochemical potentials or Fermi energies. This contact potential, $\Psi_{ij}$, is equal to the difference between the work functions of the metals so that,

$$\Psi_{ij} = \varphi_i - \varphi_j \tag{8}$$

where $\varphi_i$ and $\varphi_j$ are the work functions of the metals i and j respectively.

For a solid metal immersed in a different liquid metal a contact potential will be established. The surface of the solid metal will be at a potential $\Psi_s$, relative to the liquid. This potential will be screened by the electrons in the liquid metal giving rise to an electrical double layer at the solid surface extending some way into the liquid. As two particles of the solid metal approach one another, or if a particle approaches a plate or pipe wall of solid metal, the double layers will interact giving rise to a repulsive force. This has previously been suggested as a possible source of repulsion between metal particles in mercury (10).

At close distances of approach, as already discussed, attractive forces will arise between the solids. Just as in the classical DLVO theory, the stability and behaviour of the particles will depend on the magnitude and form with distance from the surface, of these two opposing forces.

First, to examine the electric-double layer at the surface of a infinite flat plate. By applying the semi-classical theory of liquid metals, the form of the electrical potential with distance from the surface can be calculated using the approach of Mott and

Jones (11) for the potential about a point charge in a free electron metal. This gives the usual expression,

$$\Psi_x = \Psi_s \exp. - qx \tag{9}$$

for the potential $\Psi$ at a distance x from the surface, with the boundary conditions, as $x \to 0 \Psi_x \to \Psi_s$ (the surface charge), and as $x \to \infty \Psi_2 \to 0$. q is the Mott, Jones screening length given by

$$q^2 = 4me^2 (3N_o/\pi)^{1/3}/h^2$$

m and e are electron mass and charge respectively and $N_o$ the ion number density of the liquid metal.

The repulsive force between two flat plates immersed in a large volume of liquid metal may be calculated most easily when the separation of the plates, 2d, is large compared with 1/q. Under these conditions there will be negligible interaction of the potentia of one plate at the surface of the other. At d, the centre between the plates, the potentials may be regarded as additive giving the total potential,

$$\Psi_d = 2\Psi_s \exp - qd \tag{10}$$

Following the argument of Verwey and Overbeek (12); at equilibrium for every point in the liquid phase the gradient of the hydrostatic pressure must equal the force on the electric space charge. If P is the hydrostatic pressure and $\rho$ the electric charge density then, ·

$$dP + \rho d\Psi = 0 \tag{11}$$

If the hydrostatic pressure in the liquid away from any electric charge is $P_\infty$, the force acting at d (the mid-point between the plates pushing the plates apart, is equal to the difference between the hydrostatic pressure at d, $P_d$ and $P_\infty$. This force p can be calculated using (11) since

$$p = P_d - P_\infty = \int_\infty^d dp = -\int_0^{\Psi_d} \rho d\Psi \tag{12}$$

Taking the charge density in the liquid metal between the plates as that given by the linearization of the Thomas-Fermi approximation. (as used in ref. 11 above) i.e.

$$\rho = 3/2a \ \Psi_o^{1/2}\Psi$$

(where $a = 8\pi e(2me)^{3/2}/3h^3$ and $e\Psi_o$ is the Fermi energy of the liquid metal), equ. 12 reduces to

$$P = 3a\Psi_o^{1/2} \Psi_s^2 \exp - 2 qd \tag{13}$$

Since changes of pressure are related to free energy changes, $p = P_d - P_\infty = -\frac{\partial G}{\partial d}$ where G is the Gibbs' free energy. The resultant repulsive potential energy

$$V_R = 2 (G_d - G_\infty) = -2 \int_\infty^d p\partial d \tag{14}$$

On substituting for p from (13) and integrating,

$$V_R = \frac{3a\Psi_o^{1/2} \Psi_s^2}{q} \exp - 2qd \tag{15}$$

or $\quad = \frac{q\Psi_s^2}{\pi} \exp - 2qd \tag{15a}$

This equ. applies to plates of the same material; for plates of different materials,

$$V_R = \frac{q_1 \, _1\Psi_{s2} \, \Psi_s}{\pi} \exp - 2qd \tag{16}$$

where $_1\Psi_s$ and $_2\Psi_s$ are the contact potentials between the different plates and the liquid metal.

In this derivation, equ. 8 and 8a are approximate, and only apply to weak interactions at relatively large separations. (They are of similar form to the equ. derived for aqueous electrolytes under the same conditions).

At close distances of separation, of the order of a few atomic diameters, Born type repulsive forces must be considered. If this repulsive potential takes the usual form of that between atoms, i.e. is equal to a constant, $B/r^{12}$ where r is the separation, and further more it is considered to be additive, in the same way as the dispersion forces, the attractive potential between two plates at an equilibrium separation can be shown to be reduced by at least 25 %.

TOTAL INTERACTIONS IN LIQUID SODIUM

The full interaction energy of two bodies is given by the sum of the repulsive and attractive potentials,

$$V = V_R + V_A \tag{17}$$

For flat plates the expressions given in the previous sections for $V_A$ and $V_R$ may be used. If the plates are of the same metal separated by a second liquid metal equ. 7 and 15 give,

$$V = - \frac{Aq^2}{48\pi D^2} \exp\left(\frac{-bD}{2q\delta}\right) + \frac{q\Psi_s^2}{\pi} \exp\ (-2D) \tag{18}$$

where the distance of separation is given by the dimensionless quantity $D = qd$, $d$ being the half distance.

A plot of $\log_{10}(-V_A)$ and $\log_{10}V_R$ against D provides a useful indication as to shape of the full potential energy distance curve. If the plot of $V_R$ and $-V_A$ cross, the potential energy curve has a minimum, if the plots cross twice the curve has two minima and a maximum between. In Fig. 1 such a plot is shown for iron plates in liquid sodium. Two curves for the attractive potential are shown using different estimated values of A, in equ. 7. The larger value results from using equ. 4, while the smaller value uses equ. 5 to take account of the liquid metal separating medium. It is evident from the figure that small changes in the value of A can make a large difference in the expected behaviour of the system. The value $V_A$ may also be sensitive to the value of the absorption coefficient of radiation in the separating medium. It is taken here as equal to $2\delta$, the skin depth for sodium, but may approach the screening length q, in which case the attractive potential is markedly reduced.

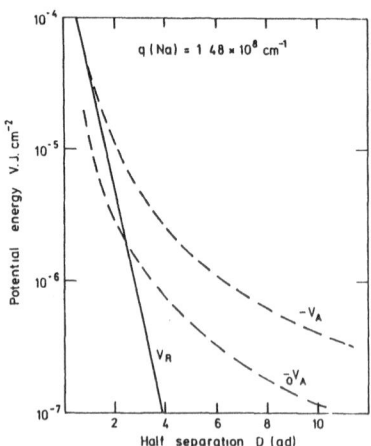

Figure 1    Interaction Energies Between Iron Plates in Liquid Sodium

Using the notions formulated above it is possible to go on to consider the potential energy distance relationship between a spherica particle and a flat plate. This system is particularly relevant to deposition of particulate corrosion products on components and pipe walls of a liquid metal system.

Assuming that the radius of the particle, a, is large compared to the closest distance of approach $H_0$, between the particle and the plate, the potential energy, V, can be calculated using equ. 7 and 15 above (see ref.10 p.137) for plate and sphere of the same metal);

$$\frac{V}{2\pi a} = -\frac{A}{12\pi}\left[\frac{\exp(-H_0 b/2\delta)}{H_0} + \frac{b}{2\delta}\left(\ln H_0 - \frac{bH_0}{2\delta} + \frac{H_0^2 b^2}{8.2!\delta^2}\right.\right.$$

$$\left.\left. - \frac{H_0^3 b^3}{24.3!\delta^3} + \dots\right)\right] + \frac{\psi_s^2}{\pi}\exp - qH_0$$

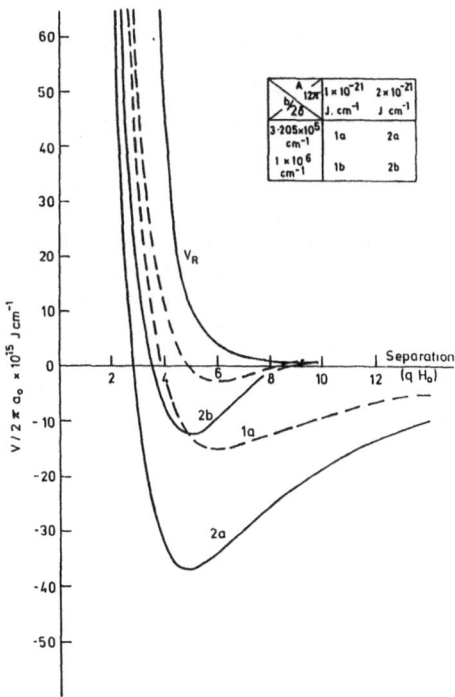

Figure 2.   Interaction Between a Metal Sphere (Radius a○) and a Plate in Liquid Sodium

If $H_0 < \frac{b}{2\delta}$, and substituting for $H_0 = D_0/q$, this equ. can be reduced to

$$\frac{V}{2\pi a} = -\frac{A}{12\pi} \left[ q \cdot \frac{\exp(-D_0 b/2\delta q)}{D_0} + \frac{b}{2\delta} \left( \ln \frac{D_0}{q} - \frac{D_0 b}{q 2\delta} \right) \right]$$

$$+ \frac{\Psi_s}{\pi} \exp - D_0 \qquad\qquad (19)$$

In Fig. 2 this equ. is plotted for a metal sphere and plate in liquid sodium. Curves are given for two values of A appropriated for the transition metals in Table 1. Curves are also shown for two different absorption coefficients, $b/2\delta$ ; one is the skin depth for sodium, and the other approximately 3 x larger. The value of $\Psi_s$ used was the contact potential between iron and sodium (2.12 volts).

Using the curves in the figure, the equilibrium separation, $_E D_0$, and the equilibrium potential energy, $V_E$ for any radius of particle may be estimated. Values of equilibrium potential energy suggest, that particles above 0.02 μm diameter deposit and stick to flat surfaces since the particles at the surface are in equilibrium with only a small number of particles in the liquid. Particles smaller than 0.02 μm may have minima in the potential energy small enough for particles at the surface to be in equilibrium with a relatively high concentration in the bulk liquid.

Table 2 gives $V_E$ values from Fig. 2, in terms of kT, and using these values are listed the number concentration of particles in the liquid $n_b$, and the equivalent concentration of iron in μg Fe/g Na, in equilibrium with a surface particle concentration, $n_s$, of one monolayer, i.e. $10^{12}$ particles per $cm^2$ for 0.02 μm diameter particles. This assumes the equilibrium constant is given by $n_s/n_b$ exp $(-V_E)/kT$. As can be seen from the table, in some instances sodium may support relatively high concentrations of small particles of iron in suspension and further more, as particle-particle interactions are similar (slightly smaller) to particle-plate interactions, in these circumstances coagulation of the particles will be slow. Equ. 19 actually gives a repulsive potential for $b/2\delta$ values as large as $10^6$ $cm^{-1}$, at separations of $D_0$ 10. This repulsive energy becomes equivalent to $-V_E$ above $2 \cdot _E D_0$. However, the energy continues to rise slowly with increasing distance whereas it might be expected that it should decrease to zero. This behaviour may be an artifact of the analysis rather than a real physical effect. If, on the other hand this repulsive potential does exist in some systems then metal colloidal particles will be stable in the liquid metal.

Table 2: $V_E$ values from Fig. 2

| Curve | 1a | 1b | 2a | 2b |
|---|---|---|---|---|
| $V_E$ (kT) | 10 | 2 | 25 | 8 |
| $n_b$ | $4.54 \times 10^7$ | $1.3 \times 10^{11}$ | 1.4 | $3.3 \times 10^8$ |
| Equivalent conc $\mu$gFe/gNa$^{x)}$ | 0.002 | 5 | - | 0.01 |

$^{x)}$ 1 $\mu$g Fe/gNa $\hat{=}$ $3 \times 10^{10}$ particles cm$^{-3}$ of 0.02 $\mu$m diameter

GENERAL DISCUSSION AND CONCLUSIONS

This work was undertaken to see if some insight could be obtained into the behaviour of metal particles in liquid metals by considering the system from the point of view of classical lyophobic colloid theory, as usually applied to aqueous ionic solutions. It has been demonstrated that there is, indeed, a parallel relationship between the two systems. A solid/liquid metal system has the characteristics of being perhaps the ultimate in concentration for strong electrolyte colloidal systems, in that the electrical double layer thickness at the solid/liquid interface is reduced to a relatively small value, namely the electronic screening length given by semi-classical free electron theory of metals. This generally implies a rapidly coagulating colloidal system since the repulsive potential between two solid metal bodies only operates over small surface separations whereas the attractive Lifshitz, Van-der-Waals' potential decays more slowly. The possibility of potential barriers and secondary minima in the interaction does not arise, therefore. However, if the attenuation of the attractive forces by the optically opaque liquid metal is taken into account, the full attractive potential can be markedly reduced, and to such an extent that strong permanent attachment of particles to surfaces does not occur. This analysis even suggests, under certain conditions, that the particles may be repelled, but as already discussed, the exact form of the attenuation, equ. 7, may not be correct and a more rigorous expression for the interaction of solid metal media separated by a liquid metal may be required.

The method used to derive the repulsive potential is essentially classical. If a full wave theory analysis is used then the electrical potential $\Psi_x$ instead of having an exponential form with distance x, (equ. 9), will go as $\cos 2k_f x/x^3$ for large x and where $k_f$ is the Fermi wave number proportional to q (13). This may lead to the possibility of a potential barrier between the bodies at certain separations and so increasing the coagulation time and deposition rate of particles. Stable sols of small iron particles ( 0.004 μm diameter) in mercury have been reported (14,15), suggesting that in this system at least a substantial potential energy barrier exists.

REFERENCES

1.  A. W. Thorley, K. C. Tyzack, B. Longson, A. C. Raine, Int.Conf. on Liquid Metal Technology in Energy Production, Champion, Penn. 1976.
2.  International Working Group on Fast Reactors. Specialist Meeting; Carbon in Sodium. IAEA, IWGFR/33. 1979.
3.  P. Richmond, Colloid Sci. 2. Chem. Soc. Sp. Pub. 1975.
4.  H. C. Hamaker, Physica 4. 1058. 1937.
5.  T. E. Faber, An Introduction to the Theory of Liquid Metals. Cambridge University Press. 1972.
6.  E. W. Lifshitz, Soviet Phys. JETP.2.73.1956.
7.  J. Heinrichs, Solid State Comm. 13. 1595. 1973.
8.  J. Gregory, Advances in Colloid and Interface Sci.2.396. 1969.
9.  V. N. Gorelkin and V. P. Smilga, Soviet Phys. JETP.36. 761.1973. (see also ref. 3 p.163).
10.  J. Popplewell, S. W. Charles and S. R. Hoon, IEE Conf. Pub. 149. 2nd Conf. on Advances in Mag. Materials and Application. 1976.
11.  N. F. Mott and H. Jones, The Theory of the Properties of Metals and Alloys. Oxford University Press. 1936.
12.  E. J. W. Verwey and J. Th. G. Overbeek, Theory of the Stability of Lyophobic Colloids. Elservier Pub. Co.1948.
13.  N. H. March, Liquid Metals. Pergamon Press. 1968.
14.  F. E. Luborsky, J. Phys. Chem. 61. 1336. 1957.
15.  J. Popplewell, S. W. Charles and R. L. Windle, IEEE Trans. on Magnetics. Mag-11. 5. 1975.

THE INFLUENCE OF HIGH TEMPERATURE SODIUM, VACUUM AND ARGON

ENVIRONMENTS ON AISI 316 STAINLESS STEEL BOLTED ASSEMBLIES

Peter Marshall, John R. Gwyther and Alan J. Hooper

Berkeley Nuclear Laboratories

United Kingdom

INTRODUCTION

Contacts involving metal alloys will be widespread in any fast reactor system. As high temperature reactor-grade sodium is expected to remove surface films and traces of grease, adhesion and self-welding of contacting surfaces could complicate the design of couplings. Alternatively, the process of self-welding would have advantages to fatigue damaged components especially if loaded under compressive stress. A number of workers, including Jerman, Williams and Leesar (1) and Bendorf (2), have found evidence of self-welding with several combinations of contacting alloys in high temperature sodium. Huber and Mattes (3) and Chang et al (4) reported self-welding in alloy couples which included combinations of AISI 304 stainless steel, Stellites 156 and 6 and Inconel 718. For in-sodium tests at temperatures > 838 K for times approaching 4,400 h under compressive loading stresses of 27.5–110 $\text{MNm}^{-2}$, subsequent separation stresses ranged from 6.9 to 30.9 $\text{MNm}^{-2}$. Although there is no reported evidence of welding below temperatures of 838 K, no tests have lasted longer than ~ 4,400 h. In general, the extent and strength of the metal bond was inversely related to material strength and increased with temperature, time and contact pressure.

This note describes experiments in which miniature AISI 316 stainless steel bolted assemblies were exposed to vacuum, argon and static or flowing sodium of controlled chemistry. The metallography, chemistry, physics and mechanical strength of the bonding process are briefly discussed.

EXPERIMENTAL

The specimens were assemblies of AISI 316 stainless steel 4BA
nuts, bolts and stacks of about twenty degreased washers, of commercial quality. The washer stacks were retained on the bolt with
either a 'finger' tightened nut (low torque) or a nut tightened
until the bolt yielded plastically (high torque). Assembly types
and exposure conditions are summarised in Table 1.

Table 1.  Conditions of Exposure of Bolts

| Medium | Temp. $^o$K | Time hours | Approx. Torque | Code | Fig. |
|--------|-------------|------------|----------------|------|------|
| Static Sodium (35 ppm $O_2$) | 906 | 2600 | High (yield stress) | A | 2 |
| Flowing Sodium (5 ppm $O_2$) | 973 | 1500 | Low | B | 3 |
| Flowing Sodium (5 ppm $O_2$) | 973 | 1500 | High (yield stress) | C | 4 |
| Static Argon | 973 | 1500 | High (yield stress) | D | - |
| Vacuum | 906 973 | 2600 1500 | 22 Nm 22 Nm | E | - |

High torque assemblies were exposed to static sodium at 906 K
containing oxygen levels controlled at    35 ppm by cold-trapping.
High and low torque assemblies were exposed to slow-flowing sodium
at 973 K containing ~ 5 ppm oxygen in a loop where the temperatures
of the cold leg and cold trap were 448 K and 398 K respectively.
(For experimental details see Gwyther, Marshall, Rowe and Withrington (5). Similar bolted assemblies were exposed in argon for 1,500 h
at 973 K and 2,600 h at 906 K.

The sodium apparatus was cooled to about 373 K, bolted assemblies were removed and carefully washed in alcohol. Both argon
and sodium exposed assemblies were mounted, ground, polished to
1/4 μm diamond and etched in picral or dilute aqua regia.

AISI 316 solution-treated tensile specimens 0.75 mm thick were
sectioned in two, degreased and fastened together by means of a
4BA nut, bolt and degreased washers. The assembly was placed in an
alignment jig and tightened to a torque of 22 Nm (Fig. 1). Assemblies
were surrounded with zirconium foil and heated in an evacuated silica
capsule for 2,600 h at 906 K and 1,500 h at 973 K (exposure conditions E in Table 1). After ageing, the room temperature torque
to release the assembly nuts and bolts was measured and the shear

load to part the self-welded surfaces at 25 °C determined by loading
the device shown in Fig. 1 in a Hounsfield tensometer.

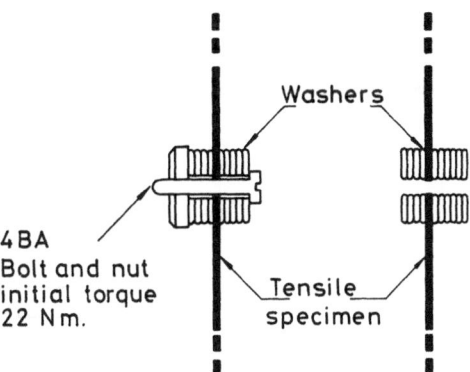

Fig. 1   Stainless Steel Assembly used for Estimating Strength of
         Self Welded Bond

DISCUSSION

     The results are discussed under five sections, namely metal-
lography, the role of sodium chemistry, self-welding model, strength
of the welded joint and finally the relevance of the work to reactor
structures.

     For definitions of the assembly types see Table 1. Important
features of the micrographs (Figs. 2 - 4) are as follows:

     (i)   Type A and C assemblies show extensive self-welding at
washer interfaces (Fig. 2).

     (ii) Type B assemblies show adherence of washer interfaces by
sodium chromite (demonstrated by X-ray crystallographic analysis
subsequent to dismantling) (Fig. 3).

     (iii) There was a lower incidence of welded contacting surfaces
in assemblies exposed in argon (type D), due to gettered surface
oxides.

     (iv) There was recrystallisation of the self-welded regions
(type C) promoted by surface working during assembly tightening
(Fig. 4).

     (v)   Bolt thread/nut interfaces were self-welded at several
points of contact, but the extent of the weld was less than that
obtained on washer interfaces.

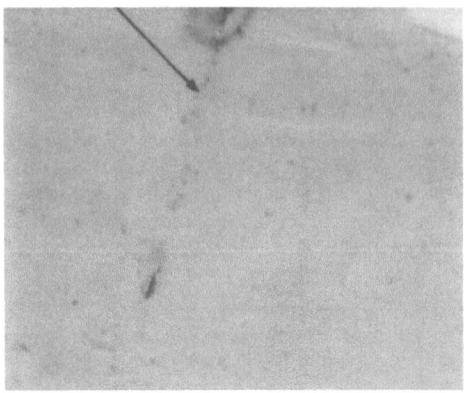

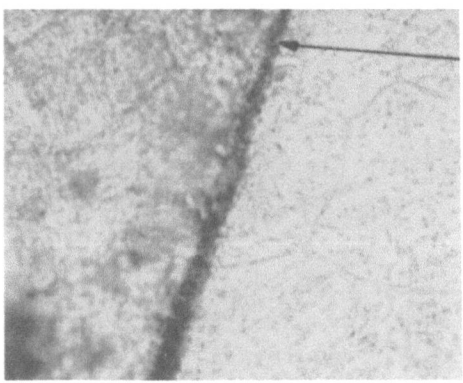

Fig. 2  Typical 'Type A'              Fig. 3  Typical 'Type B'
        Assembly (Unetched.)                  Assembly (Etched.)

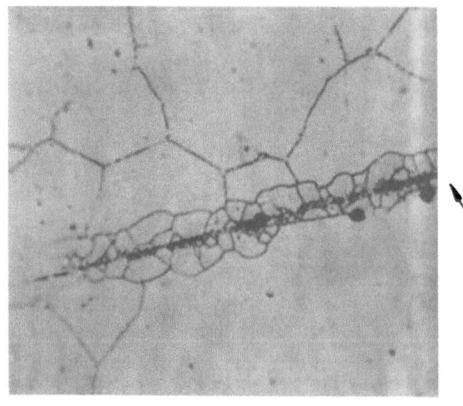

Rewelded
interface

x 2000
         Fig. 4  Typical 'Type C' Assembly
                 (Etched.)

This investigation has shown that surface oxide films can reduce the tendency for self-welding of AISI 316 stainless steel surfaces. Under the conditions prevailing in both sodium experiments, any air oxide film on the 316 washer interfaces, which was wetted by the sodium, would be degraded according to the reaction:

$$FeCr_2O_4 + 2Na \rightarrow 2NaCrO_2 + Fe$$

If fairly free sodium circulation pertained at the washer interface, as is likely where low torque was applied to the bolted assembly, formation of sodium chromite would proceed by further reaction between the chromium in the steel and the solution of oxygen in sodium. This could provide a jacking mechanism if oxide formation proceeds by the pertinent mechanism.

Alternatively, in the high torque situation where there is a relatively small space between washers, either:

(i)    Sodium may not wet the air oxide film but the low oxygen activity environment precludes further oxide growth (in which case sintering may be able to proceed in the presence of the initial oxide film) or results in reduction of the oxide film

$$FeCr_2O_4 \rightleftharpoons Fe + 2Cr + 4 \ [O] \ \text{(to system)}$$

enhancing any sintering process. Or:

(ii)   Sodium wets the air oxide film producing sodium chromite and free iron, and the competing processes of sintering and further sodium chromite formation ensue. The sintering process may be enhanced by the saturation of the sodium with iron and the production of an iron-rich surface with associated enhanced diffusion and recrystallisation properties. The process which dominates is likely to be highly dependent upon the rate of replenishment of oxygen in the sodium between the washer faces.

Bendorf (2) has hypothesised the self-welding process as:

(i)    Formation of clean surfaces following the chemical action of sodium on stainless steel, then

(ii)   Deformation of the contact areas and generation of a supporting area due to the influence of contact pressure. Huber and Mattes (3) consider that,

(iii) Self-welding of the supporting area occurs as a consequence of recrystallisation across the interface (see recrystallised grains on re-welded interface, Fig. 4).

We consider the self-welding process occurs by pore shrinkage due to a diffusion mechanism similar to that proposed by Brett and seigle (6) and Evans and Walker (7).

Brett and Seigle (6) relate sintering time (t) to pore radius (r), diffusion coefficient (D) and temperature (T) by:

$$t = \frac{k \ Tr^3}{6\gamma\Omega D} \tag{1}$$

$k$ = Boltzmann's constant = $1.3806 \times 10^{-23}$ J deg$^{-1}$,
$\Omega$ = Atomic volume = $1.15 \times 10^{-20}$ mm$^3$,
$\gamma$ = Surface energy = 1 Jm$^{-2}$.

The diffusion coefficient in equation (1) is that for nickel in stainless steel (20Cr/25Ni/1Nb) obtained by Smith and Gibbs (8).

Grain boundary diffusion is assumed for temperatures in the range
of 823 K to 973 K (note small recrystallised grains at 973 K). The
compressive stress across the washers is neglected as a simple calculation suggests that, following high temperature relaxation, this
is ~ 0.1 of yield stress. This indicates that a model based on
surface energy as the driving force is more appropriate to explain
the present test results than a creep of asperities model suggested
by Bendorf (2) or Chang et al (4) for the sintering of continuously
stressed interfaces. Although the upper bound predicted sintering
times (Fig. 5) are strongly dependent on initial pore size which
is difficult to quantify by metallography, the interface pore size
of tightened but unexposed bolted assemblies was estimated to be
~ 0.1 - 0.2 μm. At the unresolved interfaces it is reasonable to
assume therefore that pores could be typically 0.05 μm. Thus the
observation of significant self-welding following annealing for
1500 h at 973 K and 2600 h at 906 K is compatible with an initial
pore size of ~ 0.05 μm (Fig. 5).

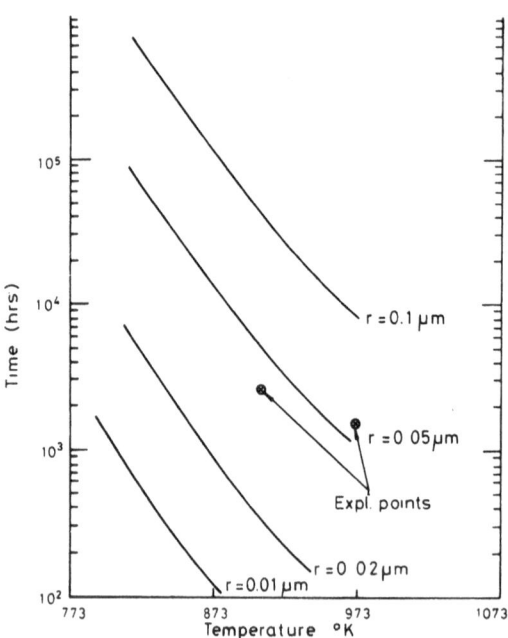

Fig. 5   The Influence of Temperature and Pore Size on the
         Rewelding Time of AISI 316 Steel Surfaces

     A further complication arises from possible changes in the
mechanism of sintering. Contributory factors could be trapped gas
or liquid retarding the collapse of voids and lower test tempera-

tures promoting grain boundary diffusion. If interfaces were highly stressed, creep of asperities could become an important mechanism of self-welding.

It was impossible to unscrew at 25 $^{\circ}$C any of the bolts from the assemblies annealed at 906 K and 973 K in vacuum. The bolts in the four assemblies tested sheared in the threads during removal at a torque of about 22 Nm. The remainder of the bolt was drilled out of the assembly. Subsequent room temperature tensile tests revealed that room temperature loads of 150 - 300 lbf. (667 - 1334 N) were required to part the tensile specimens from the washer faces (Fig. 6). Curves 'a' and 'b' refer to initial and subsequent tests on remaining ligaments of the assembly. Assuming both surfaces of the specimen were bounded to the strip specimen then the calculated shear stresses at failure were between 45 and 90 MNm$^{-2}$. (Note maximum stress is approaching the yield stress of 100 - 150 MNm$^{-2}$, Bolton et al (9).

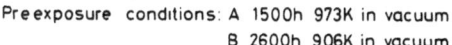

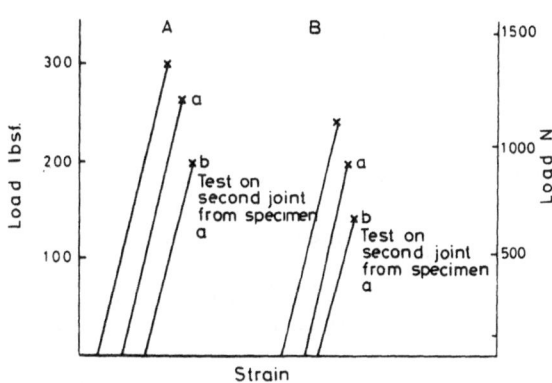

Fig. 6   Room Temperature Forces Required to Shear

Metallographic and tensile evidence, together with prediction, show that strong mechanical bonds can be formed at contacting interfaces in non-oxidising environments such as sodium at temperatures > 906 K. This agrees with work of Chang et al (4), Bendorf (2) and Huber and Mattes (3), who observed strong mechanical bonds following self-welding below 838 K. Although there is no reported evidence of self-welding below 838 K, long-term tests have never been conducted. Note Fig. 5 predicts self-welding following long-term exposure at temperatures < 838 K. The actual time is dependent on initial pore size, which is a function of the surface topography and interfacial contact pressure. Factors which increase self-diffusivity in the steel· such as sodium induced austenite to ferrite changes or irradiation could significantly reduce sintering times.

Finally, one advantage of the rewelding phenomena is that compressively loaded surface defects may reweld in sodium.

CONCLUSIONS

AISI 316 stainless steel bolted assemblies self-welded in static sodium at 906 K containing 35 ppm oxygen after 2600 h and in circulating sodium at 973 K containing 5 ppm $O_2$ after 1500 h exposure. Washer surfaces were welded by sodium chromite or metal/metal bridges. The bonded surface was recrystallised at 973 K but not at 906 K. A model based on sintering theory predicts the times for self-welding, which are compatible with experimental results. Mechanical tests on model assemblies show that high torque forces are necessary to debond bolts at room temperature and that the room temperature bond strengths are indicative of true metal bonding.

ACKNOWLEDGEMENT

This paper is published by permission of the Central Electricity Generating Board.

REFERENCES

1.  R. B. Jerman, R. C. Williams and D. O. Leesar, J. Basic Eng. 81 (1959) 213.
2.  K. Bendorf, Kernforschungszentrum, Karlsruhe, Externer Berich 8/70-1 (1970).
3.  F. Huber and K. Mattes, Liquid Alkali Metals, Proc.Int. Conf. Organised by BNES, Nottingham University, 4-6 April 1973.
4.  J. Y. Chang, P. N. Flagella and S. L. Shrock, ANS Trans. 22 (1975) 200.
5.  J. R. Gwyther, P. Marshall,. D. Rowe and D. Withrington, CEGB Report RD/B/N3232 (1974).
6.  J. Brett and L. Seigle, Acta Metl. 11 (1963) 467.
7.  H. E. Evans and G. K. Walker, Physics of Sintering 5 (1973) 133.
8.  A. F. Smith and G. B. Gibbs, Met.Sci.J. 3 (1969) 93.
9.  C. J. Bolton, J. E. Cordwell, J. A. Eades, N. S. Evans, A. J. Hooper, P. Marshall, R. D. Richardson, J. W. Steeds and L. P. Stoter, 1979, Vol. 2 ICM3, Cambridge, U.K.

# CORROSION OF AUSTENITIC STEEL IN SODIUM LOOPS

Manfred Schad

General Electric
Sunnyvale
U.S.A.

## INTRODUCTION

The object of this work was whether it is possible to predict corrosion effects for austenitic steel exposed to liquid sodium with an analytical diffusion model. The analytically predicted corrosion effects will be compared with experimental measurements of corrosion effects achieved in an accurately controlled sodium loop.

## THE DIFFUSION MODEL

The diffusion model is based on the general mass balance equation. This equation is able to account for all observed and generally recognized sodium corrosion characteristics by introducing Fick's first law for the solid wall

$$-\frac{\partial \; \rho \cdot D_r \cdot \frac{\partial x}{\partial r}}{\partial r} + \frac{\partial(\rho \cdot x \cdot \upsilon_r)}{\partial r} - \frac{\rho \cdot x \cdot \upsilon_r}{r} \cdot \frac{\partial x}{\partial r}$$

$$+\frac{\rho \cdot x \cdot \upsilon_r}{r} - \frac{\partial \; \rho \cdot D_z \cdot \frac{\partial x}{\partial x}}{\partial z} + \frac{\partial(\rho \cdot x \cdot \upsilon_z)}{\partial z} = \frac{\partial(\rho \cdot x)}{\partial t} \quad (1)$$

and for the liquid sodium by accounting additionally for the mass transport through the flowing sodium.

$$-\frac{\partial \; \rho \cdot D_r \cdot \frac{\partial x}{\partial r}}{\partial r} - \frac{\rho \cdot D_r}{r} \cdot \frac{\partial x}{\partial r} - \frac{\partial \; \rho \cdot D_z \cdot \frac{x}{z}}{\partial z} = - \frac{\partial(\rho \cdot x)}{\partial t}$$

$$(2)$$

At the solid liquid interface these two equations are inter-
connected by the condition that the allow elements absorbed by the
sodium have to be equal to the progressing wall loss and the diffusing alloy elements from the wall.

$$\rho_{Na} \cdot D_{Na} \cdot \frac{\partial x_{Na}}{\partial r} = \rho_w \cdot D_w \cdot \frac{\partial x_w}{\partial r} + \rho_w \cdot x_w \frac{\partial R}{\partial t} \tag{3}$$

The test loop's maximum and minimum sodium temperature were,
respectively, 750 $^\circ$C and 150 $^\circ$C (flow through cold trap temperature
corresponding to 2 to 3 ppm oxygen). The loop was constructed of
stainless steel 1.4571. The specimens were made of steel No. 1.4571
or 1.4988, which are steels similar to SS304 and SS316. The investigated surfaces of the sodium flow guidance tubes along the loop
heaters and the rotating disc samples were exposed to sodium at
500 $^\circ$C to 750 $^\circ$C in the hot leg of the loop. The comparison between
the experimental and analytical results shows that the model is able
to predict most characteristics of the sodium corrosion of the
austenitic steels, such as the temperature effect, the velocity
effect, and the stainless steel alloy leaching effect. The analytical
results are qualitatively correct and to a fairly good degree
quantitatively accurate as well. (Figs. 1 and 2)

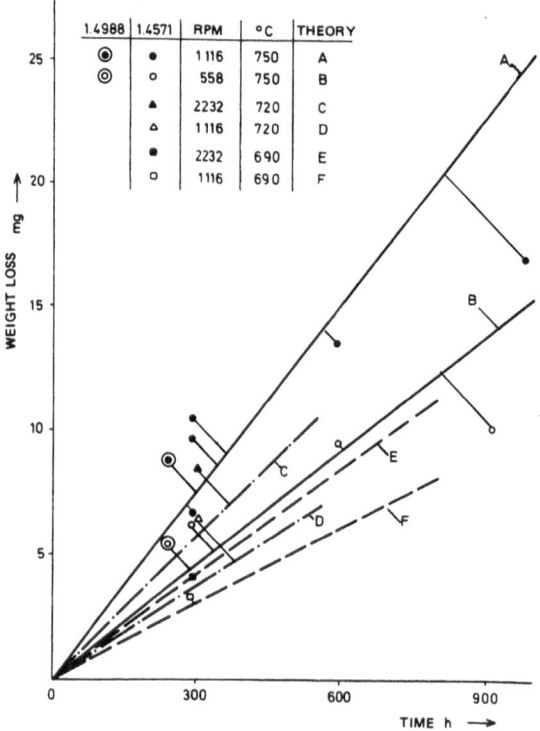

Fig. 1   Disc sample weight loss

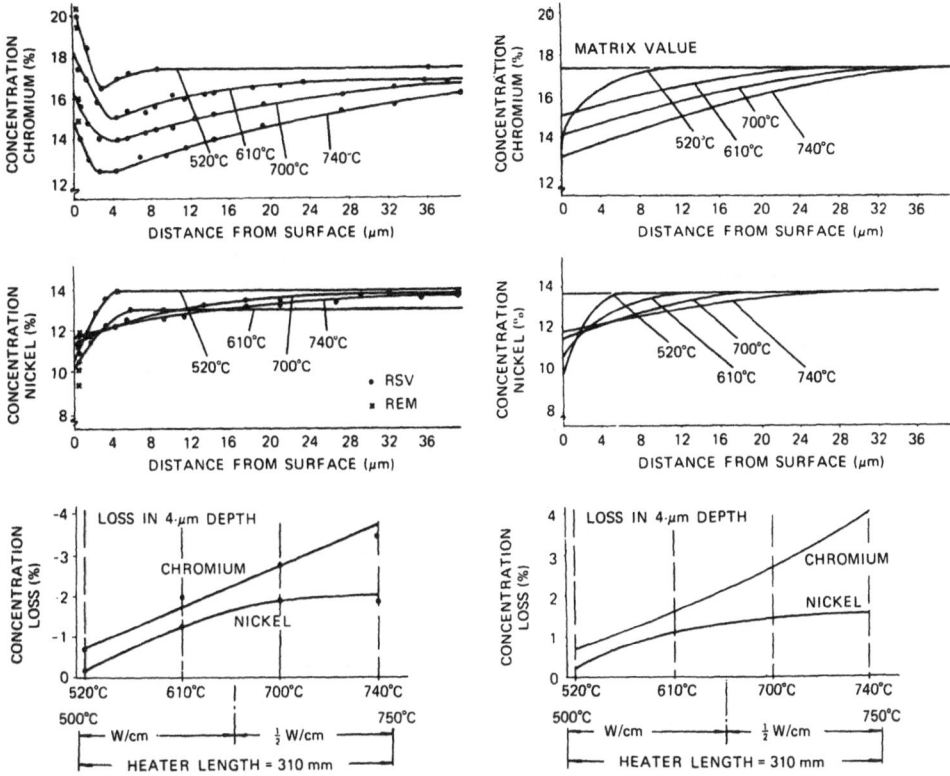

Fig. 2   Sodium flow guidance tube chromium and nickel profiles

The accuracy of the model predictions depends primarily on the degree of knowledge of the factors determining the diffusion such as the sodium saturation limits and the diffusion coefficient of the individual stainless steel elements in corroding walls.

The investigations were limited to iron, chromium, and nickel since the necessary diffusion determining factors for the other elements are not known. However, these three elements amount to > 95 % of all alloy elements in the considered steels.

Fig. 2 shows a comparison between the theoretical and actual chromium and nickel profiles in a sodium flow guidance tube. It may be noticed that the rising chromium content very close to the surface is not in agreement with the diffusion theory. This profile deviation was typical for all samples from this test series. However, it can be related to the carbon increase in parallel. Obviously, these two elements combine to precipitated chromium-carbide, which do not take part in the diffusion process.

The chosen heated loop section available for the comparison was not suited to demonstrate the "Downstream Effect" and how the individual corrosion influencing factors may alter the corrosion effects.

Therefore, a second computer program was set up for an actual fuel
rod with an isothermal section. The possible sodium mass transfer
effects along a fuel rod are shown in Fig. 3. This figure demonstrates that the model is able to predict the sodium corrosion
effects on austenitic stainless steels. However, for higher accuracy, the basic prerequisite is the exact determination of the
alloy element diffusion properties.

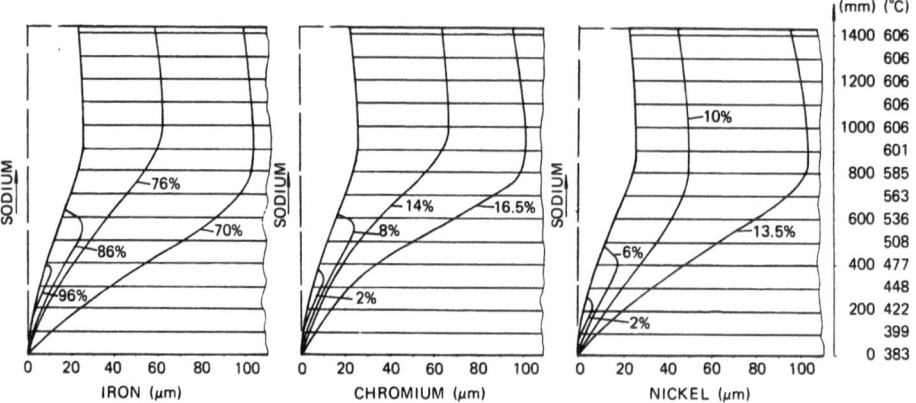

Fig. 3   Concentration profile in the fuel rod wall (diffusion model)
         after 10000 h:  $ST_{q\,max} = 450$ W/cm; G = 20 kg/s; P/D = 1.3;
         hot channel factor = 1.398

NOMENCLATURE

    D = diffusion coefficient ($m^2$/s)
    R = wall thickness loss (m)
    r = time (h)
    V = velocity (m/s)
    x = dimension less concentration (1)
    P = density (kg/$m^2$)

REFERENCES

    1.  M. K. Schad, "Corrosion of Austenitic Steel in Sodium
        Loops", Nucl. Technol. 50, 267 (1980).
    2.  M. Schad, "Zur Korrosion von Austenitischen Stählen in
        Natriumkreisläufen", KFK-2582, Gesellschaft für Kern-
        forschung Karlsruhe (Mar.1978).

PRELIMINARY OBSERVATIONS OF THE STRAIN-INDUCED CORROSION OF ALLOY-
800 IN SODIUM HYDROXIDE-SATURATED SODIUM AND ON ITS EVOLUTION IN
DYNAMIC PURIFIED SODIUM

Gianni D'Alessandro, Sergio Casadio and Giovanni Bruni

CNEN Casaccia, Rome

Italy

INTRODUCTION

Incoloy-800 of specially controlled composition and microstructure has been selected for the steam generators of sodium
cooled fast breeder reactors (1,2). The assessments involved in
such a selection were found to meet all the design requirements,
including the corrosion resistance in normal service conditions or
abnormal conditions (thermovector fluids contamined by sodium-water
reaction products) (1).

The Alloy-800 stress-corrosion resistance in caustic media
appears to be greater than that of the austenitic stainless steel
of the AISI 300 series, however under this aspect Alloy-800 is not
a still completely reliable material because of the relatively
limited available information (3) on its behaviour. Some papers
were recently published about the stress-corrosion of Alloy-800
exposed in caustic contamined liquid sodium (4,5,6). After some
tens of hours in sodium saturated by sodium hydroxide the specimens
obtained by commercial tubes (grade 1) undergo intergranular corrosion cracking at 525 $^{\circ}$C when the applied stress is greater than
130 Mpa(5). Both grades 1 and 2 of commercial tubes and plates were
found to crack under stresses greater than 145 Mpa within 284 hours
of exposure in a sodium containing about 7 - 10 % of NaOH at 550 $^{\circ}$C
(4).

The crack initiation was not observed to occur at such temperatures for a NaOH concentration less than 5 % (4, 5). In these caustic
environments an intergranular attack of few tens of micrometers was
always observed to occur, this last being assumed to be the starting
cause of the cracking under stress above the critical value (5).

479

Aim of this preliminary work is to examine the development of
this intergranular attack when the Alloy-800 is successively exposed
in purified dynamic sodium. A simple and cheap method was used to
stress plain test samples by a constant total deflection. By such
a technique it is possible to get only a qualitative picture of the
phenomenon due to the marked load relaxation induced on the strained
specimens during their exposure at high temperatures (7).

EXPERIMENTAL

Four point bending systems (7,8) were used for stress-corrosion
specimens of Alloy-800 (50 x 10 x 1 mm) whose compositions, thermal
treatment, metallurgical structure and Time-Temperature-Sensitization
diagrams are reported in detail elsewere (9). They were solution
annealed (SA) at 1010 $^\circ$C and their Ti + Al content was varying, the
carbon concentration being constant at 300 ppm. Most of the specimens were pre-aged in liquid sodium at 550 $^\circ$C (450 and 1650 hours).
By such a thermal aging they were "sensitized" and their creep
resistance was anhanced by the $\gamma'$-phase precipitation. Since by
foil (AISI-304 type) equilibration method the carbon activity of
the sodium baths was found to be about 0.04, the aged specimens were
carburized at the surfaces. 0.3 mm thick test specimens after the
1650 hours of sodium exposure were found to have a mean total carbon
of about 0.2 % against their original carbon concentration of 300 ppm.

The exposure in liquid sodium saturated with NaOH ($\sim$8 %) were
performed in all stainless steel capsule system under Argon blanket
gas. A dynamic sodium rig (Fig. 1) was then used to examine the
develope of the corrosion under the strain which was again imposed
to the specimens after their exposure in the caustic environments.
The sodium loop was managed by the technical services of the Casaccia
Nuclear Center as well as the diagnostic examinations by isotope
X-ray fluorescence spectroscopy (IXRFS), scanning electron microscopy
(SEM), Electron Probe Microanalysis (EPM) and powder X-ray analysis.

We used a WILD-M8 and an Ernst Leitz GMBH Wetzlar microscopy
and a Reichert microhardness measuring apparatus for the routine
examination of the specimens after their sectioning and polishing
with diamond paste up to 1 μm. The variation of the magnetic
susceptibility at the surface of the Alloy-800 pices was monitored
by a CRF Magnetometer.

CORROSION WITHOUT APPLIED STRAIN

In Fig. 2 the specific weight changes are reported as a function
of the exposure time for the Alloy-800 pieces exposed in liquid
sodium saturated with NaOH at 550 $^\circ$C.

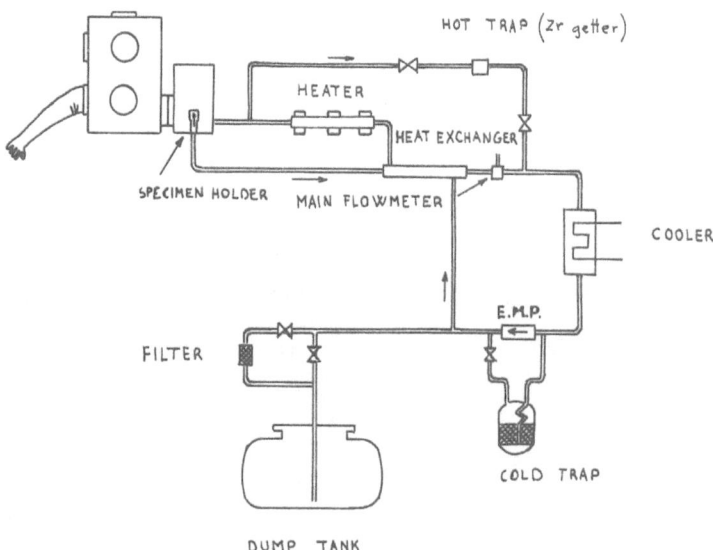

Fig. 1 Dynamic sodium rig for the corrosion studies at the Casaccia Nuclear Center.

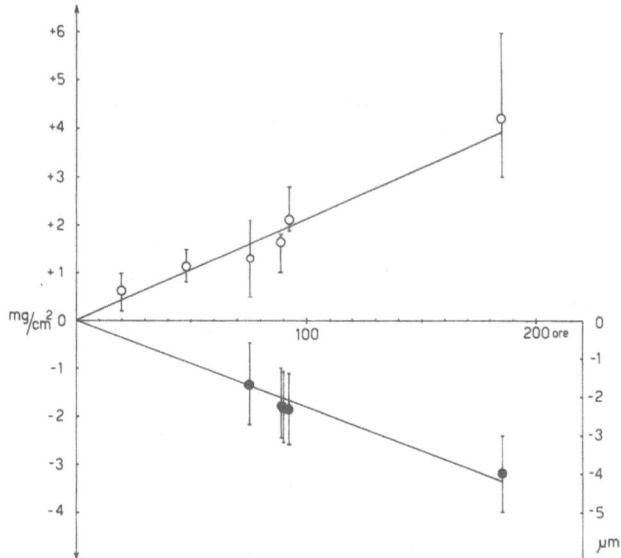

Fig. 2 Weight variation of Alloy-800 plotted vs. the exposure time in sodium saturated with NaOH at 550 °C. The full points refers to weight change observed after the mechanical removing of the formed scale. The thickness of the corresponding material loss can be evaluated by the right-side ordinate.

After about one week of exposure the specimens were found covered by an easily removable scale, whose morphology is shown by the scanning electron microscopy images of Fig. 3.

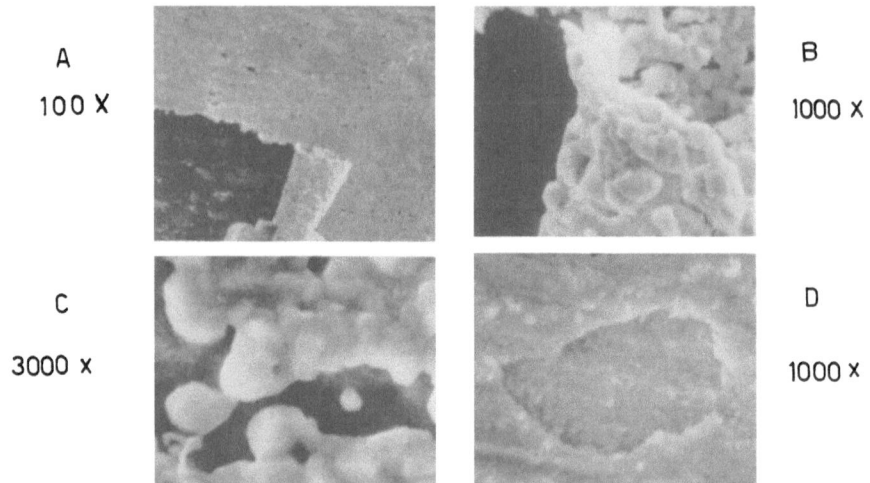

Fig. 3:   SEM surface analysis of the Alloy-800 after 90 hours of
          exposure in the caustic sodium at 550 $^{\circ}$C. Typical morphology
          of scale, partially removed (A and B), it is shown in C
          and under the scale the morphology of surface appears to
          be homogeneously flat (D).

The metallic composition of the scale was measured by isotope X-ray fluorescence spectroscopy (Fig. 4); as a result a deep de-chromization was observed. The cell parameters of such a scale were found to be compatible with the $NaMO_2$ species (M = Fe, Ni, Cr) by X-ray analysis. The presence of an intergranular attack of about 20 - 40 μm was evidenced by micrographic analysis (Fig. 5). Under steady state conditions the chemical attack at the surface seems to proceed much faster than the intergranular attack, hence this last was found generally constant with the exposure time and limited to a depth of the order of the grain size of the alloy. The magnetic susceptibility at the surface of the alloy was found deeply altered. This fact seems to be related to the corrosion damage, in particular to the Cr-depleted zones near the grain boundaries involved in the intergranular corrosion process. However also the bulk magnetic susceptibility is affected by the thermal history of the material (9).

The intergranular attack was not found to progress further by prolonged exposure (up to 700 hours) in sodium flowing at 1 m/s at 550 $^{\circ}$C and cold trapped at about 120 $^{\circ}$C (Fig. 6).

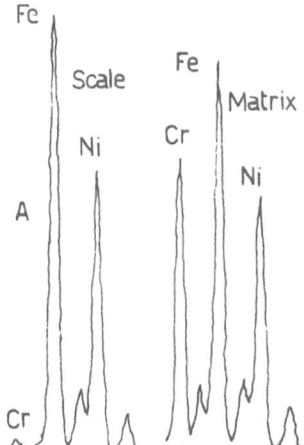

Fig. 4    IXRFS of the scale (A) and of the Alloy-800 matrix of the
          specimen of Fig. 3

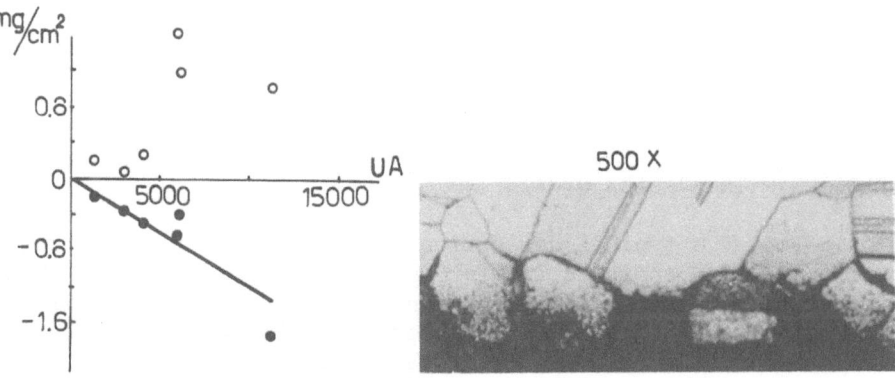

Fig. 5    Weight changes of Alloy-800 exposed 48 h in NaOH saturated
          Na (void points) and then 480 h in clean dynamic sodium
          (full points) vs. The magnetic susceptibilities (arbitrary
          units) measured at the surfaces. The micrography shows the
          typical intergranular attack observed.

No weight changes were observed in dynamic sodium without
considering the exfoliation of the scale previously formed on the
specimens exposed in the caustic media. The magnetic susceptibility
of the specimens exposed 20 h in caustic static Na and then 480 h
in clean dynamic Na was found roughly proportional to their weight
loss (Fig. 5).

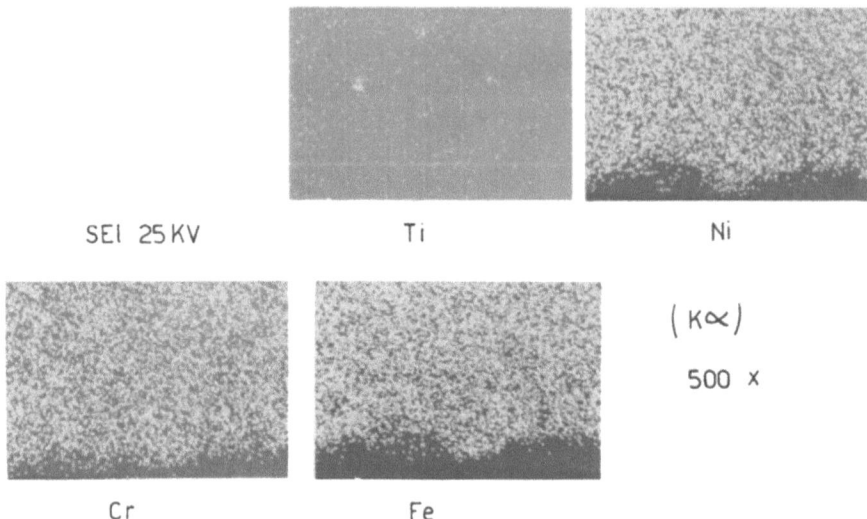

SEI 25 KV                    Ti                        Ni

( Kα )

500 x

Cr                           Fe

Fig. 6   EPM analysis of an Alloy-800 specimen exposed 20 h in
         caustic Na and 700 h in dynamic Na at 550 °C. No cross Cr
         or Ni depletion appears near the sodium exposed surface
         and in the zone where the intergranular attack is evidenced
         (SEI image).

CORROSION WITH APPLIED STRAIN

    In Fig. 7 a typical deflection-microstrain function is reported
for the solution annealed and for the 550 °C aged specimens loaded
by our systems of Fig. 1. By removing the applied load a plastic
deformation was observed, and the corresponding residual strains
after the 550 °C treatment under load were found to lie in the scatter band of Fig. 7.

    In Fig. 8 the load applied at the surface of the bent specimens
is reported as function of the imposed deflection.

    In Fig. 9 typical micrographies of sections of strained speci-
mens exposed one week in sodium hydroxide saturated sodium are reported, the initially applied load being above the yield strength.
The intergranular attack of 40 μm was still observed but we were
not able to evidence any crack initiation on the solution annealed
and on the in sodium thermal aged material. An evaluation of the
residual stress at the end of the corrosion test was made. A comparison of our results with the available data is reported in Fig. 10.
Even the critical threshold stress of 145 Mpa for the corrosion
cracking (4) was found to be conservative since we were not able to
observe crack initiations below about 180 Mpa on our solution
annealed (and 550 °C aged) bent specimens within 170 hours of ex-

posure in the caustic environment. Such a discrepancy could be ascribed: i) to the probable presence of residual cold work on commercial pieces (4, 5) that being not the case for our experimental heat specimens; ii) to the bending mode of straining as compared to the tensile constant load methods.

Some crack initiations were observed on a 13 % cold worked (by tensile strain) specimen that was found to be loaded above 180 Mpa (Fig. 10).

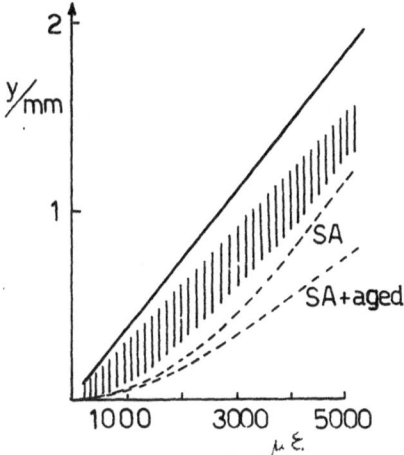

Fig. 7   Max. deflection (y) vs. microstrain (µε) at the surface of the Alloy-800 specimens loaded in the apparatus of Fig. 1. The dashed lines show the room temperature residual plastic microstrains, while these last were found to lie in the scatter-band after the 550 °C exposure under load.

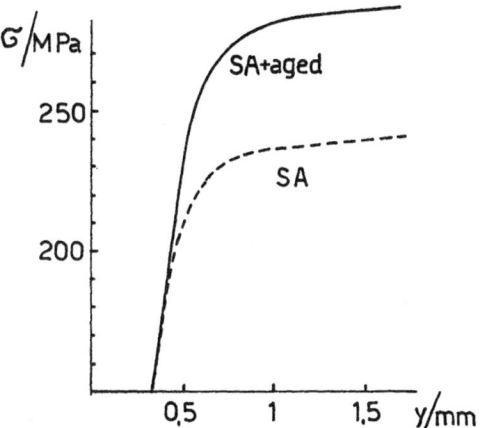

Fig. 8   Typical stress (σ) vs. y plots for the SA + aged specimens.

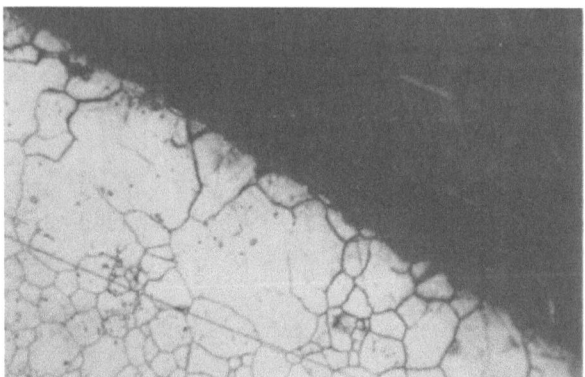

200 X

Fig. 9  Typcial intergranular attack on Alloy-800 bent specimen
        one week exposed in caustic sodium at 550 °C

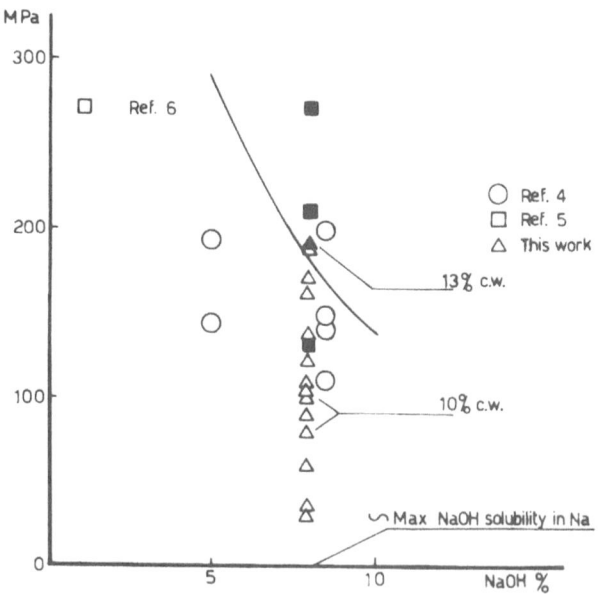

Fig. 10 Stress vs. NaOH concentration diagram showing the SCC field
        in liquid Na at 550 °C.

CONCLUSION

     Within the limits and with the cautions (7) imposed by the
straining technique used in this work, the treshold stress of about
145 Mpa obtained by Berge et al. (4) can be retained a conservative
value to avoid the stress-corrosion cracking of Alloy-800 (solution-

annealed and "sensitized") in caustic saturated liquid sodium at
550 $^{\circ}$C under total bending strain.

In fact we were not able to observe crack initiation on specimen loaded up to about 180 Mpa within one week of exposure in such
an environment. The Alloy-800 general corrosion rate was found to
be constant in the caustic sodium at 550 $^{\circ}$C, while the intergranular
attack was not found to penetrate depths greater than about the
grain size of material and it was not found to further progress in
dynamic sodium of normal technological purity, even under large
plastic bending strains. The "sensitized" state and the (Ti + Al)/C
ratio do not affect the corrosion behaviour the material.

REFERENCES

1. M. Julien; Nucl.Technol., 31, 367 (1976).
2. J. M. Duke et al; Qualification of Alloy-800 for Sodium
   Heated Steam Generators", Int.Conf.Liquid Metal Technol.
   in Energy Prod., p. 211, May 3-6, 1976, Champion,
   Pennsilvania.
3. W. E. Ray et al.; Nucl.Techn.Mat., 66, 65 (1977).
4. Ph.Berge et al.; Nucl.Mat., 66, 65 (1977).
5. M. Cappelaere et al.; Mem.Sci.Revue Métallurgie, Nov. 1977,
   p.719.
6. L. Champeix et al.; "High Temperature Corrosion and Mechanical Properties in Sodium Environment", in Alloy-800,
   W. Betteridge et al., North-Holland Pub.Comp., 1978.
7. R. N. Parkins, "Stress Corrosion Test Methods", in "Corrosion, vol. 1, L.L.Shreir, Newnes-Butterwoths (1977).
8. H. Coriou et al.; "Influence of Carbon and Nickel Content
   on Stress Corrosion of Austenitic Stainless Alloys in
   Pure and Chlorinated Water at 350 $^{\circ}$C", Proc.Conf.
   "Fundamental Aspects of Stress Corrosion Cracking",
   p. 352, the Ohio State University, Sept. 11-15, 1967.
9. A. Borello et al.; submitted for publication on Corrosion
   NACE journal.

# DISCUSSION OF SODIUM CORROSION AND MASS TRANSFER

Benjamin H. Kolster

TNO
Apeldoorn
The Netherlands

In the literature and also in papers presented at this conference attemps are made to determine the diffusion coefficients of
the alloying elements during corrosion of austenitic stainless steel
in liquid sodium. I am afraid that these approaches lack the fundamental basis. As shown in the following the diffusion coefficients
can be calculated under specific conditions, different to the conditions the above mentioned papers deal with.

The corrosion phenomenon with respect to austenitic stainless
steel comprise a number of consecutive processes. Under these conditions it is well known that, if one of the processes is definitely
the slowest process, this process determines the overall rate. If
we apply this statement to the corrosion of stainless steel at
relatively low oxygen levels (the actual situation the authors refer
to) the following consecutive processes can be distinguished:

1.  transfer of alloying elements in the liquid sodium;
2.  interaction of Fe, Ni, Cr and Mn with Na, at low oxygen level in
    such a way that the dissolution rate of Fe is less than those for
    the alloying elements;
3.  diffusion of the alloying elements in the austenitic steel;
4.  phase transformation of austenite into ferrite, if process 2 + 3
    leads to unstable austenite;
5.  diffusion of the alloying elements in the ferrite.

The situation that interests us refers to hydrodynamic conditions
for which the transfer rate in liquid sodium exceeds the processes in
the solid state. So, process 1 can be omitted.

Apart from the above mentioned, the time dependence of the overall corrosion process shows two periods: an incubation period followed by the steady state situation. For both situations the consecutive processes (except the diffusion in the austenite) are schematically represented in the proceedings of the 7th International Conference on Metallic Corrosion (Rio de Janeiro (1978), p. 1524). It should be emphasized that during both periods the interaction of Fe with liquid sodium is the rate determining step. This conclusion follows from the following experimental results:

- the leaching of the alloying elements proves that their interaction with sodium is faster than the one for Fe.

- during the steady state situation the corrosion rate is linear related to time. A diffusion process kinetically would result in a parabolic rate dependence.

Having stated this, the question is whether it is possible to determine the diffusion coefficients of the alloying elements. In general diffusion coefficients can be determined if:

- the diffusion is the rate determining step;
- the boundary conditions are known.

If we apply these conditions to our problem it is clear that they hold for the steady state situation. The first condition should be interpreted in the sense that the diffusion rate equals the inter-action rate of the alloying elements. The second condition needs no further explanation, resulting in the conclusion that by applying Fick's first law the diffusion coefficients can be determined.

However, the situation we are interested in concerns the incubation period. As stated before, under these conditions diffusion is not the determining step. Nevertheless, it is clear that the change in concentration (whether this is in the austenite or ferrite is not relevant) is partly due to the mobility of the alloying elements, that is to say the change in concentration is related to the diffusion coefficients of the alloying elements. The second cause for the concentration change are the boundary conditions, which are not constant in time. Let us first suppose that we know exactly the boundary conditions as a function of time. Even then the diffusion coefficients cannot be determined. The additional information we need to solve the problem mathematically is the exact diffusion rates as a function of time. The principle problem how is to determine the diffusion rate of all elements. Even if we have to deal with a binary alloy, say Fe and Mn, the problem is how to determine the diffusion rate of Mn. The corrosion experiments only provide the accumulated weight loss. The least we must know, thus, is the cor-rosion rate of Fe, which can only be determined from the corrosion rate during the steady state condition. However, as stated before, if we have reached the steady state condition, the diffusion coefficient for Mn can be determined directly.

On top of it we may wonder why one should determine the diffusion coefficient if one has the disposal of the diffusion rates. Diffusion coefficients are determined in order to calculate diffusion rates. So, this approach seems to be the other way round.

In conclusion, to my opinion it is not possible to determine the diffusion coefficients of the alloying elements in the solid state during the incubation period, unless one has the disposal over so many information, including the diffusion rates of the alloying elements, that the original aim makes no sense, as:

- this information comprises diffusion rates, which is in fact the quantity to be determined by calculating the diffusion coefficients;
- it needs experimental effort from which much easier the diffusion coefficients can be determined.

# NITROGEN TRANSFER IN AUSTENITIC SODIUM HEAT TRANSPORT SYSTEMS

Stuart A. Shiels and Chris Bagnall

Westinghouse Electric Corporation
Madison, Pennsylvania
USA

## INTRODUCTION

Transfer of the interstitial elements, carbon and nitrogen, in high-temperature sodium heat transport systems is a well-recognized phenomenon which has been studied extensively over the years. There are many recorded examples of carbon and nitrogen loss from, and pick-up by stainless steels exposed to high temperature sodium under a variety of conditions. Carbon transfer has, however, received the major share of attention. In-depth studies, sponsored by the Department of Energy at the Westinghouse Advanced Reactors Division and at the Argonne National Laboratories, have expanded the knowledge of this phenomenon to a stage where predictive equations are now routinely used for the prediction of end-of-life carbon levels in structural components.

The treatment of carbon relies on the ability to measure three variables: First, the carbon activity of the sodium, $A_{c,Na}$; second, the relationship between activity and the carbon equilibrium concentration in the steel, $C_e$, at any given temperature; and third, the effective diffusion coefficient of carbon in steel, $De_{ff}$, as a function of temperature. The key factor in the acceptance and application of a predictive method in the U.S. was successful development at W-ARD, of a technique for circumventing the measurement of $A_{c,Na}$. In the absence of reliable carbon activity meters, W-ARD embarked on a program to relate the carbon "potential" of the sodium, $C_s$, measured as a concentration in a standard alloy at a standard temperature, to values of $C_e$ for a variety of alloys over a wide range of temperature. Accumulation of data over several years led to the development of predictive equations for primary containment materials. Two standard alloys are now in use for measuring $C_s$, T304LSS and Fe-12Mn alloy.

Similar equations are required for nitrogen. Unfortunately, there is no technique available for measuring the nitrogen potential of sodium and no correlations exist relating nitrogen potential, $N_s$, to nitrogen equilibrium concentration in the steel, $N_c$.

The tendency has been, therefore, to assume that nitrogen will behave exactly like carbon and that $N_e$ would be equal to $C_e$. Further, it was assumed that the diffusion coefficients were comparable, and hence the same set of equations could be used to describe the movement of both species. This second assumption in particular is not valid. Carbon has a low solubility in austenite and very quickly reaches saturation causing the precipitation of carbides. The bulk of the carbon content of austenitic steels exposed to high temperatures is in fact, in the form of carbides. The presence of these carbides, not the diffusion of carbon through the austenitic matrix, controls the loss of carbon from steels exposed to decarburizing conditions. Two effective diffusion coefficients are needed to describe carbon transport, one for decarburization and one for carburization. Nitrogen, however, is highly soluble in austenitic steels and for the concentrations of interest, is always present in solid solution.

In view of the differences in behaviour, it seemed prudent to develop a separate method for predicting nitrogen transfer. This paper describes how, using available data, an estimate of nitrogen transfer in austenitic systems can be made.

NITROGEN TRANSPORT DATA ANALYTICAL APPROACH

The changes in nitrogen concentration in a specific component can be predicted if, in addition to the exposure conditions, some knowledge of $N_e$ and the nitrogen diffusion coefficient $D_N$ is available as a function of temperature. Sufficient data on $D_N$ are available to allow the derivation of an Ahrrenius equation. Nitrogen equilibrium measured are generally restricted to temperatures in excess of 650 °C. It is possible however to determine equilibrium values indirectly if sufficient information is available on the exposure conditions and geometry of the sample. Several types of data have been used in this analysis:

Bulk values, $\overline{N}$, from samples known to be at equilibrium were used directly.

$N_e$ was calculated for samples for which geometry, exposure time, exposure temperature and original concentration, $N_o$, were known. Data were used only if the process was at least 65 % toward equilibration as given by:

$$\frac{\overline{N} - Ne}{N_o - Ne} = 0.65$$

Samples which neither nitrided or de-nitrided were assumed to
be at equilibrium at that exposure temperature.

Extrapolation of nitrogen gradients, determined by incremental
analysis, to the surface to obtain $N_e$.

Values for the diffusion coefficient of nitrogen in austenitic
stainless steels were required for the determination of the equilibrium data in this paper and were also required for determination
of the nitrogen changes in structural components exposed to high
temperature sodium. The diffusion relationship used in this paper
was derived by Bagnall et al. (1), from data generated in the
temperature range of interest (1-3). These data and the calculated
line are given in Fig. 1. Values from other sources (4-7) are shown
on the same curve and generally corroborate the Bagnall curve to
such a degree that, for the purpose of this paper, it was considered
necessary  to re-calculate the diffusion relationship.

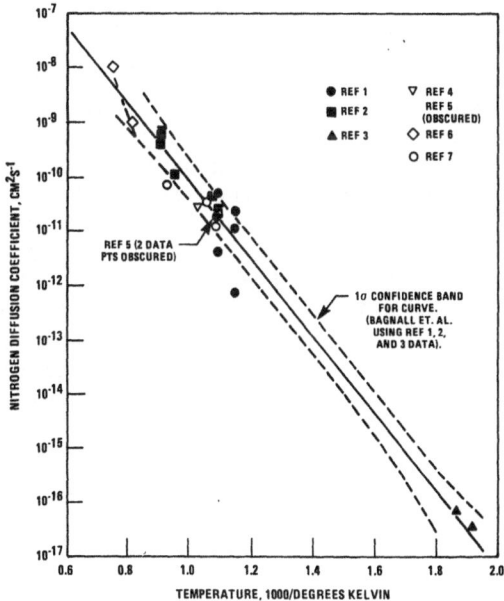

Fig. 1  Diffusion Coefficient of Nitrogen in Austenitic Stainless
        Steels

The information used to derive the diffusion relationship was
obtained from three sources:

Bagnall et.al. (1) measured matrix hardness gradients in
nitrided T304SS after various exposures at 647 °C and 593 °C.
Diffusion coefficients were determined from the hardness profiles
by computer solution of Ficks Second Law.

The second set of data (2) were derived from analysis of concentration profiles in T316SS containment tubing and tubular inserts used in a liquid potassium test loop.

Data were obtained from internal friction measurements performed on T304SS at 250 °C and 255 °C (3).

The equation of the line in Fig. 1 is:

$$D = 1.09 \times 10^{-3} \exp. \left(\frac{-32,700}{RT_K}\right) (cm \cdot s^{-1})$$

Other data (4-7) plotted in Fig. 1 are for unstabilized austenitic Fe-Cr-Ni alloys and generally fall within the $1\sigma$ limits of the Bagnall line.

The Interstitial Transfer Facility, ITF, at W-ARD, was used to study carbon and nitrogen transport in a system representing a phenomenological simulation of FFTF. The ITF has been described in detail elsewhere (8), and it is sufficient to say that it was designed in such a way that thin foil samples (152 mm x 12.5 mm x 4.08 mm) could be located almost everywhere in the loop. This allowed both nitrogen loss and pick-up to be measured in the same system during the same test run. Exposure temperatures varied between 442 °C and 740 °C. Times of exposure varied from 668 h to 12000 h. Most of the data were obtained for T316SS with some values for T304SS. The data population, shown in Fig. 2, represents results from five different runs (ref. 9 and this paper). Both steels were exposed with two different initial carbon levels. While the total population shows considerable scatter, the data for any particular run were consistent. Also, there was no discernable effect of initial nitrogen concentration, $N_o$. The variation with temperature is consistent; $N_e$ increases with decreasing temperature.

The Materials Test Loops, MTL's, which operated at W-ARD up to 1973 were destructively evaluated after each long term test run. Part of this evaluation consisted of determining nitrogen profiles through the containment walls by incremental machining and analysis (10). The profiles developed in this manner were extrapolated back to the surface and a surface value estimated. A typical profile is shown in Fig. 3.

The Pre-Exposure Loop, PEL, at W-ARD, an all-austenitic system, has been used for several years for the long term pre-exposure of mechanical property samples. Qualification samples were included in each run and were used for metallographic and chemical analysis. Bulk nitrogen values were thus obtained. The data used here were from T326SS sheet and fuel pin cladding samples. Temperatures and exposure times were sufficient to ensure that the samples had equilibrated with the sodium. Although not reported in the literature, these data are plotted in Fig. 4 with data from other sources.

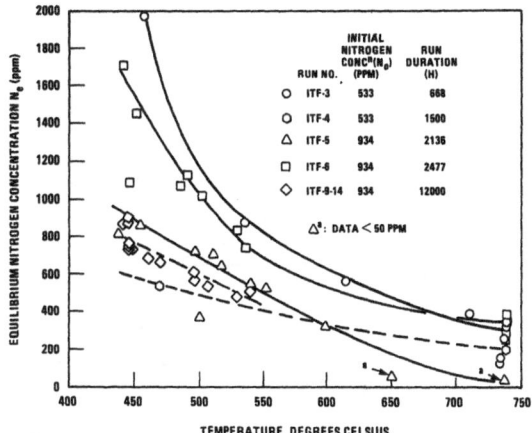

Fig. 2   Calculated Equilibrium Nitrogen Concentrations for Type
316 Stainless Steel Foils Exposed in ITF

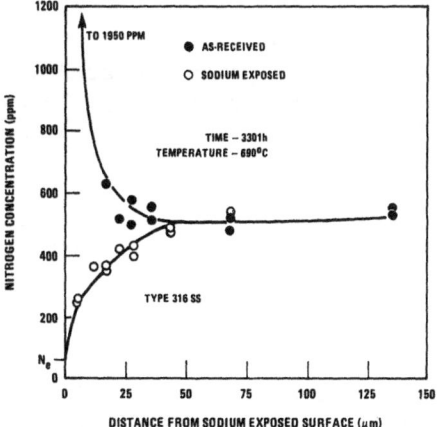

Fig. 3   Nitrogen Gradient in Type 316 Stainless Steel Main Heater
Tubing (MTL-3-14) After 3301 h

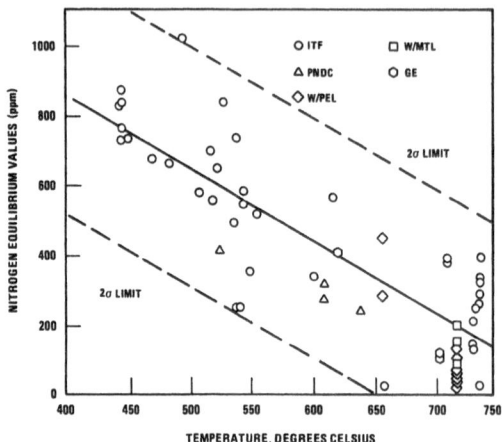

Fig. 4   Nitrogen Equilibrium Values $N_e$ as a Function of Temperature
for Type 316 Stainless Steel

An extensive mass transfer program performed by G.E. during
the late 1960's and early 1970's included some long term exposures
of T316SS and 304SS tubing (11). Post-exposure nitrogen values were
determined for these samples. Exposure times (10,000 - 15,000 h)
were of sufficient duration to ensure that the samples were at
equilibrium.

Nitrogen studies were also performed at PRNFDC as part of a
comprehensive sodium corrosion program (12). T316SS samples with
different initial nitrogen values were located throughout a heater
section in a test loop. After long-term exposure, the samples were
analyzed for bulk nitrogen. The samples were assumed to be close
to equilibrium when nitrogen loss was just detectable. Hence, an
equilibrium temperature was determined. In fact, the true equilibrium temperatures would be slightly lower than those reported.

DATA ANALYSIS

Equilibrium results are plotted for T316SS and T304SS, respectively, in Figs. 4 and 5. A least squares fit through the data produced
the following relationships:

T316SS: $N_e$ (ppm) = 1671 - 2.05T $^\circ$C; $2\sigma$ limits are approximately
$\pm$ 375 ppm; the lower band goes through 0 ppm nitrogen at 650 $^\circ$C.

T304SS: $N_e$ (ppm) = 799 - 0.85T $^\circ$C; $2\sigma$ limits are approximately
$\pm$ 170 ppm.

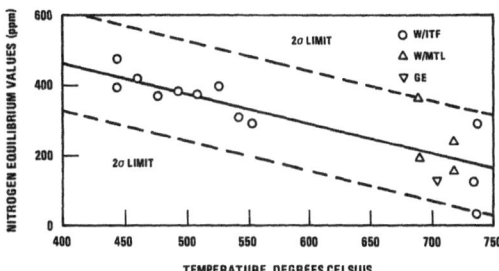

Fig. 5  Nitrogen Equilibrium Values $N_e$ as a Function of Temperature for Type 304 Stainless Steel

DISCUSSION AND CONCLUSIONS

The intent of this paper is to present the available data on nitrogen transfer in a form suitable for design purposes. The achievement of this objective is described by the equations for $N_e$ at a function of T, and by the diffusion equation for nitrogen in austenite. Some judgement in the use of the $N_e$/T equations is re- quired however, particularly since the data scatter bands are so broad. Several pertinent observations can be made regarding the use of the nitrogen equations.

The data confirm that nitrogen transport will occur in the direction of decreasing temperature which indicates that the dominant factor is not the temperature dependency of the nitrogen activity in the steel, but is the temperature dependency of the nitrogen activity in the sodium.

The data scatter is large. Run-to-run variations in ITF, for example (9), can only be attributed to variations in the nitrogen activity in the sodium. Unfortunately, these variations cannot be monitored, therefore, the $N_e$ values cannot be normalized in terms of nitrogen potential or nitrogen concentration in the sodium. The reasons for this variation are also unknown. It could be a function of dissolved nitrogen in the sodium, a function of some unknown nitrogen source, such as cover gas contamination or nitrogen dif- fusion through containment walls, or perhaps a function of a chemical interaction between nitrogen and some other species in the sodium, carbon, for example. It might be pointed out that the lowest $N_e$ values measured in ITF were obtained from a test run in which a high carbon potential was known to exist (9). Whether this was simply a coincidence is not known.

It has to be accepted, therefore, that the nitrogen transport behaviour in a reactor system cannot be defined any closer than the presented data allow. The implications of the data scatter on design must therefore be discussed.

Structural components will not generally experience temperatures above about 550 $^\circ$C, or temperatures much below 400 $^\circ$C. Referring to Fig. 4, it can be seen that T316SS can vary from a nitrogen loss situation to a nitrogen gain situation over the entire temperature range. Nitrogen pick-up, according to the data, will have relatively little impact. At 400 $^\circ$C, the predicted $N_e$ value is only 851 ppm. The 2$\sigma$ limit is about 1200 ppm. At this temperature, the diffusion coefficient is very low, about 2 x $10^{-14}$ cm$^2$/s. Bulk nitrogen changes would, therefore, be negligible. Nitrogen loss could result in levels below 200 ppm at 550 $^\circ$C if the 2$\sigma$ limit is taken as the lower bound. Nitrogen loss should, therefore, be addressed at the design stage.

The reverse argument is true for the core regions. The fuel clad will experience sodium-side temperatures in excess of 700 $^\circ$C, which could cause total loss of nitrogen. The initial nitrogen levels in U.S. fuel cladding are controlled at < 100 ppm. Loss of this small amount of nitrogen is not considered to be significant with regard to any change in mechanical properties. The importance of nitrogen pick-up is not clear. Nitrogen in the fuel cladding will undergo n$\alpha$ reactions producing helium which can lead to embrittlement. It is for this reason that the nitrogen levels are held so low in the as-received cladding. Any increase in nitrogen above the specified value might impact on fuel pin performance. The data indicate that nitriding of ex-core components in an all austenitic stainless steel system will not constitute a problem. Consideration should be given, however, to nitrogen loss above about 500 $^\circ$C.

Nitrogen loss from core components, particularly fuel cladding, is not a problem. Nitriding of fuel cladding, however, may be of concern, and should be addressed from the point of view of helium production and subsequent in-reactor embrittlement.

ACKNOWLEDGEMENTS

This work was performed under contract to the U.S. Department of Energy.

REFERENCES

1.  C. Bagnall, M. G. Cowgill, S. J. Orbon, and S. L. Schrock, "Nitrided Type 304 Stainless Steel in Sodium: Stability and Nitrogen Diffusion Data: WARD 328, November 1975.
2.  C. L. Walker, H. D. Wilstead, "Investigation of Bimetallic Liquid Metal Systems", GMAD 3643-8, September 1968.

3.  C. R. Manning, Jr., "Nitrogen Diffusion Constants in
    Austenitic Stainless Steels as Determined by Internal
    Friction Measurements", M.S. Thesis Virginia Polytechnic
    Institute, 1962.

4.  J. E. Cordwell, T. Swan, and S. P. Tyfield, "The Nitriding
    of Solution Treated AISI Type 316 Stainless Steel in
    Ammonia", RD/B/N3051 CEGB, December 1974.

5.  J. K. Stanley, "Mechanical Properties of Nitrided Austenitic
    Stainless Steel as Related to Microstructure; 1 January
    – 1 July 1969", AD-698745, November 1968.

6.  R. Hales and A. C. Hill, "The Diffusion of Nitrogen in an
    Austenitic Stainless Steel", Met.Sci., pp.241-244 (1977).

7.  "Vanadium Alloy Cladding Development. Quarterly Progress
    Report for the Period Ending March 31 1970", WARD-3791-46,
    April 1970.

8.  S. A. Shiels, C. Bagnall, S. J. Orbon and S. L. Schrock,
    "Interstitial Transfer Program. ITF Runs 2 and 3".
    WARD-NA-3045-22, July 1975.

9.  S. A. Shiels, C. Bagnall, S. J. Orbon and S. L. Schrock,
    "Interstitial Transfer Program, ITF Runs 4, 5 and 6,
    Program Conclusions", WARD-NA-3045-34. June 1976.

10. S. A. Shiels, C. Bagnall, and S. L. Schrock, "Interstitial
    Mass Transfer in Sodium Systems", Proceedings of the
    Symposium on Chemical Aspects of Carbon and Mass Transfer
    in Liquid Sodium, pp. 157-166, American Institute of
    Mining, Metallurgical and Petroleum Engineers, New York,
    1973.

11. "Effects of Sodium Exposure on the Corrosion and Strength
    of Stainless Steels Summary Report Sodium Mass Transfer
    Program", GEAP-10394, August 1971.

12. H. Atsumo et al.,"Sodium Compatibility and Corrosion Tests
    for Component Materials", "Proceedings of the Inter-
    national Conference on Liquid Metal Technology in
    Energy Production", CONF-760503-P2, November 1976.
    pp 849-859.

SOME OBSERVATIONS ON THE CARBURISATION OF TYPE 316 STAINLESS STEEL

FOIL IN A LOW CARBON ACTIVITY SODIUM ENVIRONMENT

Alan W. Thorley and Peter J. Jeffcoat

United Kingdom Atomic Energy Authority
Risley
United Kingdom

## INTRODUCTION

This paper describes work currently being undertaken at the Risley Nuclear Laboratories (RNL) to establish the equilibrium composition of carbides which form in stainless steel foils during their exposure to a low carbon activity sodium environment. The experiment is the first in a series which aim to study the equilibrium compositions of carbides which form in stainless steel foils and the time it takes the carbon to reach equilibrium during exposure to sodium of different carbon activity. The first part of the study, which is reported in this paper, deals with the lowest carbon activity measureable in our test loops where the sodium is just about carburising to stainless steel. Analytical techniques are used to determine the composition of the carbide and the austenite matrix and hence to estimate the carbon activity of the equilibrium structure. This provides a comparison with carbon activity values determined by alternative methods such as the Harwell Carbon Meter (HCM) and nickel tab techniques.

## EXPERIMENTAL

Type 316 stainless steel foils used in the experiment were obtained by mechanically working sheet material down to a thickness of 75 μm and then heat treating the steel at 1050 °C for 1 hour followed by a gas quench. The steel was then further heat treated for 1/4 hr at 1050 °C and furnace cooled. The purpose of this treatment was to remove all effects of mechanical working which might affect the diffusion kinetics of carbon in the steel while at the same time providing a single phase structure typical of those used in similar studies elsewhere (1). The composition of the steel is given in Table 1.

Table 1.   Compositon of 316 Stainless Steel Foil Specimens:

Element Wt %

| Fe  | Cr   | Ni   | C    | Mo   | Mn   | Si   | S     | P     |
|-----|------|------|------|------|------|------|-------|-------|
| Bal | 17.2 | 11.8 | .039 | 2.72 | 1.37 | .43  | .019  | .029  |

The material used to measure the carbon activity of the sodium was pure nickel (78 μm thick). This material was used in the annealed and lightly cold worked conditions.

Type 316 stainless steel and pure nickel were exposed at 650 $^{\circ}$C in a standard RNL test loop (2) fabricated from Type AISI 321 stabilised stainless steel which should minimise movement of carbon from loop pipework to the sodium. Exposure of the specimens was in the "well" of the loop which contains about 1 litre of sodium and through which sodium flows at about 0.16 m/sec. During exposure the main loop sodium was continuously cold-trapped at 120 $^{\circ}$C so as to minimise the level of soluble impurities such as oxygen. This was to ensure that the experiments were conducted below the oxygen threshold levels at 650 $^{\circ}$C required to form uniform sodium chromite films on the surface of the steel which might inhibit carbon take-up by the steel foils. In practice the measured levels (2) were generally less than 5 ppm oxygen.

Transfer of the foils into or out of the loop sodium was through an evacuable glove box which prevented contamination of the sodium. After removal the foils were washed in methanol to remove bulk sodium, then in pre-boiled distilled water and finally acetone. All foils were then lightly abraded to remove any carbonaceous material or contaminants remaining on the surface after washing. The foils were then prepared for subsequent carbon, nitrogen and microscopic analyses.

To measure the chemical activity of the carbon in the sodium, nickel tabs as well as a Harwell diffusion type carbon meter (3) were used. Nickel monitor tabs were exposed to the sodium alongside the Type 316 foil specimens and following removal, on either a weekly or monthly basis, their carbon content was determined using gamma activation analysis (4). To determine the carbon activity of the foil we have used the relationship for the terminal carbon solubility in nickel as a function of temperature recommended by Natesan and Kassner (5) and then assuming Henrian behaviour the activity of carbon in the nickel ($a_c$Ni) at 650 $^{\circ}$C, with respect to graphite as the standard state, was obtained through the use of the expression:

$$a_c\text{Ni} = \frac{\text{Wt\% of carbon in nickel}}{\text{Terminal solubility of carbon in nickel at 650}^{\circ}\text{C in wt\%}}$$

(1)

Values obtained using this expression are in reasonable agreement with alternative equations of carbon activity in nickel provided by Swartz (6).

Specimens for examination in the transmission electron microscope (JEOL 100C) were produced by two methods. First, a standard
two-stage technique (7) was used to prepare carbon extraction
replicas from the Type 316 alloy foils exposed to sodium for times
of 107, 217, 321 and 493 days. Selected area diffraction techniques.
were used to identify the various phases present and their
compositions were determined using STEM and energy dispersive X-ray
analysis facilities incorporated into the electron microscope.
Secondly, certain of the sodium exposed specimens were used to
produce conventional thin foil specimens by electrochemical polishing
in a Tenupol unit using an electrolyte of 6 % perchloric acid and
94 % glacial acetic acid at 13 $^{\circ}$C. These specimens were examined,
again using the energy dispersive techniques, to determine the
compositional variations in the austenite matrix, particularly in
areas adjacent to the precipitated carbide particles. For both
types of sample, the X-ray intensity data were converted to weight
fraction ratios using the Cliff and Lorimer technique (8).

RESULTS

Bulk carbon and nitrogen levels analysed in the steel foils are
detailed in Fig. 1. This figure shows that after the first stage of
testing the carbon level of the foil increases in roughly a parabolic
manner with exposure time whereas nitrogen losses occur in the foil
more or less at the start of the exposure period. The carbon activity
values obtained by the monitoring techniques (HCM and nickel tabs)
although giving different values, showed that the carbon activity
of the sodium during all test periods remained reasonably constant
(Fig. 1).

Transmission electron microscope studies of carbide extraction
replicas and conventional thin foil specimens produced from the 316
alloy prior to sodium exposure indicated no second phase precipitation
whatsoever. In contrast, sodium exposed specimens showed extensive
inter- and intra-granular precipitation, the size and extent of which
increased with increasing exposure time (Fig. 2). Selected area diffraction analysis indicated that there were only two different precipitate phases present in the structures irrespective of exposure
time. These were a $M_{23}C_6$ type cubic carbide and an intermetallic
Laves phase structure (Fig. 3). It should be noted that the two
phases could not be differentiated on the basis of either morphology
or position, site or size. Supporting chemical analyses of those
particles identified by electron diffraction were carried out using
the STEM and energy dispersive X-ray analysis facilities of the
microscope. The values from these analyses, which are detailed in
Table 2, represent the average as well as maximum and minimum

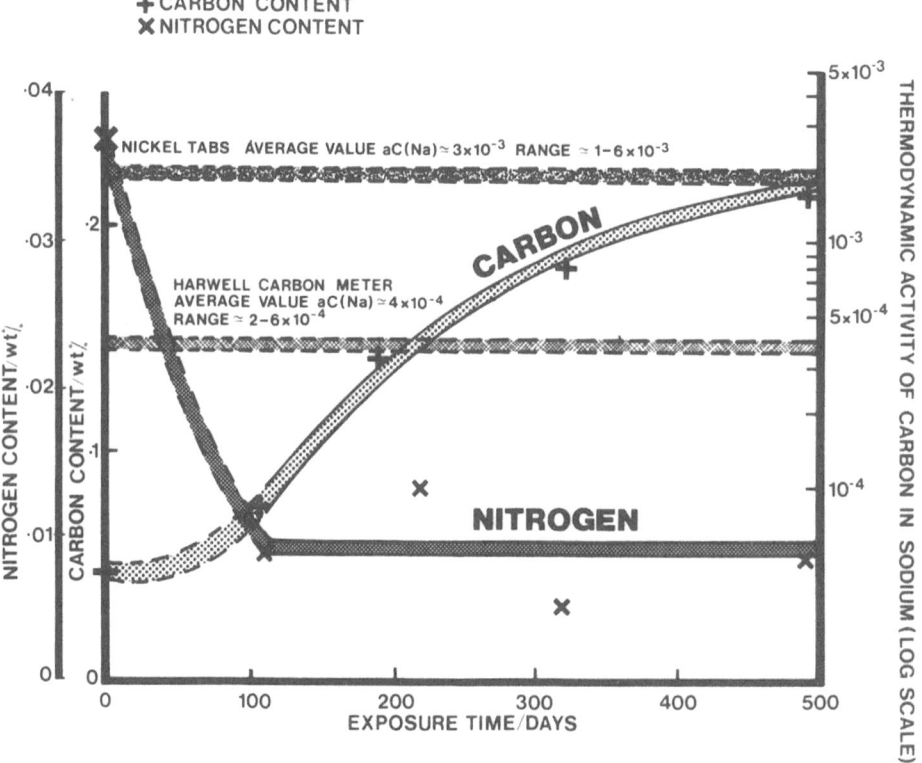

Fig. 1   Variation in Carbon and Nitrogen Levels in Heat Treated
Type 316 S/S Foils Exposed at 650 °C to low Velocity
Sodium

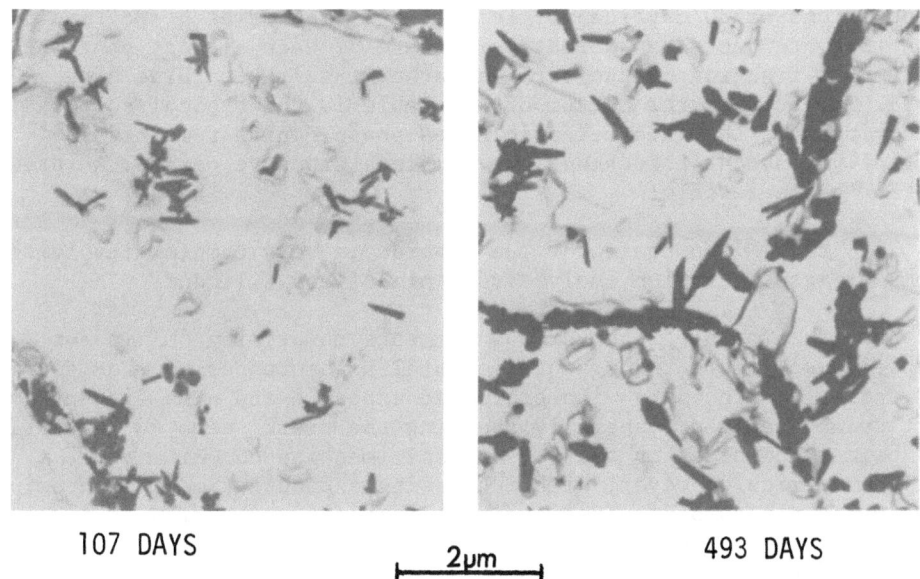

107 DAYS                    2μm                    493 DAYS

Fig. 2: Type 316 s/s foil exposed to liquid sodium at 650 $^{o}$C for
        107 days and 493 days. Carbide extraction T.E.M. studies.

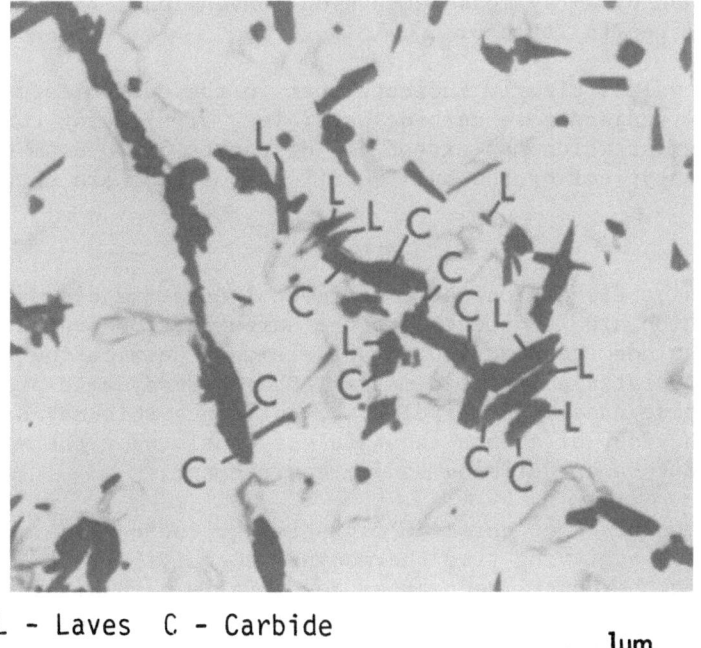

L - Laves  C - Carbide                        1μm

Fig. 3: Laves phase and carbide precipitation in Type 316 s/s fol-
        lowing exposure to liquid sodium at 650 $^{o}$C for 493 days

compositions of 30 carbide particles identified in each specimen.
Also they are quoted as weight percentage of each element contained
in the 'M' fraction of the carbide rather than the carbide as a
whole. Included at the bottom of the table is the typical composition
determined for the intermetallic Laves phase, which remained essentially constant throughout the complete exposure period. Reference
to this table reveals:

    i.  the 'M' component of the carbide is very complex involving
atoms of iron, chromium, molybdenum, nickel and silicon;

    ii. the ratio of the various elements comprising 'M' is not a
simple fixed figure throughout the total exposure period. For example
'M' varies continuously with prolonged exposure and also within any
particular specimen. Generally speaking the trend seems to be that
the primary carbides in the 107 day specimen are chromium and
molybdenum rich. However, with increasing exposure time the chromium
content of the carbides progressively increases primarily at the
expense of the molybdenum, nickel and silicon. In contrast, the iron
content of the carbides remains reasonably constant and along with
the chromium is the major component of 'M' in the 493 day specimen
(Table 2 and Fig. 4).

    Values for the variation in composition of the austenite matrix,
as determined by X-ray measurements on conventional thin foil specimens, are given in Table 3.

    These values clearly indicate that in the austenite matrix,
particularly adjacent to carbide particles, significant concentration
gradients exist which may extend for distances of $\sim 0.8$ μm before
values more typical of the bulk matrix composition are obtained.

DISCUSSION

    To assist discussion the experimental data are discussed under
headings which are fundamental to the carburisation process. These
headings include the distribution of elements between the carbide
and the associated austenitic matrix, the thermodynamic properties
of the carbide phase and thirdly, the influence this may have on the
kinetic behaviour of carbon in stainless steels when the materials
are exposed to a carburising sodium environment.

    The importance of chromium migration to the carbide phase and
its effect on achieving true thermodynamic equilibrium in the steel
foil between the constituent atoms, especially carbon, cannot be
ignored in the light of the observed compositional changes. The
experimental evidence indicates that in the initial stages of exposure to sodium, relatively metastable carbides of the $M_{23}C_6$ type
are formed, whereas over the longer term chromium replaces a proportion of elements like Ni, Si, Mo, to form an equilibrium carbide
structure containing the major elements, Fe, Cr and carbon.

Table 2. Composition of $M_{23}C_6$ carbides extracted from sodium exposed 316 alloy foils. (STEM plus energy dispersive X-ray analysis.)

| Exposure Time (days) | Cr | Fe | Mo | Ni | Si | Comments |
|---|---|---|---|---|---|---|
| 107 range | 23.8-34 | 13.7-17 | 34.8-39.5 | 9.9-14.1 | 5.8-11 | Primary carbides |
| average | 28.1 | 15.1 | 37.1 | 11.6 | 8.1 | high Mo content |
| | 46.9-53.2 | 16.1-17.5 | 19.7-27.3 | 5.7-6.9 | 3.8-5.2 | )Very few carbides |
| | 50.1 | 17 | 22.5 | 6.2 | 4.2 | )of these compo- |
| | 62.5-65.9 | 13.5-16.6 | 13.2-13.9 | 3.6-4.1 | 3.3-3.4 | )sitions found. |
| | 64.2 | 15 | 13.6 | 3.9 | 3.4 | )Note increased )Cr level |
| 217 | 47.6-67.9 | 15.8-26.8 | 10.4-20.2 | 2.6-4.7 | 0.5-2.4 | Primary carbides |
| | 61 | 17.9 | 16 | 3.9 | 1.2 | |
| | 26.9-28.7 | 15-18 | 37.2-41.7 | 11.4-12.2 | 3.2-4.4 | Very few carbides |
| | 27.9 | 16.2 | 40 | 11.9 | 4 | in this composi-tion range remaining |
| 321 | 55-70.2 | 12.5-19.8 | 12.2-18.9 | 1.6-3.9 | 0.6-4.3 | |
| | 63.4 | 15.8 | 15.8 | 2.6 | 2.4 | |
| 493 | 60.6-74.8 | 11.8-14.6 | 9.3-18 | 2.8-5.1 | 0.2-1.7 | |
| | 68.8 | 13.2 | 13.3 | 3.8 | 0.9 | |
| Laves Phase | 11.2 | 27.6 | 56.6 | 3 | 1.6 | |

Table 3. Variation in composition of austenite matrix adjacent to carbide particles. Conventional thin foil specimens examined by STEM and energy dispersive X-ray analysis.

| Exposure Time (days) | Cr | Fe | Mo | Ni | Si | Comments |
|---|---|---|---|---|---|---|
| 107 | 14.2 | 68.5 | 5.5 | 11.5 | .3 | $= 3 \times 10^{-5}$ mms from carbide particle. |
| | 14.6 | 69.5 | 5.6 | 10.0 | .3 | $= 1 \times 10^{-4}$ mms "   "   " |
| | 16.9 | 69.5 | 2.6 | 10.7 | .3 | Bulk matrix remote from ppt. |
| 217 | 13.7 | 70 | 5.4 | 10.5 | .4 | Immediately adjacent to carbide. |
| | 14.1 | 71 | 3.2 | 11 | .7 | |
| | 16.2 | 69.4 | 3.8 | 10.3 | .3 $^V$ | $\simeq 8 \times 10^{-4}$ mms. |
| | 17.5 | 68.9 | 2.8 | 10.5 | .3 | Bulk matrix remote from ppt. |
| 321 | 17 | 70 | 2.8 | 10 | .2 | Bulk matrix. |
| 493 | 14 | 71.6 | 2.3 | 11.5 | .6 | Adjacent to carbide. |
| | 15.9 | 68.1 | 5.9 | 9.5 | .6 | $\simeq 4 \times 10^{-4}$ mms. |
| | 16.5 | 69.4 | 2.8 | 10.7 | .6$^V$ | Bulk matrix. |

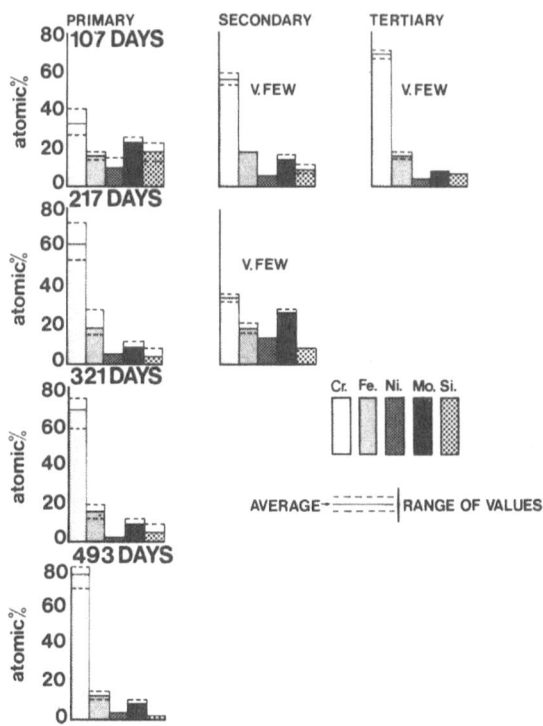

Fig. 4    Variation in Composition of $M_{23}C_6$ Carbides with Prolonged
Exposure to Sodium at 650 °C

        To provide correlations which allow one to relate carbon contents
in foils to true thermodynamic carbon activities it is apparent from
these experiments that in a carburising environment where the carbon
activities are relatively low, time is an important parameter. Time
for example affects the level of chromium and nickel distribution
between the carbide and the matrix and this in turn affects the
metallurgical stability of the austenite phase. Removal of chromium
to the carbide may also decrease the carbon activity of the carbon
in the matrix by increasing the terminal solubility of carbon in
austenite. At the same time the foil is trying to come to equilibrium
with the carbon in the sodium and the only way this can be done is
by taking up more carbon as carbides transform and precipitate.
Therefore it seems that carbon take-up by the foil is controlled, in
part, by time dependent compositional changes both in the carbide and
the austenite solid solution, and not entirely by carbon diffusion
in the austenite.

For this calculation it has been assumed that the precipitated $M_{23}C_6$ carbide is a stoichiometric carbide, minor element concentration in the carbide such as Ni, Si and Mo have not been normalised to Fe and Cr equivalents, therefore mCr + nFe = 23 atoms, the observed Laves phase does not interfere with the carburisation process and the equilibrium bulk carbon level in the foil is 0.22 w/o (Fig. 1). The composition of the carbide according to the analytical values in Table 2 is $Cr_{20}Fe_3C_6$.

To establish the carbon activity of the carbide phase we have used first of all an empirical approach using equations provided by Tuma (9) and RNL (10) which relate the equilibrium carbon content of the steel to carbon activities. These data were obtained from gas equilibration methods. The second approach calculates the free energy of formation of the carbide phase from data provided by Rand (11) for the free energies of the constituent carbides $F_{23/29}C_{6/29}$ and $Cr_{23/29}C_{6/29}$, an energy term for the ideal mixing of the constituent atoms and an excess free energy of mixing.[x)]

ie, $\Delta G_f^o = xFe \Delta G_f^o Fe_{23/29}C_{6/29} + xCr \Delta G_f^o Cr_{23/29} C_{6/29} + RT$

$(xFe \ell nxFe + xCr \ell nxCr) + G_M^{XS}$        (2)

where x denotes the atom fraction of the constituent elements which for the equilibrium $Cr_{20} Fe_3C_6$ carbide are given by:

$xCr = {}^{20}/_{29}, \quad xFe = {}^{3}/_{29} \text{ and } xC = {}^{6}/_{29}$

The estimated value for the standard free energy of formation for the precipitated carbide using this equation is - 16.1 KCals/g atom.C. This value is in line with literature values for carbides of this type (12, 13). The basic equation to describe the equilibrium situations between the precipitated carbide phase and the austenite solid solution is of the form:

$\frac{20}{6} [Cr]_{Aus} + \frac{3}{6} [Fe]_{Aus} + [C]_{Aus} \rightleftharpoons 1/6 \ (Cr_{20}Fe_3C_6)$        (3)

Through the law of Mass Action this equation can be rewritten:

$K = \frac{a^{1/6}(Cr_{20}Fe_3C_6)}{a^{20/6}_{Cr} \ a^{3/6}_{Fe} \ aC}$        (4)

where K is the equilibrium constant and is related to equ. 2 through

[x)] To apply the regular solution model, B.C.C. carbon must be used. Therefore to relate $\Delta G_f^o$ to graphite as the standard state a $\Delta G$ value for the change from B.C.C. carbon to graphite must be added to this equation.

the expression $\Delta G_f^o = -RT\ln K$. Therefore using the calculated value of $-16.1KCals/grm$ atom.C for $\Delta G_f^o$ the value for K at 650 $^o$C is 6497.

To determine the chemical activities of Fe and Cr comparisons of their thermodynamic behaviour in Fe-Cr, Fe-Ni, Ni-Cr binary alloys have been made and where possible this behaviour has been related to the constituent ternary alloys of the FeNiCrC system (13,14,15,16). This research indicates that alloying elements such as chromium, nickel and carbon have small effects on the activity of Fe in the relevant binary systems and there is only a small positive deviation from Raoultian behaviour. Therefore the deviations from ideality for iron in FeNiCrC are thought to be small ($\gamma$Fe = 1.03). The acivity of chromium in FeNiCr ternary alloys is however affected by nickel content (14,16) and according to Benz (13) and Mazandarany (14) by temperature but not by carbon at the carbon concentrations of interest (15). Therefore to account for the effect of both nickel content and temperature on the activity coefficient of chromium in FeNiCr austenite we have used a value for $\gamma_{Cr}^{Aus}$ of 4.5. This value is in reasonable agreement with an estimated value of 3.91 provided by Rand (17) for BCC Cr in FeNiCr alloys.

Substituting 6497 for K at the equilibrium temperature of 650 $^o$C and 0.85 and 0.71 for the activity of Cr and Fe respectively in equ.4, the $a_c$ value for the steel foil is estimated as $3.14 \times 10^{-4}$. This value is in better agreement with the HCM value than the nickel foil method and in reasonable agreement with predicted values for stainless steel containing 0.22 w/o carbon obtained by Tuma (9) and RNL (10) which are $5.2 \times 10^{-4}$ and $1.7 \times 10^{-4}$ respectively.

In order to try and understand how carbon transports through an austenitic matrix and how this is related to the precipitation of the carbide phase, it is convenient to consider the total quantity of carbon entering the steel as though it was composed of two hypothetical types, namely, reactive carbon and diffusing carbon. That is as the carbon penetrates the austenitic matrix the reactive fraction combines primarily with iron and chromium atoms and so precipitates as carbides whilst the remaining diffusion carbon continues on through the austenitic matrix. In many respects it is difficult to conceive a mechanism by which carbon penetrates austenite when the solid solution is already saturated with carbon. However, it seems that the carbon itself is instrumental in 'unlocking' the austenite in that once it has reacted with surface chromium it provides the necessary driving force for subsequent diffusion into the bulk austenite by locally affecting the solubility of carbon in regions of carbide precipitation.

One way of assessing the magnitude of this effect is through the use of a solubility parameter Ks which requires the solubilities of the mole concentrations of the metal and carbon components of the $Cr_{20}Fe_3C_6$ phase to be in solution in the austenite matrix in

the same ratio as they are in the carbide. Therefore Ks is related
to the equilibrium constant K. For the purposes of the calculation
we have used the estimated figure for chromium loss in the vicinity
of the carbide which is 3 % (Table 3), to see what effect this has
on the level of carbon solubility in this region.

By definition the solubility parameter Ks can be written:

$$Ks = \frac{1}{x_{Cr}^{20} \, x_{Fe}^{3} \, x_{C}^{6}} \qquad x = \text{Mole fraction} \qquad (5)$$

or rewritten

$$Ks = \frac{\gamma_{Cr}^{20} \, \gamma_{Fe}^{3} \, \gamma_{C}^{6}}{a_{Cr}^{20} \, a_{Fe}^{3} \, a_{C}^{6}} \qquad (6)$$

The denominator is recognisable as the denominator in equ. (4)
raised to the power 6. The inverse of this new denominator may be
written as K' and so

$$Ks = \gamma_{Cr}^{20} \, \gamma_{Fe}^{3} \, \gamma_{C}^{6} \, K' \qquad (7)$$

or    $ \ln Ks = 20\ln\gamma_{Cr} + 3\ln\gamma_{Fe} + 6\ln\gamma_{C} + \ln K' \qquad (8)$

Values for $\gamma_{Cr}$ and $\gamma_{Fe}$ are 4.5 and 1.03 respectively (see
earlier) and $\gamma_{C}$ can be estimated from Natesan's data (5) where
through the various interaction coefficients

$$\ln\gamma_{C} = \ln\gamma_{C}^{Cr} + \ln\gamma_{C}^{Ni} + \ln\gamma_{C}^{C} = 0.937 \qquad (9)$$

Substituting the various values in equ. 8 Ks = 2.634 x $10^{38}$.
However, because the mole fraction of iron remains sensibly constant,
equ. 5 can be rewritten:

$$Ks \, x_{Fe}^{3} = \frac{1}{x_{Cr}^{20} \cdot x_{C}^{6}} = 8.81 \times 10^{37} \qquad (10)$$

Substituting the determined values for matrix and grain boundary
Cr, 17 w/o and 14 w/o respectively, into equ. 10 it is seen that the
removal of   3 w/o Cr in the vicinity of the carbides increases the
local solubility of carbon from   15 to 32 ppm. It is felt such
differences, which are dependent on Cr diffusion from the matrix
to the carbide and subsequent precipitation of the carbide, are
sufficient to provide the necessary driving forces for further
carbon movement into the steel. In this connection it is interesting
to note that the carbon uptake values in the foil fit reasonably
well the Mehl equation which has been used in Ref. 18 to describe

the rate of precipitation of carbides in aged Type 316 stainless
steel.

CONCLUSIONS

This work has demonstrated that in low carbon activity environments it is difficult using quenched steels to obtain realistic
measurements of the carbon activity of the sodium over timescales
of 100 days at 650 $^{o}$C. The reason for this is that the quenched
structure is metastable with respect to both nitrogen and carbon
contents and in the initial stages of exposure most of the nitrogen
in the foil is released to the sodium and in the later stages
carbides start to form in the structure which are also metastable.
It seems doubtful whether during this period the steel acts as a
carbon activity monitor for the sodium.

The equilibrium carbide which eventually forms in the steel at
these low carbon activities has the composition $Cr_{20}Fe_3C_6$. Preliminary estimated carbon activities for this carbide give a value
of 3.1 x $10^{-4}$. This value is in reasonable agreement with the HCM
value and values obtained using different experimental techniques
(9, 10), However it should be noted this estimate is very sensitive
to the value used for $^{a}Cr$ and therefore better estimates are required
for this value before the activity value for carbon can be considered
to be really soundly based.

REFERENCES

1.  K. Natesan and T. F. Kassner: Nuclear Technology, Vol. 19,
    1973, pp. 46-56.
2.  A. W. Thorley and C. Tyzack: Liquid Alkali Metals. Proc.
    Nottingham, 1973, pp. 257-72.
3.  R. C. Asher et al: AERE R8020, HMSO 1977.
4.  J. S. Hislop et al: AERE R8182, HMSO 1975.
5.  K. Natesan and T. F. Kassner: Metallurgical Transactions,
    Vol. 4, 1973, pp.2557-66.
6.  J. C. Swartz: Metallurgical Transactions, Vol. 2, 1971,
    pp. 2318-9.
7.  C. J. Smithells: Metals Reference Book, 5th Edition,
    Butterworths, London, 1976.
8.  G. Cliff and G. W. Lorimer: Journal of Microscopy, Vol. 103,
    1975.
9.  H. Tuma et al: Archiv für das Eisenhüttenwesen, Vol. 40,
    1969, pp. 727 - 31.
10. A. W. Thorley and M. R. Hobdell: IAEA Specialists Meeting,
    Carbon in Sodium, Harwell, 1979.
11. I. Ansara and M. H. Rand: in The Industrial Use of Thermo-
    Chemical Data. Ed.Barry T I, Special Publication No. 34,
    Chemical Society, London, 1980.

12. V. I. Alekseev and L. A. Shvartsman: Russian Metallurgy,
    Pt. 1, 1965, pp.117-9.
13. R. Benz et al: Metallurgical Transactions, Vol. 5, 1974,
    pp.2235-40.
14. F. N. Mazandarany and R. D. Pehlke: Metallurgical Trans-
    actions, Vol. 4, 1973, pp. 2067-76.
15. J. Chipman: Metallurgical Transactions, Vol. 5, 1974, p.521.
16. W. Slough et al: Journal of Chemical Thermodynamics, Vol.2,
    1970, pp.117-24.
17. M. H. Rand: Private Communication, 1981.
18. J. K. L. Lai and M. Meshkat: Metal Science, Sept. 1978,
    pp.415-20.

STUDIES ON THE CARBURISATION OF AN AISI TYPE 316

STAINLESS STEEL IN LIQUID SODIUM

T. Jayakumar, B. Raj, D. K. Bhattacharya,
P. Rodriguez, B. Prahlad and R. D. Kale

Reactor Research Centre
Kalpakkam, India

INTRODUCTION

One of the well recognised problems in LMFBR sodium circuits
is the phenomenon of carbon transfer which takes place across sodium/
steel interface in the direction of decreasing carbon activity. The
major concern due to carbon transfer is connected with the integrity
of thin walled parts such as fuel cladding, IHX and steam generator
tubings, bellows in control rod drive mechanisms and sodium valves
etc. This is because of the fact that the property changes are more
pronounced in thin walled sections.

Carbon transfer to structural materials takes place in the
sodium circuits of an LMFBR from a number of sources (1). The effects
of carbon transport in sodium circuits are of concern since the
carbon level in the steel material governs the tensile properties
to a great extent. The deleterious effects of carburisation are
reflected in the deterioration in ductility and low cycle fatigue
life. It has been reported (2) that the tensile ductility of AISI
type 316 L stainless steel falls drastically to as low as zero at
an average carbon concentration of around 0.6 wt% at room temperature
as well as at high temperatures (773 K - 873 K).

Inspite of considerable work having been done in the field of
carbon transfer in sodium/stainless steel systems there is scope
for improving the understanding by correlating the carburised micro-
structures with tensile properties and various types of fractures.
It has been reported (3) that a 10 % maximum carburised case depth
may be allowable for safe operation of components; however, it is

desirable to investigate the general applicability of this observation. An extensive programme has been undertaken to study the
effects of carburisation of stainless steels in sodium. In the first
phase, static sodium has been chosen as the carburising medium.
Simultaneously, studies on removal of carbon from sodium by gettering
with austenitic stainless steels have also been undertaken. This
paper gives the results of tensile testing of VIRGO 14 SB (an AISI
type 316 stainless steel) exposed to static sodium at 823 and 873 K
for a maximum period of 2000 hours. The results of measurement of
carbon activity by foil equilibration technique and rates of carbon
removal using stainless steel getters are also briefly discussed.

EXPERIMENTAL

A static sodium exposure set up was fabricated. The test section
(300 mm dia) contains about 43 litres of sodium. The total surface
area of specimens and vessel is about 0.87 $m^2$ giving a surface area
to volume ratio of ca. 200 $cm^2$/1. Sodium temperature in the test
section was maintained within $\pm$ 5 K. Temperature of sodium was
measured by 3 numbers of stainless steel sheathed Chromel-Alumel
thermocouples kept in the thermowells into the vessel.

Tensile specimens of VIRGO 14 SB (unpolished, polished to 600
grit SiC, and composition (16.47 wt%-Cr, 12.43 wt%-Ni, 2.28 wt%-Mo,
1.69 wt%-Mn, 0.054 wt%-C) having gauge dimensions 32 x 6.35 x 3 mm,
and AISI type 304 stainless steel foils (60 microns thick and
composition (16.6 wt%-Cr, 8 wt%-Ni, 1.9 wt%-Mn and 0.046 wt%-C)
and wire mesh (120 microns dia) were exposed to sodium at 823 and
873 K. Filtered sodium (10 microns pore size filter), unfiltered
sodium and oil contaminated and filtered sodium were used in these
experiments. All the specimens, foils and wiremesh were ultrasonically degreased before they were exposed in sodium. After exposure, the
specimens were cleaned in ethanol and then in water.

Oil got introduced into the sodium accidentally after 1000 hours
of exposure in the system. Due to this, the data pertaining to the
exposure time beyond 1000 hours at 873 K would correspond to the
exposure in oil contaminated sodium only.

To separate out the contribution of thermal ageing on the microstructure and tensile properties, tensile specimens were thermally
aged in vacuum (better than $10^{-4}$ torr) at 823 - 873 K for similar
lengths of times.

The tension tests were carried out in an Instron testing machine
at a strain rate of 5.2 x $10^{-4}$/sec. The tests at 873 K were carried
out in vacuum (better than $10^{-4}$ torr) with a temperature measurement
accuracy of $\pm$ 2 K. Tension tests were followed by optical microscopy,
microhardness measurements and examination of fracture surfaces by
scanning electron microscope.

The foils and wire mesh, after exposure to sodium, were chemically analysed for carbon, by the total combustion method.

RESULTS AND DISCUSSION

Table 1 shows the results of carbon analysis of foils and wire
mesh, and the gettering rates of carbon in sodium. The carbon concentration of the equilibrated (1000 hours exposure) foil was taken
to determine the carbon activity in sodium using the equation developed by Natesan et al (4). The extent and rates of carburization
in commercial grade sodium (filtered/unfiltered) were observed to
be small as the carbon activity was in the range of 0.018 to 0.025
at 823 and 873 K respectively. Here, the carbon activity in sodium
appears to have increased slightly when the temperature increased
from 823 to 873 K. This may be possible in case of sodium where a
lot of particulate carbon is available which can go into solution
at higher temperature. However when sodium was contaminated with
oil, the specimens were heavily carburized and the activity of
carbon increased to about 0.28. The average rates of carburization

Table 1. Data on carbon activity and gettering rates in sodium

Abbreviations: FC% = Final carbon concentration, wt%
                ARG = Average rate of gettering, μgr-C/hr.cm$^2$

| Exposure time (hours) | Unfiltered sodium at 823 K.  ac = 0.0176 | | | | Filtered sodium at 873 K ac = 0.0247 | | | |
|---|---|---|---|---|---|---|---|---|
| | Foil 60 μm thick Ci=0.046 % | | Wire mesh 140 μm dia Ci=0.06 % | | Foil 60 μm thick Ci=0.046 % | | Wire mesh 140 μm dia. Ci=0.06 % | |
| | FC% | ARG | FC% | *ARG | FC% | ARG | FC% | *ARG |
| 400 | - | - | - | - | 0.11 | 0.0384 | - | - |
| 600 | 0.096 | 0.020 | 0.125 | 0.03 | 0.16 | 0.0456 | 0.287 | 0.1059 |
| 1000 | 0.11 | 0.0154 | 0.16 | 0.028 | 0.168 | 0.0293 | - | - |
| o 1000 | - | - | - | - | 1.51 | 0.351 | 1.59 | 0.428 |

ARG   = (C-Ci) ρss x 5000 1/t, μgmC/cm$_2^2$ hr
ARG   = (C-Ci) ρss x 2500 d/t, μgmC/cm$^2$ hr

where C   = Carbon wt% of foil/wiremesh
      Ci  = Initial carbon wt% of foil/wiremesh
      ρss = Density of stainless steel, gm/cc
      l   = Thickness of foil, cm
      d   = Diameter of wiremesh, cm
      t   = Exposure time, hrs
o Exposure in oil contaminated and filtered sodium at 873 K.
Carbon activity = 0.2817.

at 873 K were observed to be more than ten times the average rates obtained in uncontaminated sodium.

Though austenitic stainless steel appears to have the potential for gettering carbon from sodium, it requires very high temperatures, probably in the range of 923 - 973 K for quicker removal of carbon from sodium in case of accidental carbon ingress. Materials like Titanium which have greater affinity to carbon, are expected to remove carbon from sodium more efficiently at comparatively lower temperatures, and hence it may be useful to conduct experiments using such materials.

Table 2. shows some of the results obtained in tensile tests.

The data in this table supplemented by data from optical microscopy and microhardness measurements allow us to get the following observations:

(a) The carburisation in commercial unfiltered sodium is at the grain boundaries and to a small extent only. This resulted in only a slight increase in room temperature tensile strengths to the values obtained for thermally aged specimens. The reduction in room temperature ductility is more pronounced at 873 K than at 823 K. Similar observations have been obtained when the sodium is filtered. Similar grain boundary carburisations as seen by us have also been observed by others (5) at carbon activities around 0.1. It appears that the carbon burden in filtered/unfiltered commercial sodium in which the carbon activity was much less than 0.1, did not affect the tensile properties. Whatever changes in tensile properties took place can be attributed to thermal ageing effects.

(b) As expected, much more extensive carburisations were observed when accidental oil ingress took plase in the system. This caused an increase in room temperature yield stress and ultimate tensile stress and a considerable decrease in ductility. The case depths for polished and unpolished specimens as estimated from optical micrographs and microhardness profiles (Fig. 1) were found to be average 80 microns and a maximum 150 microns respectively. Figs. 2a and 2b compare the extents of carburisation in the two cases. In the case of polished specimens, the carburisations were more or less uniform and less whereas for the unpolished specimens, the extents of carburisation was non-uniform and more. Similar observation has earlier been reported (6) while studying the effect of carbonaceous impurities in Helium gas, on austenitic stainless steels. The reason given by them was an internal oxide layer with pure Fe-Ni alloy in unpolished specimens acting as a catalyst for carbon transfer, thus resulting in a larger extent of carburisation as compared in polished specimens. Since we did not have any such oxide layer, direct comparison of this study with our results may not be valid.

Table 2. Data from Tensile Tests

abbreviations: unfilt - Unfiltered Sodium, filt - Filtered Sodium, Pol - Polished tensile specimens, unpol - Unpolished tensile specimens, O/c - oil contaminated and filtered sodium.

| Treatment | YS in kg/mm² | | UTS in kg/mm² | | % Tot.Elong. | | % Unif.Elong. | |
|---|---|---|---|---|---|---|---|---|
| | Vacuum | Sodium | Vacuum | Sodium | Vacuum | Sodium | Vacuum | Sodium |
| As received | 30.0 | | 61.0 | | 65.0 | | 53.0 | |
| **Room Temperature Data** | | | | | | | | |
| 823K, 400 hrs,unfilt,pol. | 32.5 | 31.0 | 60.5 | 63.0 | 53.0 | 54.0 | 44.0 | 44.5 |
| 823K, 600 hrs,unfilt,pol. | 32.0 | 32.5 | 63.0 | 62.0 | 53.0 | 55.5 | 45.5 | 44.0 |
| 823K, 1000 hrs,unfilt,pol. | 30.0 | 31.5 | 62.5 | 65.5 | 55.5 | 43.0 | 45.5 | 37.0 |
| 873K, 600 hrs,unfilt,pol. | | 33.0 | | 64.0 | | 47.0 | | 40.0 |
| 873K, 600 hrs,filt,unpol. | | 31.0 | | 62.0 | | 54.5 | | 46.0 |
| 873K, 600 hrs,filt,pol. | 30.2 | 30.6 | 62.5 | 62.5 | 55.0 | 50.5 | 45.0 | 43.0 |
| 873K, 1000 hrs,filt,unpol. | | 31.5 | | 62.5 | | 53.0 | | 44.5 |
| 873K, 1000 hrs,filt,pol. | 30.7 | 31.5 | 63.0 | 62.0 | 54.0 | 59.0 | 46.0 | 48.0 |
| 873K, 2000 hrs,O/c,unpol. | | 33.0 | | 63.0 | | 32.0 | | 26.0 |
| 873K, 2000 hrs,O/c,pol. | 29.0 | 33.0 | 63.0 | 63.0 | 53.0 | 30.0 | 43.5 | 27.0 |
| **High Temperature Data** | | | | | | | | |
| As received | 22.7 | | 37.6 | | 38.4 | | 31.3 | |
| 873K, 600 hrs,filt,unpol. | | 21.0 | | 43.0 | | 37.0 | | 31.0 |
| 873K, 600 hrs,filt,pol. | 21.0 | 20.0 | 42.0 | 42.5 | 33.0 | 36.0 | 25.0 | 29.0 |
| 873K, 1000 hrs,filt,unpol. | | 19.0 | | 41.0 | | 40.0 | | 31.0 |
| 873K, 1000 hrs,filt,pol. | 20.0 | 19.5 | 45.0 | 41.0 | 33.0 | 39.0 | 29.0 | 32.0 |
| 873K, 2000 hrs,O/c,unpol. | | 23.0 | | 42.0 | | 27.0 | | 22.0 |
| 873K, 2000 hrs,O/c,pol. | 20.0 | 23.0 | 41.0 | 43.0 | 30.0 | 28.0 | 24.0 | 23.0 |

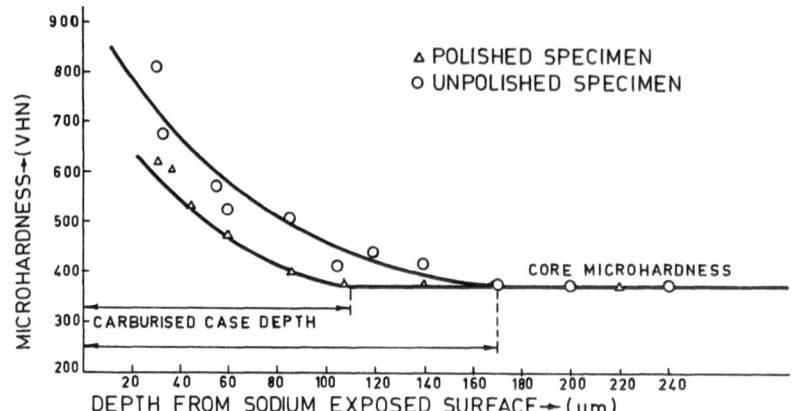

FIG.1. A TYPICAL MICROHARDNESS PROFILE ACROSS THE SECTION
OF THE SPECIMEN EXPOSED FOR 1000 hrs. IN FILTERED
SODIUM AND 1000 hrs IN OIL CONTAMINATED SODIUM
AT 873 K

(c) Tensile testing at 873 K showed comparable values for
yield stress and ultimate tensile stress and ductility for specimens
exposed to filtered sodium at 873 K as compared to thermally aged
specimens. After exposure from 1000 hours – 2000 hrs in oil contaminated sodium, there was a decrease in ductility and increase in
yield stress, ultimate tensile stress as compared to those for
thermally aged specimens.

Thermally aged specimens and specimens exposed to sodium at
823 K when tensile tested at room temperature showed a ductile type
of failure with a decrease in dimple size as exposure time increased,
indicating a decrease in ductility with time. On the other hand,
thermally aged specimens and specimens exposed to sodium at 873 K
when tensile tested at room temperature showed a mixed mode of failur
in which dimples were seen on grain facets and final failure occured
along grain boundaries. Ductile type of failure was observed when
the above specimens were tested at 873 K.

In case of specimens exposed to oil contaminated sodium at
873 K and tensile tested at room temperature and 873 K, intergranular
brittle type of failure was observed in the carburised case (Fig. 3)
whereas in the core region, the failure was found to be of mixed mode
and ductile mode for specimens tested at room temperature and 873 K
respectively.

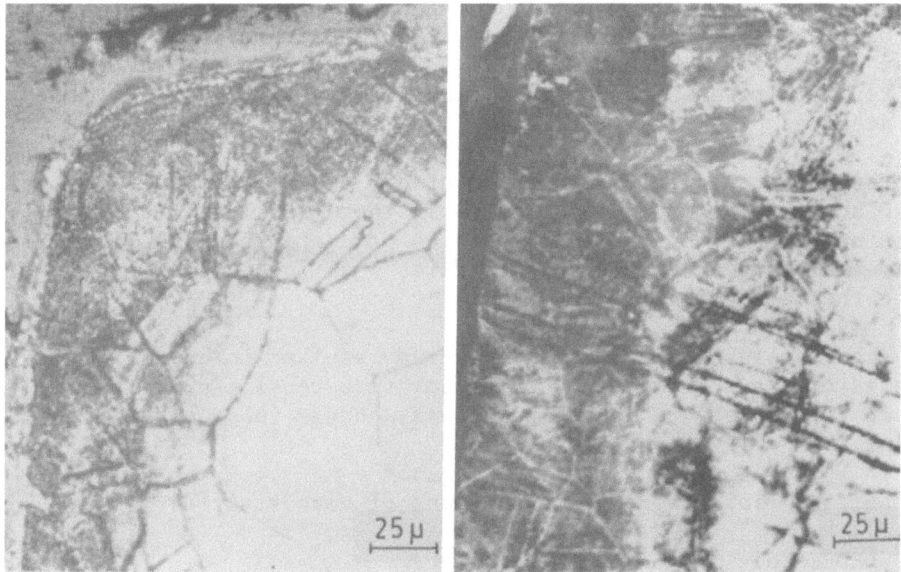

Fig. 2  Typical optical photomicrographs of the specimens exposed
        at 873 K for 1000 hours in filtered sodium and 1000 hours
        in oil contaminated sodium. (a) Polished specimen. (b) Un-
        polished specimen.

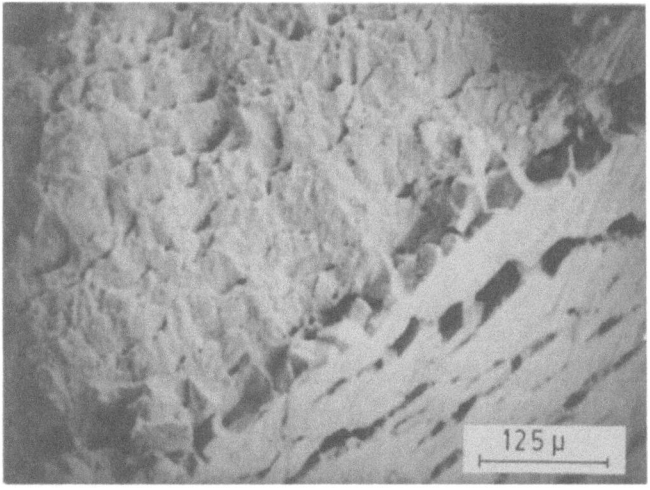

Fig. 3  A typical fractograph of the specimen exposed for 1000 hours
        in filtered sodium and 1000 hours in oil contaminated sodium
        at 873 K.

CONCLUSIONS

Oil ingress into sodium has effected substantial increase in carbon activity from about 0.025 to 0.28 which resulted in more than ten fold rise in carbon removal rates by AISI type 304 stainless steel at 873 K.

For unfiltered/filtered sodium at 823 and 873 K where the carbon activity was much less than 0.1, there was little carburisation and change in tensile properties of austenitic VIRGO 14 SB steel. The changes in tensile properties were found to be mainly due to thermal ageing.

Accidental ingress of oil in sodium resulted in extensive carburisation and change in room temperature tensile properties. Even in this case, at high temperatures, change in properties were nominal.

Surface finish affected the extent of carburisation.

ACKNOWLEDGEMENT

We thank our numerous colleagues who have helped us during different stages of this investigation. In particular, Radiochemistry Laboratory of Reactor Research Centre and Analytical Chemistry Division of Bhabha Atomic Research Centre are thanked for carrying out chemical analysis.

REFERENCES

1. R. B. Hinze, NAA-SR-Memo 12394 (May 1967).
2. A. Thorley et al, TRG Report 1909(C) 1969.
3. D. I. Sinizer et al, NAA-SR-Memo 7804 (1967).
4. K. Natesan et al, Nuclear Technology 19 (July 1973), p.46.
5. W. Charnock et al, CEGB Report RD/B/N 2980 Oct. 1974.
6. F. N. Mazandarany et al, Nuclear Technology 43(3) May 1979, p. 349.

# CABURIZATION KINETICS OF ALLOY-800 IN LIQUID SODIUM AT 550 $^\circ$C

Sergio Casadio, Giovanni Bruni, Gianni D'Alessandro
and Corrado Meloni

CNEN-Casaccia
Rome, Italy

## INTRODUCTION

The austenitic Alloy-800 steel is a structural material candidate for the steam generators of nuclear power plants, including the sodium cooled breeder reactors (1). In this case the Alloy-800 is exposed to the sodium environment of the secondary heat transfer system coupled to the AISI type 316 or 304 stainless steels generally employed for the intermediate heat exchanger components. For this all-austenitic bimetallic system the corrosion of the materials and the consequent mass transfer are not of relevant importance in the normal service conditions (good technological purity of the sodium and temperatures below 550 $^\circ$C). However a certain carbon (or nitrogen) interstitial transfer may occur because of the thermal gradients and the different nature of the alloys of the loop; a general carburization could also arise because of the presence of possible carbon sources during the plant running (2).

In this note we report our preliminary observations on the isothermal carbon transfer in low-carbon Alloy-800 exposed in sodium with AISI-304 type of steel at 550 $^\circ$C.

## MATERIALS

In Table 1 the chemical composition of the four examined Alloy-800 types are reported. We used thin specimens (thickness 0.3 and 0.5 mm) obtained by progressive rolling of the original thick plates. After each 30 % of cold work the material was solution annealed (SA) at 1050 $^\circ$C for 15 minutes in a vacuum furnace ($10^{-5}$ torr) and quenched at room temperature by an argon jet. The last SA treatment was performed at 1020 $^\circ$C, the resulting structure being intermediate between the commercial grades 1 and 2 (ASTM grain size 9).

Table 1.   Chemical composition of the alloys

| Heat | C% | Si% | Mn% | S% | P% | Cr% | Ni% | Ti% | Al% | Cu% | N ppm |
|------|------|------|------|------|------|-------|-------|------|------|------|-------|
| A | 0.029 | 0.48 | 0.63 | 0.011 | 0.007 | 21.30 | 33.40 | 0.41 | 0.18 | 0.07 | 158 |
| B | 0.028 | 0.46 | 0.56 | 0.004 | 0.008 | 21.50 | 33.20 | 0.50 | 0.05 | 0.07 | 150 |
| D | 0.029 | 0.45 | 0.59 | 0.012 | 0.008 | 21.75 | 33.25 | 0.50 | 0.28 | 0.07 | 150 |
| E | 0.030 | 0.49 | 0.61 | 0.005 | 0.007 | 21.85 | 33.25 | 0.20 | 0.20 | 0.07 | 150 |

EXPERIMENTAL

The specimens were exposed in static sodium capsule systems after a conditioning treatment to eliminate most of the oxygen impurity by foil and powder Zr getter. AISI-304 foils (thickness 150 μm of the same composition as the sodium containers systems) were placed with the Alloy-800 specimens to check the carbon activity of the environment.

The total mean carbon concentration $\overline{C}$ in the steels was determined by destructive combustion with an induction furnace and quantitative detection of the carbon oxides evolved by the LECO-WR/12 analyer.

An Ernst Leitz GmbH Wetzlar microhardness measuring apparatus was used to evaluate the carbon penetration in some thicker pieces (see Fig. 1c). Some tests were performed in a dynamic sodium (Na velocity of about 2 m/s) whose cold trap worked at  120 $^{\circ}$C.

CARBON ACTIVITY EVALUATION

Since we performed Tab-test (2) in conditions of non-equilibrium for the carbon partition between the metal foil and the sodium, the true equilibrium carbon concentration (Ce) on the AISI-304 foils (thickness = Z) was determined by measuring the amount of carbon which entered in the tab at the time t by equ. (1)

$$M = (\overline{C} - C^{o}) \times Z \tag{1}$$

$C^{o}$ being the original carbon level in the alloy.

At the equilibrium we should have (3):

$$Me = M/(1 - \sum_{n=0}^{\infty} (8/(2n+1)^2 \pi^2) \exp(-D(2n+1)^2 \pi t/Z^2) \tag{2}$$

and the corresponding Ce value in the tab is given by equ. (3):

$$Ce = Me/Z + C^{o} \tag{3}$$

We used $D = 5.7 \times 10^{-12}$ cm$^2$/s at 550 $^{\circ}$C$^2$.

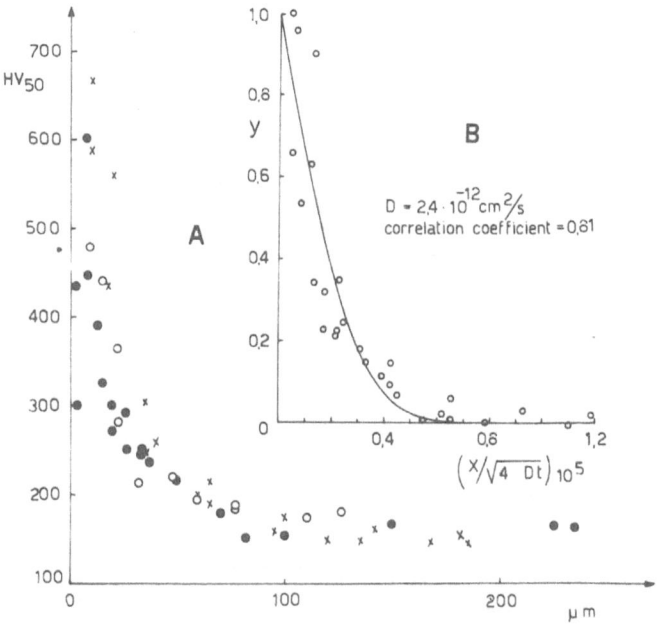

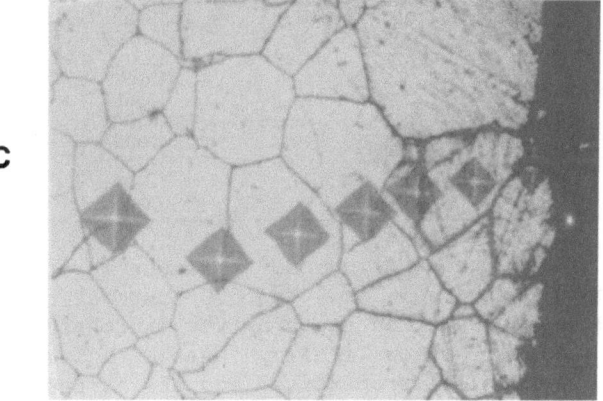

Fig. 1:   A): Microhardness vs. x for the alloy E carburized in
              sodium at 550 °C for 650 h.
          B): Best fit analysis of the experimental Y ratios (eq.(5))
              with the error function (5).
          C): Micrography of the sectioned specimen showing the
              carburized zone with the diamond indentation marks
              (gliceregia etched). (500 x)

The carbon activity in the austenitic alloy can be evaluated according to the Natesan and Kassner (4) relationship (4)

$$\ln a_C = \ln(0.048\ Ce) + (0.525 - \frac{300}{T})Ce - 1.845 + \frac{5100}{T} - \tag{4}$$

$$(0.021 - \frac{72.4}{T})C_{Ni} + (0.248 - \frac{404}{T})C_{Cr} - (0.0102 - \frac{9.422}{T})C_{Cr}^2$$

where $C_{Ni}$ and $C_{Cr}$ are the Ni and Cr weight percent concentrations in the steel.

In Table 2 the measured carbon activities are reported for the experimental tests of this work.

Table 2.   Carbon potentials in the Na environments in the systems used at 550 $^{\circ}$C to perform the carbon transfer experiments. The original carbon concentration in the AISI-304 foil was C = 0.076 %.

| Exp. | Time (hours) | $\bar{C}$ (wt %) | (M/Me) x 100 (%) | $a_C$ | Note |
|------|------|------|------|------|------|
| I | 450 | 1.02 | 45. | 0.37 | Capsule test, carb. AISI-304 SS |
| II | 450 | 0.23 | 45. | 0.053 | " " " " " " " |
| III | 1320 | 1.05 | 75. | 0.21 | " " " " " " " |
| IV | 1320 | 0.45 | 75. | 0.076 | " " " " " " " |
| V | 960 | 0.080 | 66. | 0.010 | " " " " " " " |
| VI | 960 | 0.087 | 66. | 0.011 | " " " " " " " |
| VII | 1030 | 0.077 | 67.5 | 0.0095 | " " " " " " " |
| VIII | 1030 | 0.057 | 67.5 | 0.0058 | " " " , decarb. " " " |
| IX | 2700 | 0.0785 | 94. | 0.010 | " " " , carb. " " " |
| X | 2700 | 0.023 | 94. | 0.0023 | " " " " " " " |
| XI | 2330 | 0.079 | 90. | 0.0097 | " " " " " " " |
| XII | 4130 | 0.0755 | 98 | 0.0092 | " " " , decarb. " " " |
| XIII | 5540 | 0.036 | 99.5 | 0.0044 | " " " " " " " |
| XIV | 670 | 0.070 | 55. | 0.0080 | Dynamic sodium test, decarb. " |

Some thick Alloy-800 specimens were exposed in a carburizing sodium bath ($a_C \sim 0.4$) for 650 hours at 550 $^{\circ}$C to evaluate the carbon penetration depth by the microhardness technique. For such a purpose the low Ti alloy (heat E in Table 1) was chosen since its mechanical properties (and its hardness in particular) were found to be almost unaffected by the thermal aging treatment at 550 $^{\circ}$C. That was not the case for the other alloys with a normal Ti content for which the intermetallic $\gamma'$ phase precipitation induces an important hardening process during their life in the 500 - 650 $^{\circ}$C temperature field. The microhardness measures were found widely scattered, they are plotted versus the distance from the sodium exposed surface (x) in Fig. 1A. An effective carbon diffusion coefficient D of about 2.4 x 10$^{-12}$ cm$^2$/s was found to best fit the experimental data with the error function (5)

$$Y = (Ce-C_{(x,t)})/(Ce - C^{\circ}) = erf\ (x/\ \sqrt{4\ Dt}) \tag{5}$$

the ratio Y being evaluated by using directly the microhardness values instead of the concentrations of the carbon of the central term of equ. (5) (Fig. 1B). In spite of such a very rough approximation the measured D was comparable with the values quoted at WARD (2) for the AISI-304 ($5.7 \times 10^{-12}$ cm$^2$/s) and for the AISI-316 ($2.4 \times 10^{-12}$ cm$^2$/s) at 550 °C.

In Table 2 we reported the mean carbon concentrations ($\overline{C}$) measured on the Alloy-800 specimens exposed in the sodium environments. The chemical composition of the alloys, notably the Ti content, seems not to be relevant to get different carbon transfers within our conditions. In the cases where the semiinfinite approximation (and the equ. (5)) holds, the carbon taken up is related to the carbon activity of the sodium and to the square root of the exposure time by the plot of Fig. 2.

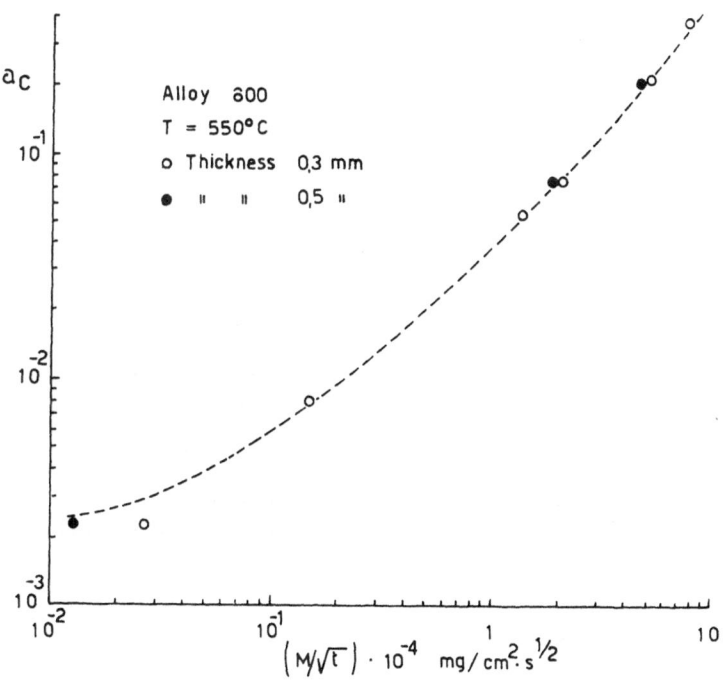

Fig. 2   Alloy-800 carburization rate constant M/$\sqrt{t}$ for the various carbon activities of the sodium at 550 °C.

We calculated the Ce values for the Alloy-800 thin specimens by using the equ. (1), (2) and (3) and they are plotted in Fig. 3 versus the carbon activity values. The empirical correlation (6)

$$Ce = 8.53 \cdot a_C^{1.26} + C^o \tag{6}$$

was found to well fit the experimental data. Most of the original
carbon content ($C^o$ = 0.03 %) was not found to move away because it
is probably present as TiC carbides (or TiCN carbonitrides) formed
during the SA treatments and as $(Cr, Fe)_{23}C_6$ carbides precipitated
at the grain boundaries during the sodium exposure. A decarburization of the Alloy-800 was infact never observed, even for carbon
cativities producing a decarburization of the AISI-304 steel
(Table 2 and Fig. 3).

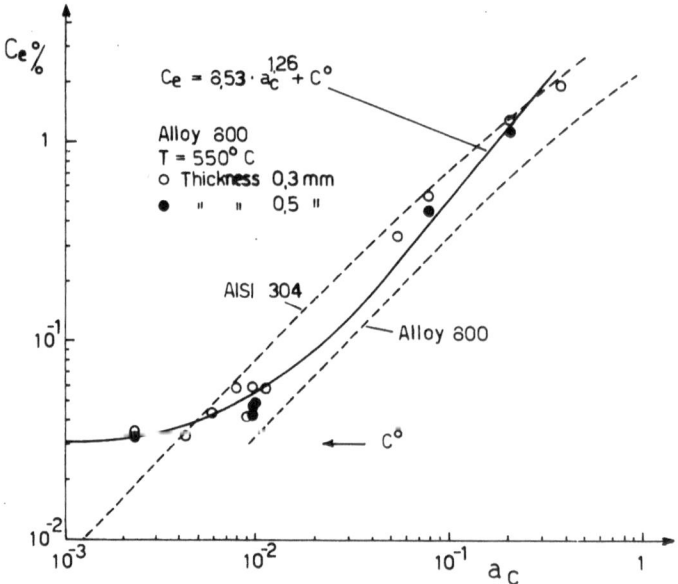

Fig. 3   Ce for Alloy-800 vs. $a_C$ at 550 °C. The
         dotted lines show the equ. (4) for the
         AISI-304 and for the Alloy-800 compositions.


        As a consequence a slight "dynamic" carburization may affect
the Alloy-800 of the steam generators by detriment of the intermediate heat exchanger stainless steel acting as a carbon source
in the secondary systems.

CONCLUSIONS

        The low-carbon Alloy-800 was found to undergo a carburization
comparable or slightly lower than that observed for the AISI-304
stainless steel for carbon activities above 6 x $10^{-3}$ at 550 °C, but
was not found to decarburize below such a carbon activity value.
In the examined range the chemical composition of the alloys does

not affect the process out of the experimental uncertainty of measure.
The effective diffusion coefficient for carbon in the Alloy-800 was
found close to those available for the 18 Cr - 8-10 Ni austenitic
steels (Fig. 1). Preliminary relationships (Fig. 2 and equ. (6))
were found for an empirical evaluation of the carburization of the
Alloy-800 in sodium of known carbon activity.

REFERENCES

1.  M. Julien; Nucl. Technol., 31, 367 (1976).
2.  J. J. McCown and C. Bagnall; HEDL-SA-1950-FP.
3.  J. Crank; The Mathematics of Diffusion, Oxford University
      Press, p. 45.
4.  K. Natesan and T. F. Kassner; Nucl. Technol., 23, 46 (1974).

# MODELLING FOR THE CARBURIZATION OF THE ALLOY - 800 IN LIQUID SODIUM

Andrea Saltelli and Sergio Casadio

CNEN-CSN-CASACCIA
Rome
Italy

## INTRODUCTION

The interstitial carbon transfer is one of the most important phenomena concerning the compatibility of metals and alloys with high temperature liquid sodium. In previous comunications we reported some methods to evaluate the extent of the carburization (or de-carburization) expected to occur on austenitic steels exposed in a given sodium environment (1, 2). The calculations were essentially based on the criteria developed by Snyder, Natesan and Kassner (3) for the AISI 304 and 316 steels, but some simplifications and minor modifications were introduced to take into account the empirical model performed by Shrock, Shiels and Bagnall (4, 1) and of the coupling of the interstitial carbon diffusion with the selective leaching of the Cr and Ni substitutional alloying elements acting on the same time at the surface of the materials exposed in the hotter zone of the sodium loops (2).

The method developed at ANL[3] make use of structural parameters to take into account the thermal-mechanical history of the material that may influence the carbide precipitation kinetics by altering the grain size and the size and distribution of the carbide particles. That was performed by evaluating the average carbon concentration in the austenitic phase as a function of time during the growth of the carbides following the Nolfi approach (5). The carbide particles were thought as spherical shaped and uniformely distributed throughout the matrix and their growth (the precipitation kinetics) controlled by the carbon diffusion in the matrix (5). Carburization of the AISI type 316 steel exposed at 650 °C in sodium with carbon activity ranging between 0.1 and 0.3 was found to proceed by $(Cr, Fe)_{23}C_6$ carbide formation solely at the

grain-boundaries (g.b.), there being no precipitated carbide in
the grains (6). This last fact has been generally assumed to occur
during the thermal aging of the austenitic steels, and was stated
to explain the "sensitization" to the intergranular corrosion of
such materials (6, 7).

Although the true carbide morphologies in sensitized steels
are known to be very thin dentritic (7), its modelling by a
continuous film enveloping the grains was found satisfactory to
describe the Time-Temperature-Sensitization (TTS) behaviour of
the austenitic steel by the "chromium depletion theory" both for
the classical 18%Cr-8%Ni steels (6, 7) and the Ni-rich Alloy-800
(8). Intragranular carbides have been also observed to occur as
a later process during the thermal aging above 650 - 700 $^\circ$C (9,
10, 11). In this work we present a mathematical analysis to
produce carbon profile concentrations in austenitic steels as a
function of both the environmental conditions and the relevant
metallurgical parameters to the characterization of the TTS field
for the alloy considered, in the starting "solution annealed"
condition. In this case the carbide precipitation kinetics is
assumed to be completely controlled by the sole chromium diffusion (6, 7, 8). The Alloy-800 has been retained to be more
resistent both to the carburization (12) and to the stress
corrosion (13) than the classical 18%Cr-8%Ni steels and it has
been chosen for the heat-exchanger tubes of the steam generators
for the Superphenix fast reactor plant (13). The present model is
then addressed to the carbon transfer analysis for such a material.
The behaviour and the modelling for the carburization-decarburization
of the AISI type 304 already stated (1, 2, 3, 4) is employed in
this work as a reference for a comparative analysis.

CARBON TRANSPORT MODELLING

Let us consider a stainless steel specimen of thickness z
exposed to the liquid sodium; if $C^{Na}$ is the carbon concentration
(ppm) in sodium and $C^{Na}_{sat}$ is the corresponding saturation level,
the Henry's law is generally assumed to hold by neglecting the
presence of poliatomic carbon species in the system (the dimeric
acetylide has been proved to be present in certain circumstances
(14), hence the carbon activity in sodium ($a_C^{Na}$) is expressed by
the eq. (1) and (2).(3)

$$a_C^{Na} = C^{Na}/C^{Na}_{sat} \tag{1}$$

$$C^{Na}_{sat} = 5.03 \times 10^7 \exp(-13.740/T) \tag{2}$$

The carbon activity in an austenitic alloy can be evaluated
according to the Natesan and Kassner (3) relationship (equ. 3)

$$\ln a_C = \ln(0.048C)+(0.525-300/T)C-1.845+5100/T - \tag{3}$$

$$(0.021 - 72.4/T)C_{Ni}+(0.248-404/T)C_{Cr}-(0.0102-9.422/T)C_{Cr}^2$$

The carbon concentration at the surface of the specimen ($C^s$) is then obtained by solving for the carbon concentration in the steel (C) the implicit equation (3) by imposing $a_C = a_C^{Na}$. Stawstrom and Hillert (7) produced a model for the temperature dipending kinetics of the chromium carbide precipitation in steels 304 or 316 types. We exstended the Stawstrom model to high Ni concent alloys (8). In the following we briefly recall such a treatment by introducing the modifications to account for the carbon, chromium and nickel transport between the sodium and the steel. A visual picture of the precipitation process is given in Fig. 1, representing the chromium profile imposed by the chromium carbide precipitation at the g.b.. By a linear approximation of the Cr gradient near the g.b. we can state

$$-J_{Cr} = D_{Cr} \ (X_{Cr}^{\gamma} - X_{Cr}^{i})/\ell \ V_m^{\gamma} \tag{4}$$

Imposing mass law to an infinitesimal carbide layer (dK) generated at the g.b. by the disappearance of a $\gamma$-phase layer (da) we have:

$$(dK/dt)/V_m^C = (da/dt)/V_m \tag{5}$$

$J_{Cr}$ can also be expressed by eq. (6)

$$-J_{Cr} = (dK/dt)X_{Cr}^C/V_m^C - (da/dt)X_{Cr}^i/V_m^{\gamma} \tag{6}$$

where the chromium concentration in the carbide ($X_{Cr}^C$) is supposed constant. By coupling equations (4) to (6) we obtain eq. (7)

$$(dK/dt)=V_m^{\gamma} \ D_{Cr}(X_{Cr}-X_{Cr}^{i})/(V_m^C \ \ell \ (X_{Cr}^C - X_{Cr}^{i})) \tag{7}$$

Inside the grain the carbon concentration gradient is supposed to be negligeable, hence the carbon molar fraction into the grain ($X_C$) will decrease uniformely with time. The rate of disappearance of carbon from the $\gamma$-phase will be equal to the rate of carbon production in the carbide phase (eq. (8))

$$(dX_C/dt)V/V_m^{\gamma} \ \text{for prec.} \ =-(dK/dt) \ 2SX_C^C/V_m^C \tag{8}$$

From (7) and (8) we have

$$(dX_C^{\gamma}/dt) = 2D_{Cr}X_C^C(X_{Cr}^{\gamma}-X_{Cr}^{i})/(h\ell(X_{Cr}^C-X_{Cr}^{i})) \tag{9}$$

where h = V/S is a measure of the grain size. As regard $\ell$ we have adopted for short times the relation $\ell = 2\sqrt{D_{Cr}t}$, interpolating for long times with the limiting value $\ell = h/2$. The chromium concentration at the austenite-carbide interface $X_{Cr}^i$ is related to the carbon concentration $X_C^i$ and to the carbon activity $a_C$ in the grain, which can be computed by eq. (3). Hence $X_{Cr}^i$ can be computed by simultaneous solving eq. (3) and (10)

$$\text{Log } C_{sat} = a(T) + b(T)(C_{Cr}-20)-0.01 \ C_{Ni})  \tag{10}$$

with

$$a(T) = -\ 5500/T + 2.92$$

and

$$b(T) = (73.36 - 3.27 \ 10^4/T - 5.03 \ 10^{-2}T)$$

Equation (10) accounts for the carbon solubility as function of Cr and Ni content of the alloy (8). To solve the system of eqs (3) and (10) with respect to carbon and chromium concentration means to find concentration values obeying to the solubility relationship (10) and which are compatible with the activity $a_C$ in the bulk of the grain. The coupling of the carbon transport from (or towards) the sodium and the carbide precipitation process inside the material is then performed by considering the concentrations referred above as "in the bulk of the grain" or "mean value in the grain" as functions of the distance from the sodium-steel interface. The boundary value problem becomes:

$$\frac{\partial C}{\partial t} = D_C \frac{\partial^2 C}{\partial x^2} - r(C, \ C_{Cr}, \ C_{Ni})  \tag{11}$$

$$\frac{\partial C_{Cr}}{\partial t} = D_{Cr} \frac{\partial^2 C_{Cr}}{\partial x^2} - f_{Cr}r(C, \ C_{Cr}, \ C_{Ni})  \tag{12}$$

$$\frac{\partial C_{Ni}}{\partial t} = D_{Ni} \frac{\partial^2 C_{Ni}}{\partial x^2}  \tag{13}$$

$$r(C,C_{Cr},C_{Ni})=f_C2D_{Cr}(X_{Cr}-X_{CR}^i)/(h\ell(X_{Cr}^c-X_{Cr}^i))  \tag{14}$$

for x=0 or x=z and t $\geq$ 0: $C=C^S$; $C_{Cr}=C_{Cr}^S$; $C_{Ni}=C_{Ni}^S$

for x$\varepsilon$ (0,z) and t=0: $C=C^o$; $C_{Cr}=C_{Cr}^o$; $C_{Ni}=C_{Ni}^o$ with

Table 1

| Exposure time, h | Na $a_C$ | | C transf.to 0.3mm thick specimens (mg/cm²) | | C transf.to 0.5 mm thick specimens (mg/cm²) | |
|---|---|---|---|---|---|---|
| | crit.ref. 19 | crit.ref. 22 | calcd. | exptl. | calcd. | exptl. |
| 445 | 0.37 | 0.36 | 0.93 | 1.04 +0.13 | | |
| 450 | 0.053 | 0.075 | 0.18 | 0.18 $\overline{+}$0.02 | | |
| 1320 | 0.21 | 0.24 | 1.12 | 1.15 $\overline{+}$0.02 | 1.12 | 1.06+0.13 |
| 1320 | 0.06 | 0.09 | 0.44 | 0.47 $\overline{+}$0.02 | 0.44 | 0.06$\overline{+}$0.13 |
| 1030 | 0.0095 | 0.0023 | 0.077 | 0.024+0.003 | | |

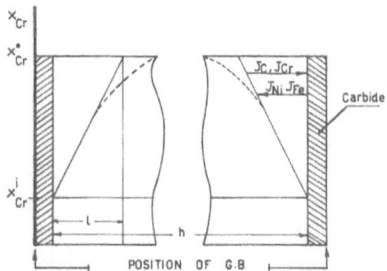

Fig. 1:   A possible schema for Cr concentration profile in a grain
          during the carbide precipitation at the grain boundaries

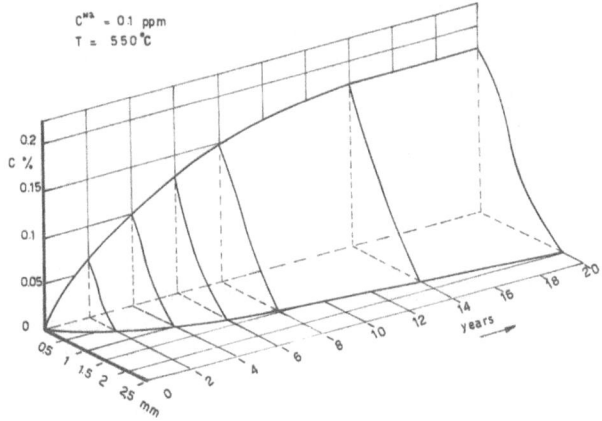

Fig. 2:   Calculated tridimensional diagram of the C profiles inside
          a 2.5 mm thick wall of Alloy-800 exposed to liquid Na at
          550 °C with a C concentration of 0.1 ppm

$$f_{Cr}=(M_{Cr}/M_C) \; 16/6 \; ; \; f_C=M_C \; f_o$$

$$f_o=C_{Cr}/M_{Cr}+C_{Ni}/M_{Ni}+C/M_C+(100-C_{Ni}-C_{Cr})/M_{Fe}$$

(the equation (14) is obviously the eq. (9) expressed in weight %).

The system of the eq.s (11) to (14) was numerically solved by a computer code written in CSMP III language. The carbon concentration profiles are given in terms of averaged carbon levels in the finite selected spatial cell $\Delta x$ at the various distances $x_i=i\Delta x$; i varying from 0 to $z/\Delta x$.

## RESULTS AND DISCUSSION

In Table 1 calculated total carbon uptake from thin Alloy-800 specimens are compared with the observed ones for various sodium environmental conditions at 550 $^o$C (19). The carbon activities of the various batches employed were determined by tab tests following the procedure described elsewere (19, 22). It can be observed how the present model gives reasonable results mainly for carbon concentrations in sodium greater than about 0.07 ppm. In both the primary and secondary sodium systems of the LMFBR's the carbon level may reach such a value (17). Snyder, Natesan and Kassner (3) performed previsional calculation for the IHX carburization by using a carbon level in the sodium of about 0.13 ppm, a value almost representative of the secondary sodium contamination when the low alloyed 2.25 % Cr - 1 % Mo ferritic steel is used for the steam generator tubing. It is well known how this material acts as a carbon source for the sodium. Nevertheless oil leakage from the sodium pumps could also occur to make possible such a carbon contamination even in a whole austenitic secondary system. By assuming a secondary sodium containing 0.1 ppm of carbon, the carburization profiles expected on a steam generator tube wall during its service life at 550 $^o$C are reported in Fig.2. In Fig.3 the corresponding average carbon contents in the wall tube are also plotted. Our calculated carbon content at the end of service life (0.13 % after 20 years of exposure) is compatible with a corresponding value determinated by the General Electric investigators (20) (0.245 % after 210,000 hours for 1 mm thick tube wall at 1000 $^o$F).

Examples of calculation of the carbon concentration as a function of both the distance from the sodium exposed surface and of exposure time are reported in Fig. 4 for thin specimens (thickness 0.5 mm) of Alloy-800 and of AISI-304 steel wetted by sodium containing 2.5 ppm of carbon at 550 $^o$C. The Alloy-800 results to undergo carburization to a lesser extent than that pertaining to the classical 18%Cr-8%Ni stainless steel, in agreement with some experimental available data (20). It has to be noticed that the diffusion rate is enhanced if the surface area

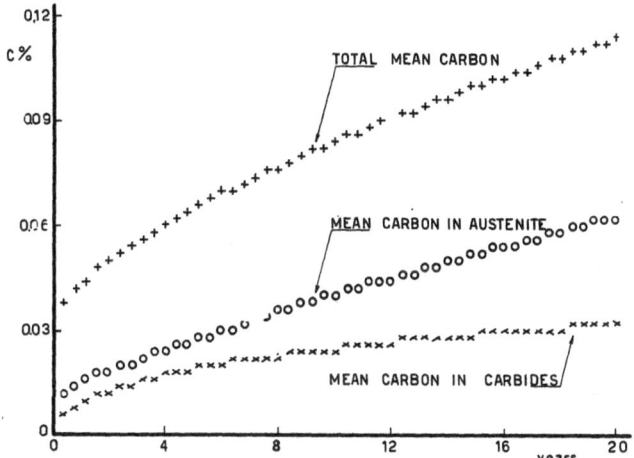

Fig. 3: Calculated average C concentration variations with time
in 2.5 mm thick wall with a surface exposed in liquid Na
with 0.1 ppm of C at 550 °C

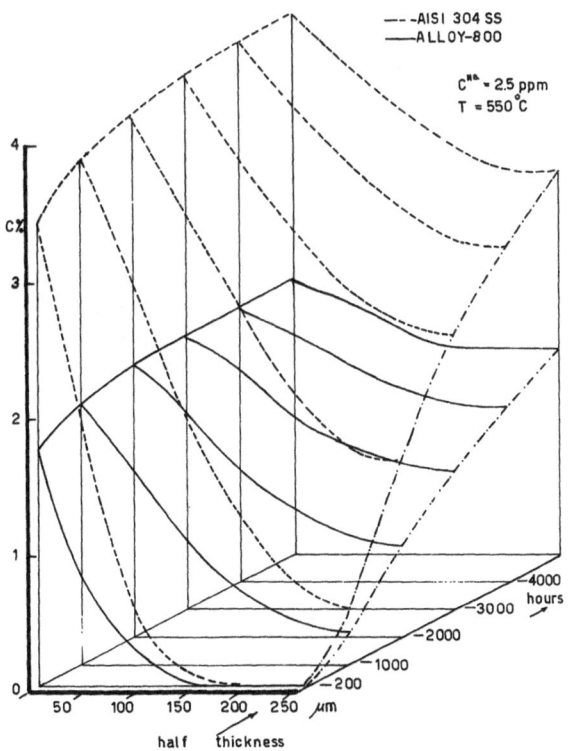

Fig. 4: Tridimensional diagram showing the evolution of C profiles
in 0.5 mm thick specimens of Alloy-800 (full lines) and
AISI-304 (dotted lines) exposed in Na containing 2.5 ppm
C at 550 °C

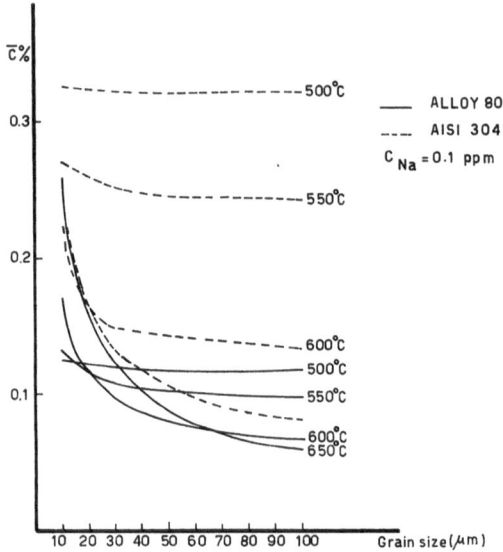

Fig. 5:  Total C taken up by 0.5 mm thick specimens of Alloy-800
         and AISI-304 steel after 5000 hours of exposure at 550 °C
         in Na with 2.5 ppm C, as a function of the grain size.

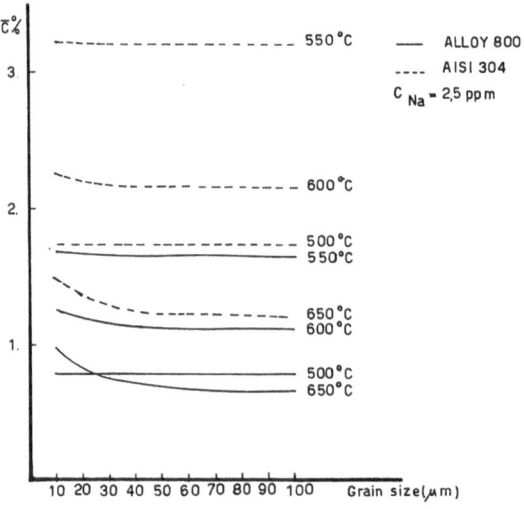

Fig. 6:  Total C taken up on the same specimens and condictions
         of Fig. 5, but for a C content in Na of 0.1 ppm

of the g.b. is increased by decreasing the grain size in the poly-
crystalline aggregates (21). By decreasing the grain size we
observe an increase (Fig.s 5 and 6) of the carbon transfer resulting
from the higher surface available for the carbide precipitation
accompanying the interstitial carbon diffusion, however this effect
becomes appreciable only above 550 $^{o}$C in a carburizing environment
(Fig. 6) representative of the in service sodium purity (3, 20).
On the other side above 650 - 700 $^{o}$C the intergranular precipitation
of the carbides should also occur (9, 10), thus reducing the relative
importance of the precipitation sites at the g.b., and the influence
of the grain size which is the sole parameter we consider in the
present simulation of the process. In all the present calculation
we used the carbon diffusion coefficients obtained by Schrock,
Shiels and Bagnall (4, 17) for the type 304 steel. Up to date
this choice seems to be reasonable looking at our experimental
data (Table 1). However longer time exposures have to be considered
to examine the evolution of the interstitial diffusion during the
in-service aging of the material.

CONCLUSIONS

    The model proposed in this work seems to describe fairly well
the carburization of the Alloy-800 exposed to a sodium environment
of known carbon activity, above a carbon concentration of 0.07 ppm
at 550 $^{o}$C.

    The grain size parameter has been included in the calculation:
its decreasing anhances the carburization process, although this
effect is not significant in highly carburizing sodium.

    The Alloy-800 resists to the carburization better than the
AISI type 304 steel because of thermodynamic reasons, the kinetic
of the process being characterized by near the same effective
diffusion coefficients for the two alloys.

    Long time extrapolations of the carburization expected at
550 $^{o}$C in sodium contamined with 0.1 ppm of carbon show a significant
carbon penetration as compared to the wall thickness of a steam
generator tube during its operating life.

REFERENCES

    1.  S. Casadio, G. Scibona; Specialists' Meeting on Carbon
        in Sodium, Harwell, U.K., 27-30 Nov. 1979, IAEA-
        IWGFR/33.
    2.  S. Casadio, A. Saltelli, G. Scibona; Second Int.Conf.on
        Liquid Metal Technology in Energy Production, Richland,
        USA, 20-24 April 1980.
    3.  B. Snyder, K. Natesan, T. F. Kassner; ANL-8015 (June 1973).

4.  S. L. Schrock, S. A. Shiels, C. Bagnall; First Int.Conf.
    on Liquid Metal Technology in Energy Production,
    Champion, USA, 3-6 May 1976.
5.  F. V. Nolfi Jr., P. G. Shewmon, J. S. Foster; Trans.Met.
    Soc. of AIME, 245, 1427 (1969).
6.  R. L. Cowan II, C. S. Tedmon Jr.; Adv.Corr.Sci.and Techn.,
    vol.3,p.317 (1973).
7.  C. Stawtrom, M. Hillert; J. of Iron and Steel Inst., Jan.
    1969, p.77.
8.  A. Borello, S. Casadio, A. Saltelli, G. Scibona; accepted
    for publication on Corrosion NACE, 1980.
9.  B. Weiss, R. Stickler; Met.Trans., 3, 851 (1972).
10. L. K. Singhal, J. W. Martin; Acta Met., 16, 1159 (1968).
11. A. Schnaas, H. J. Grable; Oxid.Met., 12, 387 (1978).
12. J. Blanchet, H. Coriou; "Alloy-800", North-Holland Pub.
    Comp., 1978, p.241.
13. M. Julien; Nucl.Techn., 31, 367 (1976).
14. M. R. Hobdell et al.; Chemical and Technological aspects
    of carbon in liquid sodium, Int.Conf.of Ref.2.
15. J. Crank; The Mathematics of Diffusion, Oxford Univ.Press
    (1970).
16. C. C. Miles; Anal.Chem., 41, 1041 (1969).
17. J. J. McCown, C. Bagnall; HEDL-SA-1950-FP (1979).
18. J. J. Goldstein, A. E. Moren; Met.Trans.A, 9A, 1515 (1978).
19. S. Casadio, G. Bruni, G. D'Alessandro, C. Meloni; Carburi-
    zation Kinetics of Alloy-800 in liquid sodium at 550 $^{o}$C,
    this Meeting on the "Material Behaviour and Physical
    Chem...", 24-26 March 1981, Karlsruhe, DFR.
20. GEAP-13919-3, March 1973, Third Quartely Report on LMFBR
    Heat Exchanger Materials Dev. Progr.
21. V. Z. Bugakov; Diffusion in metals and alloy, Israel
    Progr. for Sci. Trans., 1971.
22. A. Saltelli, S. Casadio; CNEN-RT/CHI (80)13.

CLOSING THE SEMINAR

Hans Ulrich Borgstedt

Kernforschungszentrum Karlsruhe
Karlsruhe
Fed. Rep. of Germany

The aim of the seminar on "Material Behaviour and Physical Chemistry in Liquid Metal Systems" was to discuss the forthcoming of our scientific work. We did not intend to find solutions of unsolved technological problems of nuclear power stations. After listening the presentations of the 53 contributions my impression is that the seminar gave us a platform for the desired detailed discussions. And we all have also used the contacts to colleagues to talk on the scientific work of interest.

Though large sodium-cooled reactors are already in operation without serious problems caused by the use of the alkali-metal cooling, there is still a lack of knowledge in basic understanding. The seminar treated two of such topics, the solubility of metals and non-metals in alkali metals, and the physical chemistry of multi-component systems.

New coming laboratories as those of India or Italy have contributed to the programme. Having in mind the expensive equipment for the studies one can give them the recommendations:

- They must spend money for the operation of sophisticated sodium loops to gain experience in liquid metal handling.

- However, they should assess wether they would follow the established countries in studying the corrosion resistance of certain materials or they should identify unsolved problems in this area to work on their solution or they may contribute to the elaboration of basic results.

They should use all occasions like this seminar to keep
contacts and to learn which of their contributions would
be the most valuable.

In the sodium work to support the technology one can recommend
further studies of the crack propagation in structural materials in
sodium environment, and the evaluation of the influence of sodium
under unspecified chemical conditions on the material behaviour.
Such conditions are caused by chemical contamination, for instance
by oxygen and nitrogen, water or hydroxide, and carbon hydrides.

The contributions to the liquid lithium chemistry and corrosion
have indicated that basic work is going on satisfactorily. It seems
that corrosion is limiting the application of a number of materials.
Therefore, quantitative studies under realistic parameters are
necessary to define the range of application of stainless steels.
Refractory metals should be considered as alternative materials
with potential to suppress the activation problems to a certain
degree. The state of technology of fusion reactors leaves still
time for research as well as for material development.

Only very few laboratories are studying lead and liquid lead
alloy corrosion of materials. The heavy liquid metals are much more
corrosive than alkali metals. Corrosion problems will limit the
application ranges of structural materials. Extensive studies with
the lead bismuth eutectic are to be recomended. The studies should
be performed in a realistic field of parameters.

The secretary of the programme and organization committee
is indebted to the members of this committee. Thanks are also
given to the session chairmen, the contributors and the attending
scientists. All of them have contributed to the success of the
seminar.